BIRKHÄUSER

Modern Birkhäuser Classics

Many of the original research and survey monographs in pure and applied mathematics published by Birkhäuser in recent decades have been groundbreaking and have come to be regarded as foundational to the subject. Through the MBC Series, a select number of these modern classics, entirely uncorrected, are being re-released in paperback (and as eBooks) to ensure that these treasures remain accessible to new generations of students, scholars, and researchers.

Visions in Mathematics

GAFA 2000 Special Volume, Part II

N. Alon
J. Bourgain
A. Connes
M. Gromov
V. Milman
Editors

Reprint of the 2000 Edition

Birkhäuser

Editors:

N. Alon
School of Mathematical Sciences
University of Tel Aviv
Tel Aviv 69978
Israel
e-mail: nogaa@math.tau.ac.il

J. Bourgain
School of Mathematics
Institute for Advanced Study
Princeton University
Princeton, NJ 08540
USA
e-mail: bourgain@math.ias.edu

A. Connes
Collège de France
3, rue d'Ulm
75231 Paris cedex 05
France
e-mail: alain@connes.org

M. Gromov
Institut des Hautes Études Scientifiques
35, Route de Chartres
91440 Bures-sur-Yvette
France
e-mail: gromov@ihes.fr

V. Milman
Department of Mathematics
University of Tel Aviv
Tel Aviv 69978
Israel
e-mail: milman@post.tau.ac.il

Originally published under the title "Visions in Mathematics. Towards 2000"
as Special Volume, Part II, pp. 455-983 of the Journal "GAFA Geometric And
Functional Analysis" by Birkhäuser Verlag, Switzerland, ISSN 1016-443X

2000 Mathematics Subject Classification 05-02, 00A05, 11M26, 46B07, 53D35,
53D45, 58B34, 81P68, 82B26

Library of Congress Control Number: 2010920640

Bibliographic information published by Die Deutsche Bibliothek
Die Deutsche Bibliothek lists this publication in the Deutsche Nationalbibliografie;
detailed bibliographic data is available in the Internet at <http://dnb.ddb.de>.

ISBN 978-3-0346-0424-6

P.O. Box 133, CH-4010 Basel, Switzerland
Part of Springer Science+Business Media
Printed on acid-free paper produced of chlorine-free pulp. TCF ∞

ISBN 978-3-0346-0424-6 e-ISBN 978-3-0346-0425-3

9 8 7 6 5 4 3 2 1 www.birkhauser.ch

The organizers of "Visions in Mathematics – Towards 2000" take this opportunity to express their deep gratitude to Tel Aviv University for support of the conference and for providing a most stimulating environment. Special thanks are due to the School of Mathematical Sciences and the Minkowsky Minerva Center for Geometry for their very generous support.

Table of Contents

Addendum: Discussions at the Dead Sea

Foreword

The meeting "Visions in Mathematics – Towards 2000" took place mainly at Tel Aviv University in August 25-September 3, 1999, with a few days at the Sheraton-Moriah Hotel at the Dead Sea Health Resort. The meeting included about 45 lectures by some of the leading researchers in the world, in most areas of mathematics and a number of discussions in different directions, organized in various forms.

The goals of the conference, as defined by the scientific committee, consisting of N. Alon, J. Bourgain, A. Connes, M. Gromov and V. Milman, were to discuss the importance, methods, future and unity/diversity of mathematics as we enter the 21st Century, to consider the relation between mathematics and related areas and to discuss the past and future of mathematics as well as its interaction with Science.

A new format of mathematical discussions developed by the end of the Conference into an interesting addition to the more standard form of lectures and questions. The "Addendum" to this part of the Proceedings contains the transcript of some of the discussions which took place at the Dead Sea.

We believe that the meeting succeeded in giving a wide panorama of mathematics and mathematical physics, but we did not touch upon the interaction of mathematics with the experimental sciences.

This is the second (and final) part of the proceedings of the meeting.

It is a pleasure to thank Mrs. Miriam Hercberg and Mrs. Diana Yellin for their great technical help in the preparation of this manuscript.

N. Alon J. Bourgain

A. Connes M. Gromov

V. Milman

GAFA 2000

Tel Aviv, Israel

Y. Eliashberg

N. Alon

A. Givental

R. Coifman

H. Hofer

A. Connes

H. Iwaniec

G. Kalai

P. Shor

V. Milman

T. Spencer

P. Sarnak

V. Zakharov

GAFA, Geom. funct. anal.
Special Volume – GAFA2000, 455 – 470
1016-443X/00/S20455-16 $ 1.50+0.20/0

GAFA Geometric And Functional Analysis

ALGEBRAIC AND PROBABILISTIC METHODS IN DISCRETE MATHEMATICS

Noga Alon

Abstract

Combinatorics is an essential component of many mathematical areas, and its study has experienced an impressive growth in recent years. This survey contains a discussion of two of the main general techniques that played a crucial role in the development of modern combinatorics: algebraic methods and probabilistic methods. Both techniques are illustrated by examples, where the emphasis is on the basic ideas and the connection to other areas.

1 Introduction

Mathematical Research deals with ideas that can be meaningful to everybody and there is no doubt that it also lies behind most of the major advances in Science and Technology. Yet, mathematicians often tend to formulate their questions, results and thoughts in a way that is comprehensible only to their colleagues who work in a closely related area. One of the goals of the conference "Visions in Mathematics" was to try and present the main areas in mathematics in a way that can be interesting to a general mathematical audience, and possibly even to a general scientific audience. Although this is a difficult task, it is not impossible, and I believe that many of the lectures achieved this goal.

Following the spirit of the conference, this survey is also aimed at a general mathematical audience. I try to explain two of the main techniques that played a crucial role in the development of modern combinatorics: algebraic techniques and probabilistic methods. The focus is on basic ideas, rather than on technical details, and the techniques are illustrated by examples that demonstrate the connection between combinatorics and related mathematical areas.

Research supported in part by a US-Israel BSF grant, by a grant from the Israel Science Foundation and by the Hermann Minkowski Minerva Center for Geometry at Tel Aviv University.

My choice of topics and examples is inevitably influenced by my own personal taste, and hence it is somewhat arbitrary. Still, I believe that it provides some of the flavour of the techniques, problems and results in the area, which may hopefully be appealing to researchers in mathematics, even if their main interest is not Discrete Mathematics.

2 Algebraic Techniques

Various algebraic techniques have been used successfully in tackling problems in Discrete Mathematics over the years. These include several tools that I will not discuss here, like tools from Representation Theory applied extensively in enumeration problems, or spectral techniques used in the study of highly regular structures. In this section I describe mainly two representative algebraic tools. The first one may be called Combinatorial Nullstellensatz, is based on some basic properties of polynomials, and has applications in Combinatorial Number Theory, Graph Theory and Combinatorics. The second one may be called the dimension argument, and has had numerous applications over the years. The examples given here illustrate the basic ideas. More examples can be found in various survey articles and books including [G], [Al2], [BF], [Bl].

2.1 Combinatorial Nullstellensatz. The classical Hilbert's Nullstellensatz (see, e.g., [vdW]) asserts that if F is an algebraically closed field, $f, g_1, \dots, g_m$ are polynomials in the ring of polynomials $F[x_1, \dots, x_n]$, and f vanishes over all common zeros of $g_1, \dots, g_m$, then there is an integer k and polynomials $h_1, \dots, h_m$ in $F[x_1, \dots, x_n]$ so that

$$f^k = \sum_{i=1}^{m} h_i g_i .$$

In the special case $m = n$, where each g_i is a univariate polynomial of the form $\prod_{s \in S_i}(x_i - s)$, a stronger conclusion holds, as follows.

Theorem 2.1. *Let F be an arbitrary field, and let $f = f(x_1, \dots, x_n)$ be a polynomial in $F[x_1, \dots, x_n]$. Let $S_1, \dots, S_n$ be nonempty subsets of F and define $g_i(x_i) = \prod_{s \in S_i}(x_i - s)$. If f vanishes over all the common zeros of $g_1, \dots, g_n$ (that is; if $f(s_1, \dots, s_n) = 0$ for all $s_i \in S_i$), then there are polynomials $h_1, \dots, h_n \in F[x_1, \dots, x_n]$ satisfying $\deg(h_i) \le \deg(f) - \deg(g_i)$ so that*

$$f = \sum_{i=1}^{n} h_i g_i .$$

As a consequence of the above one can prove the following,

Theorem 2.2. *Let F be an arbitrary field, and let $f = f(x_1, \dots, x_n)$ be a polynomial in $F[x_1, \dots, x_n]$. Suppose the degree $\deg(f)$ of f is $\sum_{i=1}^{n} t_i$, where each t_i is a nonnegative integer, and suppose the coefficient of $\prod_{i=1}^{n} x_i^{t_i}$ in f is nonzero. Then, if $S_1, \dots, S_n$ are subsets of F with $|S_i| > t_i$, there are $s_1 \in S_1, s_2 \in S_2, \dots, s_n \in S_n$ so that*

$$f(s_1, \dots, s_n) \neq 0.$$

These two results are proved in [Al4], where it is proposed to call them *Combinatorial Nullstellensatz*. The proofs are based on some simple properties of polynomials. It turns out that these results are related to some classical ones, and have many combinatorial applications.

One of the classical results that follow easily from Theorem 2.2 is the following theorem, conjectured by Artin in 1934, proved by Chevalley in 1935 and extended by Warning in 1935.

Theorem 2.3 (cf., e.g., [S]). *Let p be a prime, and let*

$$P_1 = P_1(x_1, \dots, x_n), P_2 = P_2(x_1, \dots, x_n), \dots, P_m = P_m(x_1, \dots, x_n)$$

be m polynomials in the ring $Z_p[x_1, \dots, x_n]$. If $n > \sum_{i=1}^{m} \deg(P_i)$ and the polynomials P_i have a common zero $(c_1, \dots, c_n)$, then they have another common zero.

The proof follows in a few lines by applying Theorem 2.2 to the polynomial

$$f = f(x_1, \dots, x_n) = \prod_{i=1}^{m} \left(1 - P_i(x_1, \dots, x_n)^{p-1}\right) - \delta \prod_{j=1}^{n} \prod_{c \in Z_p, c \neq c_j} (x_j - c),$$

where δ is chosen so that $f(c_1, \dots, c_n) = 0$.

Another classical result that follows from a similar reasoning is the Cauchy-Davenport Theorem, which is one of the fundamental results in Additive Number Theory, see, e.g., [N]. This theorem asserts that if p is a prime, and A, B are two nonempty subsets of Z_p, then

$$|A + B| \geq \min\{p, |A| + |B| - 1\}.$$

Cauchy proved this theorem in 1813, and applied it to give a new proof to a lemma of Lagrange in his well known 1770 paper that shows that any integer is a sum of four squares. Davenport formulated the theorem as a discrete analogue of a conjecture of Khintchine (proved a few years later) about the Schnirelman density of the sum of two sequences of integers. The original proofs of the theorem given by Cauchy and Davenport are purely combinatorial. As observed in [AlNR], there is a different, algebraic proof, which extends easily and gives several related results. This proof is, again, a simple application of Theorem 2.2. It readily extends to provide bounds for restricted sums in finite fields. If $h = h(x_0, x_1, \dots, x_k)$ is a polynomial over Z_p and $A_0, A_1, \dots, A_k$ are subsets of Z_p, then the method provides a lower bound (which is often tight) for the cardinality of the set

$$\{a_0 + a_1 + \ldots + a_k : a_i \in A_i,\ \ h(a_0, a_1, \dots, a_k) \neq 0\}.$$

When h is the polynomial $\prod_{k \geq i > j \geq 0}(x_i - x_j)$ the above set corresponds to sums of distinct elements. By applying Theorem 2.2 to an appropriate polynomial, and by observing that the relevant coefficient in this case can be computed from the known results about the Ballot problem (see, e.g., [M]), as well as from the known connection between this problem and the hook formula for the number of Young tableaux of a given shape, one can obtain a tight lower bound for the number of such sums. The very special case of this result in which $k = 1$, $A_0 = A$ and $A_1 = A - \{a\}$ for an arbitrary element $a \in A$, implies the following theorem, conjectured by Erdős and Heilbronn in 1964 (cf., e.g., [ErG]) and proved, after various partial results by several researchers, by Dias Da Silva and Hamidoune [DH], using some tools from linear algebra and the representation theory of the symmetric group.

Theorem 2.4 [DH]. *If p is a prime, and A is a nonempty subset of Z_p, then*

$$\big|\{a + a' : a, a' \in A, a \neq a'\}\big| \geq \min\{p, 2|A| - 3\}.$$

This special case can be proved directly by assuming it is false, taking C to be a set of cardinality $2|A| - 4$ containing all sums of distinct elements $a_1, a_2 \in A$, with $a_2 \neq a$ for some fixed $a \in A$, and then by applying Theorem 2.2 to the polynomial $f(x, y) = (x - y)\prod_{c \in C}(x + y - c)$ to get a contradiction.

Erdős, Ginzburg and Ziv [ErGZ] proved that every sequence of $2n - 1$ elements of the cyclic group Z_n contains a subsequence of exactly n terms whose sum (in Z_n) is 0. This is tight, as shown, for example, by the sequence consisting of $n - 1$ zeros and $n - 1$ ones. The main part of the proof of this statement is its proof for prime values of $n = p$, as the general case can then be easily obtained by induction. Kemnitz [Ke] conjectured that for every prime p, every sequence of $4p - 3$ elements of Z_p^2 contains a subsequence of exactly p terms whose sum (in Z_p^2) is zero. Rónyai [Ro] has proved, very recently, that $4p - 2$ elements suffice. His proof can be described as an application of Theorem 2.2. This is done by first proving the following lemma.

LEMMA 2.5 [AlD]. *If* $(a_1, b_1), \ldots, (a_{3p}, b_{3p}) \in Z_p^2$ *and* $\sum_{i=1}^{3p}(a_i, b_i) = 0$ *(in* Z_p^2*), then there is an* $I \subset \{1, 2, \ldots, 3p\}$, $|I| = p$, *such that* $\sum_{i \in I}(a_i, b_i) = 0$.

To prove the lemma, consider the polynomial

$$f(x_1, x_2, \ldots, x_{3p-1}) = \left(1 - \left(\sum_{i=1}^{3p-1} a_i x_i\right)^{p-1}\right)\left(1 - \left(\sum_{i=1}^{3p-1} b_i x_i\right)^{p-1}\right)$$
$$\left(1 - \left(\sum_{i=1}^{3p-1} x_i\right)^{p-1}\right) - \prod_{i=1}^{3p-1}(1 - x_i).$$

Then the coefficient of $\prod_{i=1}^{3p-1} x_i$ is nonzero, and hence, by Theorem 2.2 with $S_1 = S_2 \ldots = S_{3p-1} = \{0, 1\}$ there are $x_i \in \{0, 1\}$ such that $f(x_1, \ldots, x_{3p-1})$ is not zero. As $f(0, 0, \ldots, 0) = 0$, not all x_i are 0. If $\sum_{i=1}^{3p-1} x_i$ is not zero modulo p then $f(x_1, \ldots, x_{3p-1}) = 0$, hence this sum is either p or $2p$. In both cases we get the desired result, where in the second case we apply the fact that the sum of all $3p$ vectors is 0.

To prove, next, that any sequence $(a_1, b_1), (a_2, b_2), \ldots, (a_{4p-2}, b_{4p-2})$ of elements of Z_p^2 contains a subsequence of precisely p terms whose sum is 0, apply Theorem 2.2 to the polynomial

$$f(x_1, x_2, \ldots, x_{4p-2}) = \left(1 - \left(\sum_{i=1}^{4p-2} a_i x_i\right)^{p-1}\right)\left(1 - \left(\sum_{i=1}^{4p-2} b_i x_i\right)^{p-1}\right)$$
$$\left(1 - \left(\sum_{i=1}^{4p-2} x_i\right)^{p-1}\right)\left(2 - \sum_{J \subset \{1,2,\ldots,4p-2\}, |J| = p} \prod_{j \in J} x_j\right) - 2\prod_{i=1}^{4p-2}(1 - x_i),$$

with $S_1 = S_2 = \ldots = S_{4p-2} = \{0, 1\}$. As the coefficient of $\prod_i x_i$ is nonzero there are $x_i \in \{0, 1\}$ such that $f(x_1, \ldots, x_{4p-2}) \neq 0$. It is easy to check that not all x_i are zero. It also follows that $\sum_i x_i$ must be divisible by p; if it is p we are done, if it is $3p$ the desired result follows from the lemma, and the last ingredient is the fact that if it is $2p$ then the term

$$2 - \sum_{J \subset \{1,2,\ldots,4p-2\}, |J| = p} \prod_{j \in J} x_j$$

is zero and hence so is f. This completes the proof.

Theorem 2.2 has various applications in Graph Theory, including ones in Graph Coloring, which is the most popular area of the subject. We sketch below the basic approach, following [AlT]. See also [Ma] for a related method.

A *vertex coloring* of a graph G is an assignment of a color to each vertex of G. The coloring is *proper* if adjacent vertices receive distinct colors. The *chromatic number* $\chi(G)$ of G is the minimum number of colors used in a proper vertex coloring of G. An *edge coloring* of G is, similarly, an assignment of a color to each edge of G. It is *proper* if adjacent edges receive distinct colors. The minimum number of colors in a proper edge-coloring of G is the *chromatic index* $\chi'(G)$ of G. This is equal to the chromatic number of the line graph of G.

A graph $G = (V, E)$ is *k-choosable* if for every assignment of sets of integers $S(v) \subset Z$, each of size k, to the vertices $v \in V$, there is a proper vertex coloring $c : V \mapsto Z$ so that $c(v) \in S(v)$ for all $v \in V$. The *choice number* of G, denoted $ch(G)$, is the minimum integer k so that G is k-choosable. Obviously, this number is at least the chromatic number $\chi(G)$ of G. The choice number of the line graph of G, denoted here by $ch'(G)$, is usually called the *list chromatic index* of G, and it is clearly at least the chromatic index $\chi'(G)$ of G.

The study of choice numbers was introduced, independently, by Vizing [Viz] and by Erdős, Rubin and Taylor [ErRT]. There are many graphs G for which the choice number $ch(G)$ is strictly larger than the chromatic number $\chi(G)$ (a complete bipartite graph with 3 vertices in each color class is one such example). In view of this, the following conjecture, suggested independently by various researchers including Vizing, Albertson, Collins, Tucker and Gupta, which apparently appeared first in print in the paper of Bollobás and Harris ([BoH]), is somewhat surprising.

CONJECTURE 2.6 (The list coloring conjecture). *For every graph G, $ch'(G) = \chi'(G)$.*

This conjecture asserts that for *line graphs* there is no gap at all between the choice number and the chromatic number. Many of the most interesting results in the area are proofs of special cases of this conjecture, which is still wide open.

The *graph polynomial* $f_G = f_G(x_1, x_2, \dots, x_n)$ of a graph $G = (V, E)$ on a set $V = \{1, \dots, n\}$ of n vertices is defined by $f_G(x_1, x_2, \dots, x_n) = \Pi\{(x_i - x_j) : i < j\ ,\ ij \in E\}$. This polynomial has been studied by various researchers, starting already with Petersen [P] in 1891. Note that if $S_1, \dots, S_n$ are sets of integers, then there is a proper coloring assigning to each vertex i a color from its list S_i, if and only if there are $s_i \in S_i$ such that $f_G(s_1, \dots, s_n) \neq 0$. This condition is precisely the one appearing in the conclusion of Theorem 2.2, and it is therefore natural to expect that this theorem can be useful in tackling coloring problems. By applying it to line graphs of planar, cubic graphs, and by interpreting the appropriate co-

efficient of the corresponding polynomial combinatorially, it can be shown, using a known result of Vigneron [Vi] and the Four Color Theorem, that the list chromatic index of every 2-connected cubic planar graph is 3. This is a strengthening of the Four Color Theorem, which is well known to be equivalent to the fact that the chromatic index of any such graph is 3. An extension of this result appears in [ElG].
Additional results on graph coloring and choice numbers using the algebraic approach are described in the survey [Al1].

2.2 The dimension argument. In order to prove an upper bound for the cardinality of a set, it is sometimes possible to associate each member of the set with a vector in an appropriately defined vector space, and show that the set of vectors obtained in this manner is linearly independent. Thus, the cardinality of the set is at most the dimension of the vector space. This simple linear-algebra technique, which may be called the *dimension argument*, has many impressive combinatorial applications. In this subsection we describe a few representative examples.

Borsuk [Bors] asked if any set of points in R^d can be partitioned into at most $d+1$ subsets of smaller diameter. Kahn and Kalai [KK] gave an example showing that this is not the case, by applying a theorem of Frankl and Wilson [FW]. Here is a sketch of a slightly modified version of this counterexample, following Nilli [Ni]. The main part of the proof uses the the dimension argument. Let $n = 4p$, where p is an odd prime, and let $\mathcal{F}$ be the set of all vectors $\mathbf{x} = (x_1, \ldots, x_n) \in \{-1, 1\}^n$, where $x_1 = 1$ and the number of negative coordinates of $\mathbf{x}$ is even.

LEMMA 2.7. *If $\mathcal{G} \subset \mathcal{F}$ contains no two orthogonal vectors then $|\mathcal{G}| \leq \sum_{i=0}^{p-1} \binom{n-1}{i}$.*

To prove the lemma note, first, that the scalar product $\mathbf{a} \cdot \mathbf{b}$ of any two members of $\mathcal{F}$ is divisible by 4, and since there is no $\mathbf{a} \in \mathcal{F}$ for which $-\mathbf{a}$ is also in $\mathcal{F}$ the assumption implies that there are no distinct $\mathbf{a}$ and $\mathbf{b}$ in $\mathcal{G}$ so that $\mathbf{a} \cdot \mathbf{b} \equiv 0$ $(mod\ p)$. For each $\mathbf{a} \in \mathcal{G}$ define a polynomial over the finite field $GF(p)$ as follows: $P_{\mathbf{a}}(\mathbf{x}) = \prod_{i=1}^{p-1} (\mathbf{a} \cdot \mathbf{x} - i)$, where here $\mathbf{x} = (x_1, \ldots, x_n)$ is a vector of variables. Note that by the assumption

(i) $P_{\mathbf{a}}(\mathbf{b}) = 0$ (in $GF(p)$) for every two distinct members $\mathbf{a}$ and $\mathbf{b}$ of $\mathcal{G}$, and
(ii) $P_{\mathbf{a}}(\mathbf{a}) \neq 0$ for all $\mathbf{a} \in \mathcal{G}$.

Let $\overline{P}_{\mathbf{a}}$ be the multilinear polynomial obtained from the standard representation of $P_{\mathbf{a}}$ as a sum of monomials by using, repeatedly, the relations

$x_i^2 = 1$. Since $\overline{P}_{\mathbf{a}}(\mathbf{x}) = P_{\mathbf{a}}(\mathbf{x})$ for every vector $\mathbf{x}$ with $\{-1, 1\}$ coordinates, the relations (i) and (ii) above hold with every P replaced by $\overline{P}$.

It is easy to see that this implies that the polynomials $\overline{P}_{\mathbf{a}}$ for $\mathbf{a} \in \mathcal{G}$ are linearly independent. Therefore, $|\mathcal{G}|$ is bounded by the dimension of the space of multilinear polynomials of degree at most $p-1$ in $n-1$ variables (since $x_1 = 1$) over $GF(p)$, which is $\sum_{i=0}^{p-1} \binom{n-1}{i}$, completing the proof of the lemma.

For any n-vector $\mathbf{x} = (x_1, \dots, x_n)$, let $\mathbf{x} * \mathbf{x}$ denote the tensor product of $\mathbf{x}$ with itself, i.e., the vector of length n^2, $(x_{ij} : 1 \leq i, j \leq n)$, where $x_{ij} = x_i x_j$. Define $S = \{\mathbf{x} * \mathbf{x} : \mathbf{x} \in \mathcal{F}\}$, where $\mathcal{F}$ is as above. The norm of each vector in S is n and the scalar product between any two members of S is easily seen to be non-negative. Moreover, by Lemma 2.7 any set of more than $\sum_{i=0}^{p-1} \binom{n-1}{i}$ members of S contains an orthogonal pair, i.e., two points the distance between which is the diameter of S. It follows that S cannot be partitioned into less than $2^{n-2} / \sum_{i=0}^{p-1} \binom{n-1}{i}$ subsets of smaller diameter.

The vectors in S lie in an affine subspace of dimension $\binom{n}{2}$, and hence if

$$2^{n-2} \Big/ \sum_{i=0}^{p-1} \binom{n-1}{i} > \binom{n}{2} + 1,$$

the set S is a subset of R^d for $d = \binom{n}{2}$ that cannot be partitioned into at most $d+1$ subsets of smaller diameter. The smallest d for which this holds (with $n = 4p$, p an odd prime) is $d = 946 = \binom{44}{2}$ obtained by taking $p = 11$.

For an undirected graph $G = (V, E)$, let G^n denote the graph whose vertex set is V^n in which two distinct vertices $(u_1, u_2, \dots, u_n)$ and $(v_1, v_2, \dots, v_n)$ are adjacent iff for all i between 1 and n either $u_i = v_i$ or $u_i v_i \in E$. The *Shannon capacity* $c(G)$ of G is the limit $\lim_{n \to \infty} (\alpha(G^n))^{1/n}$, where $\alpha(G^n)$ is the maximum size of an independent set of vertices in G^n. This limit exists, by super-multiplicativity, and it is always at least $\alpha(G)$.

The study of this parameter was introduced by Shannon in [Sh], motivated by a question in Information Theory. Indeed, if V is the set of all possible letters a channel can transmit in one use, and two letters are adjacent if they may be confused, then $\alpha(G^n)$ is the maximum number of messages that can be transmitted in n uses of the channel with no danger of confusion. Thus $c(G)$ represents the number of distinct messages *per use* the channel can communicate with no error while used many times.

The (*disjoint*) *union* of two graphs G and H, denoted $G + H$, is the graph whose vertex set is the disjoint union of the vertex sets of G and of H and whose edge set is the (disjoint) union of the edge sets of G and H. If G and H are graphs of two channels, then their union represents the *sum* of the channels corresponding to the situation where either one of the two channels may be used, a new choice being made for each transmitted letter.

Shannon [Sh] proved that for every G and H, $c(G+H) \geq c(G)+c(H)$ and that equality holds if the vertex set of one of the graphs, say G, can be covered by $\alpha(G)$ cliques. He conjectured that in fact equality always holds. Counter examples are given in [Al3], where it is shown that there are graphs G and H satisfying $c(G) \leq k$ and $c(H) \leq k$, whereas $c(G+H) \geq k^{(1+o(1))\frac{\log k}{8 \log\log k}}$ and the $o(1)$-term tends to zero as k tends to infinity.

The construction is based on some of the ideas of Frankl and Wilson [FW], together with a method for bounding the Shannon capacity of a graph using the dimension argument. This bound, described below, is strongly related to a bound of Haemers [H].

Let $G=(V,E)$ be a graph and let $\mathcal{F}$ be a subspace of the space of polynomials in r variables over a field F. A *representation* of G over $\mathcal{F}$ is an assignment of a polynomial f_v in $\mathcal{F}$ to each vertex $v \in V$ and an assignment of a point $c_v \in F^r$ to each $v \in V$ such that the following two conditions hold:

1. For each $v \in V$, $f_v(c_v) \neq 0$.
2. If u and v are distinct nonadjacent vertices of G then $f_v(c_u)=0$.

In these notations, the following holds.

PROPOSITION 2.8. *Let $G=(V,E)$ be a graph and let $\mathcal{F}$ be a subspace of the space of polynomials in r variables over a field F. If G has a representation over $\mathcal{F}$ then $\alpha(G) \leq \dim(\mathcal{F})$.*

This is proved by associating each vertex of an independent set of maximum cardinality in a given power of G, an appropriate polynomial in the corresponding tensor power of $\mathcal{F}$, and by showing that these polynomials are linearly independent. The details can be found in [Al3].

Many additional applications of the dimension argument appear in [Bl], [BF], [G].

3 Probabilistic Methods

The discovery, demonstrated in the early work of various researchers, that deterministic statements can be proved by probabilistic reasoning, led already more than fifty years ago to several striking results in Analysis, Number Theory, Combinatorics and Information Theory. These are demonstrated in early papers of Paley, Zygmund, Kac, Shannon, Turán and Szele, and even more so in the work of Paul Erdős. It soon became clear that the method, which is now called *the probabilistic method*, is a very powerful tool for proving results in Discrete Mathematics. The early results combined combinatorial arguments with fairly elementary probabilistic techniques, whereas the development of the method in recent years required the application of more sophisticated tools from probability theory. There is, by now, a huge amount of material on the topic, and it is hopeless to try and survey it in a comprehensive manner here. My intention in this section is therefore merely to illustrate the basic ideas with a few representative examples. More material can be found in the books [AlS], [Sp] and [JLR].

The *Ramsey number* $R(k,t)$ is the minimum number n such that every graph on n vertices contains either a clique of size k or an independent set of size t. By a special case of the celebrated theorem of Ramsey (cf., e.g., [GrRS]), $R(k,t)$ is finite for every positive integers k and t, and in fact $R(k,t) \leq \binom{k+t-2}{k-1}$. In particular, $R(k,k) < 4^k$. The problem of determining or estimating the numbers $R(k,t)$ received a considerable amount of attention, and seems to be very difficult in general.

In one of the first applications of the probabilistic method in Combinatorics, Erdős [Er] proved that if $\binom{n}{k}2^{1-\binom{k}{2}} < 1$ then $R(k,k) > n$. Therefore, $R(k,k) > \lfloor 2^{k/2} \rfloor$ for all $k > 2$. The proof is (by now) extremely simple; Let $G = G(n, 1/2)$ be a *random graph* on the n vertices $\{1, 2, \ldots, n\}$, obtained by picking each pair of distinct vertices, randomly and independently, to be connected with probability $1/2$. Every fixed set of k vertices of G forms a clique or an independent set with probability $2^{1-\binom{k}{2}}$. Thus $\binom{n}{k}2^{1-\binom{k}{2}}$ (< 1) is an upper bound for the probability that G contains a clique or an independent set of size k. It follows that with positive probability G is a graph without such cliques or independent sets, and hence such a graph exists!

A proper coloring of a graph is *acyclic* if there is no two-colored cycle. The *acyclic chromatic number* of a graph is the minimum number of colors in an acyclic coloring of it. The Four Color Theorem, which is the best known result in Discrete Mathematics, asserts that the chromatic number of every planar graph is at most 4. Answering a problem of Grünbaum and improving results of various authors, Borodin [Bor] showed that every planar graph has an acyclic 5-coloring. He conjectured that for any surface but the plane, the maximum possible chromatic number of a graph embeddable on the surface, is equal to the maximum possible acyclic chromatic number of a graph embeddable on it. The Map Color Theorem proved in [RY] determines precisely the maximum possible chromatic number of any graph embeddable on a surface of genus g. This maximum is the maximum number of vertices of a complete graph embeddable on such a surface, which turns out to be

$$\left\lfloor \frac{7+\sqrt{1+48g}}{2} \right\rfloor = \Theta(g^{1/2}).$$

The following result shows that the maximum possible acyclic chromatic number of a graph on such a surface is asymptotically different, thus disproving Borodin's conjecture.

Theorem 3.1 [AlMS]. *The acyclic chromatic number of any graph embeddable on a surface of genus g is at most $O(g^{4/7})$. Moreover, for every $g > 0$ there is a graph embeddable on a surface of genus g whose acyclic chromatic number is at least $\Omega(g^{4/7}/(\log g)^{1/7})$.*

The proof of the $O(g^{4/7})$ upper bound is probabilistic, and combines some combinatorial arguments with the Lovász Local Lemma. This Lemma, proved in [ErL], is a tool for proving that under suitable conditions, with positive probability, none of a large finite collection of nearly independent, low probability events in a probability space holds. This positive probability is often extremely small, and yet the Local Lemma can be used to show it is positive. The proof of the $\Omega(g^{4/7}/(\log g)^{1/7})$ lower bound is also probabilistic, and is based on an appropriate random construction. Note that the statement of the above theorem is purely deterministic, and yet its proof relies heavily on probabilistic arguments.

The final example in this section is a recent gem; it is based on a simple result in graph theory, whose proof is probabilistic. This result has several fascinating consequences in Combinatorial Geometry and Combinatorial Number Theory. Some weaker versions of these seemingly unrelated consequences have been proved before, in a far more complicated manner

An *embedding* of a graph $G = (V, E)$ in the plane is a a planar representation of it, where each vertex is represented by a point in the plane, and each edge uv is represented by a curve connecting the points corresponding to the vertices u and v. The *crossing number* of such an embedding is the number of pairs of intersecting curves that correspond to pairs of edges with no common endpoints. The *crossing number* $cr(G)$ of G is the minimum possible crossing number in an embedding of it in the plane. The following theorem was proved by Ajtai, Chvátal, Newborn and Szemerédi [ACNS] and, independently, by Leighton [L].

Theorem 3.2. *The crossing number of any simple graph $G=(V,E)$ with $|E|\geq 4|V|$ is at least $\frac{|E|^3}{64|V|^2}$.*

The proof is by a simple probabilistic argument. By Euler's formula any simple planar graph with n vertices has at most $3n-6$ edges, implying that the crossing number of any simple graph with n vertices and m edges is at least $m-(3n-6)>m-3n$. Let $G=(V,E)$ be a graph with $|E|\geq 4|V|$ embedded in the plane with $t=cr(G)$ crossings. Let H be the random induced subgraph of G obtained by picking each vertex of G, randomly and independently, to be a vertex of H with probability p (where p will be chosen later). The expected number of vertices of H is $p|V|$, the expected number of its edges is $p^2|E|$, and the expected number of crossings in its given embedding is p^4t, implying that the expected value of its crossing number is at most p^4t. Therefore, $p^4t\geq p^2|E|-3p|V|$, implying that

$$cr(G)=t\geq \frac{|E|}{p^2}-3\frac{|V|}{p^3}.$$

Without trying to optimize the constant factor, take $p=4|V|/|E|$ (≤ 1), to get the desired result.

L. Székely [Sz] noticed that this result can be applied to obtain a surprisingly simple proof of a result of Szemerédi and Trotter in Combinatorial Geometry [SzeT]. The original proof is far more complicated.

Theorem 3.3. *Let P be a set of n distinct points in the plane, and let L be a set of m distinct lines. Then, the number of incidences between the members of P and those of L (that is, the number of pairs (p,l) with $p\in P$, $l\in L$ and $p\in l$) is at most $c(m^{2/3}n^{2/3}+m+n)$, for some absolute constant c.*

Székely's proof is short and elegant: denote the number of incidences by I. Let $G=(V,E)$ be the graph whose vertices are all members of P, where two are adjacent if and only if they are consecutive points of P on some line in L. Clearly, $|V|=n$ and $|E|=I-m$. Note that G is already given embedded in the plane, where the edges are represented by segments of the corresponding lines in L. In this embedding, every crossing is an intersection point of two members of L, implying that $cr(G)\leq\binom{m}{2}\leq m^2/2$. By Theorem 3.2, either $I-m=|E|<4|V|=4n$, that is, $I\leq m+4n$, or

$$\frac{m^2}{2}\geq cr(G)\geq \frac{(I-m)^3}{64n^2},$$

showing that $I\leq (32)^{1/3}m^{2/3}n^{2/3}+m$. In both cases $I\leq 4(m^{2/3}n^{2/3}+m+n)$, completing the proof.

G. Elekes found several applications of the last theorem to Additive Number Theory. Here, too, the proofs are amazingly simple. Here is a representative result. A related one appears in [E].

Theorem 3.4. *For any three sets A,B and C of s real numbers each,*

$$|A \cdot B + C| = \big|\{ab + c : \ a \in A, b \in B, c \in C\}\big| \geq \Omega(s^{3/2}).$$

To prove this result, define $R = A \cdot B + C$, $|R| = r$ and put

$$P = \big\{(a,t) : a \in A, t \in R\big\}, \quad L = \{y = bx + c : b \in B, c \in C\}.$$

Thus P is a set of $n = sr$ points in the plane, L is a set of $m = s^2$ lines in the plane, and each line $y = bx + c$ in L is incident with s points of P, that is, with all the points $\{(a, ab + c) : a \in A\}$. Therefore, by Theorem 3.3, $s^3 \leq 4(s^{4/3}(sr)^{2/3} + sr + s^2)$, implying that $r \geq \Omega(s^{3/2})$, as needed.

4 The Algorithmic Aspects

The rapid development of theoretical Computer Science and its tight connection to Discrete Mathematics motivated the study of the algorithmic aspects of algebraic and probabilistic techniques. Can a combinatorial structure, or a substructure of a given one, whose existence is proved by algebraic or probabilistic means, be constructed *explicitly* (that is, by an efficient deterministic algorithm)? Can the algorithmic problems corresponding to existence proofs be solved by efficient procedures? The investigation of these questions are often related to other branches of mathematics. Here we merely mention a few open problems motivated by these questions.

As mentioned in the last paragraph of subsection 2.1, the list chromatic index of any planar cubic 2-connected graph is 3. Can the corresponding algorithmic problem be solved efficiently? That is, can we color properly the edges of any given planar cubic 2-connected graph using given lists of three colors per edge, in polynomial time?

This problem, as well as several similar applications of Theorem 2.2, are widely open. Note that any efficient procedure that finds, for a given input polynomial that satisfies the assumptions of Theorem 2.2, a point $(s_1, s_2, \ldots, s_n)$ satisfying its conclusion, would provide efficient algorithms for all these algorithmic problems. It would thus be interesting to find such an efficient procedure.

Probabilistic proofs also suggest the study of the corresponding algorithmic problems. This is related to the study of randomized algorithms, a topic which has been developed tremendously during the last decade. See, e.g., [MoR] and its many references. Even the simple proof of Erdős, described in section 3, that there are graphs on more than $\lfloor 2^{k/2} \rfloor$ vertices containing neither a clique nor an independent set of size k leads to an open problem which seems very difficult. Can we construct, explicitly, a graph on $n \geq (1+\epsilon)^k$ vertices with neither a clique nor an independent set of size k, in time which is polynomial in n, where $\epsilon > 0$ is any positive absolute constant?

The above problems, as well as many related ones, could be viewed as a victory of algebraic and probabilistic techniques. They illustrate the fact that these methods often supply solutions to problems that we cannot solve constructively. I am convinced that the study of algebraic and probabilistic methods, as well as the related search for more constructive proofs, will keep playing a major role in the future development of Discrete Mathematics.

References

[ACNS] M. Ajtai, V. Chvátal, M.M. Newborn, E. Szemerédi, Crossing-free subgraphs, in "Theory and Practice of Combinatorics", North-Holland Math. Stud., 60, North-Holland, Amsterdam and New York (1982), 9–12.

[Al1] N. Alon, Restricted colorings of graphs, in "Surveys in Combinatorics", Proc. 14th British Combinatorial Conference, London Mathematical Society Lecture Notes Series 187 (K. Walker, ed.), Cambridge University Press (1993), 1-33.

[Al2] N. Alon, Tools from higher algebra, in "Handbook of Combinatorics" (R. Graham, M. Grötschel, L. Lovász, eds.), Elsevier and MIT Press (1995), 1749-1783.

[Al3] N. Alon, The Shannon capacity of a union, Combinatorica 18 (1998), 301-310.

[Al4] N. Alon, Combinatorial Nullstellensatz, Combinatorics, Probability and Computing 8 (1999), 7–29.

[AlD] N. Alon, M. Dubiner, Zero-sum sets of prescribed size, in "Combinatorics, Paul Erdős is Eighty", János Bolyai Math. Soc., Budapest (1993), 33–50.

[AlMS] N. Alon, B. Mohar, D.P. Sanders, On acyclic colorings of graphs on surfaces, Israel J. Math. 94 (1996), 273–283.

[AlNR] N. Alon, M.B. Nathanson, I.Z. Ruzsa, The polynomial method and restricted sums of congruence classes, J. Number Theory 56 (1996), 404–417.

[AlS] N. Alon, J.H. Spencer, The Probabilistic Method, 2nd edition, Wiley, New York, 2000.

[AlT] N. Alon, M. Tarsi, Colorings and orientations of graphs, Combinatorica 12 (1992), 125–134.

[BF] L. Babai, P. Frankl, Linear Algebra Methods in Combinatorics, to appear.

[Bl] A. Blokhuis, Polynomials in finite geometries and combinatorics, in "Surveys in Combinatorics", Proc. 14th British Combinatorial Conference, London Mathematical Society Lecture Notes Series 187 (K. Walker, ed.), Cambridge University Press (1993), 35–52.

[BoH] B. Bollobás, A.J. Harris, List colorings of graphs, Graphs and Combinatorics 1 (1985), 115–127.

[Bor] O.V. Borodin, On acyclic colorings of planar graphs, Discrete Math. 25 (1979), 211–236.

[Bors] K. Borsuk, Drei sätze über die n-dimensionale euklidische sphäre, Fundamenta Math. 20 (1933), 177–190.

[DH] J.A. Dias da Silva, Y.O. Hamidoune, Cyclic spaces for Grassmann derivatives and additive theory, Bull. London Math. Soc. 26 (1994), 140–146.

[E] G. Elekes, On the number of sums and products, Acta Arith. 81 (1997), 365–367.

[ElG] M.N. Ellingham, L. Goddyn, List edge colourings of some 1-factorable multigraphs, Combinatorica 16 (1996), 343–352.

[Er] P. Erdős, Some remarks on the theory of graphs, Bulletin of the Amer. Math. Soc. 53 (1947), 292–294.

[ErGZ] P. Erdős, A. Ginzburg, A. Ziv, Theorem in the additive number theory, Bull. Research Council Israel 10F (1961), 41–43.

[ErG] P. Erdős, R.L. Graham, Old and New Problems and Results in Combinatorial Number Theory, L'Enseignement Mathématique, Geneva, 1980.

[ErL] P. Erdős, L. Lovász, Problems and results on 3-chromatic hypergraphs and some related questions, in "Infinite and Finite Sets" (A. Hajnal et al., eds.), North Holland (1975), 609–628.

[ErRT] P. Erdős, A.L. Rubin, H. Taylor, Choosability in graphs, Proc. West Coast Conf. on Combinatorics, Graph Theory and Computing, Congressus Numerantium XXVI (1979), 125–157.

[FW] P. Frankl, R. Wilson, Intersection theorems with geometric consequences, Combinatorica 1 (1981), 259–286.

[G] C. Godsil, Tools from linear algebra, in "Handbook of Combinatorics" (R. Graham, M. Grötschel, L. Lovász, eds.), Elsevier and MIT Press (1995), 1705–1748.

[GrRS] R.L. Graham, B.L. Rothschild, J.H. Spencer, Ramsey Theory, 2nd edition, Wiley, New York, 1990.

[H] W. Haemers, On some problems of Lovász concerning the Shannon capacity of a graph, IEEE Trans. Inform. Theory 25 (1979), 231–232.

[JLR] S. Janson, T. Łuczak, A. Ruciński, Random Graphs, Wiley, New York, 2000.

[KK] J. KAHN, G. KALAI, A counterexample to Borsuk's conjecture, Bulletin of the AMS 29 (1993), 60–62.

[Ke] A. KEMNITZ, On a lattice point problem, Ars Combinatoria 16b (1983), 151–160.

[L] F.T. LEIGHTON, Complexity Issues in VLSI, MIT Press, 1983.

[M] M.P.A. MACMAHON, Combinatory Analysis, Chelsea Publishing Company, 1915, Chapter V.

[Ma] Y. MATIYASEVICH, A criterion for colorability of vertices of a graph stated in terms of edge orientations (in Russian), Discrete Analysis (Novosibirsk) 26 (1974), 65–71.

[MoR] R. MOTWANI, P. RAGHAVAN, Randomized Algorithms, Cambridge University Press, New York, 1995.

[N] M.B. NATHANSON, Additive Number Theory: Inverse Theorems and the Geometry of Sumsets, Springer-Verlag, New York, 1996.

[Ni] A. NILLI, On Borsuk's problem, Contemporary Mathematics, Vol. 178 (1994), AMS, 209–210.

[P] J. PETERSEN, Die theorie der regulären graphs, Acta Math. 15 (1891), 193–220.

[RY] G. RINGEL, J.W.T. YOUNGS, Solution of the Heawood map coloring problem, Proc. Nat. Acad. Sci. U.S.A. 60 (1968), 438–445.

[Ro] L. RÓNYAI, On a conjecture of Kemnitz, to appear.

[S] W. SCHMIDT, Equations over finite fields, an elementary approach, Lecture Notes in Mathematics, Vol. 536, Springer, Berlin, 1976.

[Sh] C.E. SHANNON, The zero-error capacity of a noisy channel, IRE Trans. Inform. Theory 2 (1956), 8–19.

[Sp] J.H. SPENCER, Ten Lectures on the Probabilistic Method, 2nd edition, SIAM, Philadelphia, 1994.

[Sz] L. SZÉKELY, Crossing numbers and hard Erdős problems in discrete geometry, Combin. Probab. Comput. 6 (1997), 353–358.

[SzeT] E. SZEMERÉDI, W.T. TROTTER, Extremal problems in discrete geometry, Combinatorica 3:3-4 (1983), 381–392.

[vdW] B.L. VAN DER WAERDEN, Modern Algebra, Julius Springer, Berlin, 1931.

[Vi] L. VIGNERON, Remarques sur les réseaux cubiques de classe 3 associés au probléme des quatre couleurs, C. R. Acad. Sc. Paris, t. 223 (1946), 770–772.

[Viz] V.G. VIZING, Coloring the vertices of a graph in prescribed colors (in Russian), Diskret. Analiz. No. 29, Metody Diskret. Anal. v. Teorii Kodov i Shem 101 (1976), 3–10.

NOGA ALON, School of Mathematical Sciences and School of Computer Science, Tel Aviv University, Tel Aviv 69978, Israel nogaa@post.tau.ac.il

GAFA, Geom. funct. anal.
Special Volume – GAFA2000, 471 – 480
1016-443X/00/S20471-10 $ 1.50+0.20/0

GAFA Geometric And Functional Analysis

CHALLENGES IN ANALYSIS

R. Coifman

Mathematical analysis, and in particular Harmonic Analysis, has traditionally been tied to physical modeling – providing the language to describe the infinitesimal laws of nature through calculus and partial differential expressions as well as descriptions of field effects through integral operators, spectral and functional analysis.

A variety of deep analytical methods and tools were developed enabling detailed understanding and descriptions of natural transforms of analysis. The Fourier transform, the Hilbert transform and their generalizations as Singular Integrals, pseudodifferential and Fourier Integral calculi, have played a central role in 20th century analysis.

Over the last few years, while attempting to deal computationally with the problems that existing theory was supposed to elucidate, it became clear that a large number of fundamental issues both theoretical and computational need to be addressed; and that new mathematical/algorithmic tools and languages need to be developed.

It has become obvious that major obstructions exist to the development of an effective computational harmonic analysis. Moreover, success in overcoming these difficulties will provide the scientist dealing with complex scientific structure with a language to formulate and model his science. Our goal is to describe some of these challenges, both algorithmic and theoretical, by providing a few examples, hinting at the existence of a rich field of research.

The main theme governing these examples is our lack of understanding of analysis and geometry in high dimension (> 10).

The main issue involves our ability to evaluate effectively an analytical expression. We will see that this question provides a natural mechanism to test our analytical/synthetic understanding, and leads to deep structural and organizational insights.[1]

[1]Such insights have recently led to the solution by Lacey and Thiele of Calderon's conjecture and provided a conversion of Carleson's proof of the convergence of Fourier Series into a powerful analytic method, as well as deep insights in complex function theory.

1 Digital Transcriptions of Functions, Libraries of Waveforms

For this exposition it is convenient to think of a function f as a Fourier transform of a compactly supported square integrable function $\hat{f}$, (with $\mathrm{supp}\hat{f} \subseteq [-N_0, N_0]$.

Such a function is determined by its "samples" $f(k/N_0)$. We can identify the function with the vector $f = \{f(k/N_0)\}k = 0, \pm 1, \ldots, f = (f_k)$.

We should think of $f(t)$ as a recorded sound and of f_k as digital samples of f. Unfortunately this simple-minded digitization of f is neither efficient nor very useful. Our goal is to transcribe the function to a given precision ε using a minimal or close to minimal number of parameters. Moreover, we would like to automate the transcription mode and to develop a calculus with these transcriptions. (In much the same way as the standard binary or digital notation enables the automation of a numerical computation).

The standard procedure in signal processing is to window $f(t)$ by multiplication by $\omega(t-j)$ where ω is compactly supported on $[-1,1]$ and $\sum \omega^2(t-j) = 1$ and then expand $f(t)\omega(t-j)$ as a Fourier series in t. The Fourier coefficients are kept (to some precision ε) and used to represent the function. This kind of representation is convenient for storing sound or other one dimensional signals providing a local frequency content of the function.

The following figures show the effect of various window functions. The function being digitized is digitized for various choices of window size. The third choice is more effective, revealing the full structure of the function as a sum of three sounds with linearly increasing frequency.

We now describe briefly a mode for automatic transcription of functions resembling an "orchestration" of the function as a superposition of "musical scores" for different instruments. A "score" is a superposition of notes, where each note has a location, duration, pitch and amplitude.

More precisely we consider a small basic window $\omega(t)$ which is supported on an interval $\left[-\frac{1}{2}, \frac{3}{2}\right] \Sigma\omega^2(t-j) = 1$ and $\omega(t)\omega(t+1)$ is even. Then the functions $\omega(t-j)\sin\left[\left(k+\frac{1}{2}\right)\pi(t-j)\right]$ form an orthonormal basis of $L^2(\mathbf{R})$ (see [CM]). Similarly, if we let $\omega_1(t) = [\omega^2(t) + \omega^2(t-1)]^{1/2}$ the functions

$$\omega_1(t-2j)\sin\left[\left(k+\tfrac{1}{2}\right)(t-2j)\tfrac{1}{2}\right]$$

form a basis.

Chirps – best level

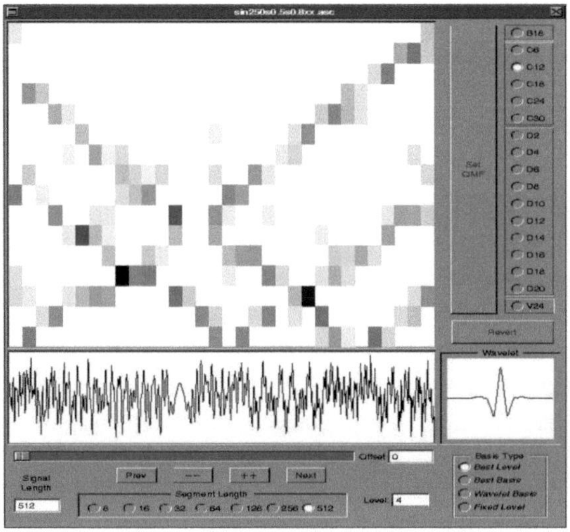

Chirps – bad level choice

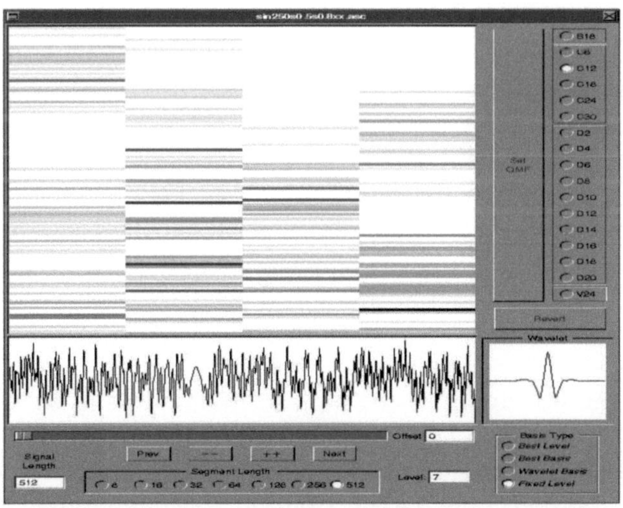

Continuing, we can associate to each dyadic interval $I = [j2^\ell, (j+1)2^\ell] = I_j^\ell$ an orthogonal set of functions

$$\omega_1(t - j2^\ell) \sin \left[\left(k + \tfrac{1}{2}\right] \pi (2^{-\ell} t - j) \right]$$

such that whenever a collection of dyadic intervals covers $\mathbf{R}$. The corresponding collection of functions is orthonormal. (The set of bases is indexed by dyadic covers of $\mathbf{R}$ with intervals of length $\geq \delta_0$).

An optimal transcription relative to this collection of bases is obtained by selecting the best basis for a given task. For example we could look for the basis providing the shortest expansion for a given error.

Another library of bases can be obtained by doing this construction (or variants) in the Fourier domain. This corresponds to the wavelet-packet libraries and is much closer to the musical score concept since in original space a function at a given location is a superposition of notes of different length scales.

An orchestration is obtained by picking a best basis in a collection of libraries (mathematical musical instruments) selecting a most efficient transcription, keeping only that portion of the function which is extracted at low entropy, and repeating the procedure on the residual, until we reach a residual whose entropy is similar to that of a random function, at which point we give up and stop.

The following two-dimensional version (developed by F. Meyer) reveals the various structures needed to synthesize the Mandrill image efficiently.

This mode of transcribing natural "mechanically" generated data sets has many practical uses such as data compression, efficient approximation, feature extraction, etc. We are mostly concerned here with analytic aspects: we want to transcribe the data so as to extract structures and attributes automatically, moreover any processing on the data should be simple and efficient in the chosen transcription parameters. We will apply this signal processing paradigm as a tool to analyze and organize complex physical transformations such as acoustic scattering, by processing the transformation as complex images that need to be orchestrated.

Before proceeding, we observe that the libraries of waveforms described above are totally inadequate to transcribe "seemingly unstructured data" having a random appearance such as the data arising in a quantized elastic mechanical system for which the eigenfunctions are uniformly spread in time frequency. The challenge for such systems is to invent analysis tools for structural detection. (Alain Connes' noncommutative geometry might be a step in the right direction.) There exist various mathematical struc-

Mandrill

Wavelet compression (watercolor)

Residual with two structures

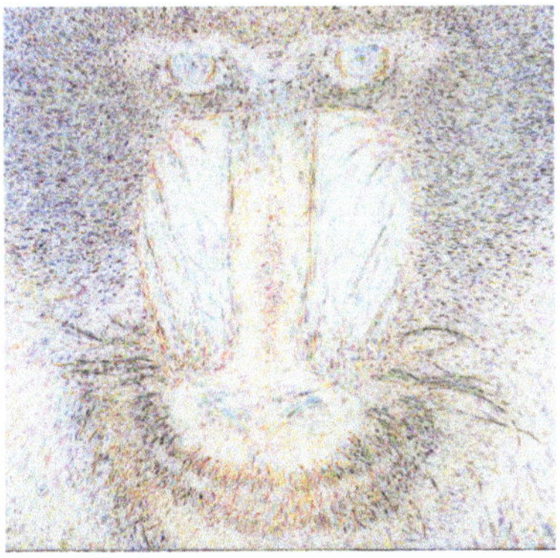

Brushstrokes

pointillistic

tures "parallel universes" equipped with their own notion of time frequency libraries which, when analyzed in conventional terms, seem like pure noise.

2 Transcribing Dense Matrices for Efficient Computations

Consider the Green function for the Helmholtz operator in $\mathbf{R}^3$,

$$k(x,y) = \frac{e^{i2\pi M|x-y|}}{|x-y|}$$

where $x, y \in S$ with S a two-dimensional surface. The computation of

$$T(f)(x) = \int_S k(x,y) f(y) dy$$

requires CN^4 operations when S is discretized at N^2 points (for obvious reasons $N \geq 4M$), which renders this computation prohibitively expensive even for $M = 100$.

We can easily discretize the integral operator to obtain a matrix representation. For simplicity we consider the case of $\mathbf{R}$ where S is a one-dimensional curve in $\mathbf{R}^2$, $k(x,y)$ can be viewed as an image and "orchestrated" as the Mandrill (see [R]). Clearly, the non oscillatory part corresponding to $2\pi|x-y|M \leq 1$ would be compressed by wavelets (like the description of the Mandrill's nose in watercolor) while the oscillatory part should be treated as brushlets using local Fourier or trig. expansions. This mode of description of $k(x,y)$ provides an unraveling of the operator by lifting T as a sparse operator $\tilde{T}$ on $\mathbf{R}^{N \log N}$ (i.e. having only $CN \log N$ entries, as opposed to N^2)

$$\begin{array}{ccc} \tilde{T} : \mathbf{R}^{N \log N} & \longrightarrow & \mathbf{R}^{N \log N} \\ \uparrow & & \downarrow \\ T : \mathbf{R}^N & \longrightarrow & \mathbf{R}^n \end{array}$$

where the vector in $\mathbf{R}^N$ is expanded in $\log N$ local trig. bases corresponding to windows of size $2^\ell \leq N$ and where $\tilde{T}$ operates on this redundant vector as in the figure.

This unwinding of T is a powerful analytic tool (usually viewed as microlocalization) enabling a fast computation of both linear and nonlinear transforms. In the example given above, it automatically selects pairs of intervals on the curve S and allocates to each pair a sparse coupling matrix corresponding to the fact that the oscillatory layer of the operator "beams" each localized cosine in a frequency dependent direction, and that beam interacts on each interval with a basis function having a frequency depending on the arrival direction. This geometric optic approach is automatically obtained from the transcription mode and provides a precise numerical way of describing complex physical phenomena, in this case acoustic scattering.

Observe that the number of parameters needed to describe the surface (or curve) could be quite large. The scattered field depends in a complex way on all of these parameters. The efficient description of the effect of the Green kernel in an incoming plane wave is quite complex, and requires much more than the simple transcription described here. The breakthrough in electromagnetic and acoustic scattering computations was done by V. Rokhlin who introduced special efficient representations for acoustic fields [R].

The point here is that we have a rudimentary mathematical algorithmic method for describing precisely relatively complex objects.

This is clearly a major difficulty faced daily by the natural scientist. While fundamental infinitesimal laws are well understood, more global effects and complex interactions are difficult to describe efficiently, this requires a detailed understanding of the organization of the Green operator, and its decompositions.

Another aspect of this higher dimensional lifting map involves our ability to provide numerically computable moving frames. This capability might simplify the description of complex phenomena such as turbulence, by viewing say a vorticity field as having a simple description in a moving frame whose description is also less complex.

Approximation in high dimensions. We have discussed briefly the challenge of computing effectively a linear transformation in $\mathbf{R}^N$. Nonlinear maps are much more baffling. The general problem that confronts us is understanding which nonlinear functions can be approximated to error $\varepsilon > 0$ using no more than $N(\log N)^\alpha[\log(1/\varepsilon)]^\beta$ terms. We call such approximations "computationally effective". We start with a simple observation that the usual classical methods for approximating a function by a trigonometric polynomial require $(1/\varepsilon)^N$ terms for precision $\varepsilon > 0$. Even in dimension 10 this is excessive for modest precision. The challenge to provide descriptions for empirical functions depending on say more than 10 parameters has led to an "industry of adhoc" methods, such as neural nets or other algorithms, for which no rate of approximation can be proved. There are however, indications that a powerful theory exists, but will require better understanding of geometry in high dimensions.

We illustrate these ideas quoting from J.O. Stromberg, who observed that under relatively simple conditions and for moderate N, useful results exist.

Theorem. *Let $P(x_1, \dots, x_N)$ be of bounded mixed variation*

$$\left| \frac{\partial^N}{\partial x_1 \dots \partial x_N} P \right| \leq M \quad \text{on} \quad x \in [0,1]^N$$

then

$$P(x_1, \dots, x_N) = \sum_{|R|>\varepsilon} \alpha_R h_R(x) + O(\varepsilon)$$

where R is a dyadic rectangle $R = I_1 \times I_2 \dots \times I_N$, I_K are dyadic intervals in $[0,1]$, $|R|$ = volume of R, and $h_R = h_{I_1}(x) h_I(x_2) \dots h_{I_N}(x_N)$, h_I is the Haar function based on I:

$$h_I(x) = |I|^{-1/2} \begin{cases} 1 & \text{on left half of } I \\ -1 & \text{on right half of } I \,. \end{cases}$$

Moreover, the number of rectangles of volume exceeding ε is $\frac{1}{\varepsilon}\left[\log\left(\frac{1}{\varepsilon}\right)\right]^{N-1}$ (as opposed to $1/\varepsilon^N$).

Observe first that any function of the form

$$\int_0^{x_N} \int_0^{x_1} p(t_1 \dots t_N) dt_1 \dots dt_N = P(x_1 \dots x_N)$$

where p is bounded satisfies the hypothesis, or any product $\prod_i f_i(x_i)$ with the f_i' bounded. (If $p(t)$ is interpreted as a probability density on $[0,1]^N$, then $P(x)$ is the probability of finding a point in the rectangle $[0,x] = [0,x_1] \times [0,x_2] \times \dots [0,x_N]$.)

Observe that since $|R| = |I_1||I_2| \dots |I_N|$, for precision $\varepsilon = 10^{-3}$ we cannot have more than ten I_i whose length is smaller than $1/2$, therefore for this precision the function of N variables is a superposition of functions of at most 10 variables, moreover the finer the resolution in $\mathbf{R}^N$ the fewer the number of variables needed to achieve the precision. The example of Stromberg shows that even in two dimensions it suffices to compute the probability $P(R)$ of a small set of well chosen rectangles to obtain precision $1/64$ as opposed to the evaluation of P at the 64^2 rectangles ending at the regular good points.

Unfortunately, this theorem is of limited use since $(\log 1/\varepsilon)^N$ grows exponentially with N. Moreover, even for moderate dimension $N \leq 10$ the assumption on the mixed derivative is not rotationally invariant. Given a function it is necessary to find a local coordinate system on which a representation like this might work. More generally, criteria for efficient description are needed as well as methods for geometric descriptions of effective domains for f.

In general, we are confronted with the problem of approximating a function of N parameters which is given empirically, usually the input parameters are not really in a box but on some lower dimensional subset of $\mathbf{R}^N$ and the question then is to parametrize this set.

More specifically, we assume that $f(x)$ is measured for a large number of points in $\mathbf{R}^N$ but that the points x are drawn from a low dimensional subset, that we can write $x = \varphi(\lambda), \lambda \in \mathbf{R}^n$ with φ bilipschitz, (i.e. $|\varphi(\lambda)-\varphi(\lambda')| \simeq |\lambda - \lambda'|$). This enables us to reduce the modeling of f to $\mathbf{R}^n$.

The remarkable theorems of Jones, David and Semmes permit the verification on the empirical domain of f whether it can be parametrized as above. This verification is obtained by performing a multiscale variance statistic on the points (see [J], [DS]).

For example, we might want to model the melting temperature of an alloy as a function of the various concentrations of constituents and their material attributes. This could involve a complex simulation using a range of feasible parameters, or could be collected from a large data base and a regression for the temperature has to be built.

Even if a simulation to compute the melting point could be performed, it would involve a large number of particles and would be a costly calculation for each value of input parameters. Assume for example that we have only the input parameters and end up with a melting point. We would like to predict directly that value without performing a simulation involving thousands of particles. The issue then is to efficiently describe dependencies.

The main area of analysis that needs attention is the development of effective approximation algorithms with weak dimensional dependence, it is clear that randomized methods generalizing Monte Carlo will play a considerable role in the actual computations and search for approximants, moreover, most quantitative results will have to be quantified modulo small sets of exceptions. Unfortunately, standard-harmonic analysis in $\mathbf{R}^N$ ignores the effectiveness issue; every single theorem say in Stein–Weiss and Stein books is not effective and could be recast in this light.

Similarly, practically all results in complex analysis related to analytic continuation and vanishing of holomorphic functions are meaningless as computational guides.

We should conclude here by observing that the question of computational effectiveness for transformations is a remarkably good test of analytic understanding. Generations of harmonic analysts have asked the question of existence of L^p estimates for various operators as a way of forcing this understanding, leading to fundamental analytic tools such as Calderon–Zygmund theory.

The kind of decomposition of operators obtained while trying to achieve computational effectiveness, provide very natural powerful generalizations of Calderon-Zygmund decompositions, providing organization for interactions and simple guidance for analytic insight.

References

[J] P.W. Jones, Rectifiable sets and the traveling salesman problem, Inventiones Math. 102 (1990), 1–15.

[DS] G. David, S. Semmes, Analysis of and on Uniformly Rectifiable Sets, Providence: American Math. Society, 1993.

[CMW] R. Coifman, Y. Meyer, V. Wickerhauser, Wavelet analysis and signal processing, in "Wavelets and Their Applications", Jones and Barlett, Boston, MA, (1992), 153–178.

[CM] R. Coifman, Y. Meyer, Remarques sur l'analyse de Fourier à fenêtre. Jour. C. R. Acad. Sci. Paris série A 312 (1991), 259–261.

[R] V. Rokhlin, Diagonal forms of translation operators for the Helmholtz equation in three dimensions, Appl. Comput. Harm. Anal. 1:1 (1993), 82–93.

Raphy Coifman, Department of Mathematics, Yale University, New Haven, CT 06520-8283, USA.

GAFA, Geom. funct. anal.
Special Volume – GAFA2000, 481 – 559
1016-443X/00/S20481-79 $ 1.50+0.20/0

GAFA Geometric And Functional Analysis

NONCOMMUTATIVE GEOMETRY YEAR 2000

Alain Connes

Abstract

Our geometric concepts evolved first through the discovery of Non-Euclidean geometry. The discovery of quantum mechanics in the form of the noncommuting coordinates on the phase space of atomic systems entails an equally drastic evolution. We describe a basic construction which extends the familiar duality between ordinary spaces and commutative algebras to a duality between Quotient spaces and Noncommutative algebras. The basic tools of the theory, K-theory, Cyclic cohomology, Morita equivalence, Operator theoretic index theorems, Hopf algebra symmetry are reviewed. They cover the global aspects of noncommutative spaces, such as the transformation $\theta \to 1/\theta$ for the noncommutative torus $\mathbb{T}^2_\theta$ which are unseen in perturbative expansions in θ such as star or Moyal products. We discuss the foundational problem of "what is a manifold in NCG" and explain the fundamental role of Poincare duality in K-homology which is the basic reason for the spectral point of view. This leads us, when specializing to 4-geometries to a universal algebra called the "Instanton algebra". We describe our joint work with G. Landi which gives noncommutative spheres S^4_θ from representations of the Instanton algebra. We show that any compact Riemannian spin manifold whose isometry group has rank $r \geq 2$ admits isospectral deformations to noncommutative geometries. We give a survey of several recent developments. First our joint work with H. Moscovici on the transverse geometry of foliations which yields a diffeomorphism invariant (rather than the usual covariant one) geometry on the bundle of metrics on a manifold and a natural extension of cyclic cohomology to Hopf algebras. Second, our joint work with D. Kreimer on renormalization and the Riemann–Hilbert problem. Finally we describe the spectral realization of zeros of zeta and L-functions from the noncommutative space of Adele classes on a global field and its relation with the Arthur–Selberg trace formula in the Langlands program. We end with a tantalizing connection between the renormalization group and the missing Galois theory at Archimedean places.

1 Introduction

There are two fundamental sources of 'bare' facts for the mathematician. These are, on the one hand the physical world which is the source of *geometry*, and on the other hand the arithmetic of numbers which is the source of *number theory*. Any theory concerning either of these subjects can be tested by performing experiments either in the physical world or with numbers. That is, there are some real things out there to which we can confront our understanding.

If one looks back at the 23 problems of Hilbert then one finds that, fortunately, the twentieth century saw very important discoveries which nobody could have foreseen by 1900. Two of them (of course by no means the only discoveries) involve Hilbert space in a crucial way and will be of particular importance for this talk: The first one is quantum mechanics, and the second, equally important in a sense, is the extension of class field theory to the non-abelian case, thanks to the Langlands program.

In this lecture I'll take both of these discoveries as a pretext and point towards the extension of our familiar geometrical concepts beyond the classical, commutative case. My aim is to discuss the foundation of noncommutative geometry.

2 Geometry

Before I do that, let me remind you, using a simple example, of the power of abstraction in mathematics. Around 1800, mathematicians wondered whether it is true that Euclid's fifth axiom is actually superfluous. For instance Legendre proved that if you have one triangle whose internal angles sum to π then that is enough to guarantee ordinary Euclidean geometry. However, as we all know Euclid's fifth axiom is not superfluous and Non-Euclidean Geometry gives a counter-example. The simplest model of Non-Euclidean Geometry is probably the Klein model. The points of the geometric space X are the points inside an ellipse,

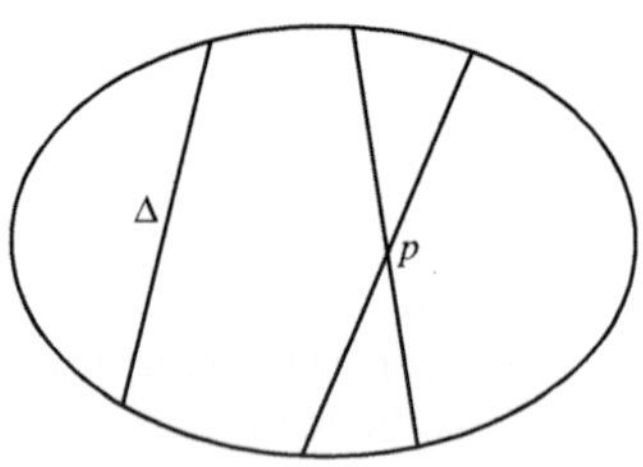

The lines are the intersections of the ordinary Euclidean lines with X. If you take a point p, outside the line Δ then there are distinct lines which don't meet Δ (*i.e.* are parallel to Δ) but meet each other at p.

At first this was considered as an esoteric example and Gauss didn't publish his discovery, but after some time it became clear that rather than just being a strange counter-example, it was something with remarkable beauty and power. The question then became "what is the source of this beauty and power?" Often in mathematics, understanding comes from generalisation, instead of considering the object *per se* what one tries to find are the concepts which embody the power of the object.

A first generalisation is the *Erlangen* program of Klein and the theory of Lie groups which attributes the beauty of this example to its symmetries, namely the group of projective transformations of the plane which preserve the ellipse.

The second conceptual generalisation is Riemannian geometry as explained in Riemann's inaugural lecture ([26]) in which he reflected on the hypotheses of geometry and introduced two key notions: the concepts of *manifold* and *line element.*

By a manifold Riemann meant 'any space you can think of whose points can vary continuously'. For example, a manifold could be a continuous collection of colours, the parameter space for some mechanical system or, of course, space. In his lecture Riemann explained that it is possible, essentially proceeding by induction, to label the points of such a space by a finite collection of real numbers.

In Riemannian geometry the distance between two points x and y is given by the following *ansatz*:

$$d(x,y) = \mathrm{Inf}\left\{ \int_\gamma ds \,|\gamma \text{ is a path between } x \text{ and } y\right\}. \tag{2.1}$$

Expanding $d(x,y)$ near the diagonal, after raising it to an even power to ensure smoothness gives a local formula for ds. The first case he considered was the quadratic case (although he explicitly mentioned the quartic case). From the Taylor expansion he obtained, in the quadratic case, the well-known formula for the metric,

$$ds^2 = g_{\mu\nu}\, dx^\mu\, dx^\nu\,. \tag{2.2}$$

Riemann's concept of geometry differs greatly from that of Klein because Klein's formulation is based on the idea of rigid motions whereas in Riemannian geometry rigid motions are no longer possible because of the variability of the curvature and the extraordinary freedom in the choice of the components $g_{\mu\nu}$.

The basic notions of ordinary geometry do make sense, for example a straight line is given by the geodesic equation,

$$\frac{d^2x^\mu}{dt^2} = -\frac{1}{2}\, g^{\mu\alpha}(g_{\alpha\nu,\rho} + g_{\alpha\rho,\nu} - g_{\nu\rho,\alpha})\, \frac{dx^\nu}{dt}\, \frac{dx^\rho}{dt} \tag{2.3}$$

but what really vindicated the point of view of Riemann, with respect to that of Klein, was another major discovery of the twentieth century, General Relativity.

One can get a glimpse of this from the following simple fact. If we take the Minkowski metric and perturb it to $dx^2+dy^2+dz^2-(1+2V(x,y,z))dt^2$ using the Newtonian potential $V(x,y,z)$, then the geodesic equation can be re-written in the obvious approximation to obtain Newton's law of motion. This makes clear that the variability of the $g_{\mu\nu}$ is precisely necessary in order to get a good geometric model of the physical universe.

It is interesting to note that Riemann was well aware of the limits of his own point of view as is clearly expressed in the last page of his inaugural lecture; ([26])

"Questions about the immeasurably large are idle questions for the explanation of Nature. But the situation is quite different with questions about the immeasurably small. Upon the exactness with which we pursue phenomenon into the infinitely small, does our knowledge of their causal connections essentially depend. The progress of recent centuries in understanding the mechanisms of Nature depends almost entirely on the exactness of construction which has become possible through the invention of the analysis of the infinite and through the simple principles discovered by Archimedes, Galileo and Newton, which modern physics makes use of. By contrast, in the natural sciences where the simple principles for such constructions are still lacking, to discover causal connections one pursues phenomenon into the spatially small, just so far as the microscope permits. Questions about the metric relations of Space in the immeasurably small are thus not idle ones.

If one assumes that bodies exist independently of position, then the curvature is everywhere constant, and it then follows from astronomical measurements that it cannot be different from zero; or at any rate its reciprocal must be an area in comparison with which the range of our telescopes can be neglected. But if such an independence of bodies from position does not exist, then one cannot draw conclusions about metric relations in the infinitely small from those in the large; at every point the curvature can have arbitrary values in three directions, provided only that the total curvature of every measurable portion of Space is not perceptibly different from zero. Still more complicated relations can occur if the line element cannot be represented, as was presupposed, by the square root of a differential expression of the second degree. Now it seems that the empirical notions on which the metric determinations of Space are based, the concept of a solid body and that of a light ray, lose their validity in the infinitely small; it is therefore quite definitely conceivable that the metric relations of Space in the infinitely small do not conform to the hypotheses of geometry; and in fact one ought to assume this as soon as it permits a simpler way of explaining phenomena.

The question of the validity of the hypotheses of geometry in the infinitely small is connected with the question of the basis for the metric relations of space. In connection with this question, which may indeed still be ranked as part of the study of Space, the above remark is applicable, that in a discrete manifold the principle of metric relations is already contained in the concept of the manifold, but in a continuous one it must come from something else. Therefore, either the reality underlying Space must form a discrete manifold, or the basis for the metric relations must be sought outside it, in binding forces acting upon it.

An answer to these questions can be found only by starting from that conception of phenomena which has hitherto been approved by experience, for which Newton laid the foundation, and gradually modifying it under the compulsion of facts which cannot be explained by it. Investigations like the one just made, which begin from general concepts, can serve only to insure that this work is not hindered by too restricted concepts, and that progress in comprehending the connection of things is not obstructed by traditional prejudices.

This leads us away into the domain of another science, the realm of physics, into which the nature of the present occasion does not allow us to enter".

3 Quantum Mechanics

In fact quantum mechanics showed that indeed the parameter space, or phase space of the mechanical system given by a single atom fails to be a manifold. It is important to convince oneself of this fact and to understand that this conclusion is indeed dictated by the experimental findings of spectroscopy. The information we get from the light coming from distant stars is of spectral nature, the spectral lines are absorption or emission lines

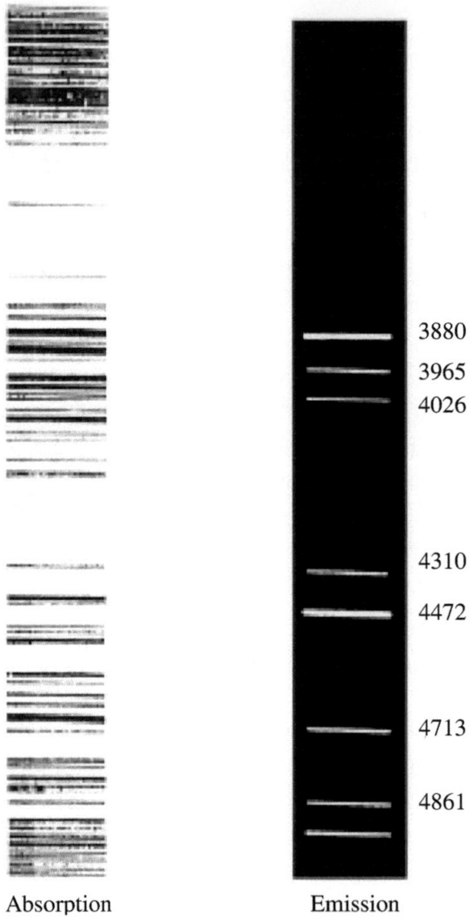

One can infer from this spectral information the chemical composition of the star since the simple elements have recognisable spectra. These spectra obey experimentally discovered laws, the most notable being the Ritz-Rydberg combination principle. The principle can be stated as follows; spectral lines are indexed by pairs of objects. These objects could be numbers, Greek letters, or any kind of labels. The statement of the principle then is that certain pairs of spectral lines, when expressed in terms of frequencies, do add up to give another line in the spectrum. Moreover, this happens precisely when the labels are of the form i, j and j, k.

What Heisenberg understood, by analogy with the classical treatment of the interaction of a mechanical system with the electromagnetic field, is that this Ritz-Rydberg combination principle actually dictates an algebraic formula for the product of any two observable physical quantities attached to the atomic system

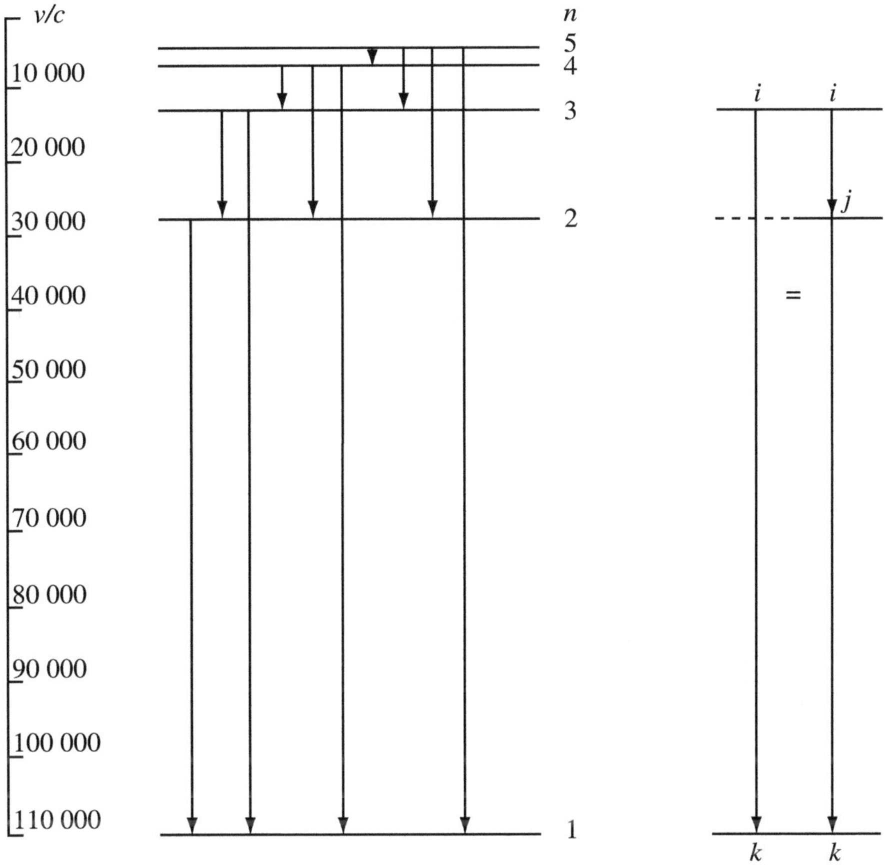

Heisenberg wrote down the formula for the product of two observables;

$$(A\,B)_{(i,k)} = A_{(i,j)}\,B_{(j,k)} \tag{3.1}$$

and he noticed of course that this algebra he had found is no longer commutative,

$$A\,B \neq B\,A\,. \tag{3.2}$$

Now Heisenberg didn't know about matrices, he just worked it out, but he was told later by Born, Jordan and Dirac that the algebra he had worked out was known to mathematicians as the algebra of matrices.

Physicists often tell jokes such as: A physicist walks down the main street of a strange town looking for a laundrette. He sees a shop with signs in the window saying 'bakery' 'grocers' 'laundrette', so he enters. However, the shop is owned by a mathematician and when the physicist asks "when will the washing be ready?" the mathematician replies "we don't clean clothes, we just sell signs!".

In the case of Heisenberg and also that of Einstein who was helped out by Riemann, this was no joke.

However, soon after Heisenberg's discovery, Schrödinger came up with his equation so physicists happily returned to the study of partial differential equations, and the message of Heisenberg was buried to a great extent. Most of my work has been an attempt to take this discovery of Heisenberg seriously. On reflection, this discovery actually clearly displays the limitation of Riemann's formulation of geometry. If we look at the phase space of an atomic system and follow Riemann's procedure to parametrize its points by finitely many real numbers, we first split the manifold into the levels on which some particular function is constant, but we then need to iterate this process and apply it to the level hypersurfaces. However, according to Heisenberg this doesn't work because as soon as we make the first measurement, we alter the situation drastically. The right way to think about this new phenomenon is to think in terms of a new kind of space in which the coordinates do not commute.

The starting point of noncommutative geometry is to take this new notion of space seriously.

4 Noncommutative Geometry

The basis of noncommutative geometry is twofold. On the one hand there is a wealth of examples of spaces whose coordinate algebra is no longer commutative but which have obvious relevance in physics or mathematics. The first examples came, as we saw above, from phase space in quantum mechanics but there are many others, such as the leaf spaces of foliations, the duals of nonabelian groups, the space of Penrose tilings, the Brillouin zone in solid state physics, the noncommutative tori which appear naturally in string theory and in M-theory compactification, and the Adele class space which as we shall see below provides a natural spectral realisation of zeros of zeta functions. Finally various recent models of space-time itself are interesting examples of noncommutative spaces.

On the other hand the stretching of geometric thinking imposed by passing to noncommutative spaces forces one to rethink about most of our familiar notions. The difficulty is not to add arbitrarily the adjective quantum to our geometric words but to develop far reaching extensions of classical concepts, ranging from the simplest which is measure theory, to the most sophisticated which is geometry itself.

Let us first discuss in greater detail the general principles that allow to construct huge classes of such spaces, it is a vital ingredient indeed since there is no way to build a satisfactory theory without being able to test it on a large variety of examples. We have two principles which allow us to construct examples.

The first is deformation theory which allows us to explore the neighborhood of the commutative world, the second is a new and very important mathematical principle; the quotient operation. Most of the spaces we are concerned with are not defined by naming every one of their points, but by giving a much bigger set and dividing it by an equivalence relation.

It turns out that there are two ways of extending the geometric-algebraic duality

$$\text{Space} \leftrightarrow \text{Commutative algebra} \tag{4.1}$$

between a space X and the algebra of functions on that space, when you want to identify two points a and b. The first way which gives the usual algebra of functions associated to the quotient is to restrict oneself to functions which have the same value at the two points.

$$\mathcal{A} = \{f; f(a) = f(b)\}\,. \tag{4.2}$$

The second way is to keep the two points a and b, but to allow them to 'speak' to each other by using matrices with off-diagonal elements. It consists, instead of taking the subalgebra given by 4-4.2, to adjoin to the algebra of functions on $\{a, b\}$ the identification of a with b. The obtained algebra is the algebra of two by two matrices

$$\mathcal{B} = \left\{f = \begin{bmatrix} f_{aa} & f_{ab} \\ f_{ba} & f_{bb} \end{bmatrix}\right\}\,. \tag{4.3}$$

When one computes the spectrum of this algebra it turns out that it is composed of only one point, so the two points a and b have been identified. As we shall see this second method is very powerful and allows one to construct thousands of very interesting examples. It allows us to refine the above duality of algebraic geometry to,

$$\text{Quotient-Space} \leftrightarrow \text{Noncommutative algebra} \tag{4.4}$$

in the situation where the space one is contemplating is obtained by the operation of quotient.

At first sight it might seem that, as far as the general theory is concerned, passing from the commutative to the noncommutative situation would just be a matter of cleverly rewriting in algebraic terms our familiar geometric notions without using commutativity anywhere. If noncommutative geometry was just that it would be boring indeed. Fortunately, even at the coarsest level which is measure theory, it became clear at the beginning of the seventies that the noncommutative world is full of beautiful totally unexpected facts which have no commutative counterpart whatsoever. The prototype of such facts is the following

$$\text{Noncommutative measure spaces evolve with time!} \tag{4.5}$$

In other words there is a 'god-given' one parameter group of automorphisms of the algebra M of measurable coordinates. It is given by the group homomorphism, ([1])

$$\delta : \mathbb{R} \to \mathrm{Out}(M) = \mathrm{Aut}(M)/\mathrm{Int}(M) \tag{4.6}$$

from the additive group $\mathbb{R}$ to the group of automorphism classes of M modulo inner automorphisms.

I discovered this fact in 1972 when working on the Tomita–Takesaki theory ([2]) and it convinced me that there are amazing features of noncommutative spaces which have no counterpart in the static commutative case.

5 A Basic Example

Let us start with a prototype example of quotient space in which the distinction between the quotient operations (4.2) and (4.3) appears clearly, and which played a key role in 1980 at the early stage of the theory ([40]). This example is the following: consider the 2-torus

$$M = \mathbb{R}^2/\mathbb{Z}^2\,. \tag{5.1}$$

The space X which we contemplate is the space of solutions of the differential equation,

$$dx = \theta dy \qquad x, y \in \mathbb{R}/\mathbb{Z} \tag{5.2}$$

where $\theta \in]0,1[$ is a fixed irrational number.

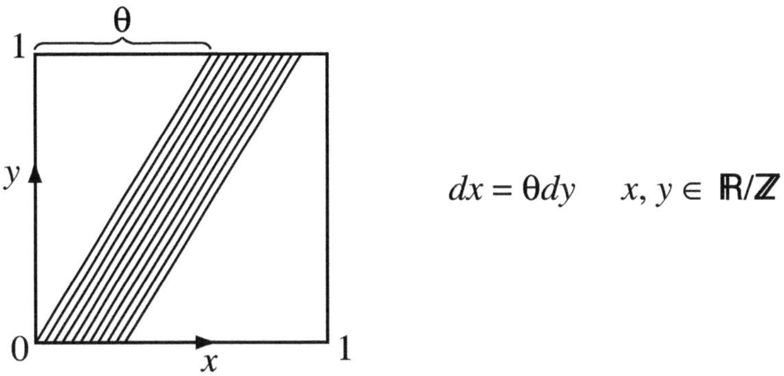

Thus the space we are interested in here is just the space of leaves of the foliation defined by the differential equation (5.2). We can label such a leaf by a point of the transversal given by $y = 0$ which is a circle $S^1 = \mathbb{R}/\mathbb{Z}$, but clearly two points of the transversal which differ by an integer multiple of θ give rise to the same leaf. Thus

$$X = S^1/\theta\mathbb{Z} \tag{5.3}$$

i.e. X is the quotient of S^1 by the equivalence relation which identifies any two points on the orbits of the irrational rotation

$$R_\theta x = x + \theta \mod 1\,. \tag{5.4}$$

When we deal with S^1 as a space in the various categories (smooth, topological, measurable) it is perfectly described by the corresponding algebra of functions,

$$C^\infty(S^1) \subset C(S^1) \subset L^\infty(S^1)\,. \tag{5.5}$$

When one applies the naive operation (4-4.2) to pass to the quotient, one finds, irrespective of which category one works with, the trivial answer

$$\mathcal{A} = \mathbb{C}\,. \tag{5.6}$$

The operation (4.3) however gives very interesting algebras, by no means reduced to $\mathbb{C}$. Elements of the algebra $\mathcal{B}$ associated to the transversal S^1 by the operation (4.3) are just matrices $a(i,j)$ where the indices (i,j) are arbitrary pairs of elements i,j of S^1 which belong to the same leaf, i.e. give the same element of X. The algebraic rules are the same as for ordinary matrices. In the above situation since the equivalence is given by a group action, the construction coincides with the crossed product familiar to algebraist from the theory of central simple algebras.

An element of $\mathcal{B}$ is given by a power series

$$b = \sum_{n \in \mathbb{Z}} b_n U^n \tag{5.7}$$

where each b_n is an element of the algebra (5.5), while the multiplication rule is given by

$$U h U^{-1} = h \circ R_\theta^{-1} . \tag{5.8}$$

Now the algebra (5.5) is generated by the function V on S^1,

$$V(\alpha) = \exp(2\pi i \alpha) \qquad \alpha \in S^1 \tag{5.9}$$

and it follows that $\mathcal{B}$ admits the generating system (U, V) with presentation given by the relation

$$VU = \lambda UV \qquad \lambda = \exp 2\pi i \theta . \tag{5.10}$$

Thus, if for instance we work in the smooth category a generic element b of $\mathcal{B}$ is given by a power series

$$b = \sum_{\mathbb{Z}^2} b_{nm} U^n V^m , \quad b \in \mathcal{S}(\mathbb{Z}^2) \tag{5.11}$$

where $\mathcal{S}(\mathbb{Z}^2)$ is the Schwartz space of sequences of rapid decay on $\mathbb{Z}^2$.

This algebra is by no means trivial and has a very rich and interesting algebraic structure. It is (canonically up to Morita equivalence) associated to the foliation 5-5.2 and the interplay between the geometry of the foliation and the algebraic structure of $\mathcal{B}$ begins by noticing that to a *closed transversal* T of the foliation corresponds canonically a *finite projective module* over $\mathcal{B}$. Elements of the module associated to the transversal T are rectangular matrices, $\xi(i,j)$ where $(i,j) \in T \times S^1$ while i and j belong to the same leaf, i.e. give the same element of X. The right action of $a(i,j) \in \mathcal{B}$ is by matrix multiplication.

From the transversal $x = 0$, one obtains the following right module over $\mathcal{B}$. The underlying linear space is the usual Schwartz space,

$$\mathcal{S}(\mathbb{R}) = \{\xi, \xi(s) \in \mathbb{C} \quad \forall s \in \mathbb{R}\} \tag{5.12}$$

of smooth functions on the real line all of whose derivatives are of rapid decay.

The right module structure is given by the action of the generators U, V

$$(\xi U)(s) = \xi(s + \theta) , \ (\xi V)(s) = e^{2\pi i s} \xi(s) \quad \forall s \in \mathbb{R} . \tag{5.13}$$

One of course checks the relation (5.10), and it is a beautiful fact that as a right module over $\mathcal{B}$ the space $\mathcal{S}(\mathbb{R})$ is *finitely generated* and *projective* (*i.e.* complements to a free module). It follows that it has the correct algebraic attributes to deserve the name of "noncommutative vector bundle" according to the dictionary,

Space	Algebra
Vector bundle	Finite projective module.

The concrete description of the general finite projective modules over $\mathcal{A}_\theta$ is obtained by combining the results of [62], [40], [63]. They are classified up to isomorphism by a pair of integers (p,q) such that $p+q\theta \geq 0$ and the corresponding modules $\mathcal{H}^\theta_{p,q}$ are obtained by the above construction from the transversals given by closed geodesics of the torus M.

The algebraic counterpart of a vector bundle is its space of smooth sections $C^\infty(X,E)$ and one can in particular compute its dimension by computing the trace of the identity endomorphism of E. If one applies this method in the above noncommutative example, one finds

$$\dim_{\mathcal{B}}(\mathcal{S}) = \theta\,. \tag{5.14}$$

The appearance of non-integral dimension is very exciting and displays a basic feature of von Neumann algebras of type II. The dimension of a vector bundle is the only invariant that remains when one looks from the measure theoretic point of view (*i.e.* when one takes the third algebra in (5.5)). The von Neumann algebra which describes the quotient space X from the measure theoretic point of view is the crossed product,

$$R = L^\infty(S^1) >\!\!\lhd_{R_\theta} \mathbb{Z} \tag{5.15}$$

and is the well-known hyperfinite factor of type II_1. In particular the classification of finite projective modules $\mathcal{E}$ over R is given by a positive real number, the Murray and von Neumann *dimension*,

$$\dim_R(\mathcal{E}) \in \mathbb{R}_+\,. \tag{5.16}$$

The next surprise is that even though the *dimension* of the above module is irrational, when we compute the analogue of the first Chern class, *i.e.* of the integral of the curvature of the vector bundle, we obtain an integer. Indeed the two commuting vector fields which span the tangent space for an ordinary (commutative) 2-torus correspond algebraically to two commuting derivations of the algebra of smooth functions. These derivations continue to make sense when the generators U and V of $C^\infty(\mathbb{T}^2)$ no longer commute but satisfy (5.10) so that they generate $\mathcal{B} = C^\infty(\mathbb{T}^2_\theta)$. They are given by the same formulas as in the commutative case,

$$\delta_1 = 2\pi i U \,\frac{\partial}{\partial U}\,, \quad \delta_2 = 2\pi i V\,\frac{\partial}{\partial V} \tag{5.17}$$

so that $\delta_1\left(\sum b_{nm} U^n V^m\right) = 2\pi i \sum n b_{nm} U^n V^m$ and similarly for δ_2. One still has of course

$$\delta_1\delta_2 = \delta_2\delta_1 \tag{5.18}$$

and the δ_j are still derivations of the algebra $\mathcal{B} = C^\infty(\mathbb{T}^2_\theta)$,

$$\delta_j(bb') = \delta_j(b)b' + b\delta_j(b') \quad \forall b, b' \in \mathcal{B}\,. \tag{5.19}$$

The analogues of the notions of connection and curvature of vector bundles are straightforward to obtain ([40]) since a connection is just given by the associated covariant differentiation ∇ on the space of smooth sections. Thus here it is given by a pair of linear operators,

$$\nabla_j : \mathcal{S}(\mathbb{R}) \to \mathcal{S}(\mathbb{R}) \tag{5.20}$$

such that

$$\nabla_j(\xi b) = (\nabla_j \xi)b + \xi\delta_j(b) \quad \forall \xi \in \mathcal{S}\ , b \in \mathcal{B}\,. \tag{5.21}$$

One checks that, as in the usual case, the trace of the curvature $\Omega = \nabla_1\nabla_2 - \nabla_2\nabla_1$, is independent of the choice of the connection. Now the remarkable fact here is that (up to the correct powers of $2\pi i$) the total curvature of $\mathcal{S}$ is an integer. In fact for the following choice of connection the curvature Ω is constant, equal to $1/\theta$ so that the irrational number θ disappears in the total curvature, $\theta \times \frac{1}{\theta}$

$$(\nabla_1\xi)(s) = -\tfrac{2\pi i s}{\theta}\,\xi(s) \quad (\nabla_2\xi)(s) = \xi'(s)\,. \tag{5.22}$$

With this integrality, one could get the wrong impression that the algebra $\mathcal{B} = C^\infty(\mathbb{T}^2_\theta)$ looks very similar to the algebra $C^\infty(\mathbb{T}^2)$ of smooth functions on the 2-torus. A striking difference is obtained by looking at the range of Morse functions. The range of a Morse function on $\mathbb{T}^2$ is of course a connected interval. For the above noncommutative torus $\mathbb{T}^2_\theta$ the range of a Morse function is the spectrum of a real valued function such as

$$h = U + U^* + \mu(V + V^*) \tag{5.23}$$

and it can be a Cantor set, *i.e.* have infinitely many disconnected pieces. This shows that the one dimensional pictures of our space $\mathbb{T}^2_\theta$ are truly different from what they are in the commutative case. The above noncommutative torus $\mathbb{T}^2_\theta$ is the simplest example of noncommutative manifold, it arises naturally not only from foliations but also from the Brillouin zone in the Quantum Hall effect as understood by J. Bellissard, and in M-theory as we shall see next. In the Quantum Hall effect, the above integrality of the total curvature corresponds to the observed integrality of the Hall

conductivity

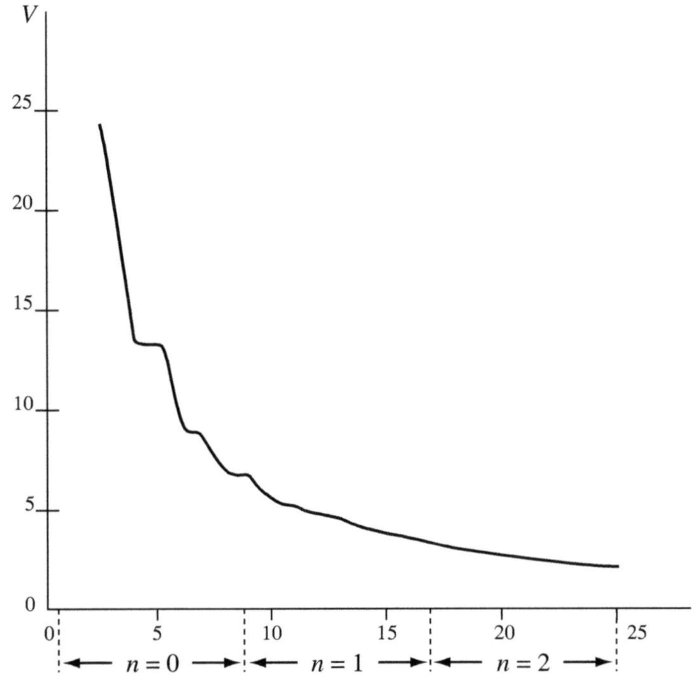

The analogue of the Yang–Mills action functional and the classification of Yang–Mills connections on the noncommutative tori was developed in [64], with the primary goal of finding a "manifold shadow" for these noncommutative spaces. These moduli spaces turned out indeed to fit this purpose perfectly, allowing for instance to find the usual Riemannian space of gauge equivalence classes of Yang–Mills connections as an invariant of the noncommutative metric.

The next surprise came from the natural occurrence (as an unexpected guest) of both the noncommutative tori and the components of the Yang–Mills connections in the classification of the BPS states in M-theory [67].

In the matrix formulation of M-theory the basic equations to obtain periodicity of two of the basic coordinates X_i turn out to be the following,

$$U_i X_j U_i^{-1} = X_j + a\delta_i^j, i = 1, 2 \tag{5.24}$$

where the U_i are unitary gauge transformations.

The multiplicative commutator $U_1 U_2 U_1^{-1} U_2^{-1}$ is then central and in the irreducible case its scalar value $\lambda = \exp 2\pi i\theta$ brings in the algebra of coordinates on the noncommutative torus. The X_j are then the components of the Yang–Mills connections. It is quite remarkable that the same picture emerged from the other information one has about M-theory concerning its relation with 11 dimensional supergravity and that string theory dualities could be interpreted using Morita equivalence. The latter relates the values of θ on an orbit of $SL(2,\mathbb{Z})$ and simply illustrates that the leaf-space of the original foliation is independent of which transversal is used to parametrize it. This type of relation between for instance θ and $1/\theta$ would be invisible in a purely deformation theoretic perturbative expansion like the one given by the Moyal product.

Nekrasov and Schwarz [74] showed that Yang–Mills gauge theory on noncommutative $\mathbb{R}^4$ gives a conceptual understanding of the non-zero B-field desingularization of the moduli space of instantons obtained by perturbing the ADHM equations.

In [75], Seiberg and Witten exhibited the unexpected relation between the standard gauge theory and the noncommutative one, and clarified the limit in which the entire string dynamics is described by a gauge theory on a noncommutative space.

One should understand from the very start that foliations provide an inexhaustible source of interesting examples of noncommutative spaces. In the above example of $\mathbb{T}^2_\theta$ we could make use of the special vector fields on the torus in order to obtain the analogues of elementary notions of differential geometry. It is quite important to develop the general theory independently of these special features and this is what we shall do in section 7. We shall start by the noncommutative analogues of topology and vector bundles which are necessary preliminary steps.

6 Topology

The development of the topological ideas was prompted by the work of Israel Gel'fand, whose C* algebras give the required framework for noncommutative topology. The two main driving forces were the Novikov conjecture on homotopy invariance of higher signatures of ordinary manifolds as well as the Atiyah–Singer Index theorem. It has led, through the work of Atiyah, Singer, Brown, Douglas, Fillmore, Miscenko and Kasparov [4], [5], [6], [7], [8], to the recognition that not only the Atiyah–Hirzebruch K-theory but more importantly the dual K-homology admit Hilbert space techniques and functional analysis as their natural framework. The cycles in the K-homology group $K_*(X)$ of a compact space X are indeed given by Fredholm representations of the C* algebra A of continuous functions on X. The central tool is the Kasparov bivariant K-theory. A basic example of C* algebra to which the theory applies is the group ring of a discrete group and this makes it clear that restricting oneself to commutative algebras is an undesirable assumption.

For a C^* algebra A, let $K_0(A)$, $K_1(A)$ be its K theory groups. Thus $K_0(A)$ is the algebraic K_0 theory of the ring A and $K_1(A)$ is the algebraic K_0 theory of the ring $A \otimes C_0(\mathbb{R}) = C_0(\mathbb{R}, A)$. If $A \to B$ is a morphism of C^* algebras, then there are induced homomorphisms of abelian groups $K_i(A) \to K_i(B)$. Bott periodicity provides a six term K theory exact sequence for each exact sequence $0 \to J \to A \to B \to 0$ of C^* algebras and excision shows that the K groups involved in the exact sequence only depend on the respective C^* algebras. As an exercise to appreciate the power of this abstract tool one should for instance use the six term K theory exact sequence to give a short proof of the Jordan curve theorem.

Discrete groups, Lie groups, group actions and foliations give rise through their convolution algebra to a canonical C^* algebra, and hence to K theory groups. The analytical meaning of these K theory groups is clear as a receptacle for indices of elliptic operators. However, these groups are difficult to compute. For instance, in the case of semi-simple Lie groups the free abelian group with one generator for each irreducible discrete series representation is contained in $K_0\, C_r^* G$ where $C_r^* G$ is the reduced C^* algebra of G. Thus an explicit determination of the K theory in this case in particular involves an enumeration of the discrete series.

We introduced with P. Baum [9] a geometrically defined K theory which specializes to discrete groups, Lie groups, group actions, and foliations. Its main features are its computability and the simplicity of its definition. In the case of semi-simple Lie groups it elucidates the role of the homogeneous space G/K (K the maximal compact subgroup of G) in the Atiyah–Schmid geometric construction of the discrete series [10]. Using elliptic operators we constructed a natural map μ from our geometrically defined K theory groups to the above analytic (*i.e.* C^* algebra) K theory groups. Much progress has been made in the past years to determine the range of validity of the isomorphism between the geometrically defined K theory groups and the above analytic (*i.e.* C^* algebra) K theory groups. We refer to the three Bourbaki seminars [11], [12], [13], for an update on this topic and for a precise account of the various contributions. Among the most important contributions are those of Kasparov and Higson who showed that the conjectured isomorphism holds for all amenable groups, thus proving the Novikov conjecture for all amenable groups and the Kadison conjecture (i.e. the absence of non-trivial idempotents in the reduced C^*-algebra) for all torsion free amenable groups. The conjectured isomorphism also holds for real semi-simple Lie groups thanks in particular to the work of A. Wassermann. Moreover the recent work of V. Lafforgue crossed the barrier of property T, showing that it holds for cocompact subgroups of rank one Lie groups and also of $SL(3,\mathbb{R})$ or of p-adic Lie groups. He also gave the first general conceptual proof of the isomorphism for real or p-adic semi-simple Lie groups (and as a corollary a direct K-theoretic proof of the construction of all discrete series representations by Dirac-induction). The proof of the isomorphism is certainly accessible for all connected locally compact groups. The proof by G. Yu of the analogue (due to J. Roe) of the conjecture in the context of coarse geometry for metric spaces which are uniformly embeddable in hilbert space, and the work of G. Skandalis, J.L. Tu, J. Roe and N. Higson on the groupoid case got very striking consequences such as the injectivity of the map μ for exact $C_r^*(\Gamma)$ due to Kaminker, Guentner and Ozawa, but recent progress due to Gromov, Higson, Lafforgue and Skandalis gives counterexamples to the general conjecture for locally compact groupoids for the simple reason that the functor $G \to K_0(C_r^*(G))$ is not half exact, unlike the functor given by the geometric group. This makes the general problem of computing $K(C_r^*(G))$ really interesting. It shows that besides determining the large class of locally compact groups for which the original conjecture is valid, one should understand how to take homological algebra into account to deal with the correct general formulation.

7 Differential Topology

The development of differential geometric ideas, including de Rham homology, connections and curvature of vector bundles, etc... took place during the eighties thanks to cyclic cohomology which came from two different horizons ([14], [15], [16], [17], [18]).

In the commutative case, for a compact space X, we have at our disposal in K-theory a tool of great relevance, the Chern character

$$\mathrm{ch} : K^*(X) \to H^*(X,\mathbb{Q}) \tag{7.1}$$

which relates the K-theory of X to the cohomology of X. When X is a smooth manifold the Chern character may be calculated explicitly by the differential calculus of forms, currents, connections and curvature. More precisely, given a smooth vector bundle E over X, or equivalently the finite projective module, $\mathcal{E} = C^\infty(X, E)$ over $\mathcal{A} = C^\infty(X)$ of smooth sections of E, the Chern character of E

$$\mathrm{ch}(E) \in H^*(X, \mathbb{R}) \tag{7.2}$$

is represented by the closed differential form:

$$\mathrm{ch}(E) = \mathrm{trace}\big(\exp(\nabla^2/2\pi i)\big) \tag{7.3}$$

for any connection ∇ on the vector bundle E. Any closed de Rham current C on the manifold X determines a map φ_C from $K^*(X)$ to $\mathbb{C}$ by the equality

$$\varphi_C(E) = \langle C, \mathrm{ch}(E)\rangle \tag{7.4}$$

where the pairing between currents and differential forms is the usual one.

One obtains in this way numerical invariants of K-theory classes whose knowledge for arbitrary closed currents C is equivalent to that of $\mathrm{ch}(E)$.

The noncommutative torus gave a striking example where it was obviously worthwhile to adapt the above construction of differential geometry to the noncommutative framework ([40]). As an easy preliminary step towards cyclic cohomology one can reformulate the essential ingredient of the construction without direct reference to derivations in the following way ([17]).

By a cycle of dimension n we mean a triple $(\Omega, d, \int)$ where (Ω, d) is a graded differential algebra, and $\int : \Omega^n \to \mathbb{C}$ is a closed graded trace on Ω.

Let $\mathcal{A}$ be an algebra over $\mathbb{C}$. Then a cycle over $\mathcal{A}$ is given by a cycle $(\Omega, d, \int)$ and a homomorphism $\rho : \mathcal{A} \to \Omega^0$.

Thus a *cycle* over an algebra $\mathcal{A}$ is a way to embed $\mathcal{A}$ as a subalgebra of a differential graded algebra (DGA). We shall see in (f) below the role of the graded trace.

The usual notions of connection and curvature extend in a straightforward manner to this context ([17]).

Let $\mathcal{A} \xrightarrow{\rho} \Omega$ be a cycle over $\mathcal{A}$, and $\mathcal{E}$ a finite projective module over $\mathcal{A}$. Then a connection ∇ on $\mathcal{E}$ is a linear map $\nabla : \mathcal{E} \to \mathcal{E} \otimes_{\mathcal{A}} \Omega^1$ such that

$$\nabla(\xi x) = (\nabla \xi)x + \xi \otimes d\rho(x)\ , \quad \forall \xi \in \mathcal{E}\ ,\ x \in \mathcal{A}\,. \tag{7.5}$$

Here $\mathcal{E}$ is a *right* module over $\mathcal{A}$ and Ω^1 is considered as a bimodule over $\mathcal{A}$ using the homomorphism $\rho : \mathcal{A} \to \Omega^0$ and the ring structure of Ω^*. Let us list a number of easy properties ([17]):

(a) Let $e \in \mathrm{End}_{\mathcal{A}}(\mathcal{E})$ be an idempotent and ∇ a connection on $\mathcal{E}$; then $\xi \mapsto (e \otimes 1)\nabla\xi$ is a connection on $e\mathcal{E}$.

(b) Any finite projective module $\mathcal{E}$ admits a connection.

(c) The space of connections is an affine space over the vector space

$$\mathrm{Hom}_{\mathcal{A}}(\mathcal{E}, \mathcal{E} \otimes_{\mathcal{A}} \Omega^1) . \tag{7.6}$$

(d) Any connection ∇ extends uniquely to a linear map of $\widetilde{\mathcal{E}} = \mathcal{E} \otimes_{\mathcal{A}} \Omega$ into itself such that

$$\nabla(\xi \otimes \omega) = (\nabla \xi)\omega + \xi \otimes d\omega\,, \quad \forall \xi \in \mathcal{E}\,,\ \omega \in \Omega\,. \tag{7.7}$$

(e) The map $\theta = \nabla^2$ of $\widetilde{\mathcal{E}}$ to $\widetilde{\mathcal{E}}$ is an endomorphism: $\theta \in \mathrm{End}_{\Omega}(\widetilde{\mathcal{E}})$ and with $\delta(T) = \nabla T - (-1)^{degT} T\nabla$, one has $\delta^2(T) = \theta T - T\theta$ for all $T \in \mathrm{End}_{\Omega}(\widetilde{\mathcal{E}})$.

(f) For n even, $n = 2m$, the equality

$$\langle [\mathcal{E}], [\tau] \rangle = \frac{1}{m!} \int \theta^m\,, \tag{7.8}$$

defines an additive map from the K-group $K_0(\mathcal{A})$ to the scalars.

Of course one can reformulate (f) by dualizing the closed graded trace $\int$, i.e. by considering the homology of the quotient $\Omega/[\Omega,\Omega]$ ([60]) and one might be tempted at first sight to assert that a noncommutative algebra often comes naturally equipped with a natural embedding in a DGA which should suffice for the Chern character. This however would be rather naive and would overlook for instance the role of *integral* cycles for which the above additive map only affects *integer* values.

The starting point of cyclic cohomology is the ability to compare different cycles on the same algebra. In fact the invariant of K-theory defined in (f) by a given cycle only depends on the multilinear form

$$\varphi(a^0, \ldots, a^n) = \int \rho(a^0) d(\rho(a^1)) d(\rho(a^2)) \ldots d(\rho(a^n)) \qquad \forall\, a^j \in \mathcal{A} \tag{7.9}$$

(called the character of the cycle) and the functionals thus obtained are exactly those multilinear forms on $\mathcal{A}$ such that

φ is *cyclic* i.e.

$$\varphi(a^0, a^1, \ldots, a^n) = (-1)^n\, \varphi(a^1, a^2, \ldots, a^0) \qquad \forall\, a_j \in \mathcal{A}\,, \tag{7.10}$$

$b\varphi = 0$ where

$$(b\varphi)(a^0, \ldots, a^{n+1}) = \sum_0^n (-1)^j \varphi(a^0, \ldots, a^j a^{j+1}, \ldots, a^{n+1}) + (-1)^{n+1} \varphi(a^{n+1} a^0, a^1, \ldots, a^n)\,. \tag{7.11}$$

This second condition means that φ is a Hochschild cocycle. In particular such a φ admits a Hochschild class

$$I(\varphi) \in H^n(\mathcal{A}, \mathcal{A}^*) \tag{7.12}$$

for the Hochschild cohomology of $\mathcal{A}$ with coefficients in the bimodule $\mathcal{A}^*$ of linear forms on $\mathcal{A}$.

The n-dimensional *cyclic cohomology* of $\mathcal{A}$ is simply the cohomology $HC^n(\mathcal{A})$ of the *subcomplex* of the Hochschild complex given by cochains which are *cyclic* i.e. fulfil (7.10). One has an obvious "forgetful" map

$$HC^n(\mathcal{A}) \xrightarrow{I} H^n(\mathcal{A}, \mathcal{A}^*) \tag{7.13}$$

but the real story starts with the following long exact sequence which allows in many cases to compute cyclic cohomology from the B operator acting on Hochschild cohomology:

Theorem 1. *The following triangle is exact:*

$$\begin{array}{ccc} & H^*(\mathcal{A}, \mathcal{A}^*) & \\ {}^{B}\swarrow & & \nwarrow^{I} \\ HC^*(\mathcal{A}) & \xrightarrow{S} & HC^*(\mathcal{A}) \end{array}$$

The operator S is obtained by tensoring cycles by the canonical 2-dimensional generator of the cyclic cohomology of $\mathbb{C}$.

The operator B is explicitly defined at the cochain level by the equality

$$B = AB_0\,,\ B_0\varphi(a^0, \ldots, a^{n-1}) = \varphi(1, a^0, \ldots, a^{n-1}) - (-1)^n \varphi(a^0, \ldots, a^{n-1}, 1)$$

$$(A\psi)(a^0, \ldots, a^{n-1}) = \sum_0^{n-1} (-1)^{(n-1)j} \psi(a^j, a^{j+1}, \ldots, a^{j-1})\,.$$

Its conceptual origin lies in the notion of cobordism of cycles which allows us to compare different inclusion of $\mathcal{A}$ in DGA as follows. By a *chain* of dimension $n+1$ we shall mean a quadruple $(\Omega, \partial\Omega, d, \int)$ where Ω and $\partial\Omega$ are differential graded algebras of dimensions $n+1$ and n with a given surjective morphism $r : \Omega \to \partial\Omega$ of degree 0, and where $\int : \Omega^{n+1} \to \mathbb{C}$ is a graded trace such that

$$\int d\omega = 0\,, \quad \forall\, \omega \in \Omega^n \text{ such that } r(\omega) = 0\,. \tag{7.14}$$

By the *boundary* of such a chain we mean the cycle $(\partial\Omega, d, \int')$ where for $\omega' \in (\partial\Omega)^n$ one takes $\int' \omega' = \int d\omega$ for any $\omega \in \Omega^n$ with $r(\omega) = \omega'$. One easily checks, using the surjectivity of r, that $\int'$ is a graded trace on $\partial\Omega$ and is closed by construction.

We shall say that two cycles $\mathcal{A} \xrightarrow{\rho} \Omega$ and $\mathcal{A} \xrightarrow{\rho'} \Omega'$ over $\mathcal{A}$ are *cobordant* if there exists a chain Ω'' with boundary $\Omega \oplus \widetilde{\Omega}'$ (where $\widetilde{\Omega}'$ is obtained from Ω' by changing the sign of $\int$) and a homomorphism $\rho'' : \mathcal{A} \to \Omega''$ such that $r \circ \rho'' = (\rho, \rho')$.

The conceptual role of the operator B is clarified by the following result,

Theorem 2. *Two cycles over $\mathcal{A}$ are cobordant if and only if their characters $\tau_1, \tau_2 \in HC^n(\mathcal{A})$ differ by an element of the image of B, where*

$$B : H^{n+1}(\mathcal{A}, \mathcal{A}^*) \to HC^n(\mathcal{A}) .$$

The operators b, B given as above by

$$(b\varphi)(a^0, \dots, a^{n+1}) = \sum_0^n (-1)^j \varphi(a^0, \dots, a^j a^{j+1}, \dots, a^{n+1}) + (-1)^{n+1} \varphi(a^{n+1} a^0, a^1, \dots, a^n)$$

$$B = AB_0 , \quad B_0\varphi(a^0, ..., a^{n-1}) = \varphi(1, a^0, ..., a^{n-1}) - (-1)^n \varphi(a^0, ..., a^{n-1}, 1)$$

$$(A\psi)(a^0, ..., a^{n-1}) = \sum_0^{n-1} (-1)^{(n-1)j} \, \psi(a^j, a^{j+1}, ..., a^{j-1})$$

satisfy $b^2 = B^2 = 0$ and $bB = -Bb$ and periodic cyclic cohomology which is the inductive limit of the $HC^n(\mathcal{A})$ under the periodicity map S admits an equivalent description as the cohomology of the (b, B) bicomplex.

With these notation one has the following formula for the Chern character of the class of an idempotent e, up to normalization one has

$$Ch_n(e) = (e - 1/2) \otimes e \otimes e \otimes ... \otimes e, \tag{7.15}$$

where $\otimes$ appears 2n times in the right-hand side of the equation.

Both the Hochschild and Cyclic cohomologies of the algebra $\mathcal{A} = C^\infty(V)$ of smooth functions on a manifold V were computed in [16] and [17].

Let V be a smooth compact manifold and $\mathcal{A}$ the locally convex topological algebra $C^\infty(V)$. Then the following map $\varphi \to C_\varphi$ is a canonical isomorphism of the continuous Hochschild cohomology group $H^k(\mathcal{A}, \mathcal{A}^*)$ with the space of k-dimensional de Rham currents on V:

$$\langle C_\varphi, f^0 \, d f^1 \wedge \dots \wedge d f^k \rangle = \frac{1}{k!} \sum_{\sigma \in S_k} \varepsilon(\sigma) \varphi(f^0, f^{\sigma(1)}, \dots, f^{\sigma(k)})$$

$\forall \, f^0, \dots, f^k \in C^\infty(V)$.

Under the isomorphism C the operator $I \circ B : H^k(\mathcal{A}, \mathcal{A}^*) \to H^{k-1}(\mathcal{A}, \mathcal{A}^*)$ is (k times) the de Rham boundary b for currents.

Theorem 3. *Let $\mathcal{A}$ be the locally convex topological algebra $C^\infty(V)$. Then*

1) *For each k, $HC^k(\mathcal{A})$ is canonically isomorphic to the direct sum*
$$\operatorname{Ker} b \oplus H_{k-2}(V,\mathbb{C}) \oplus H_{k-4}(V,\mathbb{C}) \oplus \cdots$$
where $H_q(V,\mathbb{C})$ is the usual de Rham homology of V and b the de Rham boundary.
2) *The periodic cyclic cohomology of $C^\infty(V)$ is canonically isomorphic to the de Rham homology $H_*(V,\mathbb{C})$, with filtration by dimension.*

As soon as we pass to the noncommutative case, more subtle phenomena arise. Thus for instance the filtration of the periodic cyclic homology (dual to periodic cyclic cohomology) together with the lattice $K_0(\mathcal{A}) \subset HC_{\mathrm{ev}}(\mathcal{A})$, for $\mathcal{A} = C^\infty(\mathbb{T}^2_\theta)$, gives an even analogue of the Jacobian of an elliptic curve. More precisely the filtration of HC_{ev} yields a canonical foliation of the torus HC_{ev}/K_0 and one can show that the foliation algebra associated as above to the canonical transversal segment $[0,1]$ is isomorphic to $C^\infty(\mathbb{T}^2_\theta)$.

A simple example of cyclic cocycle on a nonabelian group ring is provided by the following formula. Any *group cocycle* $c \in H^*(B\Gamma) = H^*(\Gamma)$ gives rise to a cyclic cocycle φ_c on the algebra $\mathcal{A} = \mathbb{C}\Gamma$
$$\varphi_c(g_0, g_1, \dots, g_n) = \begin{cases} 0 & \text{if} \quad g_0 \cdots g_n \neq 1 \\ c(g_1,\dots,g_n) & \text{if} \quad g_0 \cdots g_n = 1 \end{cases}$$
where $c \in Z^n(\Gamma,\mathbb{C})$ is suitably normalized, and the formula is extended by linearity to $\mathbb{C}\Gamma$. The cyclic cohomology of group rings is given by,

Theorem 4 [22]. *Let Γ be a discrete group, $\mathcal{A} = \mathbb{C}\Gamma$ its group ring.*

a) *The Hochschild cohomology $H^*(\mathcal{A},\mathcal{A}^*)$ is canonically isomorphic to the cohomology $H^*((B\Gamma)^{\mathbb{S}^1},\mathbb{C})$ of the free loop space of the classifying space of Γ.*
b) *The cyclic cohomology $HC^*(\mathcal{A})$ is canonically isomorphic to the $\mathbb{S}^1$-equivariant cohomology $H^*_{\mathbb{S}^1}((B\Gamma)^{\mathbb{S}^1},\mathbb{C})$.*

The role of the free loop space in this theorem is not accidental and is clarified in general by the equality
$$B\Lambda = BS^1$$
of the classifying space $B\Lambda$ of the *cyclic category* with the classifying space of the compact group S^1. We refer to appendix XVIII for this point.

As we saw in section 5 the integral curvature of vector bundles on $\mathbb{T}^2_\theta$ was surprisingly giving an integer, in spite of the irrationality of θ. The conceptual understanding of this type of integrality result lies in the existence of a natural lattice of *integral cycles* which we now describe.

DEFINITION. *Let $\mathcal{A}$ be an algebra, a Fredholm module over $\mathcal{A}$ is given by:*

1) *a representation of $\mathcal{A}$ in a Hilbert space $\mathcal{H}$;*
2) *an operator $F = F^*$, $F^2 = 1$, on $\mathcal{H}$ such that*

$$[F, a] \text{ is a compact operator for any } a \in \mathcal{A}.$$

Such a Fredholm module will be called *odd.* An *even* Fredholm module is given by an odd Fredholm module $(\mathcal{H}, F)$ as above together with a $\mathbb{Z}/2$ grading γ, $\gamma = \gamma^*$, $\gamma^2 = 1$ of the Hilbert space $\mathcal{H}$ such that:

a) $\gamma a = a\gamma \; \forall \, a \in \mathcal{A}$
b) $\gamma F = -F\gamma$.

The above definition is, up to trivial changes, the same as Atiyah's definition [4] of abstract elliptic operators, and the same as Kasparov's definition [8] for the cycles in K-homology, $KK(A, \mathbb{C})$, when A is a C^*-algebra.

The main point is that a Fredholm module over an algebra $\mathcal{A}$ gives rise in a very simple manner to a DGA containing $\mathcal{A}$. One simply defines Ω^k as the linear span of operators of the form,

$$\omega = a^0 \, [F, a^1] \dots [F, a^k] \qquad a^j \in \mathcal{A}$$

and the differential is given by

$$d\omega = F\omega - (-1)^k \, \omega F \qquad \forall \, \omega \in \Omega^k \, .$$

One easily checks that the ordinary product of operators gives an algebra structure, $\Omega^k \, \Omega^\ell \subset \Omega^{k+\ell}$ and that $d^2 = 0$ owing to $F^2 = 1$.

Moreover if one assumes that the size of the differential $da = [F, a]$ is controlled, i.e. that

$$|da|^{n+1} \quad \text{is trace class,}$$

then one obtains a natural closed graded trace of degree n by the formula,

$$\int \omega = \operatorname{Trace}(\omega)$$

(with the supertrace $\operatorname{Trace}(\gamma\omega)$ in the even case, see [36] for details).

Hence the original Fredholm module gives rise to a *cycle* over $\mathcal{A}$. Such cycles have the remarkable *integrality* property that when we pair them with the K theory of $\mathcal{A}$ we only get *integers* as follows from an elementary index formula ([36]).

We let $Ch_*(\mathcal{H}, F) \in HC^n(\mathcal{A})$ be the character of the cycle associated to a Fredholm module $(\mathcal{H}, F)$ over $\mathcal{A}$. This formula defines the Chern character in K-homology.

Cyclic cohomology got many applications [21], it led for instance to the proof of the Novikov conjecture for hyperbolic groups [19]. Basically, by extending the Chern–Weil characteristic classes to the general framework it allows for many concrete computations of differential geometric nature on noncommutative spaces. It also showed the depth of the relation between the classification of factors and the geometry of foliations.

Von Neumann algebras arise very naturally in geometry from foliated manifolds (V, F). The von Neumann algebra $L^\infty(V, F)$ of a foliated manifold is easy to describe, its elements are random operators $T = (T_f)$, i.e. bounded measurable families of operators T_f parametrized by the leaves f of the foliation. For each leaf f the operator T_f acts in the Hilbert space $L^2(f)$ of square integrable densities on the manifold f. Two random operators are identified if they are equal for almost all leaves f (i.e. a set of leaves whose union in V is negligible). The algebraic operations of sum and product are given by,

$$(T_1 + T_2)_f = (T_1)_f + (T_2)_f\,, \quad (T_1\, T_2)_f = (T_1)_f (T_2)_f\,, \tag{7.16}$$

i.e. are effected pointwise.

All types of factors occur from this geometric construction and the continuous dimensions of Murray and von-Neumann play an essential role in the longitudinal index theorem.

Using cyclic cohomology together with the following simple fact,

$$\text{“A connected group can only act trivially on a homotopy invariant cohomology theory”}, \tag{7.17}$$

one proves (cf. [20]) that for any codimension one foliation F of a compact manifold V with non-vanishing Godbillon–Vey class one has,

$$\mathrm{Mod}(M) \text{ has finite covolume in } \mathbb{R}_+^*\,, \tag{7.18}$$

where $\mathrm{Mod}(M)$ is the flow of weights of $M = L^\infty(V, F)$.

In the recent years J. Cuntz and D. Quillen ([23], [24], [25]) have developed a powerful new approach to cyclic cohomology which allowed them to prove excision in full generality.

8 Calculus and Infinitesimals

The central notion of noncommutative geometry comes from the identification of the noncommutative analogue of the two basic concepts in Riemann's formulation of Geometry, namely those of manifold and of infinitesimal line element. Both of these noncommutative analogues are of spectral nature and combine to give rise to the notion of spectral triple and spectral manifold, which will be described below. We shall first describe an operator theoretic framework for the calculus of infinitesimals which will provide a natural home for the line element ds.

I first have to make a little excursion, and I want it as naive as possible. I want to turn back to an extremely naive question about what is an infinitesimal. Let me first explain one answer that was proposed for this intuitive idea of infinitesimal and let me explain why this answer is not satisfactory and then give another answer which hopefully is satisfactory. So, I remember quite a long time ago to have seen an answer which was proposed by non-standard analysis. The book I was reading [78] began with the following problem:

You play a game of throwing darts at some target called Ω

and the question which is asked is: what is the probability $dp(x)$ that actually when you throw the dart it lands exactly at a given point $x \in \Omega$? Then the following argument was given: certainly this probability $dp(x)$ is smaller than $1/2$ because you can cut the target into two equal halves, only one of which contains x. For the same reason $dp(x)$ is smaller than $1/4$, and so on and so forth. So what you find out is that $dp(x)$ is smaller than any positive real number ϵ. On the other hand, if you give the answer that $dp(x)$ is 0, this is not really satisfactory, because whenever you send the dart it will land somewhere. So now, if you ask a mathematician about this naive question, he might very well answer: well, $dp(x)$ is a 2-form, or it's a measure, or something like that. But then you can try to ask him more precise questions, for instance "what is the exponential of $-1/dp(x)$". And then it will be hard for him to give a satisfactory answer, because you know that the Taylor expansion of the function $f(y) = e^{-1/y}$ is zero at $y = 0$. Now the book I was reading claimed to give an answer, and it was what is called a non-standard number. So I worked on this theory for some time, learning some logics, until eventually I realized there was a very bad obstruction preventing one to get concrete answers. It is the following: it's a little lemma that one can easily prove, that if you are given a non-standard number you can canonically produce a subset of the interval which is not Lebesgue measurable. Now we know from logic (from results of Paul Cohen and Solovay) that it will forever be impossible to produce explicitly a subset of the real numbers, of the interval $[0,1]$, say, that is not Lebesgue measurable. So, what this says is that for instance in this example, nobody will actually be able to name a non-standard number. A non-standard number is some sort of chimera which is impossible to grasp and certainly not a concrete object. In fact when you look at non-standard analysis you find out that except for the use of ultraproducts, which is very efficient, it just shifts the order in logic by one step; it's not doing much more. Now, what I want to explain is that to the above naive question there is a very beautiful and simple answer which is provided by quantum mechanics. This answer will be obtained just by going through the usual dictionary of quantum mechanics, but looking at it more closely. So, let us thus look at the first two lines of the following dictionary which translates classical notions into the language of operators in the Hilbert space $\mathcal{H}$:

Complex variable	Operator in $\mathcal{H}$
Real variable	Selfadjoint operator
Infinitesimal	Compact operator
Infinitesimal of order α	Compact operator with characteristic values μ_n satisfying $\mu_n = O(n^{-\alpha})$, $n \to \infty$
Integral of an infinitesimal of order 1	$⨍ T =$ Coefficient of logarithmic divergence in the trace of T.

The first two lines of the dictionary are familiar from quantum mechanics. The range of a complex variable corresponds to the *spectrum* of an operator. The holomorphic functional calculus gives a meaning to $f(T)$ for all holomorphic functions f on the spectrum of T. It is only holomorphic functions which operate in this generality which reflects the difference between complex and real analysis. When $T = T^*$ is selfadjoint then $f(T)$ has a meaning for all Borel functions f.

The size of the infinitesimal $T \in \mathcal{K}$ is governed by the order of decay of the sequence of characteristic values $\mu_n = \mu_n(T)$ as $n \to \infty$. In particular, for all real positive α the following condition defines infinitesimals of order α:

$$\mu_n(T) = O(n^{-\alpha}) \quad \text{when } n \to \infty \tag{8.1}$$

(i.e. there exists $C > 0$ such that $\mu_n(T) \leq Cn^{-\alpha} \quad \forall n \geq 1$). Infinitesimals of order α also form a two–sided ideal and moreover,

$$T_j \text{ of order } \alpha_j \Rightarrow T_1 T_2 \text{ of order } \alpha_1 + \alpha_2 \,. \tag{8.2}$$

Hence, apart from commutativity, intuitive properties of the infinitesimal calculus are fulfilled.

Since the size of an infinitesimal is measured by the sequence $\mu_n \downarrow 0$ it might seem that one does not need the operator formalism at all, and that it would be enough to replace the ideal $\mathcal{K}$ in $\mathcal{L}(\mathcal{H})$ by the ideal $c_0(\mathbb{N})$ of sequences converging to zero in the algebra $\ell^\infty(\mathbb{N})$ of bounded sequences. A variable would just be a bounded sequence, and an infinitesimal a sequence μ_n, $\mu_n \to 0$. However, this commutative version does not allow for the existence of variables with range a continuum since all elements of $\ell^\infty(\mathbb{N})$ have a point spectrum and a discrete spectral measure. Only *noncommutativity* of $\mathcal{L}(\mathcal{H})$ allows for the coexistence of variables with Lebesgue spectrum together with infinitesimal variables. As we shall see shortly, it is precisely this lack of commutativity between the line element and the coordinates on a space that will provide the measurement of distances.

The integral is obtained by the following analysis, mainly due to Dixmier ([28]), of the logarithmic divergence of the partial traces

$$\text{Trace}_N(T) = \sum_0^{N-1} \mu_n(T) \,, \quad T \geq 0 \,. \tag{8.3}$$

In fact, it is useful to define $\text{Trace}_\Lambda(T)$ for any positive real $\Lambda > 0$ by piecewise affine interpolation for noninteger Λ.

Define for all order 1 operators $T \geq 0$

$$\tau_\Lambda(T) = \frac{1}{\log \Lambda} \int_e^\Lambda \frac{\text{Trace}_\mu(T)}{\log \mu} \, \frac{d\mu}{\mu} \tag{8.4}$$

which is the Cesaro mean of the function $\text{Trace}_\mu(T)/\log \mu$ over the scaling group $\mathbb{R}_+^*$.

For $T \geq 0$, an infinitesimal of order 1, one has

$$\text{Trace}_\Lambda(T) \leq C \log \Lambda \tag{8.5}$$

so that $\tau_\Lambda(T)$ is bounded. The essential property is the following *asymptotic additivity* of the coefficient $\tau_\Lambda(T)$ of the logarithmic divergence (8.5):

$$|\tau_\Lambda(T_1 + T_2) - \tau_\Lambda(T_1) - \tau_\Lambda(T_2)| \leq 3C \, \frac{\log(\log \Lambda)}{\log \Lambda} \tag{8.6}$$

for $T_j \geq 0$.

An easy consequence of (8.6) is that any limit point τ of the nonlinear functionals τ_Λ for $\Lambda \to \infty$ defines a positive and linear trace on the two–sided ideal of infinitesimals of order 1,

In practice the choice of the limit point τ is irrelevant because in all important examples T is a *measurable* operator, i.e.:

$$\tau_\Lambda(T) \text{ converges when } \Lambda \to \infty\,. \tag{8.7}$$

Thus the value $\tau(T)$ is independent of the choice of the limit point τ and is denoted

$$⨍ T\,. \tag{8.8}$$

The first interesting example is provided by pseudodifferential operators T on a differentiable manifold M. When T is of order 1 in the above sense, it is measurable and $⨍ T$ is the noncommutative residue of T ([29]). It has a local expression in terms of the distribution kernel $k(x,y)$, $x, y \in M$. For T of order 1 the kernel $k(x,y)$ diverges logarithmically near the diagonal,

$$k(x,y) = -a(x) \log |x-y| + 0(1) \text{ (for } \; y \to x) \tag{8.9}$$

where $a(x)$ is a 1–density independent of the choice of Riemannian distance $|x-y|$. Then one has (up to normalization),

$$⨍ T = \int_M a(x)\,. \tag{8.10}$$

The right-hand side of this formula makes sense for all pseudodifferential operators (cf. [29]) since one can see that the kernel of such an operator is asymptotically of the form

$$k(x,y) = \sum a_k(x, x-y) - a(x) \log |x-y| + 0(1) \tag{8.11}$$

where $a_k(x,\xi)$ is homogeneous of degree $-k$ in ξ, and the 1–density $a(x)$ is defined intrinsically.

The same principle of extension of $⨍$ to infinitesimals of order < 1 works for hypoelliptic operators and more generally as we shall see below, for spectral triples whose dimension spectrum is simple.

We can now go back to our initial naive question about the target and the darts, we find that quantum mechanics gives us an obvious infinitesimal which answers the question: it is the inverse of the Dirichlet Laplacian for the domain Ω. Thus there is now a clear meaning for the exponential of $-1/dp$, that's the well-known heat kernel which is an infinitesimal of arbitrarily large order as we expected from the Taylor expansion.

From the H. Weyl theorem on the asymptotic behavior of eigenvalues of Δ it follows that dp is of order 1, and that given a function f on Ω the product $f\,dp$ is measurable, while

$$⨍ f\,dp = \int_\Omega f(x_1, x_2)\,dx_1 \wedge dx_2 \tag{8.12}$$

gives the ordinary integral of f with respect to the measure given by the area of the target.

9 Spectral Triples

In this section we shall come back to the two basic notions introduced by Riemann in the classical framework, those of *manifold* and of *line element*. We shall see that both of these notions adapt remarkably well to the noncommutative framework and this will lead us to the notion of spectral manifold which noncommutative geometry is based on.

In ordinary geometry of course you can give a manifold by a cooking recipe, by charts and local diffeomorphisms, and one could be tempted to propose an analogous cooking recipe in the noncommutative case. This is pretty much what is achieved by the general construction of the algebras of foliations and it is a good test of any general idea that it should at least cover that large class of examples.

But at a more conceptual level, it was recognized long ago by geometers that the main quality of the homotopy type of an oriented manifold is to satisfy Poincaré duality not only in ordinary homology but also in K-homology. Poincaré duality in ordinary homology is not sufficient to describe homotopy type of manifolds [30] but D. Sullivan [31] showed (in the simply connected PL case of dimension ≥ 5 ignoring 2-torsion) that it is sufficient to replace ordinary homology by KO-homology. Moreover the Chern character of the KO-homology fundamental class contains all the rational information on the Pontrjagin classes.

The characteristic property of *differentiable manifolds* which is carried over to the noncommutative case is *Poincaré duality* in KO-homology [31].

Moreover, as we saw above in the discussion of Fredholm modules, K-homology admits a fairly simple definition in terms of Hilbert space and Fredholm representations of algebras.

For an ordinary manifold the choice of the fundamental cycle in K-homology is a refinement of the choice of orientation of the manifold and in its simplest form is a choice of Spin-structure. Of course the role of a spin structure is to allow for the construction of the corresponding Dirac operator which gives a corresponding Fredholm representation of the algebra of smooth functions.

What is rewarding is that this will not only guide us towards the notion of noncommutative manifold but also to a formula, of operator theoretic nature, for the line element ds.

The infinitesimal unit of length"ds" should be an infinitesimal in the sense of section 8 and one way to get an intuitive understanding of the formula for ds is to consider Feynman diagrams which physicist use currently in the computations of quantum field theory. Let us contemplate the diagram

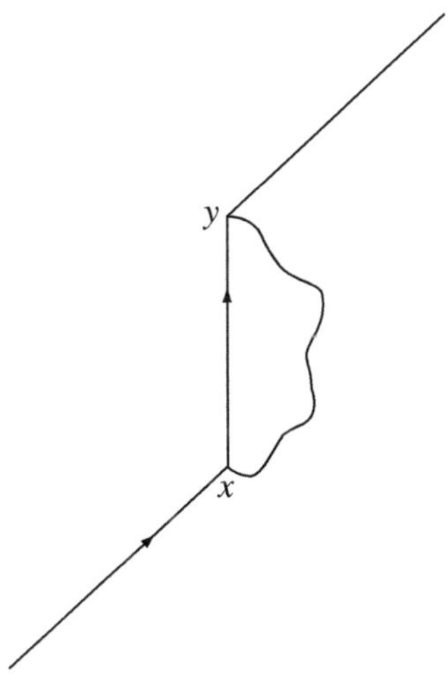

which is involved in the computation of the self-energy of an electron in QED. The two points x and y of space-time at which the photon (the wiggly line) is emitted and reabsorbed are very close by and our ansatz for ds will be at the intuitive level,

$$ds = \times\!\!\longrightarrow\!\!\times\ . \tag{9.1}$$

The right-hand side has good meaning in physics, it is called the Fermion propagator and is given by

$$\times\!\!\longrightarrow\!\!\times = D^{-1} \tag{9.2}$$

where D is the Dirac operator.

We thus arrive at the following basic ansatz,

$$ds = D^{-1}\,. \tag{9.3}$$

In some sense it is simpler than the ansatz giving ds^2 as $g_{\mu\nu}\,dx^\mu\,dx^\nu$, the point being that the spin structure allows really to extract the square root of ds^2 (as is well known Dirac found the corresponding operator as a differential square root of a Laplacian).

The first thing we need to do is to check that we are still able to measure distances with our "unit of length" ds. In fact we saw in the discussion of the quantized calculus that variables with continuous range cannot commute with "infinitesimals" such as ds and it is thus not very surprising that this lack of commutativity allows us to compute, in the classical Riemannian case, the geodesic distance $d(x,y)$ between two points. The precise formula is

$$d(x,y) = \mathrm{Sup}\big\{|f(x) - f(y)|\,;\ f \in \mathcal{A}\,,\ \|[D,f]\| \le 1\big\} \tag{9.4}$$

where $D = ds^{-1}$ as above and $\mathcal{A}$ is the algebra of smooth functions. Note that if ds has the dimension of a length L, then D has dimension L^{-1} and the above expression for $d(x,y)$ also has the dimension of a length.

Thus we see in the classical geometric case that both the fundamental cycle in K-homology and the metric are encoded in the *spectral triple* $(\mathcal{A}, \mathcal{H}, D)$ where $\mathcal{A}$ is the algebra of functions acting in the Hilbert space $\mathcal{H}$ of spinors, while D is the Dirac operator.

To get familiar with this notion one should check that we recover the volume form of the Riemannian metric by the equality (valid up to a normalization constant [36])

$$⨍ f\,|ds|^n = \int_{M_n} f\,\sqrt{g}\;d^n x \tag{9.5}$$

but the first interesting point is that besides this coherence with the usual computations there are new simple questions we can ask now such as "what is the two-dimensional measure of a four manifold" in other words "what is its area ?". Thus one should compute

$$⨍ ds^2 \tag{9.6}$$

It is obvious from invariant theory that this should be proportional to the Hilbert–Einstein action but doing the direct computation is a worthwhile exercise (cf. [52], [51]), the exact result being

$$\int\!\!\!\!\!- ds^2 = \frac{-1}{48\pi^2} \int_{M_4} r \sqrt{g}\, d^4x \tag{9.7}$$

where as above $dv = \sqrt{g}\, d^4x$ is the volume form, $ds = D^{-1}$ the length element, *i.e.* the inverse of the Dirac operator and r is the scalar curvature.

In the general framework of Noncommutative Geometry the confluence of the Hilbert space incarnation of the two notions of metric and fundamental class for a manifold led very naturally to define a geometric space as given by a *spectral triple:*

$$(\mathcal{A}, \mathcal{H}, D) \tag{9.8}$$

where $\mathcal{A}$ is a concrete algebra of coordinates represented on a Hilbert space $\mathcal{H}$ and the operator D is the inverse of the line element

$$ds = 1/D\,. \tag{9.9}$$

This definition is entirely spectral; the elements of the algebra are operators, the points, if they exist, come from the joint spectrum of operators and the line element is an operator.

The basic properties of such spectral triples are easy to formulate and do not make any reference to the commutativity of the algebra $\mathcal{A}$. They are

$$[D, a] \text{ is bounded for any } a \in \mathcal{A}\,, \tag{9.10}$$

$$D = D^* \text{ and } (D + \lambda)^{-1} \text{ is a compact operator } \forall\, \lambda \notin \mathbb{C}\,. \tag{9.11}$$

(Of course D is an *unbounded* operator).

There is no difficulty to adapt the above formula for the distance in the general noncommutative case, one uses the same, the points x and y being replaced by arbitrary states φ and ψ on the algebra $\mathcal{A}$. Recall that a state is a normalized positive linear form on $\mathcal{A}$ such that $\varphi(1) = 1$,

$$\varphi : \bar{\mathcal{A}} \to \mathbb{C}\,,\ \varphi(a^* a) \geq 0\,, \quad \forall\, a \in \bar{\mathcal{A}}\,,\ \varphi(1) = 1\,. \tag{9.12}$$

The distance between two states is given by

$$d(\varphi, \psi) = \mathrm{Sup}\left\{ |\varphi(a) - \psi(a)|\ ;\ a \in \mathcal{A}\,,\ \|[D, a]\| \leq 1 \right\}. \tag{9.13}$$

The significance of D is two-fold. On the one hand it defines the metric by the above equation, on the other hand its homotopy class represents the K-homology fundamental class of the space under consideration.

It is crucial to understand from the start the tension between the conditions (9.10) and (9.11). The first condition would be trivially fulfilled if D were bounded but condition (9.11) shows that it is unbounded. To understand this tension let us work out a very simple case. We let the algebra $\mathcal{A}$ be generated by a single unitary operator U. Let us show that if the index pairing between U and D, i.e. the index of PUP where P is the orthogonal projection on the positive eigenspace of D, *does not vanish* then the number $N(E)$ of eigenvalues of D whose absolute value is less than E grows at least like E when $E \to \infty$. This means that in the above circumstance $ds = D^{-1}$ is of order one or less.

To prove this we choose a smooth function $f \in C_c^\infty(\mathbb{R})$ identically one near 0, even and with Support $(f) \subset [-1,1]$. We then let $R(\varepsilon) = f(\varepsilon D)$. One first shows ([36]) that the operator norm of the commutator $[R(\varepsilon), U]$ tends to 0 like ε. It then follows that the trace norm satisfies

$$\left\| [R(\varepsilon), U] \right\|_1 \leq C\, \varepsilon\, N(1/\varepsilon) \tag{9.14}$$

as one sees using the control of the rank of $R(\varepsilon)$ from $N(1/\varepsilon)$. The index pairing is given by $-\frac{1}{2}\,\mathrm{Trace}\,(U^*[F,U])$ where F is the sign of D and one has,

$$\mathrm{Trace}\big(U^*[F,U]\big) = \lim_{\varepsilon\to 0} \mathrm{Trace}\big(U^*[F,U]R(\varepsilon)\big) = \lim_{\varepsilon \to 0} \mathrm{Trace}\big(U^*\, F[U, R(\varepsilon)]\big)\,. \tag{9.15}$$

Thus the limit being non-zero we get a lower bound on the trace norm of $[U, R(\varepsilon)]$ and hence on $\varepsilon\, N(1/\varepsilon)$ which shows that $N(E)$ grows at least like E when $E \to \infty$.

This shows that ds cannot be too small (it cannot be of order $\alpha > 1$). In fact when ds is of order 1 one has the following index formula,

$$\mathrm{Index}\,(PUP) = -\frac{1}{2} \int\!\!\!\!\!\!- \; U^{-1}[D,U]\, |ds|\,. \tag{9.16}$$

The simplest case in which the index pairing between D and U does not vanish, with ds of order 1, is obtained by requiring the further condition,

$$U^{-1}[D,U] = 1\,. \tag{9.17}$$

It is a simple exercise to compute the geometry on $S^1 = \mathrm{Spectrum}\,(U)$ given by an irreducible representation of condition (9.17). One obtains the standard circle with length 2π.

The above index formula is a special case of a general result ([36]) which computes the n-dimensional Hochschild class of the Chern character of a spectral triple of dimension n.

Theorem 5. *Let $(\mathcal{H}, F)$ be a Fredholm module over an involutive algebra $\mathcal{A}$. Let D be an unbounded selfadjoint operator in $\mathcal{H}$ such that D^{-1} is of order $1/n$, $\mathrm{Sign}\, D = F$, and such that for any $a \in \mathcal{A}$ the operators a and $[D,a]$ are in the domain of all powers of the derivations δ, given by $\delta(x) = [|D|, x]$. Let $\tau_n \in HC^n(\mathcal{A})$ be the Chern character of $(\mathcal{H}, F)$.*

For every n-dimensional Hochschild cycle $c \in Z_n(\mathcal{A},\mathcal{A})$, $c = \sum a^0 \otimes a^1 \ldots \otimes a^n$, one has $\langle \tau_n, c\rangle = ⨍ \sum a^0\,[D,a^1]\ldots[D,a^n]\,|D|^{-n}$.

We refer to [36] for precise normalization and to [66] for the detailed proof. By construction, this formula is scale invariant, i.e. it remains unchanged if we replace D by λD for $\lambda \in \mathbb{R}_+^*$. The operators T_c of the form

$$T_c = \sum a^0\,[D,a^1]\ldots[D,a^n]\,|D|^{-n} \tag{9.18}$$

are *measurable* in the sense of section 8.

The long exact sequence of cyclic cohomology (section 7) shows that the Hochschild class of τ_n is the obstruction to a better summability of $(\mathcal{H},F)$, indeed τ_n belongs to the image $S(HC^{n-2}(\mathcal{A}))$ (which is the case if the degree of summability can be improved by 2) if and only if the Hochschild cohomology class $I(\tau_n) \in H^n(\mathcal{A},\mathcal{A}^*)$ is equal to 0.

In particular, the above theorem implies nonvanishing of residues when the cohomological dimension of $\mathrm{ch}_*(\mathcal{H},F)$ is not lower than n:

Corollary. *With the hypothesis of Theorem 5 and if the Hochschild class of $\mathrm{ch}_*(\mathcal{H},F)$ pairs non-trivially with $H_n(\mathcal{A},\mathcal{A})$ one has*

$$⨍ |D|^{-n} \neq 0\,. \tag{9.19}$$

In other words the residue of the function $\zeta(s) = \mathrm{Trace}\,(|D|^{-s})$ at $s = n$ cannot vanish.

In higher dimension, the Hochschild class of the character suffices to determine the index pairing with the K-theory class of an idempotent e provided the lower dimensional components of $\mathrm{ch}(e)$ vanish. As we saw above these components are given, up to normalization by,

$$\mathrm{ch}_n(e) = \left(e - \tfrac{1}{2}\right) \otimes e \otimes \cdots \otimes e \tag{9.20}$$

(with $2n$ tensor signs) and as such cannot vanish. But both Hochschild and cyclic cohomology are Morita invariant, which implies that the class of $\mathrm{ch}(e)$ in the normalized (b,B) bicomplex (in homology) does not change when we project each of its components $\mathrm{ch}_n(e)$ on the commutant of a matrix algebra $M_q(\mathbb{C}) \subset \mathcal{A}$. The formula for this projection $\langle \mathrm{ch}_n(e)\rangle$ in terms of the matrix components e_{ij},

$$e = [e_{ij}]\,, \qquad e_{ij} \in M_q(\mathbb{C})' \cap \mathcal{A} \tag{9.21}$$

is the following,

$$\langle \mathrm{ch}_n(e)\rangle = \sum \left(e_{i_0 i_1} - \tfrac{1}{2}\delta_{i_0 i_1}\right) \otimes e_{i_1 i_2} \otimes e_{i_2 i_3} \otimes \cdots \otimes e_{i_{2n} i_0} \tag{9.22}$$

and there are very interesting situations in which all the lower components $\langle \mathrm{ch}_j(e)\rangle$ actually vanish,

$$\langle \mathrm{ch}_j(e)\rangle = 0 \qquad j < m\,. \tag{9.23}$$

For $m=1$ for instance we can take $q=2$ and the condition $\langle \mathrm{ch}_0(e)\rangle = 0$ means that e is of the form,

$$e = \begin{bmatrix} t & z \\ z^* & (1-t) \end{bmatrix} . \tag{9.24}$$

(The equation $e^2 = e$ then means that $t^2 + z^*z = t$, $tz + z(1-t) = z$, $z^*t + (1-t)z^* = z^*$, $z^*z + (1-t)^2 = (1-t)$ which shows that the algebra generated by the components z, z^*, t of e is abelian).

It then follows automatically that $\langle \mathrm{ch}_1(e)\rangle$ is a Hochschild cycle and hence by Theorem 5, that if $ds = D^{-1}$ is of order $1/2$ the index pairing is given by,

$$\mathrm{Index}\, D_e^+ = - \int\!\!\!\!\!\!- \gamma \left(e - \tfrac{1}{2}\right) [D,e]^2 \, ds^2 . \tag{9.25}$$

Exactly as above this shows that ds cannot be of order $\alpha > 1/2$ if the index pairing is non-zero, and we also get the analogue of equation (9.17) in the form,

$$\left\langle \left(e - \tfrac{1}{2}\right) [D,e]^2 \right\rangle = \gamma \tag{9.26}$$

where $\langle\ \rangle$ is simply the projection on the commutant of $M_2(\mathbb{C})$ in $\mathcal{L}(\mathcal{H})$.

This equation together with (9.25) implies that the area $\int\!\!\!\!-\, ds^2$ is an integer since it is given by a Fredholm index. One can show that the algebra $\mathcal{A}$ generated by the components of e is $C(S^2)$ the algebra of continuous functions on S^2 and that any Riemannian metric g on S^2 with fixed volume form gives a solution to the above equations.

There is a converse to that result ([50]) but it requires the further hypothesis that D is of order one:

$$\big[[D, e_{ij}], e_{k\ell}\big] = 0 \tag{9.27}$$

where the e_{ij} are the components of the idempotent e, i.e. are the generators of the algebra.

This order one condition is the counterpart in our operator theoretic setting of the "quadratic" nature of Riemann's equation $ds^2 = g_{\mu\nu}\, dx^\mu\, dx^\nu$. It is easier to formulate in terms of the square root which we extracted using the spin structure. We shall come later to the correct formulation of the order one condition when the algebra $\mathcal{A}$ is noncommutative.

To end this section let us move on to the four dimensional case, i.e. $n = 2$. We take $q = 4$, i.e. we deal with $M_4(\mathbb{C})$.

We first determine the C^* algebra generated by $M_4(\mathbb{C})$ and a projection $e = e^*$ such that $\left\langle e - \tfrac{1}{2}\right\rangle = 0$ as above and whose two by two matrix expression is of the form,

$$[e^{ij}] = \begin{bmatrix} q_{11} & q_{12} \\ q_{21} & q_{22} \end{bmatrix} \tag{9.28}$$

where each q_{ij} is a 2×2 matrix of the form,

$$q = \begin{bmatrix} a & \beta \\ -\beta^* & \alpha^* \end{bmatrix} . \tag{9.29}$$

Since $e = e^*$, both q_{11} and q_{22} are selfadjoint, moreover since $\langle e - \frac{1}{2} \rangle = 0$, we can find $t = t^*$ such that,

$$q_{11} = \begin{bmatrix} t & 0 \\ 0 & t \end{bmatrix} , \quad q_{22} = \begin{bmatrix} 1-t) & 0 \\ 0 & (1-t) \end{bmatrix} . \tag{9.30}$$

We let $q_{12} = \begin{bmatrix} a & \beta \\ -\beta^* & \alpha^* \end{bmatrix}$, we then get from $e = e^*$,

$$q_{21} = \begin{bmatrix} a^* & -\beta \\ \beta^* & \alpha \end{bmatrix} . \tag{9.31}$$

We thus see that the commutant $\mathcal{A}$ of $M_4(\mathbb{C})$ is generated by t, α, β and we first need to find the relations imposed by the equality $e^2 = e$.

In terms of $e = \begin{bmatrix} t & q \\ q^* & 1-t \end{bmatrix}$, the equation $e^2 = e$ means that $t^2 - t + qq^* = 0$, $t^2 - t + q^* q = 0$ and $[t, q] = 0$. This shows that t commutes with α, β, α^* and β^* and since $qq^* = q^* q$ is a diagonal matrix

$$\alpha\alpha^* = \alpha^*\alpha , \quad \alpha\beta = \beta\alpha , \quad \alpha^*\beta = \beta\alpha^* , \quad \beta\beta^* = \beta^*\beta \tag{9.32}$$

so that the C^* algebra $\mathcal{A}$ is abelian, with the only further relation (besides $t = t^*$),

$$\alpha\alpha^* + \beta\beta^* + t^2 - t = 0 . \tag{9.33}$$

This is enough to check that,

$$\mathcal{A} = C(S^4) \tag{9.34}$$

where S^4 appears naturally as quaternionic projective space,

$$S^4 = P_1(\mathbb{H}) . \tag{9.35}$$

The original C^* algebra is thus,

$$B = C(S^4) \otimes M_4(\mathbb{C}) . \tag{9.36}$$

We shall now check that the two dimensional component $\langle Ch_1(e) \rangle$ automatically vanishes as an element of the (normalized) (b,B)-bicomplex.

$$\langle Ch_n(e) \rangle = 0 , \ n = 0, 1. \tag{9.37}$$

With $q = \begin{bmatrix} \alpha & \beta \\ -\beta^* & \alpha^* \end{bmatrix}$, we get,

$$\langle Ch_1(e) \rangle = \big\langle \left(t - \tfrac{1}{2}\right) \, (dq \, dq^* - dq^* \, dq) \\ + q \, (dq^* \, dt - dt \, dq^*) + q^* \, (dt \, dq - dq \, dt) \big\rangle \tag{9.38}$$

where the expectation in the right-hand side is relative to $M_2(\mathbb{C})$ and we use the notation dx instead of the tensor notation.

The diagonal elements of $\omega = dq\, dq^*$ are

$$\omega_{11} = d\alpha\, d\alpha^* + d\beta\, d\beta^*\,, \quad \omega_{22} = d\beta^*\, d\beta + d\alpha^*\, d\alpha$$

while for $\omega' = dq^*\, dq$ we get,

$$\omega'_{11} = d\alpha^*\, d\alpha + d\beta\, d\beta^*\,, \quad \omega'_{22} = d\beta^*\, d\beta + d\alpha\, d\alpha^*\,.$$

It follows that, since t is diagonal,

$$\left\langle \left(t - \tfrac{1}{2}\right) (dq\, dq^* - dq^*\, dq)\right\rangle = 0\,. \tag{9.39}$$

The diagonal elements of $q\, dq^*\, dt = \rho$ are

$$\rho_{11} = \alpha\, d\alpha^*\, dt + \beta\, d\beta^*\, dt\,, \quad \rho_{22} = \beta^*\, d\beta\, dt + \alpha^*\, d\alpha\, dt$$

while for $\rho' = q^*\, dq\, dt$ they are

$$\rho'_{11} = \alpha^*\, d\alpha\, dt + \beta\, d\beta^*\, dt\,, \quad \rho'_{22} = \beta^*\, d\beta\, dt + \alpha\, d\alpha^*\, dt\,.$$

Similarly for $\sigma = q\, dt\, dq^*$ and $\sigma' = q^*\, dt\, dq$ one gets the required cancellations so that

$$\langle Ch_1(e)\rangle = 0\,. \tag{9.40}$$

It follows thus that $\langle Ch_2(e)\rangle$ is a Hochschild cycle and that for any $ds = D^{-1}$ of order $1/4$ commuting with $M_4(\mathbb{C})$, the index pairing of D with e is

$$\mathrm{Index} D_e^+ = \int\!\!\!\!\!- \gamma \left(e - \tfrac{1}{2}\right) [D,e]^4\, ds^4\,. \tag{9.41}$$

Exactly as above this shows that ds cannot be of order $\alpha > 1/4$ if the index pairing is non-zero, and we also get the analogue of equation 9-9.17 in the form,

$$\left\langle \left(e - \tfrac{1}{2}\right) [D,e]^4\right\rangle = \gamma \tag{9.42}$$

where $\langle\ \rangle$ is simply the projection on the commutant of $M_4(\mathbb{C})$ in $\mathcal{L}(\mathcal{H})$.

This equation together with (9.41) implies the integrality of the 4-dimensional volume,

$$\int\!\!\!\!\!- ds^4 \in \mathbb{N}\,, \tag{9.43}$$

since it is given by a Fredholm index.

One can show that the algebra $\mathcal{A}$ generated by the components of e is $C(S^4)$ the algebra of continuous functions on S^4 and that any Riemannian metric g on S^4 gives a solution to the above equations, provided its volume form is,

$$v = \frac{1}{1-2t}\, d\alpha \wedge d\overline{\alpha} \wedge d\beta \wedge d\overline{\beta}\,. \tag{9.44}$$

As in the two dimensional case there is a converse, assuming the order one condition on D.

The next question is how is D to be chosen from within the homotopy class which characterizes its K-homology class? There are two answers to this question. The first uses the naive idea of a formal metric,

$$G = \sum_{\mu,\nu=1}^{d} dx^\mu g_{\mu\nu} (dx^\nu)^* \in \Omega_+^2(\mathcal{A}) , \tag{9.45}$$

and the choice of D is performed by minimizing the action functional,

$$A = \sum_{\mu,\nu=1}^{d} ⨍ [D, x^\mu] g_{\mu\nu} \big([D, x^\nu]\big)^* |D^{-4}| , \tag{9.46}$$

among the D's which fulfil equation (9.42) holding G fixed.

The minimum is then given by the Dirac operator associated to the unique Riemannian metric with volume form v in the conformal class of $g_{\mu\nu} dx^\mu dx^\nu$.

The second way to select D from within its K-homology class is to use an action functional with the largest possible invariance group which is the unitary group of Hilbert space. The corresponding action is then spectral and only depends upon the eigenvalues of D. The simplest such action is of the form, [58]

$$S(D) = \operatorname{Trace}(f(D)) . \tag{9.47}$$

where f is an even function vanishing at ∞. If we take for f a step function equal to 1 in $[-\Lambda, \Lambda]$, the value of $S(D)$ is,

$$N(\Lambda) = \# \text{ eigenvalues of } D \text{ in } [-\Lambda, \Lambda] . \tag{9.48}$$

This step function $N(\Lambda)$ is the superposition of two terms,

$$N(\Lambda) = \langle N(\Lambda) \rangle + N_{\rm osc}(\Lambda) .$$

The oscillatory part $N_{\rm osc}(\Lambda)$ is the same as for a random matrix, governed by the statistic dictated by the symmetries of the system and does not concern us here. The average part $\langle N(\Lambda) \rangle$ is computed by a semiclassical approximation and the leading term in the asymptotic expansion is,

$$\frac{\Lambda^4}{2} ⨍ ds^4 \tag{9.49}$$

which by (43) is independent of the choice of D in its K-homology class.

If we restrict ourselves to solutions given by ordinary Riemannian metrics the next term in the asymptotic expansion is the Hilbert–Einstein action functional for the Riemannian metric,

$$-\frac{\Lambda^2}{96\pi^2} \int_{S_4} r \sqrt{g}\, d^4x . \tag{9.50}$$

Other non-zero terms in the asymptotic expansion are cosmological, Weyl gravity and topological terms.

10 Noncommutative 4-manifolds and the Instanton Algebra

In this section, based on our collaboration with G. Landi ([65]), we shall show that the basic equation for an instanton in dimension 4, namely

$$e = e^2 = e^* \tag{10.1}$$

and

$$\langle \mathrm{ch}_0(e) \rangle = 0 \ , \quad \langle \mathrm{ch}_1(e) \rangle = 0 \tag{10.2}$$

(where ch_n are the components of the Chern character,

$$\mathrm{ch}_n(e) = \left(e - \tfrac{1}{2}\right) \otimes e \otimes \ldots \otimes e \tag{10.3}$$

and $\langle\ \rangle$ is the projection onto the commutant of a 4×4 matrix algebra) do admit noncommutative solutions. In other words the algebra generated by the 16 components of the 4×4 matrix,

$$e = [e_{ij}] \tag{10.4}$$

will be noncommutative.

In fact this prompts us to introduce, a priori, the algebra $\mathcal{A}$ with 16 generators e_{ij} and whose presentation is given by the relations (10.1) and (10.2). The relation $\langle \mathrm{ch}_0(e) \rangle = 0$ just means that

$$e_{11} + e_{22} + e_{33} + e_{44} = 2 \tag{10.5}$$

and the equation $e = e^*$ defines the involution in $\mathcal{A}$. The relation $e^2 = e$ is easy to comprehend as a quadratic relation between the generators.

The relation $\langle \mathrm{ch}_1(e) \rangle = 0$ is more delicate to understand since it involves tensors and the simplest way to think about it is to represent the e_{ij} as operators in Hilbert space $\mathcal{H}$. What we ask then is that,

$$\sum \left(e_{ij} - \tfrac{1}{2}\, \delta_{ij}\right) \otimes \widetilde{e}_{jk} \otimes \widetilde{e}_{ki} = 0 \tag{10.6}$$

where the $\sim$ means that we take the class modulo the scalar multiples of 1.

This allows us to define what is a unitary representation π of the algebra $\mathcal{A}$ and we can endow its elements, i.e polynomials in the noncommuting generators e_{ij}, with the C^*-norm,

$$\|x\| = \sup_{\pi} \|\pi(x)\| \tag{10.7}$$

where π ranges through all unitary representations. It is easy to show that for $x \in \mathcal{A}$ the supremum is finite since in any unitary representation, the

e_{ij} satisfy,

$$\|\pi(e_{ij})\| \leq 1 \tag{10.8}$$

as matrix elements of a selfadjoint idempotent.

DEFINITION. *We let* $C(\mathrm{Gr})$ *be the* C^* *completion of* $\mathcal{A}$ *and* $C^\infty(\mathrm{Gr})$ *the smooth closure of* $\mathcal{A}$ *in* $C(\mathrm{Gr})$.

The letters Gr stand for the Grassmanian but our construction has little to do with the known "noncommutative Grassmanians". The really non-trivial condition is the cubic condition 10.6. In fact as we saw above the same construction in dimension 2 does give a *commutative* answer namely $P_1(\mathbb{C})$.

One should observe from the outset that the compact Lie group $SU(4)$ acts by automorphisms,

$$PSU(4) \subset \mathrm{Aut}\,(C^\infty(\mathrm{Gr})) \tag{10.9}$$

by the following operation,

$$e \to U\,e\,U^* \tag{10.10}$$

where $U \in SU(4)$ is viewed as a 4×4 matrix and $e = [e_{ij}]$ is as above.

What we saw in section 9 is that there is a surjection,

$$C(\mathrm{Gr}) \to C(S^4) \tag{10.11}$$

while the corresponding symmetry group breaks down to $SO(4)$, the isometry group of the 3-sphere from which S^4 is obtained by suspension. We shall now show that the algebra $C(\mathrm{Gr})$ is noncommutative by constructing explicit surjections,

$$C(\mathrm{Gr}) \to C(S^4_\theta) \tag{10.12}$$

whose form is dictated by natural deformations of the 4-sphere similar in spirit to the above deformation of $\mathbb{T}^2$ to $\mathbb{T}^2_\theta$.

We first determine the C^* algebra generated by $M_4(\mathbb{C})$ and a projection $e = e^*$ such that $\langle e - \frac{1}{2}\rangle = 0$ as above and whose two by two matrix expression is of the form,

$$[e^{ij}] = \begin{bmatrix} q_{11} & q_{12} \\ q_{21} & q_{22} \end{bmatrix} \tag{10.13}$$

where each q_{ij} is a 2×2 matrix of the form,

$$q = \begin{bmatrix} a & \beta \\ -\lambda\beta^* & \alpha^* \end{bmatrix}. \tag{10.14}$$

where $\lambda = \exp 2\pi i\theta$ is a complex number of modulus one, different from -1 for convenience. Since $e = e^*$, both q_{11} and q_{22} are selfadjoint, moreover since $\left\langle e-\frac{1}{2}\right\rangle=0$, we can find $t=t^*$ such that,

$$q_{11}=\begin{bmatrix} t & 0\\ 0 & t\end{bmatrix}, \quad q_{22}=\begin{bmatrix} 1-t) & 0\\ 0 & (1-t)\end{bmatrix}. \qquad (10.15)$$

We let $q_{12}=\left[\begin{smallmatrix} a & \beta\\ -\lambda\beta^* & \alpha^*\end{smallmatrix}\right]$, we then get from $e=e^*$,

$$q_{21}=\begin{bmatrix} a^* & -\bar\lambda\beta\\ \beta^* & \alpha\end{bmatrix}. \qquad (10.16)$$

We thus see that the commutant $\mathcal{B}_\theta$ of $M_4(\mathbb{C})$ is generated by t,α,β and we first need to find the relations imposed by the equality $e^2=e$.

In terms of $e=\left[\begin{smallmatrix} t & q\\ q^* & 1-t\end{smallmatrix}\right]$, the equation $e^2=e$ means that $t^2-t+qq^*=0$, $t^2-t+q^*q=0$ and $[t,q]=0$. This shows that t commutes with α, β, α^* and β^* and since $qq^*=q^*q$ is a diagonal matrix

$$\alpha\alpha^*=\alpha^*\alpha\,,\quad \alpha\beta=\lambda\beta\alpha\,,\quad \alpha^*\beta=\bar\lambda\beta\alpha^*\,,\quad \beta\beta^*=\beta^*\beta \qquad (10.17)$$

so that the C^* algebra $\mathcal{B}_\theta$ is not abelian for λ different from 1. The only further relation is (besides $t=t^*$)

$$\alpha\alpha^*+\beta\beta^*+t^2-t=0\,. \qquad (10.18)$$

We denote by S_θ^4 the corresponding noncommutative space, so that $C(S_\theta^4)=\mathcal{B}_\theta$. It is by construction the suspension of the noncommutative 3-sphere S_θ^3 whose coordinate algebra is generated by α and β as above for the special value $t=1/2$. This noncommutative 3-sphere is related by analytic continuation of the parameter q to the quantum group $SU(2)_q$ but the usual theory requires q to be real whereas we need a complex number of modulus one which spoils the unitarity of the coproduct.

We shall now check that the two dimensional component $\langle Ch_1(e)\rangle$ automatically vanishes as an element of the (normalized) (b,B)-bicomplex.

$$\langle Ch_n(e)\rangle=0\,,\quad n=0,1\,. \qquad (10.19)$$

With $q=\left[\begin{smallmatrix} \alpha & \beta\\ -\lambda\beta^* & \alpha^*\end{smallmatrix}\right]$, we get,

$$\begin{aligned}\langle Ch_1(e)\rangle=\Big\langle \left(t-\tfrac{1}{2}\right)&(dq\,dq^*-dq^*\,dq)\\ &+q(dq^*\,dt-dt\,dq^*)+q^*(dt\,dq-dq\,dt)\Big\rangle \end{aligned} \qquad (10.20)$$

where the expectation in the right-hand side is relative to $M_2(\mathbb{C})$ and we use the notation dx instead of the tensor notation.

The diagonal elements of $\omega=dq\,dq^*$ are computed as above,

$$\omega_{11}=d\alpha\,d\alpha^*+d\beta\,d\beta^*\,,\quad \omega_{22}=d\beta^*\,d\beta+d\alpha^*\,d\alpha$$

while for $\omega'=dq^*\,dq$ we get,

$$\omega'_{11}=d\alpha^*\,d\alpha+d\beta\,d\beta^*\,,\quad \omega'_{22}=d\beta^*\,d\beta+d\alpha\,d\alpha^*\,.$$

It follows that, since t is diagonal,

$$\left\langle \left(t - \tfrac{1}{2}\right) \left(dq\, dq^* - dq^*\, dq\right) \right\rangle = 0\,. \tag{10.21}$$

The diagonal elements of $q\, dq^*\, dt = \rho$ are

$$\rho_{11} = \alpha\, d\alpha^*\, dt + \beta\, d\beta^*\, dt\,, \qquad \rho_{22} = \beta^*\, d\beta\, dt + \alpha^*\, d\alpha\, dt$$

while for $\rho' = q^*\, dq\, dt$ they are

$$\rho'_{11} = \alpha^*\, d\alpha\, dt + \beta\, d\beta^*\, dt\,, \qquad \rho'_{22} = \beta^*\, d\beta\, dt + \alpha\, d\alpha^*\, dt\,.$$

Similarly for $\sigma = q\, dt\, dq^*$ and $\sigma' = q^*\, dt\, dq$ one gets the required cancellations so that,

$$\langle Ch_1(e)\rangle = 0\,, \tag{10.22}$$

It follows thus that $\langle Ch_2(e)\rangle$ is a Hochschild cycle and that for any $ds = D^{-1}$ of order $1/4$ commuting with $M_4(\mathbb{C})$, the index pairing of D with e is

$$\mathrm{Index} D_e^+ = \int\!\!\!\!\!- \gamma \left(e - \tfrac{1}{2}\right) [D, e]^4\, ds^4\,. \tag{10.23}$$

Exactly as above this shows that ds cannot be of order $\alpha > 1/4$ if the index pairing is non-zero, and we also get the analogue of equation (9.17) in the form,

$$\left\langle \left(e - \tfrac{1}{2}\right) [D, e]^4 \right\rangle = \gamma \tag{10.24}$$

where $\langle\ \rangle$ is simply the projection on the commutant of $M_4(\mathbb{C})$ in $\mathcal{L}(\mathcal{H})$.

This equation together with (10.23) implies the integrality of the 4-dimensional volume,

$$\int\!\!\!\!\!- ds^4 \in \mathbb{N}\,, \tag{10.25}$$

since it is given by a Fredholm index. We shall refer to [65] for the explicit construction of solutions of (10.24). It should be clear to the reader that this amply justifies the clarification of the notion of a manifold in Noncommutative Geometry, to which we turn next.

11 Noncommutative Spectral Manifolds

In our discussion in section 9 of the K-homology fundamental class of a manifold we skipped over the nuance between K-homology and KO-homology. This nuance turns out to be essential in the noncommutative case. Thus to describe the fundamental class of a noncommuative space by a spectral triple $(\mathcal{A}, \mathcal{H}, D)$, will require an additional "real structure" on the Hilbert space $\mathcal{H}$ given by an antilinear isometry J. The anti-linear isometry J is given in Riemannian geometry by the charge conjugation operator and in the noncommutative case by the Tomita–Takesaki antilinear conjugation operator [2].

The action of $\mathcal{A}$ satisfies the commutation rule, $[a, b^0] = 0$, $\forall\, a, b \in \mathcal{A}$ where

$$b^0 = Jb^*J^{-1} \qquad \forall b \in \mathcal{A} \tag{11.1}$$

so $\mathcal{H}$ becomes an $\mathcal{A}$-bimodule using the representation of $\mathcal{A} \otimes \mathcal{A}^0$, where $\mathcal{A}^0$ is the opposite algebra, given by,

$$a \otimes b^0 \to aJb^*J^{-1} \qquad \forall a, b \in \mathcal{A} \tag{11.2}$$

This allows us to overcome the main difficulty of the noncommutative case which is that the diagonal in the square of the space no longer corresponds to an algebra homomorphism (the map $x \otimes y \to xy$ is no longer an algebra homomorphism),

The *fundamental class* of a noncommutative space is a class μ in the KR–homology of the algebra $\mathcal{A} \otimes \mathcal{A}^0$ equipped with the involution

$$\tau(x \otimes y^0) = y^* \otimes (x^*)^0 \qquad \forall\, x, y \in \mathcal{A} \tag{11.3}$$

where $\mathcal{A}^0$ denotes the algebra opposite to $\mathcal{A}$. The KR-homology cycle representing μ is given by a spectral triple, as above, equipped with an anti-linear isometry J on $\mathcal{H}$ which implements the involution τ,

$$JwJ^{-1} = \tau(w) \qquad \forall\, w \in \mathcal{A} \otimes \mathcal{A}^0\,, \tag{11.4}$$

KR-homology ([8] [55]) is periodic with period 8 and the dimension modulo 8 is specified by the following commutation rules. One has $J^2 = \varepsilon$, $JD = \varepsilon' DJ$, $J\gamma = \varepsilon''\gamma J$ where $\varepsilon, \varepsilon', \varepsilon'' \in \{-1, 1\}$ and with n the dimension modulo 8,

n	0	1	2	3	4	5	6	7
ε	1	1	-1	-1	-1	-1	1	1
ε'	1	-1	1	1	1	-1	1	1
ε''	1		-1		1		-1	

The class μ specifies only the stable homotopy class of the spectral triple $(\mathcal{A}, \mathcal{H}, D)$ equipped with the isometry J (and $\mathbb{Z}/2$–grading γ if n is even). The non-triviality of this homotopy class shows up in the intersection form

$$K_*(\mathcal{A}) \times K_*(\mathcal{A}) \to \mathbb{Z} \tag{11.5}$$

which is obtained from the Fredholm index of D with coefficients in $K_*(\mathcal{A} \otimes \mathcal{A}^0)$. Note that it is defined without using the diagonal map $m : \mathcal{A} \otimes \mathcal{A} \to \mathcal{A}$, which is not a homomorphism in the noncommutative case. This form is quadratic or symplectic according to the value of n modulo 8.

The Kasparov intersection product [8] allows us to formulate the Poincaré duality in terms of the invertibility of μ,

$$\exists\,\beta \in KR_n(\mathcal{A}^0 \otimes \mathcal{A})\,, \quad \beta \otimes_{\mathcal{A}} \mu = \mathrm{id}_{\mathcal{A}^0}\,, \quad \mu \otimes_{\mathcal{A}^0} \beta = \mathrm{id}_{\mathcal{A}}\,. \tag{11.6}$$

It implies the isomorphism $K_*(\mathcal{A}) \xrightarrow{\cap\mu} K^*(\mathcal{A})$.

The condition that D is an operator of order one becomes

$$\big[[D,a],b^0\big] = 0 \qquad \forall\, a,b \in \mathcal{A}\,. \tag{11.7}$$

(Notice that since a and b^0 commute this condition is equivalent to $[[D,a^0],b] = 0$, $\forall\, a,b \in \mathcal{A}$.)

One can show that the von Neumann algebra $\mathcal{A}''$ generated by $\mathcal{A}$ in $\mathcal{H}$ is automatically finite and hyperfinite and there is a complete list of such algebras up to isomorphism. The algebra $\mathcal{A}$ is stable under smooth functional calculus in its norm closure $A = \bar{\mathcal{A}}$ so that $K_j(\mathcal{A}) \simeq K_j(A)$, i.e. $K_j(\mathcal{A})$ depends only on the underlying topology (defined by the C^* algebra A). The integer $\chi = \langle \mu, \beta \rangle \in \mathbb{Z}$ gives the Euler characteristic in the form

$$\chi = \mathrm{Rang}\, K_0(\mathcal{A}) - \mathrm{Rang}\, K_1(\mathcal{A}) \tag{11.8}$$

and the general operator theoretic index formula of section 13 below, gives a local formula for χ.

We gave in [50] the necessary and sufficient conditions that a spectral triple (with real structure J) should fulfil in order to come from an ordinary compact Riemannian spin manifold. These conditions extend in a straightforward manner to the noncommutative case ([50]). To appreciate the richness of examples which fulfil them we shall just quote the following result ([65]),

Theorem 6. *Let M be a compact Riemannian spin manifold. Then if the isometry group of M has rank $r \geq 2$, M admits a non-trivial one parameter isospectral deformation to noncommutative geometries M_θ.*

The group $\mathrm{Aut}^+(\mathcal{A})$ of automorphisms α of the involutive algebra $\mathcal{A}$, which are implemented by a unitary operator U in $\mathcal{H}$ commuting with J,

$$\alpha(x) = U\,x\,U^{-1} \qquad \forall\, x \in \mathcal{A}\,, \tag{11.9}$$

plays the role of the group $\mathrm{Diff}^+(M)$ of diffeomorphisms preserving the K-homology fundamental class for a manifold M.

In the general noncommutative case, parallel to the normal subgroup $\mathrm{Int}\,\mathcal{A} \subset \mathrm{Aut}\,\mathcal{A}$ of inner automorphisms of $\mathcal{A}$,

$$\alpha(f) = u f u^* \qquad \forall\, f \in \mathcal{A} \tag{11.10}$$

where u is a unitary element of $\mathcal{A}$ (i.e. $uu^* = u^*u = 1$), there exists a natural foliation of the space of spectral geometries on $\mathcal{A}$ by equivalence classes of *inner deformations* of a given geometry. To understand how they arise we need to understand how to transfer a given spectral geometry to a Morita equivalent algebra. Given a spectral triple $(\mathcal{A},\mathcal{H},D)$ and the Morita equivalence [56] between $\mathcal{A}$ and an algebra $\mathcal{B}$ where

$$\mathcal{B} = \mathrm{End}_{\mathcal{A}}(\mathcal{E}) \tag{11.11}$$

where $\mathcal{E}$ is a finite, projective, hermitian right $\mathcal{A}$–module, one gets a spectral triple on $\mathcal{B}$ by the choice of a *hermitian connection* on $\mathcal{E}$. Such a connection ∇ is a linear map $\nabla : \mathcal{E} \to \mathcal{E} \otimes_{\mathcal{A}} \Omega^1_D$ satisfying the rules ([36])

$$\nabla(\xi a) = (\nabla\xi)a + \xi \otimes da \qquad \forall\, \xi \in \mathcal{E}\ ,\ a \in \mathcal{A} \tag{11.12}$$

$$(\xi,\nabla\eta) - (\nabla\xi,\eta) = d(\xi,\eta) \qquad \forall\, \xi,\eta \in \mathcal{E} \tag{11.13}$$

where $da = [D,a]$ and where $\Omega^1_D \subset \mathcal{L}(\mathcal{H})$ is the $\mathcal{A}$–bimodule of operators of the form

$$A = \Sigma\, a_i[D,b_i]\,, \quad a_i, b_i \in \mathcal{A}\,. \tag{11.14}$$

Any algebra $\mathcal{A}$ is Morita equivalent to itself (with $\mathcal{E} = \mathcal{A}$) and when one applies the above construction in the above context one gets the inner deformations of the spectral geometry.

Such a deformation is obtained by the following formula (with suitable signs depending on the dimension mod 8) without modifying either the representation of $\mathcal{A}$ in $\mathcal{H}$ or the anti-linear isometry J

$$D \to D + A + JAJ^{-1} \tag{11.15}$$

where $A = A^*$ is an arbitrary selfadjoint operator of the form (11.14). The action of the group $\mathrm{Int}(\mathcal{A})$ on the spectral geometries is simply the following gauge transformation of A

$$\gamma_u(A) = u[D,u^*] + uAu^*\,. \tag{11.16}$$

The required unitary equivalence is implemented by the following representation of the unitary group of $\mathcal{A}$ in $\mathcal{H}$,

$$u \to uJuJ^{-1} = u(u^*)^0\,. \tag{11.17}$$

The transformation (11.15) is the identity in the usual Riemannian case. To get a non-trivial example it suffices to consider the product of a Riemannian triple by the unique spectral geometry on the finite-dimensional algebra $\mathcal{A}_F = M_N(\mathbb{C})$ of $N \times N$ matrices on $\mathbb{C}$, $N \geq 2$. One then has $\mathcal{A} = C^\infty(M) \otimes \mathcal{A}_F$, $\mathrm{Int}(\mathcal{A}) = C^\infty(M, PSU(N))$ and inner deformations of the geometry are parameterized by the gauge potentials for the gauge theory of the group $SU(N)$. The space of pure states of the algebra $\mathcal{A}$, $P(\mathcal{A})$, is the product $P = M \times P_{N-1}(\mathbb{C})$ and the metric on $P(\mathcal{A})$ determined by the formula (9.13) depends on the gauge potential A. It coincides with the

Carnot metric [57] on P defined by the horizontal distribution given by the connection associated to A. The group $\mathrm{Aut}(\mathcal{A})$ of automorphisms of $\mathcal{A}$ is the following semi-direct product

$$\mathrm{Aut}(\mathcal{A}) = \mathcal{U} >\!\!\lhd \mathrm{Diff}^{+}(M) \tag{11.18}$$

of the local gauge transformation group $\mathrm{Int}(\mathcal{A})$ by the group of diffeomorphisms.

12 Test with Space-time

What we have done so far is to stretch the usual framework of ordinary geometry beyond its commutative restrictions (set theoretic restrictions) and of course now it's not perhaps a bad idea to test it with what we know about physics and to try to find a better model of space-time within this new framework. The best way is to start with the hard core information one has from physics and that can be summarized by a Lagrangian. This Lagrangian is the Einstein Lagrangian plus the standard model Lagrangian. I am not going to write it down, it's a very complicated expression since just the standard model Lagrangian comprises five types of terms. But one can start understanding something by looking at the symmetry group of this Lagrangian. Now, if it were just the Einstein theory, the symmetry group of the Lagrangian would just be, by the equivalence principle, the diffeomorphism group of the space-time manifold. But because of the standard model piece the symmetry group of this Lagrangian is not just the diffeomorphism group, because the gauge theory has another huge symmetry group which is the group of maps from the manifold to the small gauge group, namely $U_1 \times SU_2 \times SU_3$ as far as we know. Thus, the symmetry group G of the full Lagrangian is neither the diffeomorphism group nor the group of gauge transformations of the second kind nor their product, but it is their semi-direct product. It is exactly like what happens with the Poincaré group where you have translations and Lorentz transformations, so it is the semi-direct product of these two subgroups. Now we can ask a very simple question: would there be some space X so that this group G would be equal to $\mathrm{Diff}(X)$? If such a space would exist, then we would have some chance to actually geometrize the theory completely, namely to be able to say that it's pure gravity on the space X. Now, if you look for the space X among ordinary manifolds, you have no chance since by a result of John Mather the diffeomorphism group of a (connected) manifold is a simple group. A simple group cannot have a non-trivial normal subgroup, so you cannot have this structure of semi-direct product.

However, we can use our dictionary, and in this dictionary, if we browse through it, we find that what corresponds to diffeomorphisms for a noncommutative space is just the group $\mathrm{Aut}^+(\mathcal{A})$ of automorphisms of the algebra of coordinates $\mathcal{A}$, which preserve the fundamental class in K-homology, as described above in section 11.

Now there is a beautiful fact which is that when an algebra is not commutative, then among its automorphisms there are very trivial ones, there are automorphisms which are there for free, I mean the inner ones, which associate to an element x of the algebra the element uxu^{-1}. Of course uxu^{-1} is not, in general equal to x because the algebra is not commutative, and these automorphisms form a normal subgroup of the group of automorphisms. Thus you see that the group $\mathrm{Aut}^+(\mathcal{A})$ has the same type of structure, namely it has a normal subgroup of internal automorphisms and it has a quotient. Now it turns out that there is one very natural noncommutative algebra $\mathcal{A}$ whose group of internal automorphisms corresponds to the group of gauge transformations and the quotient $\mathrm{Aut}^+(\mathcal{A})/\mathrm{Int}(\mathcal{A})$ corresponds exactly to diffeomorphisms [54]. It is amusing that the physics vocabulary is actually the same as the mathematical vocabulary. Namely, in physics you talk about internal symmetries and in mathematics you talk about inner automorphisms, you could call them internal automorphisms. Now the corresponding space is a product $M \times F$ of an ordinary manifold M by a finite noncommutative space F. The corresponding algebra $\mathcal{A}_F$ is the direct sum of the algebras $\mathbb{C}, \mathbb{H}$ (the quaternions), and $M_3(\mathbb{C})$ of 3×3 complex matrices.

The algebra $\mathcal{A}_F$ corresponds to a *finite* space where the standard model fermions and the Yukawa parameters (masses of fermions and mixing matrix of Kobayashi Maskawa) determine the spectral geometry in the following manner. The Hilbert space is finite-dimensional and admits the set of elementary fermions as a basis. For example, for the first generation of quarks, this set is

$$u_L, u_R, d_L, d_R, \bar{u}_L, \bar{u}_R, \bar{d}_L, \bar{d}_R \,. \tag{12.1}$$

The algebra $\mathcal{A}_F$ admits a natural representation in $\mathcal{H}_F$ (see [53]) and the Yukawa coupling matrix Y determines the operator D.

The detailed structure of Y (and in particular the fact that color is not broken) allows us to check the axioms of noncommutative geometry.

The next step consists of the computation of internal deformations

$$D \to D + A + JAJ^{-1} \tag{12.2}$$

(cf. section 11), of the product geometry $M \times F$ where M is a 4–dimensional Riemannian spin manifold. The computation gives the standard model gauge bosons $\gamma, W^{\pm}, Z$, the eight gluons and the Higgs fields φ with accurate quantum numbers.

Now the next question that arises is how do we recover the original action functional which contained both the Einstein–Hilbert term as well as the standard model? The answer is very simple: the Fermionic part of this action is there from the start and one recovers the bosonic part as follows. Both the Hilbert–Einstein action functional for the Riemannian metric, the Yang–Mills action for the vector potentials, the self interaction and the minimal coupling for the Higgs fields all appear with the correct signs in the asymptotic expansion for large Λ of the number $N(\Lambda)$ of eigenvalues of D which are $\leq \Lambda$ (cf. [58]),

$$N(\Lambda) = \# \text{ eigenvalues of } D \text{ in } [-\Lambda, \Lambda] . \tag{12.3}$$

Exactly as above, this step function $N(\Lambda)$ is the superposition of two terms,

$$N(\Lambda) = \langle N(\Lambda) \rangle + N_{\mathrm{osc}}(\Lambda) .$$

The oscillatory part $N_{\mathrm{osc}}(\Lambda)$ is the same as for a random matrix, governed by the statistic dictated by the symmetries of the system and does not concern us here. The average part $\langle N(\Lambda) \rangle$ is computed by a semiclassical approximation from local expressions involving the familiar heat equation expansion and delivers the correct terms. We showed above in section 9, that if one studies natural presentations of the algebra generated by $\mathcal{A}$ and D one naturally gets only metrics with a fixed volume form so that the bothering cosmological term does not enter in the variational equations associated to the spectral action $\langle N(\Lambda) \rangle$. It is tempting to speculate that the phenomenological Lagrangian of physics, combining matter and gravity appears from the solution of an extremely simple operator theoretic equation along the lines described above in sections 9 and 10.

13 Operator Theoretic Index Formula

The power of the general theory comes from deeper general theorems such as the local computation of the analogue of Pontrjagin classes: *i.e.* of the components of the cyclic cocycle which is the Chern character of the K-homology class of D and which make sense in general. This result allows, using the infinitesimal calculus, to go from local to global in the general framework of spectral triples $(\mathcal{A}, \mathcal{H}, D)$.

The Fredholm index of the operator D determines (in the odd case) an additive map $K_1(\mathcal{A}) \xrightarrow{\varphi} \mathbb{Z}$ given by the equality

$$\varphi([u]) = \text{Index}(PuP)\,, \quad u \in GL_1(\mathcal{A}) \tag{13.1}$$

where P is the projector $P = \frac{1+F}{2}$, $F = \text{Sign}\,(D)$.

This map is computed by the pairing of $K_1(\mathcal{A})$ with the following cyclic cocycle

$$\tau(a^0, \dots, a^n) = \text{Trace}\left(a^0[F, a^1] \dots [F, a^n]\right) \quad \forall\, a^j \in \mathcal{A} \tag{13.2}$$

where $F = \text{Sign}\, D$ and we assume that the dimension p of our space is finite, which means that $(D+i)^{-1}$ is of order $1/p$, also $n \geq p$ is an odd integer. There are similar formulas involving the grading γ in the even case, and it is quite satisfactory ([33], [34]) that both cyclic cohomology and the Chern character formula adapt to the infinite dimensional case in which the only hypothesis is that $\exp(-D^2)$ is a trace class operator.

It is difficult to compute the cocycle τ in general because the formula (13.2) involves the ordinary trace instead of the local trace $⨍$ and it is crucial to obtain a local form of the above cocycle.

This problem is solved by a general formula [35] which we now describe.

Let us make the following regularity hypothesis on $(\mathcal{A}, \mathcal{H}, D)$

$$a \text{ and } [D, a] \in \cap\, \text{Dom}\, \delta^k\,, \quad \forall\, a \in \mathcal{A} \tag{13.3}$$

where δ is the derivation $\delta(T) = [|D|, T]$ for any operator T.

We let $\mathcal{B}$ denote the algebra generated by $\delta^k(a)$, $\delta^k([D, a])$. The usual notion of *dimension* of a space is replaced by the *dimension spectrum* which is a subset of $\mathbb{C}$. The precise definition of the dimension spectrum is the subset $\Sigma \subset \mathbb{C}$ of singularities of the analytic functions

$$\zeta_b(z) = \text{Trace}\,(b|D|^{-z}) \qquad \text{Re}\, z > p\,,\ b \in \mathcal{B}\,. \tag{13.4}$$

The dimension spectrum of a manifold M is the set $\{0, 1, \dots, n\}$, $n = \dim M$; it is simple. Multiplicities appear for singular manifolds. Cantor sets provide examples of complex points $z \notin \mathbb{R}$ in the dimension spectrum.

We assume that Σ is discrete and simple, i.e. that ζ_b can be extended to $\mathbb{C}/\Sigma$ with simple poles in Σ.

We refer to [35] for the case of a spectrum with multiplicities. Let $(\mathcal{A}, \mathcal{H}, D)$ be a spectral triple satisfying the hypothesis (13.3) and (13.4). The local index theorem is the following, [35]:

Theorem 7. 1. *The equality*

$$⨍ P = \text{Res}_{z=0}\, \text{Trace}\,(P|D|^{-z})$$

defines a trace on the algebra generated by $\mathcal{A}$, $[D,\mathcal{A}]$ and $|D|^z$, where $z \in \mathbb{C}$.

2. *There is only a finite number of non-zero terms in the following formula which defines the odd components $(\varphi_n)_{n=1,3,\ldots}$ of a cocycle in the bicomplex (b,B) of $\mathcal{A}$,*

$$\varphi_n(a^0,\ldots,a^n) = \sum_k c_{n,k} \int\!\!\!\!\!- \, a^0[D,a^1]^{(k_1)}\ldots[D,a^n]^{(k_n)}\,|D|^{-n-2|k|} \quad \forall\, a^j \in \mathcal{A}$$

where the following notation are used: $T^{(k)}=\nabla^k(T)$ and $\nabla(T)=D^2T-TD^2$, k is a multi-index, $|k| = k_1+\ldots+k_n$,

$$c_{n,k} = (-1)^{|k|}\sqrt{2i}(k_1!\ldots k_n!)^{-1}\big((k_1{+}1)\ldots(k_1{+}k_2{+}\ldots{+}k_n{+}n)\big)^{-1}\Gamma\left(|k|+\tfrac{n}{2}\right).$$

3. *The pairing of the cyclic cohomology class $(\varphi_n) \in HC^*(\mathcal{A})$ with $K_1(\mathcal{A})$ gives the Fredholm index of D with coefficients in $K_1(\mathcal{A})$.*

For the normalization of the pairing between HC^* and $K(\mathcal{A})$ see [36]. In the even case, i.e. when $\mathcal{H}$ is $\mathbb{Z}/2$ graded by γ,

$$\gamma=\gamma^*,\quad \gamma^2=1,\quad \gamma a = a\gamma \quad \forall\, a\in\mathcal{A},\ \gamma D = -D\gamma,$$

there is an analogous formula for a cocycle (φ_n), n even, which gives the Fredholm index of D with coefficients in K_0. However, φ_0 is not expressed in terms of the residue $\int\!\!\!\!\!-$ because it is not local for a finite dimensional $\mathcal{H}$.

14 Diffeomorphism Invariant Geometry

The power of the above operator theoretic local trace formula lies in its generality and in the existence of really new geometric examples to which it applies.

In this section we shall explain how the transverse structure of foliations is described by a spectral triple $(\mathcal{A},\mathcal{H},D)$ with simple dimension spectrum. This allows moreover to give a precise meaning to diffeomorphism invariant geometry on a manifold M, by the construction of a spectral triple $(\mathcal{A},\mathcal{H},D)$ where the algebra $\mathcal{A}$ is the crossed product of the algebra of smooth functions on the finite dimensional bundle P of metrics on M by the natural action of the diffeomorphism group of M. While ordinary geometric constructions are "covariant" with respect to diffeomorphisms, our construction ([37]) is "invariant" inasmuch as the algebra now incorporates the full group of diffeomorphisms and the metrics involved are canonical.

The operator D is an hypoelliptic operator ([38]) which is directly associated to the reduction of the structure group of the manifold P to a group of triangular matrices whose diagonal blocks are orthogonal. By construction the fiber of $P \xrightarrow{\pi} M$ is the quotient $F^+/SO(n)$ of the $GL^+(n)$–principal bundle F^+ of oriented frames on M by the action of the orthogonal group $SO(n) \subset GL^+(n)$. The space P admits a canonical foliation: the vertical foliation $V \subset TP$, $V = \operatorname{Ker} \pi_*$ and on the fibers V and on $N = (TP)/V$ the following Euclidean structures. A choice of $GL^+(n)$–invariant Riemannian metric on $GL^+(n)/SO(n)$ determines a metric on V. The metric on N is defined tautologically: for every $p \in P$ one has a metric on $T_{\pi(p)}(M)$ which is isomorphic to N_p by π_*.

We first consider the hypoelliptic signature operator Q on F^+. It is not a scalar operator but it acts in the tensor product

$$\mathcal{H}_0 = L^2(F^+, v) \otimes E \tag{14.1}$$

where E is a finite dimensional representation of $SO(n)$ specifically given by

$$E = \wedge P_n \otimes \wedge \mathbb{R}^n \,, \quad P_n = S^2\mathbb{R}^n \,. \tag{14.2}$$

The operator Q is the graded sum,

$$Q = (d_V^* \, d_V - d_V \, d_V^*) \oplus (d_H + d_H^*) \tag{14.3}$$

where the horizontal (resp. vertical) differentiation d_H (resp. d_V) is a matrix in the horizontal and vertical vector fields $\mathbf{X}_i$ and $\mathbf{Y}_\ell^k$ as well as their adjoints (which also involve scalars). When n is equal to 1 or 2 modulo 4 one has to replace F^+ by its product by S^1 so that the dimension of the vertical fiber is even (it is then $1 + \frac{n(n+1)}{2}$) and the vertical signature operator makes sense. The longitudinal part is not elliptic but only transversally elliptic with respect to the action of $SO(n)$. Thus to get an hypoelliptic operator one restricts Q to the Hilbert space,

$$\mathcal{H} = \left(L^2(F^+, v) \otimes E\right)^{SO(n)} \tag{14.4}$$

and one takes the following algebra $\mathcal{A}$,

$$\mathcal{A} = C_c^\infty(P) {>\!\!\!\lhd} \mathrm{Diff}^+ \,, \quad P = F^+/SO(n) \,. \tag{14.5}$$

Let us note that the operator Q is in fact the image under the right regular representation of the affine group G_{affine} of a (matrix valued) hypoelliptic symmetric element in the enveloping algebra $\mathcal{U}(G_{affine})$. By an easy adaptation of a theorem of Nelson and Stinespring, it then follows that Q is essentially selfadjoint (with core any dense G_{affine}-invariant subspace of the space of C^∞ vectors of the right regular representation of G_{affine}).

Theorem 8 [[37]]. *Let $\mathcal{A}$ be the crossed product $C_c^\infty(P) {>\!\!\!\lhd} \mathit{Diff}^+$ acting in $\mathcal{H}$ as above.*

1. *The equality $D|D| = Q$ defines a spectral triple $(\mathcal{A}, \mathcal{H}, D)$ which satisfies the hypotheses of Theorem 7; its dimension spectrum is simple and given by $\Sigma = \left\{0, 1, ..., 2n + \frac{n(n+1)}{2}\right\}$.*

2. *The cocycle φ_j given by the local index formula (Theorem 7) is the image by the characteristic map of a universal Gelfand-Fuchs cohomology class.*

The equality $D|D| = Q$ defining D while Q is a differential operator of second order, is characteristic of "quartic" geometries.

The computation of the local index formula for diffeomorphism invariant geometry [37] was quite complicated even in the case of codimension 1 foliations: there were innumerable terms to be computed; this could be done by hand, by 3 weeks of eight hours per day tedious computations, but it was of course hopeless to proceed by direct computations in the general case. Henri and I finally found how to get the answer for the general case after discovering that the computation generated a Hopf algebra $\mathcal{H}(n)$ which only depends on n= codimension of the foliation, and which allows us to organize the computation provided cyclic cohomology is suitably adapted to Hopf algebras as in the next section.

The Hopf algebra $\mathcal{H}(n)$ only depends upon the integer n and is neither commutative nor cocommutative. We proved in [37] that it is isomorphic to the bicrossed product Hopf algebra ([70], [69], [71]) associated to the following pair of subgroups of $G = \mathrm{Diff}(\mathbb{R}^n)$.

We let $G_1 \subset G$ be the subgroup of affine diffeomorphisms,

$$k(x) = Ax + b \qquad \forall\, x \in \mathbb{R}^n \tag{14.6}$$

and we let $G_2 \subset G$ be the subgroup,

$$\varphi \in G\,, \quad \varphi(0) = 0\,, \quad \varphi'(0) = 1\,. \tag{14.7}$$

Given $\varphi \in G$ it has a unique decomposition $\varphi = k\,\psi$ where $k \in G_1$, $\psi \in G_2$ which allows us to perform the bicrossed product construction.

15 Characteristic Classes for Actions of Hopf Algebras

Hopf algebras arise very naturally from their actions on noncommutative algebras [39]. Given an algebra A, an action of the Hopf algebra $\mathcal{H}$ on A is given by a linear map,

$$\mathcal{H} \otimes A \to A, \quad h \otimes a \to h(a) \tag{15.1}$$

satisfying $h_1(h_2 a) = (h_1 h_2)(a)$, $\forall h_i \in \mathcal{H}$, $a \in A$ and

$$h(ab) = \sum h_{(1)}(a) h_{(2)}(b) \qquad \forall\, a, b \in A\,,\ h \in \mathcal{H}\,. \tag{15.2}$$

where the coproduct of h is,

$$\Delta(h) = \sum h_{(1)} \otimes h_{(2)} \tag{15.3}$$

In concrete examples, the algebra A appears first, together with linear maps

$A \to A$ satisfying a relation of the form (15.2) which dictates the Hopf algebra structure. This is exactly what occurred in the above example (see [37] for the description of $\mathcal{H}(n)$ and its relation with Diff($\mathbb{R}^n$)).

The theory of characteristic classes for actions of $\mathcal{H}$ extends the construction [40] of cyclic cocycles from a Lie algebra of derivations of a C^* algebra A, together with an *invariant trace* τ on A.

This theory was developed in [37] in order to solve the above computational problem for diffeomorphism invariant geometry but it was shown in [41] that the correct framework for the cyclic cohomology of Hopf algebras is that of modular pairs in involution. It is quite satisfactory that exactly the same structure emerged from the analysis of locally compact quantum groups. The resulting cyclic cohomology appears to be the natural candidate for the analogue of Lie algebra cohomology in the context of Hopf algebras. We fix a group-like element σ and a character δ of $\mathcal{H}$ with $\delta(\sigma) = 1$. They will play the role of the module of locally compact groups.

We then introduce the twisted antipode,

$$\widetilde{S}(y) = \sum \delta(y_{(1)}) S(y_{(2)}) \,, \quad y \in \mathcal{H} \,, \quad \Delta y = \sum y_{(1)} \otimes y_{(2)} \,. \tag{15.4}$$

We shall say that the modular pair (σ, δ) is in involution if the (σ, δ)-twisted antipode is an involution,

$$(\sigma^{-1}\widetilde{S})^2 = I. \tag{15.5}$$

We associate a cyclic complex (in fact a Λ-module, where Λ is the cyclic category), to any Hopf algebra together with a modular pair in involution. More precisely the following graded vector space $\mathcal{H}^\natural_{(\delta,\sigma)} = \{\mathcal{H}^{\otimes n}\}_{n\geq 1}$ equipped with the operators given by the following formulas (15.6)–(15.8) defines a module over the cyclic category Λ. First, by transposing the standard simplicial operators underlying the Hochschild homology complex of an algebra, one associates to $\mathcal{H}$, viewed only as a coalgebra, the natural cosimplicial module $\{\mathcal{H}^{\otimes n}\}_{n\geq 1}$, with face operators $\delta_i : \mathcal{H}^{\otimes n-1} \to \mathcal{H}^{\otimes n}$,

$$\begin{aligned}
&\delta_0(h^1 \otimes \ldots \otimes h^{n-1}) = 1 \otimes h^1 \otimes \ldots \otimes h^{n-1} \\
&\delta_j(h^1 \otimes \ldots \otimes h^{n-1}) = h^1 \otimes \ldots \otimes \Delta h^j \otimes \ldots \otimes h^n, \ \forall 1 \leq j \leq n-1, \\
&\delta_n(h^1 \otimes \ldots \otimes h^{n-1}) = h^1 \otimes \ldots \otimes h^{n-1} \otimes \sigma
\end{aligned} \tag{15.6}$$

and degeneracy operators $\sigma_i : \mathcal{H}^{\otimes n+1} \to \mathcal{H}^{\otimes n}$,

$$\sigma_i(h^1 \otimes ... \otimes h^{n+1}) = h^1 \otimes ... \otimes \varepsilon(h^{i+1}) \otimes ... \otimes h^{n+1} \,, \ 0 \leq i \leq n \,. \tag{15.7}$$

The remaining two essential features of a Hopf algebra – *product* and *antipode* – are brought into play, to define the *cyclic operators* $\tau_n : \mathcal{H}^{\otimes n} \to \mathcal{H}^{\otimes n}$,

$$\tau_n(h^1 \otimes \ldots \otimes h^n) = \left(\Delta^{n-1}\widetilde{S}(h^1)\right) \cdot h^2 \otimes \ldots \otimes h^n \otimes \sigma \,. \tag{15.8}$$

The theory of characteristic classes applies to actions of the Hopf algebra on an algebra endowed with a δ-invariant σ-trace. A linear form τ on A is a σ-trace under the action of $\mathcal{H}$ iff one has,

$$\tau(ab) = \tau(b\sigma(a)) \qquad \forall\, a, b \in A\,. \tag{15.9}$$

A σ-trace τ on A is δ-invariant under the action of $\mathcal{H}$ iff

$$\tau(h(a)b) = \tau\big(a\widetilde{S}(h)(b)\big) \qquad \forall\, a, b \in A\,,\ h \in \mathcal{H}\,. \tag{15.10}$$

Note that equation (15.9) is an excellent guide in order to construct Hopf algebra actions, since by the modular theory any positive linear functional τ on an algebra A gives rise to an (unbounded) automorphism σ of its weak closure fulfilling equation (15.9).

The theory of characteristic classes for actions of Hopf algebras is governed by the following general result:

Theorem 9 ([41]). *Let $\mathcal{H}$ be a Hopf algebra endowed with a modular pair in involution Then $\mathcal{H}^{\natural}_{\delta,\sigma} = \{\mathcal{H}^{\otimes n}\}_{n\geq 1}$ equipped with the operators given by (15.6)–(15.8) defines a module over the cyclic category Λ. Let $\mathcal{H}$ act on an algebra A endowed with a δ-invariant σ-trace τ , then the following defines a canonical map from $HC^*_{\delta,\sigma}(\mathcal{H})$ to $HC^*(A)$,*

$$\gamma(h^1 \otimes \ldots \otimes h^n) \in C^n(A)\,,$$

$$\gamma(h^1 \otimes \ldots \otimes h^n)(x^0, \ldots, x^n) = \tau\big(x^0 h^1(x^1) \ldots h^n(x^n)\big)\,.$$

We refer to [41] for the discussion of the remarkable agreement of this theory with the standard theory of quantum groups and their locally compact versions.

16 Hopf Algebras, Renormalization and the Riemann–Hilbert Problem

We describe in this section our joint work with Dirk Kreimer. Perturbative renormalization is by far the most successful technique for computing physical quantities in quantum field theory. It is well known for instance that it accurately predicts the first ten decimal places of the anomalous magnetic moment of the electron.

The physical motivation behind the renormalization technique is quite clear and goes back to the concept of effective mass in nineteen century hydrodynamics. To appreciate it, one should dive under water with a ping-pong ball and start applying Newton's law,

$$F = m\,a \tag{16.1}$$

to compute the initial acceleration of the ball B when we let it loose (at zero speed relative to the water). If one naively applies (16.1), one finds (see the QFT course by Sidney Coleman) an unrealistic initial acceleration of about 20g! In fact as explained in loc. cit. due to the interaction of B with the surrounding field of water, the inertial mass m involved in (16.1) is not the bare mass m_0 of B but is modified to

$$m = m_0 + \tfrac{1}{2}\, M \tag{16.2}$$

where M is the mass of the water occupied by B.

It follows for instance that the initial acceleration a of B is given, using the Archimedean law, by

$$-(M - m_0)g = \left(m_0 + \tfrac{1}{2}\, M\right) a \tag{16.3}$$

and is always of magnitude less than $2g$.

The additional inertial mass $\delta\, m = m - m_0$ is due to the interaction of B with the surrounding field of water and if this interaction could not be turned off (which is the case if we deal with an electron instead of a ping-pong ball) there would be no way to measure the bare mass m_0.

The analogy between hydrodynamics and electromagnetism led (through the work of Thomson, Lorentz, Kramers... [80]) to the crucial distinction between the bare parameters, such as m_0, which enter the field theoretic equations, and the observed parameters, such as the inertial mass m.

A quantum field theory in $D = 4$ dimensions, is given by a classical action functional,

$$S\,(A) = \int \mathcal{L}\,(A)\, d^4x \tag{16.4}$$

where A is a classical field and the Lagrangian is of the form,

$$\mathcal{L}\,(A) = (\partial A)^2/2 - \tfrac{m^2}{2}\, A^2 + \mathcal{L}_{\rm int}(A) \tag{16.5}$$

where $\mathcal{L}_{\rm int}(A)$ is usually a polynomial in A and possibly its derivatives.

One way to describe the quantum fields $\phi(x)$, is by means of the time ordered Green's functions,

$$G_N(x_1, \ldots, x_N) = \langle 0|T\,\phi(x_1)\ldots\phi(x_N)|0\rangle \tag{16.6}$$

where the time ordering symbol T means that the $\phi(x_j)$'s are written in order of increasing time from right to left.

The probability amplitude of a classical field configuration A is given by,

$$e^{i\,\frac{S(A)}{\hbar}} \tag{16.7}$$

and if one could ignore the renormalization problem, the Green's functions would be computed as,

$$G_N(x_1,\ldots,x_N) = \mathcal{N}\int e^{i\,\frac{S(A)}{\hbar}}\, A(x_1)\ldots A(x_N)\,[dA] \tag{16.8}$$

where $\mathcal{N}$ is a normalization factor required to ensure the normalization of the vacuum state,

$$\langle\, 0 \mid 0 \,\rangle = 1\,. \tag{16.9}$$

If one could ignore renormalization, the functional integral (16.8) would be easy to compute in perturbation theory, i.e. by treating the term $\mathcal{L}_{\rm int}$ in (16.5) as a perturbation of

$$\mathcal{L}_0(A) = (\partial A)^2/2 - \tfrac{m^2}{2}\,A^2\,. \tag{16.10}$$

With obvious notation the action functional splits as

$$S(A) = S_0(A) + S_{\rm int}(A) \tag{16.11}$$

where the free action S_0 generates a Gaussian measure $\exp(iS_0(A))[dA] = d\mu$.

The series expansion of the Green's functions is then given in terms of Gaussian integrals of polynomials as,

$$G_N(x_1,\ldots,x_N) = \left(\sum_{n=0}^{\infty} i^n/n! \int A(x_1)\ldots A(x_N)(S_{\rm int}(A))^n d\mu\right) \left(\sum_{n=0}^{\infty} i^n/n! \int S_{\rm int}(A)^n d\mu\right)^{-1}.$$

The various terms of this expansion are computed using integration by parts under the Gaussian measure μ. This generates a large number of terms $U(\Gamma)$, each being labelled by a Feynman graph Γ, and having a numerical value $U(\Gamma)$ obtained as a multiple integral in a finite number of space-time variables. As a rule the unrenormalized values $U(\Gamma)$ are given by nonsensical divergent integrals.

The conceptually really nasty divergences are called ultraviolet and are associated to the presence of arbitrarily large frequencies or equivalently to the unboundedness of momentum space on which integration has to be carried out. Equivalently, when one attempts to integrate in coordinate space, one confronts divergences along diagonals, reflecting the fact that products of field operators are defined only on the configuration space of distinct space-time points.

The physics resolution of this problem is obtained by first introducing a cut-off in momentum space (or any suitable regularization procedure) and then by cleverly making use of the unobservability of the bare parameters, such as the bare mass m_0. By adjusting, term by term of the perturbative expansion, the dependence of the bare parameters on the cut-off parameter, it is possible for a large class of theories, called renormalizable, to eliminate the unwanted ultraviolet divergences.

The main calculational complication of this subtraction procedure occurs for diagrams which possess non-trivial subdivergences, i.e. subdiagrams which are already divergent. In that situation the procedure becomes very involved since it is no longer a simple subtraction, and this for two obvious reasons: i) the divergences are no longer given by local terms, and ii) the previous corrections (those for the subdivergences) have to be taken into account.

To have an example for the combinatorial burden imposed by these difficulties consider the problem below of the renormalization of a two-loop four-point function in massless scalar ϕ^4 theory in four dimensions, given by the following Feynman graph:

$\Gamma^{(2)} =$ 

It contains a divergent subgraph:

$\Gamma^{(1)} =$ 

We work in the Euclidean framework and introduce a cut-off λ which we assume to be always much bigger than the square of any external momentum p_i. Analytic expressions for these Feynman graphs are obtained by utilizing a map Γ_λ which assigns integrals to them according to the Feynman rules and employs the cut-off λ to the momentum integrations. Then $\Gamma_\lambda[\Gamma^{[1,2]}]$ are given by

$$\Gamma_\lambda[\Gamma^{[1]}](p_i) = \int d^4k \frac{\Theta(\lambda^2 - k^2)}{k^2} \frac{1}{(k+p_1+p_2)^2},$$

and

$$\Gamma_\lambda[\Gamma^{[2]}](p_i) = \int d^4l \frac{\Theta(\lambda^2 - l^2)}{l^2(l+p_1+p_2)^2} \Gamma_\lambda\left[\Gamma^{[1]}(p_1,l,p_2,l)\right].$$

It is easy to see that $\Gamma_\lambda[\Gamma^{[1]}]$ decomposes into the form $b \log \lambda$ (where b is a real number) plus terms which remain finite for $\lambda \to \infty$, and hence will produce a divergence which is a non-local function of external momenta

$$\sim \log\lambda \int d^4 l \frac{\Theta(\lambda^2 - l^2)}{l^2 (l + p_1 + p_2)^2} \sim \log\lambda \, \log(p_1 + p_2)^2 \,.$$

Fortunately, the counterterm $\mathcal{L}_{\Gamma^{[1]}} \sim \log\lambda$ generated to subtract the divergence in $\Gamma_\lambda[\Gamma^{[1]}]$ will precisely cancel this non-local divergence in $\Gamma^{[2]}$.

That this type of cancellation occurs at any order of perturbation theory, i.e. that the two diseases above actually cure each other in general is a very non-trivial fact that took decades to prove [79].

The detailed combinatorics is governed by the $\bar{R}$ operation of Bogoliubov and Parasiuk (for a 1PI graph Γ)

$$\bar{R}(\Gamma) = U(\Gamma) + \sum_{\gamma \subsetneq \Gamma} C(\gamma)\, U(\Gamma/\gamma) \tag{16.12}$$

which prepares a given graph with unrenormalized value $U(\Gamma)$ by adding the counterterms $C(\gamma)$. The latter are constructed by induction using

$$C(\Gamma) = -T\Big(U(\Gamma) + \sum_{\gamma \subsetneq \Gamma} C(\gamma) U(\Gamma/\gamma)\Big) \tag{16.13}$$

where, using dimensional regularization T is just the extraction of the pole part in $D = 4 - \epsilon$. The renormalized graph is then given by

$$R(\Gamma) = U(\Gamma) + C(\Gamma) + \sum_{\gamma \subsetneq \Gamma} C(\gamma) U(\Gamma/\gamma)\,. \tag{16.14}$$

For a mathematician the intricacies of the detailed combinatorics and the lack of any obvious mathematical structure underlying it make it totally inaccessible, in spite of the existence of a satisfactory formal approach to the problem [81].

This situation was drastically changed by the discovery by Dirk Kreimer ([42]) who understood that the formula for the $\bar{R}$ operation in fact dictates a Hopf algebra coproduct on the free commutative algebra $\mathcal{H}_K$ generated by the 1PI graphs Γ,

$$\Delta\Gamma = \Gamma \otimes 1 + 1 \otimes \Gamma + \sum_{\gamma \subsetneq \Gamma} \gamma \otimes \Gamma/\gamma \,, \tag{16.15}$$

(In fact he first formulated it in terms of rooted trees, but the graph formulation is easier to explain.)

This Hopf algebra is commutative as an algebra and we showed in [44], [46], that it is the dual Hopf algebra of the enveloping algebra of a Lie algebra $\underline{G}$ whose basis is labelled by the one particle irreducible Feynman graphs. The Lie bracket of two such graphs is computed from insertions of one graph in the other and vice versa. The corresponding Lie group G is the group of characters of $\mathcal{H}$.

The next breakthrough came from our joint discovery [46] that identical formulas to equations (16.12), (16.13), (16.14) occur in the solution of the Riemann–Hilbert problem for an arbitrary pronilpotent Lie group G!

This really unveils the true nature of this seemingly complicated combinatorics and shows that it is a special case of a general extraction of finite values based on the Riemann–Hilbert problem.

The Riemann–Hilbert problem comes from Hilbert's 21st problem which he formulated as follows:

> "Prove that there always exists a Fuchsian linear differential equation with given singularities and given monodromy."

In this form it admits a positive answer due to Plemelj and Birkhoff (cf. [47] for a careful exposition). When formulated in terms of linear systems of the form,

$$y'(z) = A(z)y(z)\,, \quad A(z) = \sum_{\alpha \in S} \frac{A_\alpha}{z-\alpha}\,, \tag{16.16}$$

(where S is the given finite set of singularities, $\infty \notin S$, the A_α are complex matrices such that

$$\sum A_\alpha = 0 \tag{16.17}$$

to avoid singularities at ∞), the answer is not always positive [48], but the solution exists when the monodromy matrices M_α are sufficiently close to 1. It can then be explicitly written as a series of polylogarithms [47].

Another formulation of the Riemann–Hilbert problem, intimately tied up to the classification of holomorphic vector bundles on the Riemann sphere $P_1(\mathbb{C})$, is in terms of the Birkhoff decomposition

$$\gamma\,(z) = \gamma_-(z)^{-1}\,\gamma_+(z) \qquad z \in C \tag{16.18}$$

where we let $C \subset P_1(\mathbb{C})$ be a smooth simple curve, C_- the component of the complement of C containing $\infty \notin C$ and C_+ the other component. Both γ and $\gamma_\pm$ are loops with values in $\mathrm{GL}_n(\mathbb{C})$,

$$\gamma\,(z) \in G = \mathrm{GL}_n(\mathbb{C}) \qquad \forall\, z \in \mathbb{C} \tag{16.19}$$

and $\gamma_\pm$ are boundary values of holomorphic maps (still denoted by the same symbol)

$$\gamma_\pm : C_\pm \to \mathrm{GL}_n(\mathbb{C})\,. \tag{16.20}$$

The normalization condition $\gamma_-(\infty) = 1$ ensures that, if it exists, the decomposition (16.18) is unique (under suitable regularity conditions).

The existence of the Birkhoff decomposition (16.18) is equivalent to the vanishing,

$$c_1\,(L_j) = 0 \tag{16.21}$$

of the Chern numbers $n_j = c_1\,(L_j)$ of the holomorphic line bundles of the Birkhoff-Grothendieck decomposition,

$$E = \oplus\, L_j \tag{16.22}$$

where E is the holomorphic vector bundle on $P_1(\mathbb{C})$ associated to γ, i.e. with total space:

$$(C_+ \times \mathbb{C}^n) \cup_\gamma (C_- \times \mathbb{C}^n)\,. \tag{16.23}$$

The above discussion for $G = \mathrm{GL}_n(\mathbb{C})$ extends to arbitrary complex Lie groups.

When G is a simply connected nilpotent complex Lie group the existence (and uniqueness) of the Birkhoff decomposition (16.18) is valid for any γ. When the loop $\gamma : C \to G$ extends to a holomorphic loop: $C_+ \to G$, the Birkhoff decomposition is given by $\gamma_+ = \gamma$, $\gamma_- = 1$. In general, for $z_0 \in C_+$ the evaluation,

$$\gamma \to \gamma_+(z_0) \in G \tag{16.24}$$

is a natural principle to extract a finite value from the singular expression $\gamma(z_0)$. This extraction of finite values is a multiplicative removal of the pole part for a meromorphic loop γ when we let C be an infinitesimal circle centered at z_0.

We are now ready to apply this procedure in Quantum Field Theory. First, using dimensional regularization, the bare (unrenormalized) theory gives rise to a meromorphic loop,

$$\gamma(z) \in G\,, \qquad z \in \mathbb{C} \tag{16.25}$$

Our main result [45] [46] is that the renormalized theory is just the evaluation at the integer dimension $z_0 = D$ of space-time of the holomorphic part γ_+ of the Birkhoff decomposition of γ.

In fact, the original loop $d \to \gamma(d)$ not only depends upon the parameters of the theory but also on the additional "unit of mass" μ required by dimensional analysis. We showed in [49] that the mathematical concepts developed in our earlier papers provide very powerful tools to lift the usual concepts of the β-function and renormalization group from the space of coupling constants of the theory to the complex Lie group G.

We first observed that even though the loop $\gamma(d)$ does depend on the additional parameter μ,

$$\mu \to \gamma_\mu(d)\,, \tag{16.26}$$

the negative part γ_{μ^-} in the Birkhoff decomposition,

$$\gamma_\mu(d) = \gamma_{\mu^-}(d)^{-1}\,\gamma_{\mu^+}(d) \tag{16.27}$$

is actually independent of μ,

$$\frac{\partial}{\partial\mu}\,\gamma_{\mu^-}(d) = 0\,. \tag{16.28}$$

This is a restatement of a well-known fact and follows immediately from dimensional analysis. Moreover, by construction, the Lie group G turns out to be graded, with grading,

$$\theta_t \in \operatorname{Aut} G\,, \quad t \in \mathbb{R}\,, \tag{16.29}$$

inherited from the grading of the Hopf algebra $\mathcal{H}$ of Feynman graphs given by the loop number,

$$L(\Gamma) = \text{loop number of } \Gamma \tag{16.30}$$

for any 1PI graph Γ.

The straightforward equality,

$$\gamma_{e^t\mu}(d) = \theta_{t\varepsilon}(\gamma_\mu(d)) \qquad \forall\, t \in \mathbb{R}\,,\ \varepsilon = D - d \tag{16.31}$$

shows that the loops γ_μ associated to the unrenormalized theory satisfy the striking property that the negative part of their Birkhoff decomposition is unaltered by the operation,

$$\gamma(\varepsilon) \to \theta_{t\varepsilon}(\gamma(\varepsilon))\,, \tag{16.32}$$

In other words, if we replace $\gamma(\varepsilon)$ by $\theta_{t\varepsilon}(\gamma(\varepsilon))$ we don't change the negative part of its Birkhoff decomposition. We settled now for the variable,

$$\varepsilon = D - d \in \mathbb{C}\backslash\{0\}\,. \tag{16.33}$$

We give in [49] a complete characterization of the loops $\gamma(\varepsilon) \in G$ fulfilling the above striking invariance. This characterization only involves the negative part $\gamma_-(\varepsilon)$ of their Birkhoff decomposition which by hypothesis fulfils,

$$\gamma_-(\varepsilon)\,\theta_{t\varepsilon}(\gamma_-(\varepsilon)^{-1}) \text{ is convergent for } \varepsilon \to 0\,. \tag{16.34}$$

It is easy to see that the limit of (34) for $\varepsilon \to 0$ defines a one parameter subgroup,

$$F_t \in G\,,\ t \in \mathbb{R} \tag{16.35}$$

and that the generator $\beta = \left(\frac{\partial}{\partial t}\,F_t\right)_{t=0}$ of this one parameter group is related to the *residue* of γ

$$\operatorname*{Res}_{\varepsilon=0}\,\gamma = -\left(\frac{\partial}{\partial u}\,\gamma_-\left(\frac{1}{u}\right)\right)_{u=0} \tag{16.36}$$

by the simple equation,

$$\beta = Y \operatorname{Res} \gamma\,, \tag{16.37}$$

where $Y = \left(\frac{\partial}{\partial t}\,\theta_t\right)_{t=0}$ is the grading.

This is straightforward but our result is the following formula (16.39) which gives $\gamma_-(\varepsilon)$ in closed form as a function of β. We shall for convenience introduce an additional generator in the Lie algebra of G (i.e. primitive elements of $\mathcal{H}^*$) such that,

$$[Z_0, X] = Y(X) \qquad \forall\, X \in \operatorname{Lie} G\,. \tag{16.38}$$

The scattering formula for $\gamma_-(\varepsilon)$ is then,

$$\gamma_-(\varepsilon) = \lim_{t\to\infty} e^{-t\left(\frac{\beta}{\varepsilon}+Z_0\right)}\, e^{tZ_0}\,. \tag{16.39}$$

Both factors in the right-hand side belong to the semi direct product,

$$\widetilde{G} = G >\!\!\lhd_\theta \mathbb{R} \tag{16.40}$$

of the group G by the grading, but of course the ratio (16.39) belongs to the group G.

This shows ([49]) that the higher pole structure of the divergences is uniquely determined by the residue and gives a strong form of the t'Hooft relations, which will come as an immediate corollary.

The main new result of [49], specializing to the massless case and taking φ_6^3 as an illustrative example to fix ideas and notation, is that the formula for the bare coupling constant,

$$g_0 = g\, Z_1\, Z_3^{-3/2} \tag{16.41}$$

where both $g\,Z_1 = g + \delta g$ and the field strength renormalization constant Z_3 are thought of as power series (in g) of elements of the Hopf algebra $\mathcal{H}$, does define a Hopf algebra homomorphism,

$$\mathcal{H}_{CM} \xrightarrow{g_0} \mathcal{H}_K\,, \tag{16.42}$$

from the Hopf algebra $\mathcal{H}_{CM}$ of coordinates on the group of formal diffeomorphisms of $\mathbb{C}$ such that,

$$\varphi(0) = 0\,,\ \varphi'(0) = \mathrm{id} \tag{16.43}$$

to the Hopf algebra $\mathcal{H}_K$ of the massless theory. We had already constructed in [46] a Hopf algebra homomorphism from $\mathcal{H}_{\rm CM}$ to the Hopf algebra of rooted trees, but the physical significance of this construction was unclear.

The homomorphism (16.42) is quite different in that for instance the transposed group homomorphism,

$$G \xrightarrow{\rho} \operatorname{Diff}(\mathbb{C}) \tag{16.44}$$

lands in the subgroup of *odd* diffeomorphisms,

$$\varphi(-z) = -\varphi(z) \qquad \forall\, z\,. \tag{16.45}$$

Moreover its physical significance is transparent. In particular the image by ρ of $\beta = Y\,\mathrm{Res}\,\gamma$ is the usual β-function of the coupling constant g.

We discovered the homomorphism (16.42) by lengthy concrete computations which were an excellent test for the explicit ways of handling the coproduct, coassociativity, symmetry factors... that underly the theory.

As a corollary of the construction of ρ we get an *action* by (formal) diffeomorphisms of the group G on the space X of (dimensionless) coupling constants of the theory. We can then in particular formulate the Birkhoff decomposition *directly* in the group,

$$\mathrm{Diff}\,(X) \tag{16.46}$$

of formal diffeomorphisms of the space of coupling constants.

Theorem 10 ([49]). *Let the unrenormalized effective coupling constant $g_{\mathrm{eff}}(\varepsilon)$ be viewed as a formal power series in g and let $g_{\mathrm{eff}}(\varepsilon) = g_{\mathrm{eff}_+}(\varepsilon)(g_{\mathrm{eff}_-}(\varepsilon))^{-1}$ be its (opposite) Birkhoff decomposition in the group of formal diffeomorphisms. Then the loop $g_{\mathrm{eff}_-}(\varepsilon)$ is the bare coupling constant and $g_{\mathrm{eff}_+}(0)$ is the renormalized effective coupling.*

This result is now, in its statement, no longer depending upon our group G or the Hopf algebra $\mathcal{H}$. But of course the proof makes heavy use of the above ingredients. It is a challenge to physicists to find a direct proof of this result.

Finally the Birkhoff decomposition of a loop,

$$\delta(\varepsilon) \in \mathrm{Diff}\,(X)\,, \tag{16.47}$$

admits a beautiful geometric interpretation. If we let X be a complex manifold and pass from formal diffeomorphisms to actual ones, the data (16.47) is the initial data to perform, by the clutching operation, the construction of a complex bundle,

$$P = (S^+ \times X) \cup_\delta (S^- \times X) \tag{16.48}$$

over the sphere $S = P_1(\mathbb{C}) = S^+ \cup S^-$, and with fiber X,

$$X \longrightarrow P \xrightarrow{\ \pi\ } S\,. \tag{16.49}$$

The meaning of the Birkhoff decomposition,

$$\delta(\varepsilon) = \delta_-(\varepsilon)^{-1}\,\delta_+(\varepsilon) \tag{16.50}$$

is then exactly captured by an isomorphism of the bundle P with the trivial bundle,

$$S \times X\,. \tag{16.51}$$

17 Number Theory

I shall conclude these notes by giving a brief glimpse at the connection between noncommutative geometry and number theory. There are two points of contact of the two subjects, the first gives a spectral interpretation of zeros of zeta and L-functions in terms of a construction involving adeles, more specifically the noncommutative space of adele classes. The second has to do with the missing Galois theory at Archimedean places. For the specialists of quantum chaos looking for a spectral realization of the non-trivial zeros of the Riemann zeta function from the quantization of classical mechanical systems, the adeles might look rather exotic at first sight and we first need to explain briefly (for non-specialists) why Ideles and Adeles are natural and important in number theory.

Let us start with the reciprocity law (Gauss 1801)

$$\left(\frac{\ell}{p}\right) = (-1)^{\varepsilon(p)\varepsilon(\ell)}\left(\frac{p}{\ell}\right), \quad \varepsilon(p) = \frac{p-1}{2} \pmod 2 \tag{17.1}$$

where ℓ and p are odd primes and (ℓ/p) is the Legendre symbol whose value is $+1$ if the equation

$$x^2 = \ell \qquad \pmod p \tag{17.2}$$

admits a solution, and is -1 if it does not.

For instance, with $\ell = 5$ we see that whether the equation $x^2 = 5 (\operatorname{mod} p)$ admits a solution only depends upon $p (\operatorname{mod} 5)$, i.e. only on the last digit of p. Thus the answer is the same for $p = 7$ and $p = 1997$ or for $p = 19$ and $p = 1999$. It follows that the primes p thus fall into *classes*. The language of Adeles and Ideles extends this simple notion of *classes* of primes to those of *ideal classes* and then of *Idele classes*.

To the proof of Dirichlet of the existence of infinitely many primes in an arithmetic progression corresponds the construction of an L-function associated to a character χ modulo m,

$$L(s,\chi) = \prod \frac{1}{1 - \chi(p)\, p^{-s}}\,. \tag{17.3}$$

More generally a Hecke L-function is associated to a character of the ideal class group modulo m and in fact also to a Grössencharakter which is a character of the Idele class group of a number field k.

The quickest way to introduce the Idele class group of a number field k is to understand (cf. Iwasawa, Ann. of Math. 57 (1953)) that such a field sits uniquely as a discrete cocompact subfield of a unique locally compact (semi-simple and non-discrete) ring A

$$k \subset A\,, \quad k \text{ cocompact} \tag{17.4}$$

called the ring of *Adeles* of k. One then has,

$$\text{Idele class group of } k = \mathrm{GL}_1(k)\backslash \mathrm{GL}_1(A)\,, \tag{17.5}$$

and a Grössencharakter is a character of this locally compact group. Iwasawa and Tate showed how to use analysis on adeles to prove the basic properties of the Hecke L-functions which were then extended to L-functions associated to automorphic forms which appear in the action of $\mathrm{GL}_n(A)$ on the Hilbert space

$$L^2(\mathrm{GL}_n(k)\backslash \mathrm{GL}_n(A))\,. \tag{17.6}$$

To understand the other language involved in the basic dictionary which underlies the Langlands program let us go back to the equation (17.2) say with $\ell = 5$ and simply adjoin $\sqrt{5}$ to the field $\mathbb{Q}$ of rational numbers which gives an algebraic extension $K = \mathbb{Q}(\sqrt{5})$ of $k = \mathbb{Q}$. The Galois group $\mathrm{Gal}(\mathbb{Q}(\sqrt{5}) : \mathbb{Q}) = \mathrm{Gal}(K : k)$ is of course $\mathbb{Z}/2$ in this case and admits an obvious non-trivial one dimensional representation π. In general, the *Artin* L-function associated to a representation

$$\mathrm{Gal}(K : k) \to \mathrm{GL}(n, \mathbb{C}) \tag{17.7}$$

of the Galois group of a finite Galois extension K of k, is

$$L(s,\pi) = \prod_p L_p(s,\pi) \tag{17.8}$$

where p runs through the prime ideals of k and the local L factor is given at unramified p by,

$$L_p(s,\pi) = \det\left(1 - \pi(\sigma)\, N(p)^{-s}\right) \tag{17.9}$$

where σ is the Frobenius automorphism of p.

When K/k is an abelian extension and π a one dimensional representation it follows from class field theory that π defines a character modulo the conductor of K/k and that the Artin L-function equals the Hecke L-function. This Artin reciprocity law is a far reaching extension of the Gauss reciprocity law (17.1).

The Langlands program extends Hecke's theory of Euler products associated to automorphic forms on GL(2) to arbitrary reductive groups G and gives a correspondence, extending Artin's reciprocity law to the non-Abelian case, between automorphic representations of G and representations,

$$\mathrm{Gal}(K : k) \to {}^L G \tag{17.10}$$

in the Langlands dual ${}^L G$ of G.

A basic tool of the theory is the trace formula [76] which extends to the adelic context the Selberg trace formula. The trace formula is the equality obtained by computing in two different ways the trace of operators of the form,

$$\text{Trace}\,(C_\Lambda\,\pi(f)) \tag{17.11}$$

where (for $G = \text{GL}_n$), π is the natural representation of $\text{GL}_n(A)$ in the Hilbert space

$$L^2_\chi(\text{GL}_n(k)\backslash\text{GL}_n(A)) \tag{17.12}$$

where χ is a Grössencharakter, and C_Λ is a cutoff. The Grössencharakter χ allows one to restrict to vectors with a fixed behaviour relative to GL_1.

The spectral side of the trace formula is obtained from the harmonic analysis of the representation π. The geometric side expresses the trace as a sum of orbital integrals.

The restriction imposed in (17.12) by the Grössencharakter χ shows that the case $n = 1$ becomes trivial and concentrates essentially on the SL_n aspect for $n \geq 2$. So far the zeros of L-functions do not appear in this language.

Our contribution to this subject is to show that both the zeros of L-functions and the Riemann–Weil explicit formulas appear directly in a refinement of the trace formula obtained as follows. Instead of restricting the Hilbert space,

$$L^2(\text{GL}_n(k)\backslash\text{GL}_n(A)) \tag{17.13}$$

by the choice of Grössencharakter χ as above, one introduces on the full Hilbert space (13) a finer cutoff operator Q_Λ taking care of the "GL_1" behaviour of vectors.

To understand in which way the corresponding trace formula refines the Arthur trace formula, it is simplest to restrict to the case of GL_1. In order to simplify even further we shall replace the number field k by a function field of positive characteristic. This allows for a straightforward definition of the cutoff operators Q_Λ as the orthogonal projection on the subspace,

$$Q_\Lambda \subset L^2(\text{GL}_1(k)\backslash\text{GL}_1(A)) \tag{17.14}$$

spanned by the vectors $\xi \in \mathcal{S}(A)$ (averaged on $\text{GL}_1(k)$) which vanish as well as their Fourier transform for $|x| > \Lambda$. Note that we use Fourier transform on the *additive* group of adeles so that the space $\text{GL}_1(k)\backslash A$ of Adele classes is implicit in this construction. To define this Fourier transformation we needed to choose a basic character $\alpha = \prod \alpha_v$ of the additive group A for which the lattice k is selfdual.

The spectral computation of the trace of $Q_\Lambda\pi(f)$ involves all the nontrivial zeros of Hecke L-functions and is given by the following formula

([73]),

$$\mathrm{Trace}(Q_\Lambda \pi(f)) = 2\Big(\sum_{\mathrm{GL}_1} f(k)\Big)\log' \Lambda$$

$$+ \widehat{f}(0) + \widehat{f}(1) - \sum_{\substack{L(\chi, \frac{1}{2}+\rho)=0 \\ \rho \in B/N^\perp}} N\left(\chi, \tfrac{1}{2}+\rho\right) \int_{i\mathbb{R}} \widehat{f}(\chi, z)\, d\mu_\rho(z) + o(1) \quad (17.15)$$

where B is the open strip $B = \{\rho \in \mathbb{C}\,; |\mathrm{Re}\,\rho| < \frac{1}{2}\}$, $N\left(\chi, \frac{1}{2}+\rho\right)$ is the multiplicity of $\frac{1}{2}+\rho$ as a zero of the L function $L(\chi, s)$, χ varying through Grössencharakters (modulo principal ones), N being the module,

$$N = \mathrm{Mod}(k)\,. \quad (17.16)$$

Also $2\log' \Lambda = \int_{|\lambda|\in[\Lambda^{-1},\Lambda]} d^*\lambda$, and the measure $d\mu_\rho(z)$ is the harmonic measure of $\rho \in \mathbb{C}$ with respect to the line $i\mathbb{R}$. In particular if the zero $\frac{1}{2}+\rho$ is on the critical line $d\mu_\rho(z)$ is just the Dirac mass at $z = \rho$. Finally the Fourier transform of f is given by,

$$\widehat{f}(\chi, z) = \int_{\mathrm{GL}_1(A)} f(u^{-1})\, \chi(u)\, |u|^z\, d^*u\,. \quad (17.17)$$

The geometric side of the trace formula has so far only be fully justified in the simplified situation where only finitely many places are used. It is then given by the following formula ([73])

$$\mathrm{Trace}(Q_\Lambda \pi(f)) = 2\Big(\sum_{\mathrm{GL}_1} f(k)\Big)\log' \Lambda + \sum_{v,k} \int'_{k_v^*} \frac{f(ku)}{|1-u|}\, d^*u + o(1)\,; \quad (17.18)$$

where each k_v^* is embedded in $(\mathrm{GL}_1(k)\backslash \mathrm{GL}_1(A))$ by the map $u \to (1, 1, \ldots, u, \ldots, 1)$ and the principal value $\int'$ is uniquely determined by the pairing with the unique distribution on k_v which agrees with $du/|1-u|$ for $u \neq 1$ and whose Fourier transform relative to α_v vanishes at 1.

By proving that it entails the positivity of the Weil distribution, we showed in [73] that the validity of the geometric side, i.e. the global trace formula, is equivalent to the Riemann Hypothesis for all L-functions with Grössencharakter.

Theorem 10. *The following two conditions are equivalent:*

a) *When $\Lambda \to \infty$, one has, for all $f \in \mathcal{S}(\mathrm{GL}_1(k)\backslash \mathrm{GL}_1(A))$ with compact support,*

$$\mathrm{Trace}(Q_\Lambda \pi(f)) = 2\Big(\sum_{\mathrm{GL}_1} f(k)\Big)\log' \Lambda + \sum_{v,k} \int'_{k_v^*} \frac{f(ku)}{|1-u|}\, d^*u + o(1)\,; \quad (17.19)$$

b) *All L functions with Grössencharakter on k satisfy the Riemann Hypothesis.*

We have thus obtained a spectral interpretation of the zeros of zeta and L-functions as an absorption spectrum, i.e. as missing spectral lines. All zeros do play a role in the spectral side of the trace formula, but while the critical zeros do appear per se, the noncritical ones appear as resonances and enter in the trace formula through their harmonic potential with respect to the critical line. The spectral side is entirely canonical, and its validity is justified in the global case [73]. It is quite important to understand why a crucial negative sign in the analysis of the statistical fluctuations of the zeros of zeta indicated from the start that the spectral interpretation should be as an absorption spectrum, or equivalently should be of a cohomological nature.

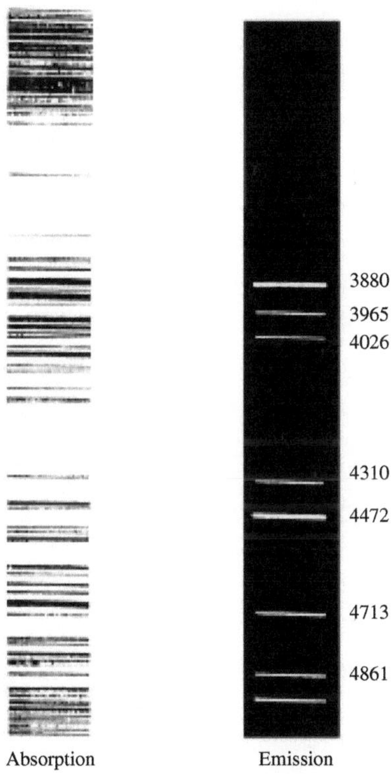

The number of zeros of zeta whose imaginary part is less than $E > 0$,

$$N(E) = \# \text{ of zeros } \rho \ , \ 0 < \operatorname{Im} \rho < E \tag{17.20}$$

has an asymptotic expression ([26]) given by

$$N(E) = \tfrac{E}{2\pi}\left(\log\left(\tfrac{E}{2\pi}\right) - 1\right) + \tfrac{7}{8} + o(1) + N_{\rm osc}(E) \quad (17.21)$$

where the oscillatory part of this step function is

$$N_{\rm osc}(E) = \tfrac{1}{\pi}\,\mathrm{Im}\,\log\zeta\left(\tfrac{1}{2} + iE\right) \quad (17.22)$$

which is of the order of $\mathrm{Log}(E)$. (We assume that E is not the imaginary part of a zero and take for the logarithm the branch which is 0 at $+\infty$). The Euler product formula for the zeta function yields (cf. [72]) a heuristic asymptotic formula for $N_{\rm osc}(E)$,

$$N_{\rm osc}(E) \simeq \frac{-1}{\pi} \sum_p \sum_{m=1}^{\infty} \frac{1}{m}\,\frac{1}{p^{m/2}}\,\sin\left(m\,E\,\log p\right). \quad (17.23)$$

One can compare this formula with what appears in the direct attempt [72] to construct a spectral realization of zeros of zeta from quantization of a classical dynamical system. In this theory the quantization of the classical dynamical system given by the phase space X and hamiltonian h gives rise to a Hilbert space $\mathcal{H}$ and a selfadjoint operator H whose spectrum is the essential physical observable of the system. For complicated systems the only useful information about this spectrum is that, while the average part of the counting function,

$$N(E) = \#\ \text{eigenvalues of } H \text{ in } [0,E] \quad (17.24)$$

is computed by a semiclassical approximation mainly as a volume in phase space, the oscillatory part,

$$N_{\rm osc}(E) = N(E) - \langle N(E)\rangle \quad (17.25)$$

is the same as for a random matrix, governed by the statistic dictated by the symmetries of the system.

One can then ([72]) write down an asymptotic semiclassical approximation to the oscillatory function $N_{\rm osc}(E)$

$$N_{\rm osc}(E) = \frac{1}{\pi}\,\mathrm{Im} \int_0^{\infty} \mathrm{Trace}(H - (E + i\eta))^{-1} i d\eta \quad (17.26)$$

using the stationary phase approximation of the corresponding functional integral. For a system whose configuration space is 2-dimensional, this gives ([72] (15)),

$$N_{\rm osc}(E) \simeq \frac{1}{\pi} \sum_{\gamma_p} \sum_{m=1}^{\infty} \frac{1}{m}\,\frac{1}{2\mathrm{sh}\left(\frac{m\lambda_p}{2}\right)}\,\sin(S_{\rm pm}(E)) \quad (17.27)$$

where the γ_p are the primitive periodic orbits, the label m corresponds to the number of traversals of this orbit, while the corresponding instability exponents are $\pm\lambda_p$. The phase $S_{\rm pm}(E)$ is up to a constant equal to $m\,E\,T_\gamma^\#$ where $T_\gamma^\#$ is the period of the primitive orbit γ_p.

Comparing the formulas one sees a fundamental mismatch (cf. [72]) which is the overall *minus sign* in front of formula (17.23) as opposed to the plus sign of (17.27). This problem is resolved in our spectral interpretation by the minus sign present in the spectral side of the trace formula (17.15). The point is that the spectral analysis of the action of the Idele class group on the Adele class space shows ([73]) white light with dark absorption lines labelled by the zeros of zeta and L-functions. This also provides the correct explanation for the asymptotic form of the formula for the average number of zeros

$$\langle N(E)\rangle \sim (E/2\pi)\big(\log(E/2\pi)-1\big)+7/8+o(1) \tag{17.28}$$

from a semiclassical computation for the number of quantum mechanical states in one degree of freedom which fulfil the conditions

$$|q|\leq\Lambda\,,\quad |p|\leq\Lambda\,,\quad |H|\leq E\,, \tag{17.29}$$

where $H=2\pi qp$ is the Hamiltonian which generates the group involved in the action of the Idele class group namely the scaling transformations (see [73] for precise normalization). We are thus computing the area of,

$$D=\big\{(p,q);pq\geq 0,|q|\leq\Lambda,|p|\leq\Lambda,|pq|\leq E/2\pi\big\}\,. \tag{17.30}$$

(since we deal with zeta alone we restrict ourselves to even functions so that we exclude the region $pq\leq 0$ of the semiclassical (p,q) plane). The computation yields

$$\frac{1}{2}\int_D dp\,dq=\frac{2E}{2\pi}\log\Lambda-\frac{E}{2\pi}\left(\log\frac{E}{2\pi}-1\right)\,. \tag{17.31}$$

In this formula we see in the right-hand side the overall term $\langle N(E)\rangle$ which appears with a *minus* sign. This shows that the number of quantum mechanical states is equal to $\frac{4E}{2\pi}\log\Lambda$ minus the first approximation to the number of zeros of zeta whose imaginary part is less than E in absolute value (one just multiplies by 2 the equality (17.31)). Now $\frac{1}{2\pi}(2E)(2\log\Lambda)$ is the number of quantum states in the Hilbert space $L^2(\mathbb{R}_+^*,d^*x)$ which are localized in $\mathbb{R}_+^*$ between Λ^{-1} and Λ and are localized in the dual group $\mathbb{R}$ (for the pairing $\langle\lambda,t\rangle=\lambda^{it}$) between $-E$ and E. Thus we see clearly that the first approximation to $N(E)$ appears as the lack of surjectivity of the map which associates to quantum states ξ with support in D the function on $\mathbb{R}_+^*$,

$$E(\xi)(x)=|x|^{1/2}\sum_{n\in\mathbb{Z}}\xi(nx) \tag{17.32}$$

where we assume the additional conditions $\xi(0) = \int \xi(x)dx = 0$.

A finer analysis, which is just what the trace formula is doing, would yield the additional terms $7/8+o(1)+N_{osc}(E)$. The above discussion yields an explicit construction of a large matrix whose spectrum approaches the zeros of zeta as $\Lambda \to \infty$.

While the above discussion clearly gives the sought for spectral interpretation of zeros of zeta it is unclear that one can expect to justify the (geometric side of) trace formula without a deeper understanding of the symmetries of the situation, which might well involve quantum groups.

As we mentioned earlier, the second point of contact between noncommutative geometry and number theory has to do with the missing Galois theory at Archimedean places.

Let k be a *global* field, when the characteristic of k is $p > 1$ so that k is a function field over $\mathbb{F}_q$, one has

$$k \subset k_{\mathrm{un}} \subset k_{\mathrm{ab}} \subset k_{\mathrm{sep}} \subset \overline{k}\,,$$

where $\overline{k}$ is an algebraic closure of k, k_{sep} the separable algebraic closure, k_{ab} the maximal abelian extension and k_{un} is obtained by adjoining to k all roots of unity of order prime to p.

One defines the Weil group W_k as the subgroup of $\mathrm{Gal}(k_{\mathrm{ab}} : k)$ of those automorphisms which induce on k_{un} an integral power of the Frobenius automorphism σ,

$$\sigma(\mu) = \mu^q \qquad \forall\, \mu \text{ root of 1 of order prime to } p\,.$$

The main theorem of global class field theory asserts the existence of a canonical isomorphism,

$$W_k \simeq C_k = GL_1(A)/GL_1(k)\,,$$

of locally compact groups.

When k is of characteristic 0, i.e. is a number field, one has a canonical isomorphism,

$$\mathrm{Gal}(k_{\mathrm{ab}} : k) \simeq C_k/D_k\,,$$

where D_k is the connected component of identity in the Idele class group C_k, but because of the Archimedean places of k there is no interpretation of C_k analogous to the Galois group interpretation for function fields. According to A. Weil [77], “La recherche d’une interprétation pour C_k si k est un corps de nombres, analogue en quelque manière à l’interprétation par un groupe de Galois quand k est un corps de fonctions, me semble constituer l’un des problèmes fondamentaux de la théorie des nombres à l’heure actuelle ; il se peut qu’une telle interprétation renferme la clef de l’hypothèse de Riemann …”.

Galois groups are by construction projective limits of the finite groups attached to finite extensions. To get connected groups one clearly needs to relax this finiteness condition which is the same as the finite dimensionality of the central simple algebras of the Brauer theory. Since Archimedean places of k are responsible for the non-triviality of D_k it is natural to ask the following preliminary question,

"Is there a non-trivial Brauer theory of central simple algebras over $\mathbb{C}$."

We showed in [3] that the *approximately finite dimensional* simple central algebras over $\mathbb{C}$ (called factors) provide a satisfactory answer to this question. They are classified by their module,

$$\mathrm{Mod}(M) \underset{\sim}{\subset} \mathbb{R}_+^* ,$$

which is a virtual closed subgroup of $\mathbb{R}_+^*$.

One can in fact go much further and understand that the renormalization group, once properly formulated mathematically as we did in section 16, really appears as a perfect ambiguity group between solutions to a (physics) problem. It hence plays a role very similar to that of the Galois group of an algebraic equation and is an ideal candidate for the missing Galois group at the Archimedean place. One can explore this idea further by making use of the relation between the (conjectural) Hopf algebra of Euler–Zagier numbers ([82], [83]) and the Kreimer Hopf algebra.

18 Appendix, the Cyclic Category

At the conceptual level, cyclic cohomology is a way to embed the non-additive category of algebras and algebra homomorphisms in an additive category of modules. The latter is the additive category of Λ-modules where Λ is the cyclic category. Cyclic cohomology is then obtained as an *Ext* functor ([14]).

The cyclic category is a small category which can be defined by generators and relations. It has the same objects as the small category Δ of totally ordered finite sets and increasing maps which plays a key role in simplicial topology. Let us recall that Δ has one object $[n]$ for each integer n, and is generated by faces $\delta_i, [n-1] \to [n]$ (the injection that misses i), and degeneracies $\sigma_j, [n+1] \to [n]$ (the surjection which identifies j with $j+1$), with the relations,

$$\delta_j \delta_i = \delta_i \delta_{j-1} \text{ for } i < j, \ \sigma_j \sigma_i = \sigma_i \sigma_{j+1} \qquad i \leq j \tag{18.1}$$

$$\sigma_j \delta_i = \begin{cases} \delta_i \sigma_{j-1} & i < j \\ 1_n & \text{if } i = j \text{ or } i = j+1 \\ \delta_{i-1} \sigma_j & i > j+1. \end{cases} \tag{18.2}$$

To obtain the cyclic category Λ one adds for each n a new morphism $\tau_n, [n] \to [n]$ such that,

$$\begin{aligned} &\tau_n \delta_i = \delta_{i-1} \tau_{n-1} \quad 1 \le i \le n, \quad \tau_n \delta_0 = \delta_n \\ &\tau_n \sigma_i = \sigma_{i-1} \tau_{n+1} \quad 1 \le i \le n, \quad \tau_n \sigma_0 = \sigma_n \tau_{n+1}^2 \\ &\tau_n^{n+1} = 1_n \,. \end{aligned} \tag{18.3}$$

The original definition of Λ (cf. [14]) used homotopy classes of non-decreasing maps from S^1 to S^1 of degree 1, mapping $\mathbb{Z}/n$ to $\mathbb{Z}/m$ and is trivially equivalent to the above.

Given an algebra A one obtains a module over the small category Λ by assigning to each integer $n \ge 0$ the vector space C^n of $n+1$-linear forms $\varphi(x^0, \dots, x^n)$ on A, while the basic operations are given by

$$\begin{aligned} &(\delta_i \varphi)(x^0, \dots, x^n) = \varphi(x^0, \dots, x^i x^{i+1}, \dots, x^n)\,, \quad i = 0, 1, \dots, n-1 \\ &(\delta_n \varphi)(x^0, \dots, x^n) = \varphi(x^n x^0, x^1, \dots, x^{n-1}) \\ &(\sigma_j \varphi)(x^0, \dots, x^n) = \varphi(x^0, \dots, x^j, 1, x^{j+1}, \dots, x^n)\,, \quad j = 0, 1, \dots, n \\ &(\tau_n \varphi)(x^0, \dots, x^n) = \varphi(x^n, x^0, \dots, x^{n-1})\,. \end{aligned} \tag{18.4}$$

These operations satisfy the relations (18.1) (18.2) and (18.3). This shows that any algebra A gives rise canonically to a Λ-module and allows [14, 21] to interpret the cyclic cohomology groups $HC^n(A)$ as Ext^n functors. All of the general properties of cyclic cohomology such as the long exact sequence relating it to Hochschild cohomology are shared by Ext of general Λ-modules and can be attributed to the equality of the classifying space $B\Lambda$ of the small category Λ with the classifying space BS^1 of the compact one-dimensional Lie group S^1. One has

$$B\Lambda = BS^1 = P_\infty(\mathbb{C})\,. \tag{18.5}$$

References

[1] A. Connes, Une classification des facteurs de type III. Ann. Sci. Ecole Norm. Sup. 6:4 (1973), 133–252.

[2] M. Takesaki, Tomita's theory of modular Hilbert algebras and its applications. Springer Lecture Notes in Math. 28 (1970).

[3] A. Connes, Noncommutative Geometry and the Riemann Zeta Function, Mathematics: Frontiers and Perspectives, IMU 2000 volume, 35–55.

[4] M.F. ATIYAH, Global theory of elliptic operators, Proc. Internat. Conf. on Functional Analysis and Related Topics (Tokyo, 1969), University of Tokyo Press, Tokyo (1970), 21–30.

[5] I.M. SINGER, Future extensions of index theory and elliptic operators, Ann. of Math. Studies 70 (1971), 171–185.

[6] L.G. BROWN, R.G. DOUGLAS, P.A. FILLMORE, Extensions of C^*-algebras and K-homology, Ann. of Math. 2:105 (1977), 265–324.

[7] A.S. MISCENKO, C^* algebras and K theory, Algebraic Topology, Aarhus 1978, Springer Lecture Notes in Math. 763 (1979), 262–274.

[8] G.G. KASPAROV, The operator K-functor and extensions of C^* algebras, Izv. Akad. Nauk SSSR, Ser. Mat. 44 (1980), 571–636; Math. USSR Izv. 16 (1981), 513–572.

[9] P. BAUM, A. CONNES, Geometric K-theory for Lie groups and foliations, Preprint IHES (M/82/), 1982; l'Enseignement Mathematique, t. 46 (2000), 1–35 (to appear).

[10] M.F. ATIYAH, W. SCHMID, A geometric construction of the discrete series for semisimple Lie groups, Inventiones Math. 42 (1977), 1–62.

[11] G. SKANDALIS, Approche de la conjecture de Novikov par la cohomologie cyclique, in "Séminaire Bourbaki, 1990-91", Expose 739, 201-202-203 (1992), 299–316.

[12] P. JULG, Travaux de N. Higson et G. Kasparov sur la conjecture de Baum–Connes, in "Séminaire Bourbaki, 1997-98", Expose 841, 252 (1998), 151–183.

[13] G. SKANDALIS, Progrés récents sur la conjecture de Baum–Connes, contribution de Vincent Lafforgue, in "Séminaire Bourbaki, 1999-2000", Expose 869.

[14] A. CONNES, Cohomologie cyclique et foncteur Ext^n, C.R. Acad. Sci. Paris, Ser. I Math. 296 (1983), 963–968.

[15] A. CONNES, Spectral sequence and homology of currents for operator algebras, Math. Forschungsinst. Oberwolfach Tagungsber. 41/81; Funktionalanalysis und C^*-Algebren, 27-9/3-10, 1981.

[16] A. CONNES, Noncommutative differential geometry. Part I: The Chern character in K-homology, Preprint IHES, M/82/53, 1982; Part II: de Rham homology and noncommutative algebra, Preprint IHES, M/83/19, 1983.

[17] A. CONNES, Noncommutative differential geometry, Inst. Hautes Etudes Sci. Publ. Math. 62 (1985), 257–360.

[18] B.L. TSYGAN, Homology of matrix Lie algebras over rings and the Hochschild homology, Uspekhi Math. Nauk. 38 (1983), 217–218.

[19] A. CONNES, H. MOSCOVICI, Cyclic cohomology, the Novikov conjecture and hyperbolic groups, Topology 29 (1990), 345–388.

[20] A. CONNES, Cyclic cohomology and the transverse fundamental class of a foliation, in "Geometric Methods in Operator Algebras, (Kyoto, 1983)", Pitman Res. Notes in Math. 123, Longman, Harlow (1986), 52–144.

[21] J.L. LODAY, Cyclic Homology, Springer, Berlin–Heidelberg–New York, 1998.

[22] D. BURGHELEA, The cyclic homology of the group rings, Comment. Math. Helv. 60 (1985), 354–365.

[23] J. CUNTZ, D. QUILLEN, Cyclic homology and singularity, J. Amer. Math. Soc. 8 (1995), 373–442.

[24] J. CUNTZ, D. QUILLEN, Operators on noncommutative differential forms and cyclic homology, J. Differential Geometry, to appear.

[25] J. CUNTZ, D. QUILLEN, On excision in periodic cyclic cohomology, I and II, C.R. Acad. Sci. Paris, Ser. I Math., 317 (1993), 917–922; 318 (1994), 11–12.

[26] B. RIEMANN, Mathematical Werke, Dover, New York, 1953.

[27] S. WEINBERG, Gravitation and Cosmology, John Wiley and Sons, New York–London, 1972.

[28] J. DIXMIER, Existence de traces non normales, C.R. Acad. Sci. Paris, Ser. A-B, 262 (1966).

[29] M. WODZICKI, Noncommutative residue, Part I. Fundamentals, in "K-theory, Arithmetic and Geometry", Springer Lecture Notes in Math. 1289 (1987).

[30] J. MILNOR, D. STASHEFF, Characteristic classes, Ann. of Math. Stud. Princeton University Press, Princeton, N.J. 1974.

[31] D. SULLIVAN, Geometric periodicity and the invariants of manifolds, Springer Lecture Notes in Math. 197 (1971).

[32] B. LAWSON, M.L. MICHELSON, Spin Geometry, Princeton, 1989.

[33] A. CONNES, Entire cyclic cohomology of Banach algebras and characters of θ summable Fredholm modules, K-theory 1 (1988), 519–548.

[34] A. JAFFE, A. LESNIEWSKI, K. OSTERWALDER, Quantum K-theory: I. The Chern character, Commun. Math. Phys. 118 (1988), 1–14.

[35] A. CONNES, H. MOSCOVICI, The local index formula in noncommutative geometry, GAFA 5 (1995), 174–243.

[36] A. CONNES, Noncommutative Geometry, Academic Press, 1994.

[37] A. CONNES, H. MOSCOVICI, Hopf algebras, cyclic cohomology and the transverse index theorem, Commun. Math. Phys. 198 (1998), 199–246.

[38] M. HILSUM, G. SKANDALIS, Morphismes K-orientés d'espaces de feuilles et fonctorialité en théorie de Kasparov, Ann. Sci. Ecole Norm. Sup. (4), 20 (1987), 325–390.

[39] Y. MANIN, Quantum groups and noncommutative geometry, Centre Recherche Math. Univ. Montréal, 1988.

[40] A. CONNES, C^* algèbres et géométrie differentielle, C.R. Acad. Sci. Paris, Ser. A-B 290 (1980), 599–604.

[41] A. CONNES, H. MOSCOVICI, Cyclic cohomology and Hopf algebras, Letters Math. Phys. 48:1 (1999), 97–108.

[42] D. KREIMER, On the Hopf algebra structure of perturbative Quantum Field Theory, Adv. Theor. Math. Phys. 2.2, 303 (1998); q-alg/9707029.

[43] D. Kreimer, On overlapping divergencies, Commun. Math. Phys. 204, 669 (1999); hep-th/9810022.

[44] A. Connes, D. Kreimer, Hopf algebras, renormalization and noncommutative geometry, Commun. Math. Phys. 199 (1998), 203–242.

[45] A. Connes, D. Kreimer, Renormalization in quantum field theory and the Riemann–Hilbert problem, J. High Energy Phys. 9, Paper 24 (1999), 8pp; hep-th/9909126.

[46] A. Connes, D. Kreimer, Renormalization in quantum field theory and the Riemann–Hilbert problem I: the Hopf algebra structure of graphs and the main theorem. Commun. Math. Phys. 210 (2000), 249–273; hep-th/9912092.

[47] A. Beauville, Monodromie des systèmes différentiels linéaires à pôles simples sur la sphère de Riemann, in "Séminaire Bourbaki, 45ème année", 1992-1993, n. 765.

[48] A. Bolibruch, Fuchsian systems with reducible monodromy and the Riemann–Hilbert problem, Springer Lecture Notes in Math. 1520 (1992), 139–155.

[49] A. Connes, D. Kreimer, Renormalization in quantum field theory and the Riemann–Hilbert problem II: The β function, diffeomorphisms and the renormalization group, hep-th/0003188.

[50] A. Connes, Gravity coupled with matter and foundation of noncommutative geometry, Commun. Math. Phys. 182 (1996), 155–176.

[51] W. Kalau, M. Walze, Gravity, noncommutative geometry and the Wodzicki residue, J. of Geom. and Phys. 16 (1995), 327–344.

[52] D. Kastler, The Dirac operator and gravitation, Commun. Math. Phys. 166 (1995), 633–643.

[53] A. Connes, Noncommutative geometry and reality, Journal of Math. Physics 36:11 (1995), 6194–6231.

[54] T. Schucker, Spin group and almost commutative geometry, hep-th/0007047.

[55] M.F. Atiyah, K-theory and reality. Quart. J. Math. Oxford (2) 17 (1966), 367–386.

[56] M.A. Rieffel, Morita equivalence for C^*-algebras and W^*-algebras, J. Pure Appl. Algebra 5 (1974), 51–96.

[57] M. Gromov, Carnot–Caratheodory spaces seen from within, Preprint IHES/M/94/6.

[58] A. Chamseddine, A. Connes, Universal formulas for noncommutative geometry actions, Phys. Rev. Letters 77, 24 (1996), 4868–4871.

[59] A. Connes, Noncommutative Geometry: The Spectral Aspect, in "Les Houches Session LXIV", Elsevier (1998), 643–685.

[60] M. Karoubi, Homologie cyclique et K-théorie, Asterisque 149 (1987).

[61] M.A. Rieffel, C^*-algebras associated with irrational rotations, Pacific J. Math. 93 (1981), 415–429.

[62] M. PIMSNER, D. VOICULESCU, Exact sequences for K groups and Ext group of certain crossed product C^*-algebras, J. Operator Theory 4 (1980), 93–118.
[63] M.A. RIEFFEL, The cancellation theorem for projective modules over irrational rotation C^*-algebras, Proc. London Math. Soc. 47 (1983), 285–302.
[64] A. CONNES, M. RIEFFEL, Yang–Mills for noncommutative two tori, in "Operator Algebras and Mathematical Physics (Iowa City, Iowa, 1985), Contemp. Math. Oper. Algebra Math. Phys. 62, Amer. Math. Soc., Providence, RI (1987), 237–266.
[65] A. CONNES, G. LANDI, Noncommutative manifolds, the Instanton algebra and isospectral deformations, Math-QA/0011194.
[66] J.M. GRACIA-BONDIA, J.C. VARILLY, H. FIGUEROA, Elements of Noncommutative Geometry, Birkhauser, 2000.
[67] A. CONNES, M. DOUGLAS, A. SCHWARZ, Noncommutative geometry and Matrix theory: compactification on tori, J. High Energy Physics 2, Paper 3 (1998), 35pp.
[68] A. CONNES, A short survey of noncommutative geometry, J. Math. Physics 41 (2000), 3832–3866.
[69] S. BAAJ, G. SKANDALIS, Unitaires multiplicatifs et dualité pour les produits croisés de C^*-algèbres, Ann. Sci. Ec. Norm. Sup., 4 série, t. 26 (1993), 425-488.
[70] G.I. KAC, Extensions of Groups to Ring Groups, Math. USSR Sbornik 5:3 (1968).
[71] S. MAJID, Foundations of Quantum Group Theory, Cambridge University Press, 1995.
[72] M. BERRY, Riemann's zeta function: a model of quantum chaos, Springer Lecture Notes in Physics 263 (1986).
[73] A. CONNES Trace formula in Noncommutative Geometry and the zeros of the Riemann zeta function, Selecta Mathematica New Ser. 5 (1999), 29–106.
[74] N. NEKRASOV, A. SCHWARZ, Instantons in noncommutative $\mathbb{R}^4$ and (2,0) superconformal six dimensional theory, hep-th/9802068.
[75] N. SEIBERG, E. WITTEN, String theory and noncommutative geometry, J. High Energy Physics 9 (1999).
[76] J. ARTHUR, The invariant trace formula II. Global theory, J. of the AMS I (1988), 501, 554.
[77] A. WEIL, Sur la théorie du corps de classes, J. Math. Soc. Japan 3 (1951), 1–35.
[78] A.R. BERNSTEIN, F. WATTENBERG, Non standard measure theory, in "Applications of Model Theory to Algebra Analysis and Probability" (W.A.J. Luxenburg Halt, ed.), Rinehart and Winstin, 1969.
[79] N.N. BOGOLIUBOV, D.V. SHIRKOV, Introduction to the theory of quantized fields, 3rd ed., Wiley 1980;
K. HEPP, Comm. Math. Phys. 2 (1966), 301–326;
W. ZIMMERMANN, Convergence of Bogoliubov's method of renormalization

in momentum space, Comm. Math. Phys. 15 (1969), 208–234.

[80] M. Dresden, Renormalization in historical perspective - The first stage, in "Renormalization" (L. Brown, ed.) Springer-Verlag, New York–Berlin–Heidelberg, 1994.

[81] H. Epstein, V. Glaser, The role of locality in perturbation theory, Ann. Inst. H. Poincaré A 19 (1973), 211–295.

[82] A. Goncharov, Polylogarithms in arithmetic and geometry, Proc. of ICM-94 (Zürich), 1,2, Birkhäuser (1995), 374–387.

[83] D. Zagier, Values of zeta functions and their applications, First European Congress of Mathematics, Vol. II, Birkhauser, Boston (1994), 497–512

Alain Connes, Collège de France, 3, rue Ulm, 75005 Paris, France
and
I.H.E.S., 35, route de Chartres, 91440 Bures-sur-Yvette, France

GAFA, Geom. funct. anal.
Special Volume – GAFA2000, 560 – 673
1016-443X/00/S20560-114 $ 1.50+0.20/0

GAFA Geometric And Functional Analysis

INTRODUCTION TO SYMPLECTIC FIELD THEORY

Y. Eliashberg, A. Givental and H. Hofer

Abstract

We sketch in this article a new theory, which we call *Symplectic Field Theory* or SFT, which provides an approach to Gromov-Witten invariants of symplectic manifolds and their Lagrangian submanifolds in the spirit of topological field theory, and at the same time serves as a rich source of new invariants of contact manifolds and their Legendrian submanifolds. Moreover, we hope that the applications of SFT go far beyond this framework.[1]

Contents

[1]The research is partially supported by the National Science Foundation.

Disclosure. Despite its length, the current paper presents only a very sketchy overview of Symplectic Field Theory. It contains practically no proofs, and in a few places where the proofs are given their role is just to illustrate the involved ideas, rather than to give complete rigorous arguments.

The ideas, the algebraic formalism, and some of the applications of this new theory were presented and popularized by the authors at several conferences and seminars (e.g. [E2]). As a result, currently there exists a significant mathematical community which is in some form familiar with the subject. Moreover, there are many mathematicians, including several former and current students of the authors, who are actively working on foundational aspects of the theory and its applications, and even published papers on this subject. Their results show that already the simplest versions of the theory have some remarkable corollaries (cf. [U]). We hope that the present paper will help attracting even more people to SFT.

Of course, our ideas give just a small new twist to many other active directions of research in Mathematics and Physics (Symplectic topology, Gromov-Witten invariants and quantum cohomology, Floer homology theory, String theory, just to mention few), pioneered by V.I. Arnold, C. Conley and E. Zehnder, M. Gromov, S.K. Donaldson, E. Witten, A. Floer, M. Kontsevich and others (see [Ar1], [CoZ], [Gro1], [D1], [F], [Ru1], [Wi1], [Wi2], [Ko1], [KoM]). Many people independently contributed results and ideas, which may be considered as parts of SFT. Let us just mention here the work of Yu. Chekanov [C], K. Fukaya, K. Ono, Y.-G. Oh and H. Ohta [FuOOO], A. Gathmann [G], E. Ionel and T. Parker [IP2], [I] , Y. Ruan and A.-M. Li [RuL]. It also draws on other results of the current authors and their coauthors (see [EHS2], [EH], [Giv3], [Giv1], [Giv2], [H], [HWZ1], [AH], [HWZ2], [HWZ3]). The contact-geometric ingredient of our work is greatly motivated by two outstanding conjectures in contact geometry: Weinstein's conjecture about periodic orbits of Reeb fields [W], and Arnold's chord conjecture [Ar3].

Presently, we are working on a series of papers devoted to the foundations, applications, and further development of SFT. Among the applications, some of which are mentioned in this paper, are new invariants of contact manifolds and Legendrian knots and links, new methods for computing Gromov-Witten invariants, new restrictions on the topology of Lagrangian submanifolds, new non-squeezing type theorems in contact geometry etc. We are expecting new links with the low-dimensional topology and, possibly, Physics. It seems, however, that what we see at the moment is just a tip of an iceberg. The main body of Symplectic Field Theory and its applications is yet to be discovered.

Guide for an impatient reader. The paper consists of two parts. The first part, except section 1.9 and the end of section 1.8, contains some necessary background symplectic-geometric and analytic information. An impatient reader can try to begin reading with section 1.9, and use the rest of the first part for the references. The second part begins with its own introduction (section 2.1) where we present a very rough sketch of SFT. At the end of section 2.1 we describe the plan of the remainder of the paper.

Acknowledgements. The authors benefited a lot from discussions with many mathematicians, and from the ideas which they corresponded to us. We are especially grateful to C. Abbas, P. Biran, F. Bourgeois, K. Cieliebak,

T. Coates, T. Ekholm, K. Fukaya, E. Getzler, M. Hutchings, E.-N. Ionel, V.M. Kharlamov, K. Mohnke, L. Polterovich, D. Salamon, M. Schwarz, K. Wysocki and E. Zehnder. A part of the paper was written, when the first author visited RIMS at Kyoto University. He wants to thank RIMS and K. Fukaya and K. Saito, the organizers of a special program in Geometry and String Theory, for the hospitality. He also thanks T. Tsuboi and J.-L. Brylinski for organizing cycles of lectures on SFT at the University of Tokyo and the Pennsylvania State University. The second author is thankful to A. Kirillov, R. Donagi, and the Department of Mathematics of the University of Pennsylvania for the possibility to present SFT in a series of lectures. The third author acknowledges the hospitality of FIM at the ETH Zürich, where some of the work was carried out. Our special thanks to N.M. Mishachev for drawing the pictures and to J. Sabloff for the computer verification of the formula (116).

1 Symplectic and Analytic Setup

1.1 Contact preliminaries. A 1-form α on a $(2n-1)$-dimensional manifold V is called *contact* if the restriction of $d\alpha$ to the $(2n-2)$-dimensional tangent distribution $\xi = \{\alpha = 0\}$ is non-degenerate (and hence symplectic). A codimension 1 tangent distribution ξ on V is called a *contact structure* if it can be locally (and in the co-orientable case globally) defined by the Pfaffian equation $\alpha = 0$ for some choice of a contact form α. The pair (V, ξ) is called a *contact manifold.* According to Frobenius' theorem the contact condition is a condition of maximal non-integrability of the tangent hyperplane field ξ. In particular, all integral submanifolds of ξ have dimension $\leq n-1$. On the other hand, $(n-1)$-dimensional integral submanifolds, called *Legendrian*, always exist in abundance. We will be dealing in this paper only with co-orientable, and moreover co-oriented contact structures. Any non-coorientable contact structure can be canonically double-covered by a coorientable one. If a contact form α is fixed then one can associate with it the *Reeb vector field* R_α, which is transversal to the contact structure $\xi = \{\alpha = 0\}$. The field R_α is uniquely determined by the equations $R_\alpha \lrcorner d\alpha = 0$; $\alpha(R_\alpha) = 1$. The flow of R_α preserves the contact form α.

The $2n$-dimensional manifold $M = (T(V)/\xi)^* \setminus V$, called the *symplectization* of (V, ξ), carries a natural symplectic structure ω induced by an embedding $M \to T^*(V)$ which assigns to each linear form $T(V)/\xi \to \mathbb{R}$ the corresponding form $T(V) \to T(V)/\xi \to \mathbb{R}$. A choice of a contact form α (if

ξ is co-orientable) defines a splitting $M = V \times (\mathbb{R} \setminus 0)$. As ξ is assumed to be co-oriented we can pick the positive half $V \times \mathbb{R}_+$ of M, and call it symplectization as well. The symplectic structure ω can be written in terms of this splitting as $d(\tau\alpha), \tau > 0$. It will be more convenient for us, however, to use additive notation and write ω as $d(e^t\alpha)$, $t \in \mathbb{R}$, on $M = V \times \mathbb{R}$. Notice that the vector field $T = \frac{\partial}{\partial t}$ is conformally symplectic: we have $\mathcal{L}_T\omega = \omega$, as well as $\mathcal{L}_T(e^t\alpha) = e^t\alpha$, where $\mathcal{L}_T$ denotes the Lie derivative along the vector field T. All the notions of contact geometry can be formulated as the corresponding symplectic notions, invariant or equivariant with respect to this conformal action. For instance, any contact diffeomorphism of V lifts to an equivariant symplectomorphism of M; contact vector fields on V (i.e. vector fields preserving the contact structure) are projections of $\mathbb{R}$-invariant symplectic (and automatically Hamiltonian) vector fields on M; Legendrian submanifolds in M correspond to cylindrical (i.e. invariant with respect to the $\mathbb{R}$-action) Lagrangian submanifolds of M.

Notice that the Hamiltonian vector field on $V \times \mathbb{R}$, defined by the Hamiltonian function $H = e^t$ is invariant under translations $t \mapsto t+c$, and projects to the Reeb vector field R_α under the projection $V \times \mathbb{R} \to \mathbb{R}$.

The symplectization of a contact manifold is an example of a symplectic manifold with *cylindrical* (or rather conical) ends. We mean by that a possibly non-compact symplectic manifold (W, ω) with ends of the form $E^+ = V^+ \times [0, \infty)$ and $E^- = V^- \times (-\infty, 0]$, such that $V^\pm$ are compact manifolds, and $\omega|_{V^\pm} = d(e^t\alpha^\pm)$, where $\alpha^\pm$ are contact forms on $V^\pm$. In other words, the ends $E^\pm$ of (W, ω) are symplectomorphic, respectively, to the positive or negative halves of the symplectizations of contact manifolds $(V^\pm, \xi^\pm = \{\alpha^\pm = 0\})$. We will consider the splitting of the ends and the the contact forms $\alpha^\pm$ to be parts of the structure of a symplectic manifold with cylindrical ends. We will also call (W, ω) a *directed symplectic cobordism* between the contact manifolds (V^+, ξ^+) and (V^-, ξ^-), and denote it by $\overrightarrow{V^-V^+}$.

Sometimes we will have to consider the compact part $W^0 = W \setminus (\mathrm{Int} E^+ \cup \mathrm{Int} E^-)$ of a directed symplectic cobordism $\overrightarrow{V^-V^+}$. If it is not clear from the context we will refer to W^0 as a *compact*, and to W as a *completed* symplectic cobordism.

Let us point out that "symplectic cobordism" *is not an equivalence relation, but rather a partial order.* Existence of a directed symplectic cobordism $\overrightarrow{V^-V^+}$ does not imply the existence of a directed symplectic cobordism $\overrightarrow{V^+V^-}$, even if one does not fix contact forms for the contact structures $\xi^{\pm}$. On the other hand, directed symplectic cobordisms $\overrightarrow{V_0V_1}$ and $\overrightarrow{V_1V_2}$ can be glued, in an obvious way, into a directed symplectic cobordism $\overrightarrow{V_0V_2} = \overrightarrow{V_0V_1} \circledcirc \overrightarrow{V_1V_2}$.

Contact structures have no local invariants. Moreover, any contact form is locally isomorphic to the form $\alpha_0 = dz - \sum_1^{n-1} y_i dx_i$ (Darboux' normal form). The contact structure ξ_0 on $\mathbb{R}^{2n-1}$ given by the form α_0 is called *standard*. The standard contact structure on S^{2n-1} is formed by complex tangent hyperplanes to the unit sphere in $\mathbb{C}^n$. The standard contact structure on S^{2n-1} is isomorphic in the complement of a point to the standard contact structure on $\mathbb{R}^{2n-1}$. According to a theorem of J. Gray (see [Gra]) contact structures on closed manifolds have the following stability property: *Given a family ξ_t, $t \in [0,1]$, of contact structures on a closed manifold M, there exists an isotopy $f_t : M \to M$, such that $df_t(\xi_0) = \xi_t; t \in [0,1]$.* Notice that for contact *forms* the analogous statement is wrong. For instance, the topology of the 1-dimensional foliation determined by the Reeb vector field R_α is very sensitive to deformations of the contact form α.

The conformal class of the symplectic form $d\alpha|_\xi$ depends only on the cooriented contact structure ξ and not on the choice of the contact form α. In particular, one can associate with ξ an almost complex structure $J : \xi \to \xi$, compatible with $d\alpha$ which means that $d\alpha(X, JY); X, Y \in \xi$, is an Hermitian metric on ξ. The space of almost complex structures J with this property is contractible, and hence the choice of J is homotopically canonical. Thus a co-oriented contact structure ξ defines on M a *stable almost complex structure* $\widetilde{J} = \widetilde{J}_\xi$, i.e. a splitting of the tangent bundle $T(V)$ into the Whitney sum of a complex bundle of (complex) dimension $(n-1)$ and a trivial 1-dimensional real bundle. The existence of a stable almost complex structure is necessary for the existence of a contact structure on V. If V is open (see [Gro2]) or $\dim V = 3$ (see [M], [Lu]) this property is also sufficient for the existence of a contact structure in the prescribed homotopy class. It is still unknown whether this condition is sufficient for the existence of a contact structure on a closed manifold of dimension > 3. However, a positive answer to this question is extremely unlikely. The homotopy class of $\widetilde{J}_\xi$, which we denote by $[\xi]$ and call the *formal* homotopy class of ξ, serves as an invariant of ξ. For an open V it is a complete invariant (see [Gro2]) up to homotopy of contact structures, but not up to a contact diffeomorphism. For closed manifolds this is known to be false in many, but not all dimensions. The theory discussed in this paper serves as a rich source of contact invariants, both of closed and open contact manifolds.

1.2 Dynamics of Reeb vector fields. Let (V, ξ) be a $(2n-1)$-dimensional manifold with a co-orientable contact structure with a fixed contact form α. For a generic choice of α there are only countably many periodic trajectories of the vector field R_α. Moreover, these trajectories can be assumed *non-degenerate* in the sense that the linearized Poincaré return map A_γ along any closed trajectory γ, including multiples, has no eigenvalues equal to 1. Let us denote by $\mathcal{P} = \mathcal{P}_\alpha$ the set of all periodic trajectories of R_α, including multiples.[2]

The reason for a such choice is discussed in section 1.8 below. We will also fix a point m_γ on each *simple* orbit from $\mathcal{P}$. Non-degenerate trajectories can be divided into *odd* and *even* depending on the sign of the Lefshetz number $\det(I - A_\gamma)$. Namely, we call γ odd if $\det(I - A_\gamma) < 0$, and even otherwise. The parity of a periodic orbit γ agrees with the parity of a certain integer grading which is defined if certain additional choices are made, as it is described below.

If $H_1(V) = 0$ then for each $\gamma \in \mathcal{P}$ we can choose and fix a surface F_γ spanning the trajectory γ in V. We will allow the case $H_1(V) \neq 0$, but will require in most of the paper that the torsion part is trivial.[3] In this case we choose a basis of $H_1(V)$, represent it by oriented curves $C_1, \dots, C_K$, and choose a symplectic trivialization of the bundle $\xi|_{C_i}$ for each chosen curve. We recall that the bundle ξ is endowed with the symplectic form $d\alpha$ whose conformal class depends only on ξ. For any periodic orbit $\gamma \in \mathcal{P}$ let us choose a surface F_γ with $[\partial F_\gamma] = [\gamma] - \sum n_i [C_i]$. The coefficients n_i are uniquely defined because of our assumption that $H_1(V)$ is torsion-free.

The above choices enable us to define the *Conley-Zehnder index* $\mathrm{CZ}(\gamma)$ of γ as follows. Choose a homotopically unique trivialization of the symplectic vector bundle $(\xi, d\alpha)$ over each trajectory $\gamma \in \mathcal{P}$ which extends to $\xi|_{F_\gamma}$ (and coincides with a chosen trivialization of $\xi|_{C_i}$ if C_i is not homologically trivial). The linearized flow of R_α along γ defines then a path in the group $Sp(2n-2, \mathbb{R})$ of symplectic matrices, which begins at the unit matrix and ends at a matrix with all eigenvalues different from 1. The

[2] As it is explained below in section 1.8 the orientation issues require us to exclude certain multiple periodic orbits out of consideration. Namely, let us recall that real eigenvalues of symplectic matrices different from ± 1 come in pairs λ, λ^{-1}. Let $\gamma \in \mathcal{P}$ be a simple periodic orbit and A_γ its linearized Poincaré return map. If the total multiplicity of eigenvalues of A_γ from the interval $(-1, 0)$ is odd, then we exclude from $\mathcal{P}$ all even multiples of γ.

[3] The case when $H_1(V)$ has torsion elements is discussed in section 2.9.1 below.

Maslov index of this path (see [Ar2], [RS]) is, by the definition, the Conley-Zehnder index $\mathrm{CZ}(\gamma)$ of the trajectory γ. See also [HWZ2], section 3, for an axiomatic description of the Conley-Zehnder index using our normalization conventions.

Notice that by changing the spanning surfaces for the trajectories from $\mathcal{P}$ one can change Conley-Zehnder indices by the value of the cohomology class $2c_1(\xi)$, where $c_1(\xi)$ is the first Chern class of the contact bundle ξ. In particular, mod 2 indices can be defined independently of any spanning surfaces, and even in the case when $H_1(V) \neq 0$. In fact,

$$(-1)^{\mathrm{CZ}(\gamma)} = (-1)^{n-1}\mathrm{sign}\big(\det(I - A_\gamma)\big).$$

1.3 Splitting of a symplectic manifold along a contact hypersurface. Let V be a hypersurface of contact type, or in a different terminology, a symplectically convex hypersurface in a symplectic manifold (W, ω). This means that ω is exact, $\omega = d\beta$, near V, and the restriction $\alpha = \beta|_V$ is a contact form on V. Equivalently, one can say that the conformally symplectic vector field X, ω-dual to β, is transversal to V. Let us assume that V divides W, $W = W_+ \cup W_-$, where the notation of the parts are chosen in such a way that X serves as an inward transversal for W_+, and an outward transversal for W_-. The manifolds $W_\pm$ can be viewed as compact directed symplectic cobordisms such that W_- has only positive contact boundary (V, α), while the same contact manifold serves as a negative boundary of W_+.

Let

$$(W_-^\infty, \omega_-^\infty) = (W_-, \omega) \cup \big(V \times [0, \infty), d(e^t\alpha)\big)$$

and

$$(W_+^\infty, \omega_+^\infty) = \big(V \times (-\infty, 0], d(e^t\alpha)\big) \cup (W_+, \omega)$$

be the completions, and

$$(W_-^\tau, \omega_-^\tau) = (W_-, \omega) \cup \big(V \times [0, \tau], d(e^t\alpha)\big)$$

and

$$(W_+^\tau, \omega_+^\tau) = \big(V \times [-\tau, 0], d(e^t\alpha)\big) \cup (W_+^0, \omega)$$

partial completions of $W_\pm$. Let us observe that the symplectic manifolds

$$(W_-, e^{-\tau}\omega_-^\tau), \quad (V \times [-\tau, \tau], d(e^t\alpha)) \quad \text{and} \quad (W_+, e^\tau\omega_+^\tau)$$

fit together into a symplectic manifold (W^τ, ω^τ), so that $W^0 = W$. Hence when $\tau \to \infty$ the deformation (W^τ, ω^τ) can be viewed as a decomposition

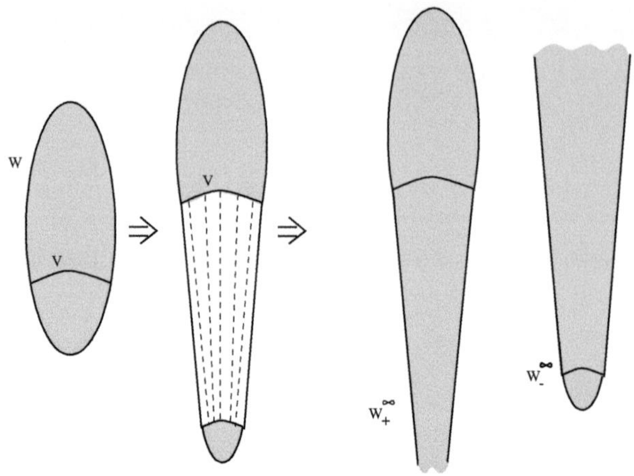

Figure 1: Splitting of a closed symplectic manifold W into two completed symplectic cobordisms W_-^∞ and W_+^∞

of the symplectic manifold $W = W^0$ into the union of two completed symplectic cobordisms W_+^∞ and W_-^∞ . We will write $W = W_- \circledcirc W_+$ and also $W^\infty = W_-^\infty \circledcirc W_+^\infty$.

Let us give here two important examples of the above splitting construction.

EXAMPLE 1.3.1. Suppose $L \subset W$ is a Lagrangian submanifold. Its neighborhood is symplectomorphic to a neighborhood of the 0-section in the cotangent bundle $T^*(L)$. The boundary V of an appropriately chosen neighborhood has contact type, and thus we can apply along V the above splitting construction. As the result we split W into W_+^∞ symplectomorphic to $W \setminus L$, and W_-^∞ symplectomorphic to $T^*(L)$.

EXAMPLE 1.3.2. Let M be a hyperplane section of a Kähler manifold W, or more generally a symplectic hyperplane section of a symplectic manifold W, in the sense of Donaldson (see [D2]). Then M has a neighborhood with a contact boundary V. The affine part $W \setminus M$ is a Stein manifold in the Kählerian case, and in any case has a structure of a symplectic Weinstein manifold $\widetilde{W}$ (notice that the symplectic structure of $\widetilde{W}$ does not coincide with the induced symplectic structure on $W \setminus M$ but contains $W \setminus M$ as an open symplectic submanifold). The Weinstein manifold $W \setminus M$ contains an isotropic deformation retract Δ. The splitting of W along V produces W_-^∞ symplectomorphic to $\widetilde{W}$, and W_+^∞ symplectomorphic to $W \setminus \Delta$. If Δ is a smooth Lagrangian submanifold, then we could get the same decomposition by splitting along the boundary of a tubular neighborhood of L, as in Example 1.3.1.

1.4 Compatible almost complex structures. According to M. Gromov (see [Gro1]) an almost complex structure J is called *tamed* by a symplectic form ω if ω is positive on complex lines. If, in addition, one adds the calibrating condition that ω is J-invariant, then J is said *compatible* with ω. For symplectic manifolds with cylindrical ends one needs further compatibility conditions at infinity, as it is described below.

At each positive, or negative end $\big(V \times \mathbb{R}_{\pm}, d(e^t\alpha)\big)$ we require J to be invariant with respect to translations $t \mapsto t \pm c$, $c > 0$ at least for sufficiently large t. We also require the contact structure $\xi^{\pm}|_{V\times t}$ to be invariant under J, and define $J\frac{\partial}{\partial t} = R_\alpha$, where R_α is the Reeb vector field (see 1.2 above) of the contact form α. In the case when $W = V \times \mathbb{R}$ is the symplectization of a manifold V, i.e. W is a cylindrical manifold, we additionally require J to be globally invariant under all translations along the second factor.

To define a compatible almost complex structure J in the above Examples 1.3.1 and 1.3.2 one needs to specify a contact form α on the contact manifold V. In the case of the boundary of a tubular neighborhood of a Lagrangian submanifold L a natural choice of a contact form is provided by a Riemannian metric on L. The Reeb vector field for such a form α generates on V the geodesic flow of the metric.

When V is the boundary of a neighborhood of a hyperplane section M then there exists another good choice of a contact form. It is a S^1-invariant connection form α on the principal S^1-bundle $V \to M$, whose curvature equals the symplectic form $\omega|_M$. The contact manifold $(V, \xi = \{\alpha = 0\})$ is called the *pre-quantization* of the symplectic manifold (M, ω). Orbits of the Reeb field R_α are all closed and coincide with the fibers of the fibration, or their multiples. Notice that though the Reeb flow in this case looks extremely nice and simple, all its periodic orbits are highly degenerate, see section 2.9.2 below. Notice that the symplectization W of V can be viewed as the total space of a complex line bundle L associated with the S^1-fibration $V \to M$ with the zero-section removed. It is possible and convenient to choose J compatible with the structure of this bundle, and in such a way that the projection $W \to M$ becomes holomorphic with respect to a certain almost complex structure J_M on M compatible with ω.

Let us describe now what the symplectic splitting construction from section 1.3 looks like from the point of view of a compatible almost complex structure.

First, we assume that the original almost complex structure J on W is chosen in such a way that the contact structure $\xi = \{\alpha = 0\}$ on V consists of complex tangencies to V, and that $JX = R_\alpha$, where X is a conformally symplectic vector field, ω-dual to α, and R_α is the Reeb vector field of α. Next we define an almost complex structure J^τ on $W^\tau = W_- \cup V \times [-\tau, \tau] \cup W_+$ by setting $J^\tau|_{W_\pm} = J$ and requiring J^τ to be independent of $t \in [-\tau, \tau]$ on $V \times [-\tau, \tau]$. When $\tau \to \infty$ the almost complex structure J^τ_- on $W^\tau_- = W_- \cup V \times [-\tau, \tau]$ converges to an almost complex structure J^∞_- on W^∞_- compatible with ω^∞_-, and J^τ_+ on $W^\tau_+ = V \times [-\tau, \tau] \cup W_+$ converges to an almost complex structure J^∞_+ on W^∞_+ compatible with ω^∞_+.

1.5 Holomorphic curves in symplectic cobordisms. Let (V, α) be a contact manifold with a fixed contact form and $(W = V \times \mathbb{R}, \omega = d(e^t \alpha))$ its symplectization. Let us denote by $\pi_\mathbb{R}$ and π_V the projections $W \to \mathbb{R}$ and $W \to V$, respectively. For a map $f : X \to W$ we write $f_\mathbb{R}$ and f_V instead of $\pi_\mathbb{R} \circ f$ and $\pi_V \circ f$.

Notice that given a trajectory γ of the Reeb field R_α, the cylinder $\mathbb{R} \times \gamma \subset W$ is a J-holomorphic curve. Let us also observe that

PROPOSITION 1.5.1. *For a J-holomorphic curve $C \subset W$ the restriction $d\alpha|_C$ is non-negative, and if $d\alpha|_C \equiv 0$ then C is a (part of a) cylinder $\mathbb{R} \times \gamma$ over a trajectory γ of the Reeb field R_α.*

Given a J-holomorphic map f of a punctured disk $D^2 \setminus 0 \to W$ we say that the map f is *asymptotically cylindrical* over a periodic orbit γ of the Reeb field R_α at $+\infty$ (resp. at $-\infty$) if $\lim_{r\to 0} f_\mathbb{R}(re^{i\theta}) = +\infty$ (resp. $= -\infty$), and $\lim_{r\to 0} f_V(re^{i\theta}) = \bar f(\theta)$, where the map $\bar f : [0, 2\pi] \to V$ parameterizes the trajectory γ.

The almost complex manifold (W, J) is bad from the point of view of the theory of holomorphic curves: it has a *pseudo-concave* end $V \times (-\infty, 0)$, or using Gromov's terminology its geometry at this end is not bounded. However, it was shown in [H] that Gromov compactness theorem can be modified to accommodate this situation, see Theorems 1.6.2 and 1.6.3 below. We will mention in this section only the following fact related to compactness, which motivates the usage of holomorphic curves asymptotically cylindrical over orbits from $\mathcal{P}_\alpha$.

PROPOSITION 1.5.2. *Suppose that all periodic orbits of the Reeb field R_α are non-degenerate. Let C be a non-compact Riemann surface without boundary and $f : C \to W$ a proper J-holomorphic curve. Suppose that* *there exists a constant $K > 0$ such that $\int_C f^*d\alpha < K$. Then C is conformally equivalent to a compact Riemann surface S_g of genus g with $s^+ + s^-$ punctures*

$$x_1^+, \dots, x_{s^+}^+, x_1^- \dots, x_{s^-}^- \in S_g,$$

such that near the punctures $\mathbf{x}^+ = (x_1^+, \dots, x_{s^+}^+)$ the map f is asymptotically cylindrical over periodic orbits $\Gamma^+ = \{\gamma_1^+, \dots, \gamma_{s^+}^+\}$ at $+\infty$, and near the punctures $\mathbf{x}^- = \{x_1^-, \dots, x_{s^-}^-\}$ the map f is asymptotically cylindrical over periodic orbits $\Gamma^- = \{\gamma_1^-, \dots, \gamma_{s^-}^-\}$ at $-\infty$.

Thus holomorphic maps of punctured Riemann surfaces, asymptotically cylindrical over periodic orbits of the Reeb vector field R_α, form a natural class of holomorphic curves to consider in symplectizations as well as more general symplectic manifolds with cylindrical ends. We will define now moduli spaces of such curves.

Let $W = \overrightarrow{V^-V^+}$ be a (completed) directed cobordism, $\alpha^\pm$ corresponding contact forms on V^- and V^+, $\mathcal{P}^\pm$ the sets of all periodic orbits (including multiple ones) of the Reeb vector fields $R_{\alpha^\pm}$. We assume that $\alpha^\pm$ satisfies the genericity assumptions from section 1.2. Choose a compatible almost complex structure J on W. Let $\Gamma^\pm$ be ordered sets of trajectories from $\mathcal{P}^\pm$ of cardinality $s^\pm$. We also assume that every *simple* periodic orbit γ from $\mathcal{P}^\pm$ comes with a fixed marker $m_\gamma \in \gamma$.

Let $S = S_g$ be a compact Riemann surface of genus g with a conformal structure j, with s^+ punctures $\mathbf{x}^+ = \{x_1^+, \dots, x_{s^+}^+\}$, called positive, s^- punctures $\mathbf{x}^- = \{x_1^-, \dots, x_{s^-}^-\}$, called negative, and r marked points $\mathbf{y} = \{y_1, \dots, y_r\}$. We will also fix an *asymptotic marker* at each puncture. We mean by that a ray originating at each puncture. Alternatively, if one takes the cylinder $S^1 \times [0, \infty)$ as a conformal model of the punctured disk $D^2 \setminus 0$ then an asymptotic marker can be viewed as a point on the circle at infinity. If a holomorphic map $f : D^2 \setminus 0 \to V^\pm \times \mathbb{R}_\pm$ is asymptotically cylindrical over a periodic orbit γ, we say that a marker $\mu = \{\theta = \theta_0\}$ is mapped by f to the marker $m_\gamma \in \overline{\gamma}$, where $\overline{\gamma}$ is the simple orbit which underlines γ, if $\lim_{r \to 0} f_{V^\pm}(re^{i\theta_0}) = m_\gamma$. Let us recall (see section 1.2 above) that we provided each periodic orbit from $\mathcal{P}^\pm$ with a "capping" surface. This surface bounds $\gamma \in \mathcal{P}^\pm$ in $V_\pm$ if γ is homologically trivial, or realizes a homology between γ and the corresponding linear combination of basic curves $C_i^\pm$. We will continue to rule out torsion elements in the first homology (see the discussion of torsion in section 2.9.1 below) and choose curves $C_k \subset W$ which represent a basis of the image $H_1(V^- \cup V^+) \to H(W)$ and for each curve $C_i^\pm$ fix a surface $G_i^\pm$ which realizes a homology in W

between $C_i^\pm$ and the corresponding linear combination of curves C_k. All the choices enable us to associate with a relative homology class $A' \in H_2(W, \Gamma^- \cup \Gamma^+)$, $\Gamma^\pm \subset \mathcal{P}^\pm$, an absolute integral class $A \in H_2(W)$, which we will view as an element of $H_2(W; \mathbb{C})$.

Let us denote by $\mathcal{M}^A_{g,r}(\Gamma^-, \Gamma^+; W, J)$ the moduli space of (j, J)-holomorphic curves $S_g \setminus (\mathbf{x}^- \cup \mathbf{x}^+) \to W$ with r marked points, which are asymptotically cylindrical over the periodic orbit γ_i^+ from Γ^+ at the positive end at the puncture x_i^+, and asymptotically cylindrical over the periodic orbit γ_i^- from Γ^- at the negative end at the puncture x_i^-, and which send asymptotic markers to the markers on the corresponding periodic orbits. The curves from $\mathcal{M}^A_{g,r}(\Gamma^-, \Gamma^+; W, J)$ are additionally required to satisfy a stability condition, discussed in the next section. We write $\mathcal{M}^A_g(\Gamma^-, \Gamma^+; W, J)$ instead of $\mathcal{M}^A_{g,0}(\Gamma^-, \Gamma^+; W, J)$, and $\mathcal{M}^A_{g,r}(\Gamma^-, \Gamma^+)$ instead of $\mathcal{M}^A_{g,r}(\Gamma^-, \Gamma^+; W, J)$ if it is clear from the context which target almost complex manifold (W, J) is considered.

Notice, that we are not fixing j, and the configurations of punctures, marked points or asymptotic markers. Two maps are called equivalent if they differ by a conformal map $S_g \to S_g$ which preserves all punctures, marked points and asymptotic markers. When the manifold $W = V \times \mathbb{R}$ is cylindrical, and hence the almost complex structure J is invariant under translations along the second factor, then all the moduli spaces $\mathcal{M}^A_{g,r}(\Gamma^-, \Gamma^+; W, J)$ inherit the $\mathbb{R}$-action. We will denote the quotient moduli space by $\mathcal{M}^A_{g,r}(\Gamma^-, \Gamma^+; W, J)/\mathbb{R}$, and by $\mathcal{M}^A_{g,r,s^-,s^+}(W, J)$ the union $\bigcup \mathcal{M}^A_{g,r}(\Gamma^-, \Gamma^+; W, J)$ taken over all sets of periodic orbits $\Gamma^\pm \subset \mathcal{P}^\pm$ with the prescribed numbers $s^\pm$ of elements. We will also need to consider the moduli space of disconnected curves of Euler characteristic $2-2g$, denoted by $\widetilde{\mathcal{M}}^A_{g,r}(\Gamma^-, \Gamma^+)$.

1.6 Compactification of the moduli spaces $\mathcal{M}^A_{g,r}(\Gamma^-, \Gamma^+)$. To describe the compactification we need an appropriate notion of a *stable holomorphic curve.*

Given a completed symplectic cobordism $W = \overrightarrow{V^- V^+}$ we first define a *stable curve of height* 1, or a 1-*story stable curve* as a "usual" stable curve in a sense of M. Kontsevich (see [Ko1]), i.e. a collection of of holomorphic curves $h_i : S_i \to W$ from moduli spaces $\mathcal{M}^A_{gi,r}(\Gamma_i^-, \Gamma_i^+)$ for various genera g_i which realize homology classes A_i, and collections of periodic orbits $\Gamma_i^\pm$, such that certain pairs of marked points (called special) on these curves are required to be mapped to one point in W. The stability condition means the absence of infinitesimal symmetries of the moduli space. Let us point out, however, that in the case when W is a cylindrical cobordism, and in particular the almost complex structure J is translationally invariant, we would need to consider along with the above moduli space its quotient under the $\mathbb{R}$-action. The stability for this new moduli space still means an absence of infinitesimal deformations, but it translates into an additional restriction on holomorphic curves. Namely, in the first case the stability condition means that each constant curve has, after removal of the marked points, a negative Euler characteristic. In the second case it additionally requires that when *all* connected components of the curve are straight cylinders $\gamma \times \mathbb{R}, \gamma \in \mathcal{P}$ then at least one of these cylinders should have a marked point.

One can define an arithmetic genus g of the resulting curve, the total sets $\mathbf{x}^{\pm}$ and $\mathbf{y}$ (equal to the union of sets $\mathbf{x}_i^{\pm}$ and $\mathbf{y}_i$ for the individual curves of the collection), and the total absolute homology class $A \in H_2(W)$ (see the discussion in section 1.5 above), realized by the union of all curves of the collection.

Moduli of stable curves of height 1, denoted by ${}_1\mathcal{M}^A_{g,r}(\Gamma^-,\Gamma^+)$, form a part of the compactification of the moduli space $\mathcal{M}^A_{g,r}(\Gamma^-,\Gamma^+)$. However, unlike the case of closed symplectic manifolds, the stable curves of height 1 are not sufficient to describe the compactification of the moduli space $\mathcal{M}^A_{g,r}(\Gamma^-,\Gamma^+)$.

A finite sequence $(W_1,\dots,W_k)$ of symplectic manifolds with cylindrical ends is called a *chain* if the positive end of W_i matches with the negative end of W_{i+1}, $i = 1,\dots,k-1$. This means that all data, assigned to an end, i.e. a contact form, marking of periodic orbits, and an almost complex structure, are the same for the matching ends.

Let us first suppose that none of the cobordisms which form a chain $(W_1,\dots,W_k)$ is cylindrical. Then a *stable curve of height* k, or a k-story stable curve in the chain $(W_1,\dots,W_k)$ is a k-tuple $f = (f_1,\dots,f_k)$, where $f_i \in {}_1\widetilde{\mathcal{M}}^{A_i}_{g_i,r_i}(\Gamma_i^-,\Gamma_i^+;W_i,J_i)$, such that the boundary data of the curve f_i at the positive end match the boundary data of f_{i+1} on the negative one. One also needs to impose the following additional equivalence relation regarding the asymptotic markers on multiple orbits. Suppose that γ is a k-multiple periodic orbit, so that the holomorphic curve f_i is asymptotically cylindrical over γ at the positive end at a puncture x^+, and f_{i+1} is asymptotically cylindrical over γ at the negative end at a puncture x^-. There are k possible positions $\mu_1^+,\dots,\mu_k^+$ and $\mu_1^-,\dots,\mu_k^-$ of asymptotic markers at each of the punctures $x^\pm$. We assume here that the markers are numbered cyclically with respect to the orientation defined by the Reeb vector field at each of the punctures, and that the markers μ_1^+ and μ_1^- are chosen for the curves f_i and f_{i+1}. Then we identify $f = \{\dots, f_i, f_{i+1}, \dots\}$ with $(k-1)$ other stable curves of height k obtained by simultaneous cyclic shift of the asymptotic markers at the punctures x^+ and x^-.

The curves f_i, which form a k-story stable curve $f = (f_1, \dots, f_k)$ are called *floors*, or *levels* of f.

If some of the cobordisms which form the chain $(W_1, \dots, W_k)$, say $W_{i_1}, \dots, W_{i_l}$, are cylindrical then we will assume that the corresponding floors of a k-story curve in $W = (W_1, \dots, W_k)$ are defined *only up to translation.* In other words, if W_i is cylindrical for some $i = 1, \dots, k$ (i.e. $W_i = V_i \times \mathbb{R}$ and J_i is translationally invariant) then f_i should be viewed as an element of ${}_1\widetilde{\mathcal{M}}^{A_i}_{g_i,r_i}(\Gamma_i^-, \Gamma_i^+; W_i, J_i)/\mathbb{R}$, rather than ${}_1\widetilde{\mathcal{M}}^{A_i}_{g_i,r_i}(\Gamma_i^-, \Gamma_i^+; W_i, J_i)$. It will be convenient for us, however, to introduce the following convention. When speaking about stable holomorphic curves in chains which contain cylindrical cobordisms, we will treat the corresponding floors as curves representing their equivalence classes from ${}_1\widetilde{\mathcal{M}}^{A_i}_{g_i,r_i}(\Gamma_i^-, \Gamma_i^+; W_i, J_i)/\mathbb{R}$. Any statement about such curves should be understood in the sense, that *there exist* representatives for which the statement is true.

Let us define now the meaning of convergence of a sequence of holomorphic curves to a stable curve of height l. For $l = 1$ this is Gromov's standard definition (see [Gro1]). Namely, with each stable curve $h = \{S_i, h_i\} \in {}_1\mathcal{M}^A_{g,r}(\Gamma^-, \Gamma^+)$ of height 1 we associate a nodal surface $\widehat{S}$ obtained by identifying special pairs of marked points on $\coprod S_i$. The maps h_i fits together to a continuous map $\widehat{S} \to W$ for which we will keep the notation h. Let us consider also a smooth surface S obtained by smoothing the nodes of $\widehat{S}$. There exist a partitioning of S by circles into open parts diffeomorphic to surfaces S_i with removed special points, and a map $g : S \to \widehat{S}$ which is a diffeomorphism from the complement $\tilde{S}$ of the dividing circles in S to the complement of the double points in $\hat{S}$, and which collapses the partitioning circles to double points. A sequence of holomorphic $\varphi_l : (S, j_l) \to (W, J)$ is said to converge to a stable curve h if the sequence $\varphi_l|_{\tilde{S}}$ converges to $h_i \circ g|_{\tilde{S}}$, and j_l converges to $g^*(j)$ uniformly on compact sets, where j is the conformal structure on the stable curve. Of course, we also require convergence of marked points and asymptotic markers. A sequence of stable curves $h^j = \{S_i^j, h_i^j\}_{i=1,\dots,k} \in {}_1\mathcal{M}^A_{g,r}(\Gamma^-, \Gamma^+), j = 1, \dots,$ is said to converge to a stable curve h, if h can be presented as a collection of stable curves h_i, $i = 1, \dots, k$, such that h_i^j converges to h_i in the above sense for each $i = 1, \dots, k$.

The convergence of a sequence of smooth curves to a stable curve of height $l > 1$ is understood in a similar sense. Let us assume here for simplicity that $l = 2$ and that the floors $f_1 : S_1 \to W_1$ and $f_2 : S_2 \to W_2$ of a stable curve f in a chain (W_1, W_2) are smooth, i.e. have no special marked points. As in the height 1 case let us consider

- the smooth surface S partitioned according to the combinatorics of our stable curve by circles into two open (possibly disconnected) parts U_1 and U_2 diffeomorphic to the punctured surfaces S_1 and S_2,
- the surface $\hat{S}$ with double points obtained by collapsing these circles to points, and
- the projection $g : S \to \hat{S}$.

Let $(W, J) = (W_1, J_1) \circledcirc (W_2, J_2)$ be the composition of (completed) directed symplectic cobordisms W_1 and W_2 with compatible almost complex structures J_1 and J_2. This means that

- there exists a contact hypersurface $V \subset W$ which splits W into two cobordisms W_1^0 and W_2^0;
- $W_1 = W_1^0 \cup V \times [0, \infty)$, $W_2 = V \times (-\infty, 0] \cup W_2^0$;
- $J|_{W_j^0} = J_1|_{W_j^0}$, $j = 1, 2$;
- J_1 and J_2 are translationally invariant at the ends $V \times [0, \infty)$ and $(-\infty, 0] \times V$.

We denote by W^k, $k = 1, \dots,$ the quotient space of the disjoint union

$$W_1^0 \coprod V \times [-k, k] \coprod W_2^0$$

obtained by identifying $V = \partial W_1^0$ with and $V \times (-k)$ and $V = \partial W_2^0$ with and $V \times k$, and extend the almost complex structures $J_1|_{W_1^0}$ and $J_2|_{W_2^0}$ to the unique almost complex structure J^k on W^k which is translationally invariant on $V \times [-k, k]$. We also consider W_1^k obtained by gluing W_1^0 and $V \times [0, k]$ along $V = \partial S_1^0 = V \times 0$, and W_2^k glued in a similar way from $V \times [-k, 0]$ and W_2^0. We have $W_j = \bigcup_{k=0}^{\infty} W_j^k$, $j = 1, 2$. On the other hand, W_1^k and W_2^k can be viewed as submanifolds of W^k.

Definition 1.6.1. Suppose that we are given a sequence j^k of conformal structures on the surface S and a sequence of 1-story (j^k, J^k)-holomorphic curves $f^k : S \to W^k$. We say that this sequence converges to a stable curve $f = (f_1, f_2)$ of height 2 in (W_1, W_2) if there exist two sequences of domains

$U_1^1 \subset \cdots \subset U_1^i \subset \cdots \subset U_1$ and $U_2^1 \subset \cdots \subset U_2^i \subset \cdots \subset U_2$, such that

$$\bigcup_{k=1}^{\infty} U_i^k = U_i,\ i = 1,2;$$

$$f^k(U_i^k) \subset W_i^k \quad \text{for} \quad i = 1,2, \quad k = 1,\dots;$$

for $i = 1,2$ the holomorphic curves $f^k|_{U_i^k}$ converges to $f_i \circ g : U_i \to W_i$, and the conformal structures $j^k|_{U_1^k}$ converge to $g^* j_i$ when $k \to \infty$ uniformly on compact sets. As in the case of stable curves of level 1 we also require convergence of marked points and asymptotic markers.

Let us emphasize that when some of the cobordisms are cylindrical then according to the convention which we introduced above one is allowed to compose the corresponding curves with translations to satisfy the above definition.

Notice that if the cobordism W_2 is cylindrical, i.e. $W_2 = V \times \mathbb{R}$ and J_2 is translationally invariant, then $W_1 \circledcirc W_2$ can be identified with W_1, and thus one can talk about convergence of a sequence of curves $f^k \in {}_1\mathcal{M}_{g,r}^A(\Gamma^-,\Gamma^+;W,J_1)$ (where the almost complex structure J_1 is fixed!) to a 2-story curve (f_1,f_2), where $f_1 \in {}_1\widetilde{\mathcal{M}}_{g_1,r_1}^{A_1}(\Gamma^-,\Gamma;W_1,J_1)$, $f_2 \in {}_1\widetilde{\mathcal{M}}_{g_2,r_2}^{A_2}(\Gamma,\widetilde{\Gamma}^+;V\times\mathbb{R},J_2)/\mathbb{R}$, $g = g_1+g_2$, $r = r_1+r_2$, $A = A_1+A_2$, and J_2 is translationally invariant. It is important to stress the point that the curve f_2 is defined only up to translation.

Theorem 1.6.2. *Let $f_k \in {}_1\mathcal{M}_g^A(\Gamma^-,\Gamma^+)$, $k = 1,\dots,$ be a sequence of stable holomorphic curves in a (complete) directed symplectic cobordism W. Then there exists a chain of directed symplectic cobordisms*

$$A_1,\dots,A_a,W,B_1,\dots,B_b,$$

where all cobordisms A_i and B_i are cylindrical, and a stable curve f_∞ of height $a+b+1$ in this chain such that a subsequence of $\{f_i\}$ converges to f_∞. See Fig. 2.

Theorem 1.6.3. *Let W be a completed directed symplectic cobordism, $V \subset W$ a contact hypersurface, and J_k a sequence of compatible almost complex structures on W which realizes the splitting of W along V into two directed symplectic cobordisms W_-^∞ and W_+^∞* (see section 1.4 above). *Let f_k be a sequence of stable J_k-holomorphic curves from ${}_1\mathcal{M}_g^A(\Gamma^-,\Gamma^+;W,J_k)$. Then there exists a chain of directed symplectic cobordisms*

$$A_1,\dots,A_a,W_-,B_1,\dots,B_b,W_+,C_1,\dots,C_c$$

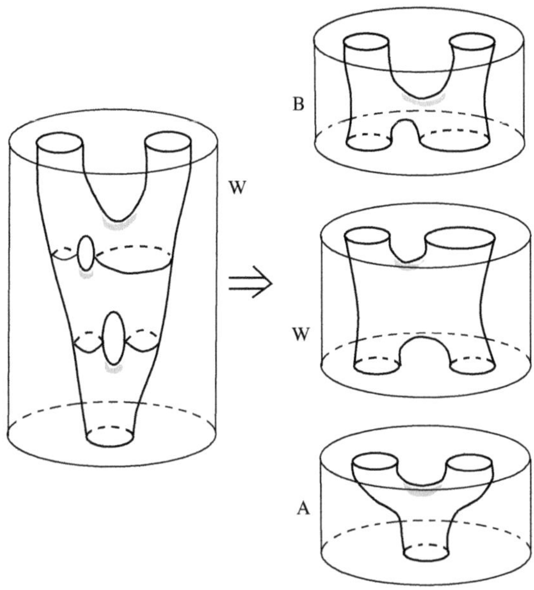

Figure 2: A possible splitting of a sequence of holomorphic curves in a completed symplectic cobordism

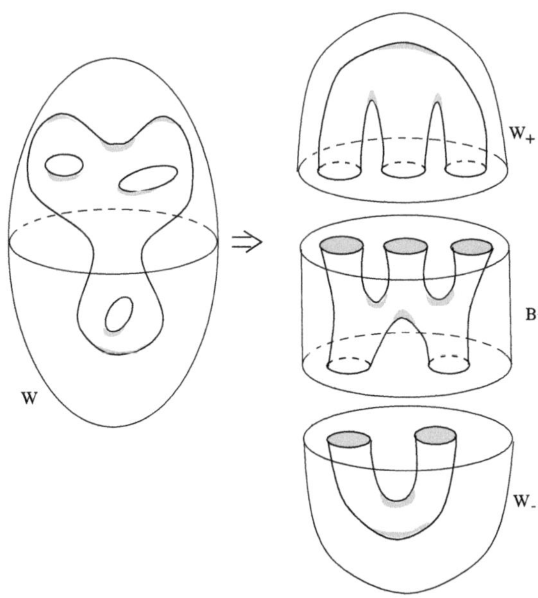

Figure 3: A possible splitting of a sequence of holomorphic curve when $J_k \to J_\infty$

where all cobordisms A_i, B_j, C_l *are cylindrical, such that a subsequence of* $\{f_i\}$ *converges to a stable curve of height* $a+b+c+2$ *in the chain*

$$A_1, \ldots, A_a, W_-, B_1, \ldots, B_b, W_+, C_1, \ldots, C_c.$$

See Fig. 3. The reader may consult [HWZ3] for the analysis of splitting $\mathbb{C}P^2$ along the boundary of a tubular neighborhood of $\mathbb{C}P^1 \subset \mathbb{C}P^2$.

The definition of convergence can be extended in an obvious way to a sequences of stable curves of height $l > 1$. Namely, we say that a sequence of l-story curves $f^k = (f_1^k, \ldots, f_l^k)$, $k = 1, \ldots, \infty$, in a chain $(W_1, \ldots, W_l)$ converge to a stable L-story, $L = m_1 + \cdots + m_l$, curve $f = (f_{11}, \ldots, , f_{1m_1}, \ldots, f_{l_1} \ldots f_{lm_l})$ in a chain

$$(W_{11}, \ldots, W_{1m_1}, \ldots, W_{l_1} \ldots W_{lm_l})$$

if for each $i = 1, \ldots, l$ the cobordism W_i splits into the composition

$$W_i = W_{i1} \circledcirc \cdots \circledcirc W_{im_i}$$

and the sequence f_i^k, $k = 1, \ldots, \infty$, of stable curves of height 1 converges to the m_i-story curve $f_i = (f_{i1}, \ldots, f_{im_i})$ in the chain $(W_{i1}, \ldots, W_{im_i})$ in the sense of Definition 1.6.1.

It is important to combine Theorems 1.6.2 and 1.6.3 with the following observation which is a corollary of Stokes' theorem combined with Proposition 1.5.1.

PROPOSITION 1.6.4. *A holomorphic curve in an exact directed symplectic cobordism (and in particular in a cylindrical one) must have at least one positive puncture.*

In particular, we have

COROLLARY 1.6.5. *Let* $f^n \in \mathcal{M}_0(W, J)$ *be a sequence of rational holomorphic curves with one positive, and possibly several negative punctures. Suppose that the sequence converges to a stable curve*

$$F = \{g_1, \ldots, g_a, f, h_1, \ldots, h_b\}$$

of height $a+b+1$ *in a chain*

$$A_1, \ldots, A_a, W, B_1, \ldots, B_b.$$

Then the W*-component* $f \in \mathcal{M}_0(W, J)$ *of the stable curve* F *has precisely one positive puncture as well.*

1.7 Dimension of the moduli spaces $\mathcal{M}^A_{g,r}(\Gamma^-, \Gamma^+)$**.** One has the following index formula for the corresponding $\bar\partial$-problem which compute the dimension of the moduli space $\mathcal{M}^A_{g,r}(\Gamma^-, \Gamma^+; W, J)$ for a *generic* choice of J.

PROPOSITION 1.7.1.

$$\dim \mathcal{M}^A_{g,r}(\Gamma^-,\Gamma^+;W,J) = \sum_1^{s^+} \mathrm{CZ}(\gamma_i^+) - \sum_1^{s^-} \mathrm{CZ}(\gamma_k^-) + (n-3)(2-2g-s^+-s^-) + 2c_1(A) + 2r, \quad (1)$$

where $s^\pm$ *are the cardinalities of the sets* $\Gamma^\pm$, *and* $c_1 \in H^2(W)$ *is the first Chern class of the almost complex manifold* (W,J)

Making the moduli spaces non-singular by picking generic J is needed for the purpose of curve counting but does not always work properly. It is therefore crucial that the moduli spaces of stable J-holomorphic curves are non-singular *virtually*. This means that for *any* J the moduli spaces, being generally speaking singular, can be equipped with some canonical additional structure that make them *function in the theory the same way as if they were orbifolds with boundary and had the dimension prescribed by the Fredholm index*. In particular, the moduli spaces come equipped with rational fundamental cycles relative to the boundary (called *virtual fundamental cycles*) which admit pairing with suitable de Rham cochains and allow us to use the Stokes integration formula.

Technically the virtual smoothness is achieved by a finite-dimensional reduction of the following picture: a singular moduli space is the zero locus of a section defined by the Cauchy-Riemann operator in a suitable orbi-bundle over a moduli orbifold of stable C^∞-maps. More general *virtual transversality* properties for families of J's also hold true (cf. [FuO1], [FuO2], [LiT], [LT], [Ru2], [Si], [Mc] et al.). We are reluctant to provide in this quite informal exposition precise formulations because of numerous not entirely innocent subtleties this would entail. Fortunately, what we intend to say in the rest of this paper does not depend much on the details we are omitting.

As it was explained in section 1.6 above the moduli space $\mathcal{M}^A_{g,r}(\Gamma^-,\Gamma^+;W,J)$ can be compactified by adding strata which consist of stable holomorphic curves of different height. This compactification looks quite similar to the Gromov-Kontsevich compactification of moduli spaces of holomorphic curves in a closed symplectic manifold with a compatible almost complex structure. There is, however, a major difference. In the case of a closed manifold all the strata which one needs to add to compactify the moduli space of smooth holomorphic curves have (modulo virtual cycle complications) codimension ≥ 2. On the other hand in our case the codimension one strata are present *generically*. Thus in this case the boundary of the compactified moduli space, rather than the moduli space itself, carries the fundamental cycle.

In particular, this boundary is tiled by codimension one strata represented by stable curves (f_-, f_+) of *height two.* Each such a stratum can be described by the constraint matching the positive ends of f_- with the negative ends of f_+ in the Cartesian product of the moduli spaces $\mathcal{M}_\pm$ corresponding to the curves $f_\pm$ separately. Proposition 1.7.2 below describes these top-dimensional boundary strata more precisely in two important for our purposes situations. Let us point out that 1.7.2 literally holds only under certain transversality conditions. Otherwise it should be understood only virtually.

PROPOSITION 1.7.2. 1. *Let $(W = V \times \mathbb{R}, J)$ be a cylindrical cobordism. Then any top-dimensional stratum $\mathcal{S}$ on the boundary of the compactified moduli space $\overline{\mathcal{M}^A_{g,r}(\Gamma^-,\Gamma^+; W, J)/\mathbb{R}}$ consists of stable curves (f_-, f_+) of height two, $f_\pm \in \mathcal{M}_\pm/\mathbb{R}$, where*

$$\mathcal{M}_- = \widetilde{\mathcal{M}}^{A_-}_{g_-,r_-}(\Gamma^-,\Gamma; W, J)/\mathbb{R}, \quad \mathcal{M}_+ = \widetilde{\mathcal{M}}^{A_+}_{g_+,r_+}(\Gamma,\Gamma^+; W, J)/\mathbb{R},$$

$$g = g_- + g_+, \quad r = r_- + r_+, \quad A = A_- + A_+, \quad \Gamma = \{\gamma_1, \dots, \gamma_l\} \subset \mathcal{P}.$$

All but one connected components of each of the curves f_- and f_+ are trivial cylinders (i.e. have the form $\gamma \times \mathbb{R}, \gamma \in \mathcal{P}$) without marked points.
2. *Let $(W = \overrightarrow{V^-V^+}, J)$ be any cobordism, and $(W_\pm, J_\pm) = (V^\pm \times \mathbb{R}, J_\pm)$ be the cylindrical cobordisms associated to its boundary. Then any top-dimensional strata $\mathcal{S}$ on the boundary of the compactified moduli space $\overline{\mathcal{M}^A_{g,r}(\Gamma^-,\Gamma^+; W, J)}$ consists of stable curves (f_-, f_+) of height two, $f_\pm \in \mathcal{M}_\pm$, where either*

$$\mathcal{M}_-=\widetilde{\mathcal{M}}^{A_-}_{g_-,r_-}(\Gamma^-,\Gamma; W_-, J_-)/\mathbb{R},\ \mathcal{M}_+=\widetilde{\mathcal{M}}^{A_+}_{g_+,r_+}(\Gamma,\Gamma^+; W, J) \ \text{ and } \Gamma \subset \mathcal{P}^-,$$

or

$$\mathcal{M}_+=\widetilde{\mathcal{M}}^{A_+}_{g_-,r_-}(\Gamma,\Gamma^+; W_+, J_+)/\mathbb{R},\ \mathcal{M}_-=\widetilde{\mathcal{M}}^{A_-}_{g_-,r_-}(\Gamma^-,\Gamma; W, J) \ \text{ and } \Gamma \subset \mathcal{P}^+.$$

In both cases we have

$$g = g_- + g_+,\ r = r_- + r_+, \ \text{ and } A = A_- + A_+.$$

The part of the stable curve (f_-, f_+) which is contained in in $W_\pm$ must have precisely one non-cylindrical connected component, while there are no restrictions on the number and the character of connected components or the other part of the stable curve.

In both cases the stratum $\mathcal{S} = \mathcal{S}(\Gamma, g_-|g_+, r_-|r_+, A_-|A_+)$ is diffeomorphic to a κ-multiple cover of the product $\mathcal{M}_- \times \mathcal{M}_+$, where the multiplicity κ is determined by the multiplicities of periodic orbits from Γ.

Proposition 1.7.2 is not quite sufficient for our purposes, as we also needs to know the structure of the boundary of moduli spaces of 1-parametric families of holomorphic curves. However, we are not formulating the corresponding statement in this paper, because it is intertwined in a much more serious way with the virtual cycle techniques and terminology. An algebraic description of this boundary is given in Theorem 2.4.2 below.

Let us consider some special cases of the formula (1). Suppose, for instance, that W is the cotangent bundle of a manifold L. Then W is a symplectic manifold which has only a positive cylindrical end. If L is orientable then there is a canonical way to define Conley-Zehnder indices. Namely, one takes any trivialization along orbits, which is tangent to vertical Lagrangian fibers. The resulting index is independent of a particular trivialization. For this trivialization, and a choice of a contact form corresponding to a metric on L we have

PROPOSITION 1.7.3. *Periodic orbits of the Reeb flow are lifts of closed geodesics, and if L is orientable their Conley-Zehnder indices are equal to Morse indices of the corresponding geodesics and we have*

$$\dim \mathcal{M}_g^A(\Gamma^+) = \sum_i \mathrm{Morse}(\gamma_i^+) + (n-3)(2-2g-s^+).$$

Notice that for a metric on L of non-positive curvature we have $s^+ > 1$, because in this case there are no contractible geodesics. Moreover, if the metric has negative curvature then all geodesics have Morse indices equal to 0. Hence, we get

COROLLARY 1.7.4. *In the cotangent bundle of a negatively curved manifold of dimension > 2 there could be only isolated holomorphic curves. If, in addition, $n \neq 3$ then these curves are spheres with two positive punctures. Each of these curves is asymptotically cylindrical at punctures over lifts of the same geodesic with opposite orientations.*

Let us point out that the orientability is not required in Corollary 1.7.4. The corresponding result for a non-orientable manifold follows from 1.7 applied to its orientable double cover.

Absence of hyperbolic Lagrangian submanifolds in uniruled manifolds. As the first application of the above compactness theorems let us prove here the following theorem of C. Viterbo. Let us recall that a complex projective manifold W is called uniruled, if there is a rational holomorphic curve through each point of W. For instance, according to Y. Myaoka–S. Mori [MyM] and J. Kollar [K] Fano manifolds are uniruled.

Theorem 1.7.5. (C. Viterbo, [Vi]) *Let W be a uniruled manifold of complex dimension > 2, ω its Kähler symplectic form, and $L \subset W$ an embedded Lagrangian submanifold. Then L does not admit a Riemannian metric of negative sectional curvature.*

Proof. J. Kollar [K] and in a more general case Y. Ruan [Ru2] proved that there exists a homology class $A \in H_2(W)$, such that for any almost complex structure compatible with ω and any point $z \in W$ there exists $f \in \mathcal{M}^A_{0,1}(W, J)$ with $f(y) = z$, where y is the marked point. Let us identify a neighborhood U of L in W with a neighborhood of the zero-section in $T^*(L)$. Suppose L admits a Riemannian metric of negative curvature. We can assume that U is the round neighborhood of radius 1 in $T^*(L)$. Let us consider a sequence J^m of almost complex structures on W, which realizes the splitting along the contact type hypersurface $(V = \partial U, \alpha = pdq|_V)$. (see section 1.4). Then according to Example 1.3.1 W splits into $W_- = T^*(L)$ and $W_+ = W \setminus L$. The almost complex structure J_- on $T^*(L)$ is compatible at infinity with the contact 1-form $\alpha = pdq|_V$. According to Corollary 1.7.4 for any choice $\Gamma = \{\gamma_1, \dots, \gamma_k\}$ and any $g \geq 0$ the moduli spaces $\mathcal{M}_g(\Gamma; W_-, J_-)$ are empty, or 0-dimensional. One the other hand, Theorem 1.6.3 together with Ruan's theorem guarantee the existence of a rational holomorphic curve with punctures through every point of L. This contradiction proves that L cannot admit a metric of negative curvature.

□

1.8 Coherent orientation of the moduli spaces of holomorphic curves. To get started with the algebraic formalism, one first needs to orient moduli spaces $\mathcal{M}(\Gamma^-, \Gamma^+)$ of holomorphic curves with punctures. This problem is much easier in the case of moduli spaces of closed holomorphic curves, because in that case moduli spaces are even-dimensional and carry a canonical almost complex structure (see section 1.8.2 below). In our case we have to adapt the philosophy of *coherent orientations* of the moduli spaces borrowed from Floer homology theory (see [FH]). We sketch this approach in this section.

1.8.1 Determinants. In order to separate the problems of orientation and transversality we are going to orient the determinant line bundles of the linearized $\bar\partial$-operators, rather than the moduli spaces themselves.

For a linear Fredholm operator $F : A \to B$ between Banach spaces we can define its determinant line $\det(F)$ by

$$\det(F) = \left(\Lambda^{\max}\mathrm{Ker}(F)\right) \otimes \left(\Lambda^{\max}\mathrm{Coker}(F)\right)^*.$$

We note that for the trivial vector space $\{0\}$ we have $\Lambda^{\max}\{0\} = \mathbb{R}$. An orientation for F is by definition an orientation for the line $\det(F)$. In particular, given an isomorphism F we can define a canonical orientation given through the vector $1 \otimes 1^* \in \mathbb{R} \otimes \mathbb{R}^*$.

Given a continuous family $F = \{F_y\}_{y \in Y}$ of Fredholm operators, parameterized by a topological space Y, the determinants of operators F_y form a line bundle $\det(F) \to Y$. The fact that this is a line bundle in a natural way might be surprising since the dimensions of kernel and cokernel vary in general. This is however a standard fact, see for example [FH].

1.8.2 Cauchy-Riemann type operators on closed surfaces. Let (S, j) be a closed, not necessarily connected Riemann surface and $E \to S$ a complex vector bundle. Denote by $X_E \to S$ the complex n-dimensional vector bundle whose fiber over $z \in S$ consists of all complex ant-linear maps

$$\phi : T_z S \to E_z,\ z \in S,\ \text{i. e. } J \circ \phi + \phi \circ j = 0,$$

where J is the complex structure on E. Fixing a connection ∇ and a smooth $a \in \mathrm{Hom}_{\mathbb{R}}(E, X_E)$ we can define a Cauchy-Riemann type operator

$$L : C^\infty(E) \to C^\infty(X_E)$$

by the formula

$$(Lh)(X) = \nabla_X h + J \nabla_{jX} h + (ah)(X),$$

where X is an arbitrary vector field on S. Since the space of connections is an affine space we immediately see that the set $\mathcal{O}_E$ of all Cauchy-Riemann type operators on E is convex. For a proper functional analytic set-up, where we may chose Hölder or Sobolev spaces, the operator L is Fredholm. By elliptic regularity theory the kernel and cokernel would be spanned always by the same smooth functions, regardless which choice we have made. The index of L is given by the Riemann-Roch formula

$$\mathrm{ind}(L) = (1 - g)\dim_{\mathbb{R}}(E) + 2c(E),$$

where $c(E)$ the first Chern number $c_1(E)(S)$ of E. Here we assume S to be a connected closed surface of genus $g = g(S)$.

Let $\phi : (S, j) \to (T, i)$ be a biholomorphic map and $\Phi : E \to F$ a $\mathbb{C}$-vector bundle isomorphism covering ϕ. Then Φ induces an isomorphism

$$\Phi_* : \mathcal{O}_E \to \mathcal{O}_F$$

in the obvious way. The operators (E, L) and (F, K) are called isomorphic if there exists $\Phi : E \to F$, so that $\Phi_*(L) = K$. We will denote by $[E, L]$ the equivalence class of an operator (E, L) which consists of operators (F, K), equivalent to (E, L) under isomorphisms, *isotopic to the identity*, and by

$[[E, L]]$ the equivalence class under the action of the full group of isomorphisms. The moduli space of equivalence classes $[[E, L]]$ will be denoted by $\mathcal{CR}$, and the "Teichmuller space" which consists of classes $[E, L]$ will be denoted by $\widetilde{\mathcal{CR}}$. An isomorphism Φ induces an isomorphism between the kernel (cokernel) of L and $\Phi_* L$ for every $L \in \mathcal{O}_E$, and hence one can canonically associate the determinant line to an isomorphism class, and thus define the *determinant line bundle* $\mathcal{V}$ over the moduli space $\mathcal{CR}$. Given an orientation o for L we obtain an induced orientation $\Phi_*(o)$. Let us note the following

LEMMA 1.8.1. *The bundle $\mathcal{V}$ is orientable.*

Proof. The lift $\widetilde{\mathcal{V}}$ of the bundle $\mathcal{V}$ to the Teichmuller space $\widetilde{\mathcal{CR}}$ is obviously orientable, because each connected component of the space $\widetilde{\mathcal{CR}}$ is contractible. However, one should check that an arbitrary isomorphism $\Phi : (E, L) \to (F, K)$ preserves the orientation. This follows from the following observation. Any connected component of $\widetilde{\mathcal{CR}}$ contains an isomorphism class of a complex linear operator (E, L_0), and any two complex linear operators representing points in a given component of $\widetilde{\mathcal{CR}}$ are homotopic in the class of complex linear operators. The determinant of (E, L_0) can be oriented canonically by observing that its kernel and cokernel are complex spaces. Any isomorphism maps a complex linear operator to a complex linear operator and preserves its complex orientation. Hence, it preserves an orientation of the determinant line of any operator (E, L). □

We will call an orientation of $\mathcal{V}$ *complex* if it coincides with the complex orientation of determinants of complex linear operators.

The components of the space $\mathcal{CR}$ are parameterized by the topological type of the underlying surface S and the isomorphism class of the bundle E. It turns out that the complex orientation of $\mathcal{V}$ satisfies three *coherency* Axioms A1–A3 which we formulate below. They relate orientations of $\mathcal{V}$ over different components of $\mathcal{CR}$. Conversely, we will see that these axioms determine the orientation uniquely up to a certain normalization.

Given (E, L) and (F, K) over surfaces Σ_0 and Σ_1 we define a *disjoint union*

$$[E, L] \mathbin{\dot{\cup}} [F, K] := [G, M]$$

of (E, L) and (F, K) as a pair (G, M), where G is a bundle over the disjoint union $\Sigma = \Sigma_0 \coprod \Sigma_1$, so that $(G, M)|_{\Sigma_0}$ is isomorphic to (E, L) and $(G, M)_{\Sigma_1}$ is isomorphic to (F, K). Clearly, the isomorphism class of a disjoint union is uniquely determined by the classes of (E, L) and (F, K). Thus, we have a well-defined construction called *disjoint union*: The determinant $\det \Sigma$ is canonically isomorphic to $\det L \otimes \det K$, and hence the orientations o_K and o_L define an orientation $o_K \otimes o_L$ of Σ. Our first axiom reads

AXIOM C1. For any disjoint union $[G, M] = [E, L] \dot{\cup} [F, K]$ the orientation o_M equals $o_K \otimes o_L$.

Given (E, L) and (F, K), where E and F are bundles over S of possibly different rank, we can define an operator $(E \oplus F, L \oplus K)$. There is a canonical map

$$\det(L) \otimes \det(K) \to \det(L \oplus K),$$

and thus given orientations o_L and o_K we obtain $o_L \oplus o_K$.

AXIOM C2.

$$o_{L \oplus K} = o_L \oplus o_K.$$

To formulate the third axiom, we need a construction, called *cutting and pasting.*

Let (E, L) be given and assume that $\gamma_1, \gamma_2 : S^1 \to S$ be real analytic embeddings with mutually disjoint images. Assume that $\Phi : E|_{\gamma_1} \to E|_{\gamma_2}$ is a complex vector bundle isomorphism covering $\sigma = \gamma_2 \circ \gamma_1^{-1}$. The maps γ_1 and γ_2 extends as holomorphic embeddings $\bar{\gamma}_j : [-\varepsilon, \varepsilon] \times S^1 \to S$ for a suitable small $\varepsilon > 0$, so that the images are still disjoint. Locally, near γ_j we can distinguish the left and the right side of γ_j. These sides correspond to the left or the right part of the annulus $[-\varepsilon, \varepsilon] \times S^1$. Cutting S along the curves γ_j we obtain a compact Riemann surface $\bar{S}$ with boundary. Its boundary components are $\gamma_j^{\pm}$, $j = 1, 2$, where $\gamma_j^{\pm}$ is canonically isomorphic to γ_j. The vector bundle E induces a vector bundle $\bar{E} \to \bar{S}$. We define a space of smooth sections $\Gamma_{\Delta}(\bar{E})$ as follows. It consists of all smooth sections $\bar{h}$ with the property that

$$\bar{h}|_{\gamma_j^-} = \bar{h}|_{\gamma_j^+} \text{ for } j = 1, 2.$$

Then L induces an operator $\bar{L} : \Gamma_{\Delta}(\bar{E}) \to \Gamma(X_{\bar{E}})$. The operators L and $\bar{L}$ have naturally isomorphic kernel and cokernel. So an orientation o of $\det(L)$ induces one of $\det(\bar{L})$. The boundary condition Δ can be written in the form

$$\begin{bmatrix} 1 & 0 \\ 0 & 1 \end{bmatrix} \cdot \begin{bmatrix} \Phi(\gamma_1(t)\bar{h} \circ \gamma_1^-(t) \\ \bar{h} \circ \gamma_2^-(t) \end{bmatrix} = \begin{bmatrix} \Phi(\gamma_1(t))\bar{h} \circ \gamma_1^+(t) \\ \bar{h} \circ \gamma_2^+(t) \end{bmatrix}.$$

We introduce a parameter depending boundary condition by

$$\begin{bmatrix} \cos(\tau) & \sin(\tau) \\ -\sin(\tau) & \cos(\tau) \end{bmatrix} \cdot \begin{bmatrix} \Phi(\gamma_1(t)\bar{h} \circ \gamma_1^-(t) \\ \bar{h} \circ \gamma_2^-(t) \end{bmatrix} = \begin{bmatrix} \Phi(\gamma_1(t))\bar{h} \circ \gamma_1^+(t) \\ \bar{h} \circ \gamma_2^+(t) \end{bmatrix}$$

for $\tau \in [0, \frac{\pi}{2}]$. For all these boundary conditions L induces an operator, which is again Fredholm of the same index. For every τ we obtain a Cauchy-Riemann type operator from $\Gamma_{\Delta_\tau}(\bar{E})$ to $\Gamma(X_{\bar{E}})$. Note that for a section h satisfying the boundary condition Δ_τ the section ih satisfies the same boundary condition. On the other hand for $\tau = \frac{\pi}{2}$ we obtain a Fredholm operator whose kernel and cokernel naturally isomorphic to the kernel on cokernel of a Fredholm operator on a new closed surface. Namely identify γ_1^+ with γ_2^- and γ_2^+ with γ_1^-. For the bundle $\bar{E}$ we identify the part above γ_1^+ via Φ with the part over γ_2^- and we identify the part above γ_1^- with $-\Phi$ to the part above γ_2^+. The latter surface and bundle we denote by $E_\Phi \to S_\Phi$ and the corresponding operator by L_Φ. Letting the parameter run we obtain starting with an orientation o for L an orientation o_Φ for L_Φ. If o is the complex orientation it is easily verified that o_Φ is the complex orientation as well. We say that the operator L_Φ is an operator obtained from L by cutting and pasting. This operator L_Φ has the same index as L, and the component of $[L, \Phi]$ in $\widetilde{\mathcal{CR}}$ depends only the isotopy classes of the embeddings γ_1 and γ_2.

Axiom C3.

$$o_{L_\Phi} = o_\Phi \,.$$

Note that we have to require here that the parts of L over the curves γ_1 and γ_2 are isomorphic via the gluing data. It is straightforward to check that

Theorem 1.8.2. *The complex orientation of $\mathcal{V}$ is coherent, i.e. it satisfies Axioms C1–C3.*

Let us point out a simple

Lemma 1.8.3. *Let (E, L) be an isomorphism then the orientation by $1 \otimes 1^*$ of the $\det L = \mathbb{R} \otimes \mathbb{R}^*$ defines the complex orientation of $\mathcal{V}$ over the component of $[E, L]$.*

The following theorem gives the converse of Theorem 1.8.2.

Theorem 1.8.4. *Suppose that a coherent orientation of $\mathcal{V}$ coincides with the complex orientation for the trivial line bundle over S^2 and for the line bundle over S^2 with Chern number 1. Then the orientation is complex.*

Proof. Let us first observe that according to Theorem 1.8.2 the disjoint union, direct sum and cutting and pasting procedures preserve the class of complex orientations. Consider the pair (E_0, L_0), where E_0 is the trivial bundle $S^2 \times \mathbb{C} \to S^2$ and L_0 is the standard Cauchy-Riemann operator. Then the $\operatorname{ind} L_0 = 2$. Take small loops around north pole and south pole on S^2 and identify the trivial bundles over these loops. Now apply the cutting and pasting procedure and Axiom C3 to obtain the disjoint union of the trivial bundle over the torus and the trivial bundle over S^2. Hence we can use Axiom C1 to obtain an induced orientation for the Cauchy-Riemann operator on the trivial bundle over T^2. Taking appropriate loops we obtain orientations for all trivial line bundles over Riemann surfaces of arbitrary genus. Using direct sums and disjoint unions constructions, and applying Axioms C1 and C2 we see that the orientation of all trivial bundles of arbitrary dimensions over Riemann surfaces of arbitrary genus are complex. Let E_1 be the bundle over S^2 with Chern number 1. Then we can use C3 to glue two copies of (E_1, L_1) to obtain the complex orientation of the disjoint union of a Cauchy Riemann operator on the trivial bundle and one on the bundle with Chern number 2. Now it is clear that the given coherent orientation has to be complex over all components of the moduli space $\mathcal{CR}$. □

In the next section we extend the coherent orientation from Cauchy-Riemann type operators over closed surfaces to a special class of Cauchy-Riemann type operators on Riemann surfaces with punctures.

1.8.3 A special class of Cauchy-Riemann type operators on punctured Riemann surfaces. Let us view $\mathbb{C}^n$ as a real vector space equipped with the Euclidean inner product which is the real part of the standard Hermitian inner product. We define a class of self-adjoint operators as follows. Their domain in $L^2(S^1, \mathbb{C}^n)$ is $H^{1,2}(S^1, \mathbb{C}^n)$ of Sobolev maps $h : S^1 \to \mathbb{C}^n$. The operators have the form

$$(Ax)(t) = -i\frac{dx}{dt} - a(t)x, \tag{2}$$

where $a(t)$ is a smooth loop of real linear self-adjoint maps. We assume that A is non-degenerate in the sense that $Ah = 0$ only has the trivial solution, which just means that the time-one map $\psi(1)$ of the Hamiltonian flow

$$\begin{aligned} \dot{\psi}(t) &= ia(t)\psi(t), \\ \psi(0) &= \mathrm{Id} \end{aligned} \tag{3}$$

has no eigenvalues equal to 1. In particular, $A : H^{1,2} \to L^2$ is an isomorphism.

Given a smooth vector bundle $E \to S^1$ we can define $H^{1,2}(E)$ and $L^2(E)$ and a class of operators B by requiring that $A = \Phi B \Phi^{-1}$ for an Hermitian trivialization Φ of the bundle E. We shall call such operators *asymptotic*, for reasons which will become clear later.

As it was defined in section 1.5 above, an asymptotically marked punctured Riemann surface is a triplet $(S, \mathbf{x}, \mu)$, where $S = (S, j)$ is a closed Riemann surface, $\mathbf{x} = \{x_1, \dots, x_s\}$ is the set of punctures, some of them called positive, some negative, and $\boldsymbol{\mu} = \{\mu_1, \dots, \mu_s\}$ is the set of asymptotic markers, i.e. tangent rays, or equivalently oriented tangent lines at the punctures.

One can introduce near each puncture $x_k \in \mathbf{x}$ a holomorphic parameterization, i.e. a holomorphic map $h_k : D \to S$ of the unit disk D such that $h_k(0) = x_k$ and the asymptotic marker μ_k is tangent to the ray $h_k(r),\ r \geq 0$. We assume that the coordinate neighborhoods $\mathcal{D}_k = h_k(D)$ of all the punctures are disjoint. Then we define $\sigma_k : \mathbb{R}^+ \times S^1 \to \mathcal{D} \setminus \{0\}$ by

$$\sigma_k(s,t) = h_k\big(e^{\pm 2\pi(s+it)}\big),$$

where the sign $-$ is chosen if the puncture x_k is positive, and the sign $+$ for the negative puncture. We will refer to σ_k as holomorphic polar coordinates adapted to (x_k, μ_k). Given two adapted polar coordinate systems σ and σ' near the same puncture $x \in \mathbf{x}$ we observe that the transition map (defined for R large enough)

$$\sigma^{-1} \circ \sigma' : [R, \infty) \times S^1 \to [0, \infty) \times S^1$$

satisfies for every multi-index α

$$D^\alpha\big[\sigma^{-1} \circ \sigma'(s,t) - (c+s, t)\big] \to 0$$

uniformly for $s \to \infty$, where c is a suitable constant. The main point is the fact that there is no phase shift in the t-coordinate.

Given $(S, \mathbf{x}, \mu)$ we associate to it a smooth surface $\bar{S}$ with boundary compactifying the punctured Riemann surface $S \setminus \mathbf{x}$ by adjoining a circle for every puncture. Each circle has a distinguished point $0 \in S^1 = \mathbb{R}/\mathbb{Z}$. Namely for each positive puncture we compactify $\mathbb{R}^+ \times S^1$ to $[0, \infty] \times S^1$, where $[0, \infty]$ has the smooth structure making the map

$$[0, \infty] \to [0, 1] : s \to s(1+s^2)^{-\frac{1}{2}}, \ \infty \to 1$$

a diffeomorphism. We call $S_k^+ = \{\infty\} \times S^1$ the circle at infinity associated to (x_k, μ_k). For negative punctures we compactify at $-\infty$ in a similar way.

Definition 1.8.5. *A smooth complex vector bundle $E \to (S, \mathbf{x}, \mu)$ is a smooth vector bundle over $\hat{S}$ together with Hermitian trivializations*

$$\Phi_k : E|_{S_k} \to S^1 \times \mathbb{C}^n.$$

An isomorphism between two bundles E and F over surfaces S and T is a a complex vector bundle isomorphism $\Psi : E \to F$ which covers a biholomorphic map $\phi : (S, j) \to (T, i)$, preserves punctures and the asymptotic *markers (their numbering and signs) and respects the asymptotic trivializations.*

Define as in section 1.8.2 above the bundle $X_E \to \bar{S}$. Set $\dot{S} = S \setminus \Gamma$. We introduce the Sobolev space $H^1(E)$ which consists of all sections h of $E \to \dot{S}$ of class $H^{1,2}_{loc}$ with the following behavior near punctures. Suppose, that x is a positive puncture and σ is an adapted system of holomorphic polar coordinates. Pick a smooth trivialization ψ of $E \to \bar{S}$ over $[0, \infty] \times S^1$ (in local coordinates) compatible with the given asymptotic trivialization. Then the map $(s,t) \to \psi(s,t) h \circ \sigma(s,t)$ is assumed to belong to $H^{1,2}(\mathbb{R}^+ \times S^1, \mathbb{C}^n)$. A similar condition is required for negative punctures. In a similar way we define the space $L^2(X_E)$. Observe that de-facto we use measures which are infinite on $\dot{S}$ and that the neighborhoods of punctures look like half-cylinders.

A Cauchy-Riemann type operator L on E has the form

$$(Lh)X = \nabla_X h + J \nabla_{jX} h + (ah)X,$$

where X is a vector field on S. We require, however a particular behaviour of L near the punctures. Namely, regarding E as a trivial bundle $[0, \infty] \times \mathbb{C}^n$ with respect to the chosen polar coordinates and trivialization near say a positive puncture we require that

$$(Lh)(s,t)\left(\frac{\partial}{\partial s}\right) = \frac{\partial h}{\partial s} - A(s)h,$$

where $A(s) \to A_\infty$ for an asymptotic operator A_∞, as it was previously introduced.

Theorem 1.8.6. *The operator L is Fredholm.*

The index of L can be computed in terms of Maslov indices of the asymptotic operators (and, of course, the first Chern class of E and the topology of S).

Similar to the case of closed surfaces we define the notion of isomorphic pairs (E, L) and (F, K), where we emphasize the importance of the compatibility of the asymptotic trivializations, define the moduli space $\mathcal{CR}_{\text{punct}} \supset \mathcal{CR}_{\text{closed}}$ and the Teichmuller spaces $\widetilde{\mathcal{CR}}_{\text{punct}} \supset \widetilde{\mathcal{CR}}_{\text{closed}}$, and extend the determinant line bundle $\mathcal{V}$ to $\mathcal{CR}_{\text{punct}}$ and $\widetilde{\mathcal{V}}$ to $\widetilde{\mathcal{CR}}_{\text{punct}}$. The bundle $\widetilde{\mathcal{V}}$ is orientable by the same reason as in the case of closed surfaces: each component of the space $\widetilde{\mathcal{CR}}_{\text{punct}}$ is contractible. However, unlike the closed case, there is no canonical (complex) orientation of $\widetilde{\mathcal{V}}$. Still due to the requirement that isomorphisms preserves the end structure of the operators, one can deduce the fact that even isotopically non-trivial isomorphisms preserve the orientation of $\widetilde{\mathcal{V}}$, which shows that the bundle $\mathcal{V}$ over $\mathcal{CR}_{\text{punct}}$ is orientable.

Let us review now Axioms C1–C3 for the line bundle $\mathcal{V}$ over $\mathcal{CR}_{\text{punct}}$. The formulation of Disjoint Union Axiom C1 should be appended by the following requirement. Let (E,L) and (F,K) be operators over the punctured Riemann surfaces $(S, \mathbf{x} = \{x_1,\dots,x_s\})$ and $(T, \mathbf{y} = \{y_1,\dots,y_t\})$, respectively. Then $(E,L)\dot{\cup}(F,K)$ is an operator over the surface $S \coprod T$ with the set of punctures $\mathbf{z} = \{x_1,\dots,x_s,y_1,\dots,y_t\}$. The disjoint union operation is associative, but not necessarily commutative (unlike the case of closed surfaces). Axioms C2 and C3 we formulate without any changes compared to the closed case. By a coherent orientation of the bundle $\mathcal{V}$ over $\mathcal{CR}_{\text{punct}}$ we will mean any orientation of $\mathcal{V}$ which satisfies Axioms C1–C3.

Take the trivial (and globally trivialized) line bundle $E_0 = \mathbb{C} \times \mathbb{C}$ over the 1-punctured Riemann sphere $\mathbb{C} = \mathbb{C}P^1 \setminus \infty$. For any admissible asymptotic operator A we choose a Cauchy-Riemann operator $L_A^{\pm}$ on E_0 which has A as its asymptotics at ∞. The superscript $\pm$ refers to the choice of ∞ as the positive or negative puncture. Note, that the component of $([E_0, L_A^{\pm})$ in the moduli spaces $\mathcal{CR}$ is uniquely determined by the homotopy class $[A]$ of the asymptotic operator A in the space of *non-degenerate* asymptotic operators.

The following theorem describes all possible coherent orientations of the line bundle $\mathcal{V}$ over $\mathcal{CR}$.

Theorem 1.8.7. *Let us choose an orientation $o_A^{\pm}$ of the operator $(E_0, L_A^{\pm})$ for a representative A of each homotopy class $[A]$ of non-degenerate asymptotic operators. Then this choice extends to the unique coherent orientation of the bundle $\mathcal{V}$ over $\mathcal{CR}_{\text{punct}}$, which coincide with the complex orientation over $\mathcal{CR}_{\text{closed}}$.*

Thus there are infinitely many coherent orientations of $\mathcal{V}$ over $\mathcal{CR}_{\text{punct}}$ unlike the case of closed surfaces, when there are precisely four.

We sketch below the proof of Theorem 1.8.7. First, similar to the case of closed surfaces, it is sufficient to consider only operators on the trivial, and even globally trivialized bundles. Next take the disjoint union of (E_0, L_A^-) and (E_0, L_A^+), consider two circles $\gamma^{\pm}$ around the punctures in the two copies of $\mathbb{C}$ and apply the cutting/pasting construction along these circles. As the result we get a disjoint union of an operator $\widetilde{L}_A$ on the trivial line bundle over the closed Riemann sphere, and an operator $\overline{L}_A$ over the cylinder $C = S^1 \times \mathbb{R}$, which we view as the Riemann sphere with two punctures $x_1 = \infty$ and $x_2 = 0$ and consider x_1 as a positive puncture and

x_2 as a negative one. The operator $\overline{L}_A$ has the same asymptotic operator A at both punctures. Then Axioms C1 and C3 determine the orientation of $\overline{L}_A$, because for the operator $\widetilde{L}_A$ we have chosen the complex orientation. Notice that if one glue $L_A^{\pm}$ in the opposite order, then we get an operator $\overline{L}'_A$ which has the reverse numbering of the punctures. The orientation of $\overline{L}'_A$ determined by the gluing may be the same, or opposite as for the operator $\overline{L}_A$, depending on the parity of the Conley-Zehnder index of the asymptotic operator A.[4]

Consider now an arbitrary operator (E, L) acting on sections of a complex line bundle E over a punctured Riemann surface $(S, \mathbf{x}, \boldsymbol{\mu})$ with $\mathbf{x} = \{x_1, \dots, x_s\}$, $E = S \times \mathbb{C}^n$, and the asymptotic operators $A_1, \dots, A_s$ at the corresponding punctures. For each $i = 1, \dots, s$ consider an operator $(E_0, L_i = L_{A_i}^{\pm})$, where $E_0 = \mathbb{C} \times \mathbb{C}^n$, the sign $+$ is chosen if the puncture x_i is negative, and the sign $+$ is chosen otherwise. Using Axiom C1 we orient the operator $(E, L)\dot{\cup}(E_0, L_s)$, and then choosing circles around the puncture x_s and ∞ apply the cutting/pasting procedure. As the result we get the disjoint union of an operator L' over the Riemann surface with punctures $(x_1, \dots, x_{s-1})$ and the operator $\overline{L}_A$, or $\overline{L}'_A$ depending on whether the puncture x_s was negative, or positive. Hence Axioms C1 and C3 determine the orientation of L' in terms of the orientation of L. Repeating the procedure for the punctures $x_{s-1}, \dots, x_1$ we express the orientation of L in terms of the complex orientation of an operator over the closed surface.

It remains to observe that if E is a trivial complex bundle of rank $r > 1$, then any asymptotic operator A can be deformed through non-degenerate asymptotic operators to an operator $\widetilde{A}$ which is split into the direct sum of asymptotic operators on the trivial complex line. Hence we can use the direct sum axiom C2 to orient determinants of operators acting on bundles of arbitrary rank.

1.8.4 Remark about the coherent orientation for asymptotic operators with symmetries. Let A be an asymptotic operator given by the formula (2), where the loop $a(t)$, $t \in S^1 = \mathbb{R}/\mathbb{Z}$, of symmetric matrices has a symmetry $a(t + 1/2) = a(t)$, $t \in \mathbb{R}/\mathbb{Z}$. Let L be a Cauchy-Riemann type operator on a bundle $E \to S$, which has A as its asymptotic operator at a puncture $x \in S$ with an asymptotic marker μ. Let L' be an operator which differs from L by rotating by the angle π the marker μ to a marker

[4]The operator $\overline{L}_A$ is homotopic to an isomorphism, and thus has a canonical orientation $1 \otimes 1^*$. If we insist on that normalization, than our construction would determine the orientation of L_A^- in terms of L_A^+.

μ', with the corresponding change of the trivialization near the puncture. Let $h : S \to S$ be a diffeomorphism which rotates the polar coordinate neighborhood $\mathcal{D}$ of the punctures x by π, and is fixed outside a slightly larger neighborhood. Then the operator $h_* L'$ has the same asymptotic data as L and the isomorphism classes $[E, L]$ and $[E, h_* L']$ belongs to the same component of the space $\widetilde{\mathcal{CR}}$. Given a coherent orientation of $\mathcal{V}$, do the orientations o_L and $o_{h_* L}$ coincide? It turns out that

LEMMA 1.8.8. *Let Ψ be the time-one map of the linear Hamiltonian flow $\psi(t)$, defined by the equation* (3). *The orientations o_L and $o_{h_* L}$ coincide if and only if the number of real eigenvalues of Ψ (counted with multiplicities) from the interval* $(-1, 0)$ *is even.*

This lemma is the reason why we excluded certain periodic orbits from $\mathcal{P}$ in section 1.2 above. See also Remarks 1.9.2 and 1.9.6.

1.8.5 Coherent orientations of moduli spaces. The moduli spaces of holomorphic curves which we need to orient are zero sets of non-linear Cauchy-Riemann type operators, whose linearizations are related to operators of the kind we described (see below for more details). In general, the moduli spaces are neither manifolds nor orbifolds, due to the fact that Fredholm sections cannot be made transversal to the zero section by changing natural parameters like the almost complex structure or the contact form. Such a transversality will only be achievable by making abstract perturbations, leading to virtual moduli spaces. Those virtual spaces will be the moduli spaces which will provide us with the data for our constructions. Nevertheless the Fredholm operators occurring in the description of the virtual moduli spaces will only be compact perturbations of the Cauchy-Riemann type operators, and hence the orientation scheme for these virtual moduli spaces does not differ from the case of moduli spaces of holomorphic curves.

A moduli space $\mathcal{M}(\Gamma^+, \Gamma^-; W, J)$ of holomorphic curves in a directed symplectic cobordism $(W = \overrightarrow{V^- V^+}, J)$ is a fiber bundle over the corresponding moduli space of Riemann surfaces. Its base is a complex orbifold, and hence it is canonically oriented, while the fiber over a point S, where S is a Riemann surface with a fixed conformal structure and positions of punctures, can be viewed as the space solutions of the $\overline{\partial}_J$-equation. If the transversality is achieved than the tangent bundle of a moduli space $\mathcal{M}(\Gamma^+, \Gamma^-; W, J)$ arise as the kernel of the linearized surjective operator $\overline{\partial}_J$. The linearization of $\overline{\partial}_J$ at a point $f \in \mathcal{M}(\Gamma^+, \Gamma^-; W, J)$ is a Fredholm operator in a suitable functional analytic setting. This set-up involves Sobolev spaces with suitable asymptotic weights derived from the non-degeneracy properties of the periodic orbits. It is a crucial observation, again a corollary of the behaviour near the punctures, that up to a compact perturbation, the operator L splits into two operators L' and L'', where L' is a complex linear operator acting on the complex line bundle $T(S)$ of the Riemann surface S, and L'' is a Cauchy-Riemann type on the the bundle E, such that $T(S) \oplus E = f^*(TM)$. This operator is usually only real linear, but most importantly it is of the kind we just described in our linear theory. The trivialization of E near the punctures is determined by the chosen in 1.5 trivialization of the contact structure near periodic orbit of the Reeb vector fields on $V^{\pm}$, and the asymptotic operators are determined by the linearized Reeb flow near the periodic orbits. We have $\det L = \det L' \otimes \det L''$. But $\det L'$ has a canonical complex orientation, and hence the orientation for $\det L$ is determined by the orientation of $\det L''$. Therefore, a choice of a coherent orientation of $\mathcal{V}$ over $\mathcal{CR}$ determines in the transversal case the orientation of all the moduli spaces $\mathcal{M}(\Gamma^+, \Gamma^-; W, J)$.

1.9 First attempt of algebraization: Contact Floer homology.

1.9.1 Recollection of finite-dimensional Floer theory. Let us first recall the basic steps in defining a Floer homology theory in the simplest case of a Morse function f on a finite-dimensional orientable closed manifold M. We refer the reader to Floer's original papers (see, for instance, [F]), as well as an excellent exposition by D. Salamon [S] for the general theory.

First, one forms a graded complex $C(f, g)$ generated by critical points $c_1, \dots, c_N$ of f, where the grading is given by the Morse index of critical points. Next, we choose a generic Riemannian metric g on M which satisfy the Morse-Smale condition of transversality of stable and unstable varieties of critical points. This enables us to define a differential $d = d_{f,g} : C(f, g) \to C(f, g)$ by counting gradient trajectories connecting critical points of neighboring indices:

$$d(c_i) = \sum L_i^j c_j,$$

where the sum is taken over all critical points c_j with $\operatorname{ind} c_j = \operatorname{ind} c_j - 1$. The coefficient L_i^j is the *algebraic number* of trajectories connecting c_i and c_j. This means that the trajectories are counted with signs. In the finite-dimensional case the signs could be determined as follows. For each critical point we orient arbitrarily its stable manifold. Together with the orientation of M this allows us to orient all unstable manifolds, as well as the intersections of stable and unstable ones. If $\operatorname{ind} c_j = \operatorname{ind} c_i - 1$ then the stable manifold of c_i and the unstable manifold of c_j intersect along finitely many trajectories which we want to count, and hence each of these trajectories gets an orientation. Comparing this orientation with the one given by the direction of the gradient ∇f we can associate with every trajectory a sign.[5]

To show that $d^2 = 0$, which then would allow us to define the homology group $H_*(C(f,g), d)$, we proceed as follows. Let us observe that the coefficients K_i^j in the expansion $d^2(c_i) = \sum K_i^j c_j$ count the algebraic number of broken gradient trajectories $(\delta_{il}, \delta_{lj})$ passing through an intermediate critical point c_l, $l = 1, \dots N$. But each broken trajectory $(\delta_{il}, \delta_{lj})$, which connects critical points whose indices differ by 2, is a boundary point of the 1-dimensional manifold of smooth trajectories connecting c_i and c_j. The algebraic number of boundary points of a compact 1-dimensional manifold is, of course, equal to 0. Hence $K_i^j = 0$, and thus $d^2 = 0$.

Next we want to show that the homology group $H_*(C(f,g), d)$ is an invariant of the manifold M (of course, in the case we consider it is just $H_*(M)$), i.e. it is independent of the choice of the function f and the Riemannian metric g. The proof of the invariance consists of three steps.

Step 1. Let us show that given a homotopy of functions $F = \{f_t\}_{t\in[0,1]}$, and a homotopy of Riemannian metrics $G = \{g_t\}_{t\in[0,1]}$, one can define a homomorphism $\Phi = \Phi_{F,G} : C(f_1, g_1) \to C(f_0, g_0)$ which commutes with the boundary homomorphisms $d_0 = d_{f_0,g_0}$ and $d_1 = d_{f_1,g_1}$, i.e.

$$\Phi \circ d_1 - d_0 \circ \Phi = 0\,. \tag{4}$$

To construct Φ we consider the product $W = M \times \mathbb{R}$ and, assuming that the homotopies $\{f_t\}$ and $\{g_t\}$ are extended to all $t \in \mathbb{R}$ as independent of t on $(-\infty, -1] \cup [1, \infty)$, we define on W a function, still denoted by F, by the formula

$$F(x,t) = \begin{cases} f_0(x) + ct, & t \in (\infty, 0); \\ f_t(x) + ct, & t \in [0,1]\ ; \\ f_1(x) + ct, & t \in (0, \infty), \end{cases}$$

[5] The generalization of this procedure to an infinite-dimensional case is not straightforward, because stable and unstable manifolds not only can become infinite-dimensional, but in most interesting cases cannot be defined at all. On the other hand, the moduli spaces of gradient trajectories connecting pairs of critical points (which in the finite-dimensional case coincide with the intersection of stable and unstable manifolds of the critical points) are often defined, and one can use the coherent orientation scheme, similar to the one described in section 1.8 above for the moduli spaces of holomorphic curves, to define their orientation.

where the constant c is chosen to ensure that $\frac{\partial F}{\partial t} > 0$. Similarly, we use the family of Riemannian metrics g_t to define a metric G on W which is equal to g_t on $M \times t$ for all $t \in \mathbb{R}$, and such that $\frac{\partial}{\partial t}$ is the unit vector field orthogonal to the slices $M \times t, t \in \mathbb{R}$. The gradient trajectories of ∇F converge to critical points of f_1 at $+\infty$, and to the critical points of f_0 at $-\infty$. For a generic choice of G the moduli space of the (unparameterized) trajectories connecting two critical points, c^1 of f_1 and c^0 of f_0, is a compact k-manifold with boundary with corners, where $k = \operatorname{ind} c^1 - \operatorname{ind} c^0$. Hence, similarly to the above definition of the differential d, we can define a homomorphism $\Phi : C(f_1, g_1) \to C(f_0, g_0)$ by taking an algebraic count of gradient trajectories between the critical point of f_1 and f_0 of the same Morse index, i.e. $\Phi(c^1_j) = \sum \widetilde{L}^i_j c^0_j$. The identity (4) comes from the description of the boundary of the 1-dimensional moduli spaces of trajectories of ∇F. Notice that the function F has no critical points, and hence a family of gradient trajectories cannot converge to a broken trajectory in a usual sense. However, this can happen *at infinity.* Let us recall that the function F and the metric G are cylindrical outside of $M \times [-1, 1]$. Hence away from a compact set a gradient trajectory of F projects to a gradient trajectory of f_0 or f_1. When the projection, say at $+\infty$, of a sequence $\delta_n : \mathbb{R} \to W$ of trajectories of ∇F converges to a broken trajectory of ∇f_1 this can be interpreted as a splitting at $+\infty$. This phenomenon is very similar to the one described for the moduli spaces of holomorphic curves in section 1.6. Namely, there exist gradient trajectories $\delta : \mathbb{R} \to W$ of ∇F, and $\delta' : \mathbb{R} \to M_1$ of ∇f_1, such that

- $\delta_n \to \delta$ uniformly on $(-\infty, C]$ for all C;
- there exists a sequence $C_n \to +\infty$ such that $\delta'_n(t) = \delta_n(t + C_n)$ converges to $(\delta'(t), t)$ uniformly on all subsets $[-C, \infty)$.

In this sense broken trajectories of the form (δ, δ') and (δ'', δ), where δ'' is a trajectory of ∇f_0 form the boundary of the 1-dimensional moduli spaces of trajectories of ∇F connecting critical points c^1 of f_1 and c^0 of f_0 with $\operatorname{ind} c^1 - \operatorname{ind} c^0 = 1$. Therefore the algebraic number of these trajectories equals 0. On the other hand, this number is equal to $\Phi \circ d_1 - d_0 \circ \Phi$ which yields the identity (4).

Step 2. Our next goal is to check that if $(F_u, G_u), u \in [0, 1]$, is a homotopy of homotopies which is constant outside of a compact subset of W, then the homomorphisms $\Phi_0 = \Phi_{F_0,G_0}$ and $\Phi_1 = \Phi_{F_1,G_1}$ are related via the chain homotopy formula

$$\Phi_1 - \Phi_0 = K \circ d_1 + d_0 \circ K, \tag{5}$$

for a homomorphism $K : C(f_1, g_1) \to C(f_0, g_0)$. The space of all homotopies (F, G) connecting given pairs (f_0, g_0) and (f_1, g_1) is contractible, and hence (5) implies that the homomorphism $\Phi_* : H_*(C(f_1, g_1), d_1) \to H_*(C(f_0, g_0), d_0)$ is independent of the choice of a homotopy (F, G).

To prove (5) one studies moduli spaces of gradient trajectories of the whole 1-parametric family of functions F_u. For a generic choice of the homotopy one has isolated critical values of the parameter u when appear *handle-slides*, i.e. gradient connections between critical points with the index difference -1. By counting these trajectories one can then define a homomorphism $K : C(f_1, g_1) \to \mathbb{C}\,(f_0, g_0)$ in exactly the same way as the homomorphism Φ was defined in Step 1 by counting trajectories with the index difference 0.

The identity (5) expresses the fact that the broken trajectories of the form (δ, δ') and (δ'', δ), where δ is a handle-slide trajectory and δ' is a trajectory of ∇f_1, form the boundary of the moduli space of index 0 trajectories in the family (F_u, G_u). The difference in signs in formulas (4) and (5) is a reflection of the fact that the homomorphism K raises the grading by 1, while Φ leaves it unchanged.

Step 3. Finally we need to show that

$$(\Phi_{F,G})_* = (\Phi_{F',G'})_* \circ (\Phi_{F'',G''})_*, \tag{6}$$

if $(F, G) = \{f_t, g_t\}_{t \in [0,2]}$ is the composition of homotopies $(F'', G'') = \{f''_t, g''_t\}_{t \in [0,1]}$ and $(F', G') = \{f'_t, g'_t\}_{t \in [1,2]}$. To prove this we view, as in Step 1, the homotopy (F, G) as a function and a metric on the cylinder $W = M \times \mathbb{R}$. Consider a deformation (F_T, G_T) of F, G, by cutting W open along $M \times 1$ and inserting a cylinder $M \times [0, T]$ of growing height T with the function and the metric independent of the coordinate t. When $T \to +\infty$ the gradient trajectories of F_T with respect to G_T split in a sense, similar to the one explained in Step 2,[6] into a "broken trajectory" (δ'', δ'), where δ' (resp. δ") is a trajectory of $\nabla_{G'} F'$ (resp. $\nabla_{G''} F''$). Consider the 1-dimensional moduli space $\mathcal{M}$ of trajectories of $\nabla_{G_T} F_T, T \in [0, \infty)$, connecting a fixed critical point $c = c^2$ of f_2 with an arbitrary critical point c^0 of f_0 with $\operatorname{ind} c - \operatorname{ind} c^0 = 1$. Then the boundary of $\mathcal{M}$ consists of

a) all the trajectories of $\nabla_{G_0} F_0 = \nabla_G F$ connecting c^0 and c; they are given by the expression $\Phi(c)$;
b) all the broken trajectories (δ'', δ') described above, such that δ'' begins at c^0 and ends at a critical point c^1 of f^1 which is, necessarily, of the

[6] See also the discussion of a similar phenomenon for the moduli spaces of holomorphic curves in section 1.6 above.

same Morse index as c^0 and c^2, δ' begins at c^1 and ends at c; these broken trajectories are described by the expression $\Phi_{F',G'}(\Phi_{F'',G''}(c))$;

c) broken trajectories defined according to Step 2 for the 1-dimensional family $F_T, T \in [0,\infty)$; they are described by the expression $K(d_0(c))+d_2(K(c))$ for some homomorphism $K : C(f_2, g_2) \to C(f_0, g_0)$.

Thus the sum (taken with appropriate signs) of the three expressions defined in a)–c) equals 0, and thus we get

$$\Phi_{F',G'}(\Phi_{F'',G''}(c)) - \Phi(c) = K \circ d_0(c) + d_2 \circ K(c),$$

i.e. the homomorphisms Φ and $\Phi_{F',G'} \circ \Phi_{F'',G''}$ are chain homotopic, which yields formula (6).

We can finish now the proof that the homology group $H_*(C(f,g),d)$ is independent of the choice of f and g as follows. Given two pairs (f_0, g_0) and (f_1, g_1) we first take any homotopy (F,G) connecting (f_0, g_0) with (f_1, g_1), and also take the inverse homotopy $(\overline{F}, \overline{G})$ connecting (f_1, g_1) with (f_0, g_0). The composition $(\widetilde{F}, \widetilde{G})$ of the homotopies (F,G) and $(\overline{F}, \overline{G})$ connects the pair (f_0, g_0) with itself. According to Step 3 we have $(\Phi_{\widetilde{F},\widetilde{G}})_* = (\Phi_{F,G})_* \circ (\Phi_{\overline{F},\overline{G}})_*$. On the other hand, we have shown in Step 3 that the homomorphism $(\Phi_{\widetilde{F},\widetilde{G}})_*$ is independent of the choice of a homotopy, connecting (f_0, g_0) with itself, and hence it equals the identity. Therefore, we conclude that $(\Phi_{F,G})_*$ is surjective, while $(\Phi_{\overline{F},\overline{G}})_*$ is injective. Taking the composition of homotopies $(\overline{F}, \overline{G})$ and (F,G) in the opposite order we prove that both homomorphisms are bijective.

A. Floer discovered that the finite-dimensional scheme which we explained in this section works, modulo some analytic complications, for several geometrically interesting functional on infinite-dimensional spaces. For instance, in the symplectic Floer homology theory one deals with critical points of the action functional. Its critical points are periodic orbits of a Hamiltonian system, while for an appropriate choice of a metric and an almost complex structure the gradient trajectories can be interpreted as holomorphic cylinders which connect these trajectories. The role of broken trajectories is played here by split holomorphic cylinders, and finite-dimensional compactness theorems are replaced by the highly non-trivial Gromov compactness theorem for holomorphic curves.

In the rest of this section we explore the Floer-theoretic approach for the problem of defining invariants of contact manifolds. We will see that this approach works only in a very special and restrictive situation. However, the general algebraic formalism of SFT, though quite different, has a distinctive flavor of a Floer homology theory.

1.9.2 Floer homology for the action functional. Let us make an attempt to define invariants of contact manifolds in the spirit of Floer homology theory. Let (V,ξ) be a contact manifold with a fixed contact form α and an almost complex structure $J:\xi\to\xi$, compatible with the symplectic form $d\alpha|_\xi$. Then J and $d\alpha$ define a Riemannian metric on the vector bundle ξ by the formula $g(X,Y)=d\alpha(X,JY)$ for any vectors $X,Y\in\xi$. We extend g to the whole tangent bundle $T(V)$ by declaring the vector field R_α to be the unit normal field to ξ. Consider the free loop space

$$\Lambda(V)=\{u:S^1=\mathbb{R}/\mathbb{Z}\to V\},$$

and define the *action functional*

$$S:\Lambda(V)\to\mathbb{R}\quad\text{by the formula}\quad S(\gamma)=\int_\gamma\alpha. \tag{7}$$

The least action principle tells us that the critical points of the functional S are, up to parameterization, the periodic orbits of the Reeb field R_α.

The metric g on $T(V)$ defines a metric on $\Lambda(V)$ and thus allows us to consider gradient trajectories of the action functional connecting critical points of V. The gradient direction $\nabla S(u)$, $u\in\Lambda(V)$, is given by the vector field $J\pi(\frac{du}{dt})$, where $\pi:T(V)\to\xi$ is the projection along the Reeb direction, so that a gradient trajectory $u(t,s)$, $t\in\mathbb{R}/\mathbb{Z}$, $s\in\mathbb{R}$, is given by the equation

$$\frac{\partial u}{\partial s}(t,s)=J\pi\left(\frac{\partial u}{\partial t}(t,s)\right). \tag{8}$$

Equation (8) has a flavor of a Cauchy-Riemann equation. We want to modify it into a genuine one. Namely, consider the Cauchy-Riemann equation

$$\frac{\partial U}{\partial t}(t,s)=J\frac{\partial U}{\partial s}(t,s)$$

for $U(s,t)=\big(u(s,t),\varphi(s,t)\big)\in V\times\mathbb{R}$. It can be rewritten as a system

$$\begin{aligned}\frac{\partial u}{\partial s}(t,s)&=J\pi\left(\frac{\partial u}{\partial t}(t,s)\right)+\frac{\partial\varphi}{\partial t}(t,s)R_\alpha\big(u(t,s)\big)\\ \frac{\partial\varphi}{\partial s}(s,t)&=-\left\langle\frac{\partial u}{\partial t}(t,s),R_\alpha\big(u(t,s)\big)\right\rangle.\end{aligned} \tag{9}$$

Notice that $dS(\nabla S+\psi R_\alpha)\geq 0$ for any function $\psi(t,s)$. Hence, the first equation of the system (9) can be viewed as the flow equation of the gradient-like vector-field $\nabla S+\frac{\partial\varphi}{\partial t}R_\alpha$. Trajectories of this gradient like field connecting critical points γ^-,γ^+ of the action functional correspond to elements of the moduli space $\mathcal{M}_0(\gamma^-,\gamma^+;W,J)$, and therefore the Floer homology philosophy ([F]), which we described above in the finite-dimensional case, suggests the following construction.

Let us associate a variable q_γ with every periodic orbit $\gamma \in \mathcal{P}_\alpha$ and assign to it the grading

$$\deg q_\gamma = \mathrm{CZ}(\gamma) + (n-3).$$

The choice of the constant $n-3$ is not important for purposes of this definition, but it will become important for generalizations considered in the second part of this paper.

Let A be the group algebra $\mathbb{C}\,[H_2(V)]$. We will fix a basis $A_1, \dots, A_N$ of $H_2(V;\mathbb{C}\,)$ and identify each homology class $\sum d_i A_i$ with its *degree* $d = (d_1, \dots, d_N)$. Thus we can view the algebra A as the algebra of Laurent polynomials of N variables $z_1, \dots, z_N$ with complex coefficients, and write its elements in the form $\sum a_d z^d$, where $z^d = z_1^{d_1} \cdots z_N^{d_N}$. The variables z_i are also considered graded, deg $z_i = -2c_1(A_i)$, $i = 1, \dots, N$. Consider a complex $\mathfrak{F}$ generated by the (infinitely many) graded variables q_γ with coefficients in the graded algebra A, and define a differential $\partial : \mathfrak{F} \to \mathfrak{F}$ by the formula:

$$\partial q_\gamma = \sum_{\gamma', d} \frac{n_{\gamma,\gamma',d}}{\kappa_{\gamma'}} z^d q_{\gamma'}, \tag{10}$$

where $\kappa_{\gamma'}$ denotes the multiplicity of the orbit γ', the sum is taken over all trajectories $\gamma' \in \mathcal{P}_\alpha$ and $d = (d_1, \dots, d_N)$ with

$$\mathrm{CZ}(\gamma') = \mathrm{CZ}(\gamma) + 2\langle c_1, d\rangle - 1,$$

and the coefficient $n_{\gamma,\gamma',d}$ counts the algebraic number of components of the 0-dimensional moduli space $\mathcal{M}_0^d(\gamma', \gamma; W, J)/\mathbb{R}$.[7] Notice that the Liouville flow of the vector field $\frac{\partial}{\partial t}$ defines a $\mathbb{R}$-action on the moduli spaces $\mathcal{M}_0^d(\gamma', \gamma; W, J)$, which makes the 1-dimensional components of the moduli spaces canonically oriented. Comparing this orientation with the coherent orientation we produce signs which we use in the formula (10).

To simplify the assumptions in the propositions which we formulate below we will assume for the rest of this section that $c_1|_{\pi_2(V)} = 0$. This assumption allows us to define for any contractible periodic orbit γ the Conley-Zehnder index $\mathrm{CZ}_{\mathrm{disk}}(\gamma)$ computed with respect to *any disk* Δ spanned by γ in V. We denote $\deg_{\mathrm{disk}}(\gamma) = \mathrm{CZ}_{\mathrm{disk}}(\gamma) + n - 3$.

PROPOSITION 1.9.1. *If for a contact form α the Reeb field R_α has no contractible periodic orbits $\gamma \in \mathcal{P}_\alpha$ with $\deg_{\mathrm{disk}}(\gamma) = 1$, then $\partial^2 = 0$.*

[7] Let us recall that according to our definition of the moduli space $\mathcal{M}_0^d(\gamma', \gamma; W, J)/\mathbb{R}$ the coefficient $n_{\gamma,\gamma',d}$ counts equivalence classes of holomorphic curves *with asymptotic markers*, and hence each holomorphic cylinder connecting γ and γ' is counted $\kappa_\gamma \kappa_{\gamma'}$ times, unless the cylinder itself is multiply covered. The role of the denominators $\kappa_{\gamma'}$ in formula (10), as well in a similar formula (11) below, is to correct this "over-counting".

Sketch of the proof. Similarly to the finite-dimensional case considered in section 1.9.1 above the identity $\partial^2 = 0$ in Floer homology is equivalent to the fact that the codimension 1 stratum of the compactified moduli spaces $\mathcal{M}_0^d(\gamma', \gamma)$ consists of broken trajectories, which in our case are represented by the height 2 stable curves (f_1, f_2), $f_1 \in \mathcal{M}_0^{d'}(\gamma', \gamma'')/\mathbb{R}$, $f_2 \in \mathcal{M}_0^{d''}(\gamma'', \gamma)/\mathbb{R}$, where $d = d' + d''$. However, in the general case a sequence of holomorphic cylinders in $\mathcal{M}_0^d(\gamma', \gamma)$ can split into curves different from cylinders, as it is stated in Proposition 1.7.2 and Corollary 1.6.5. But if this happens then the first-floor curve f_1 must have a component which is conformally equivalent to $\mathbb{C}$ and asymptotically cylindrical over a contractible orbit at $+\infty$. Moreover, if (f_1, f_2) belongs to a top-dimensional stratum of the boundary of the moduli space $\mathcal{M}_0^d(\gamma', \gamma)$, then $\deg_{\mathrm{disk}}(\gamma) = 1$, which contradicts our assumption.

REMARK 1.9.2. Let us recall that we excluded from $\mathcal{P}$ certain "bad" periodic orbits (see the footnote in section 1.2). However on the boundary of the moduli space $\mathcal{M}_0^d(\gamma', \gamma)$ there could be a stratum which consists of height 2 stable curves (f_1, f_2), $f_1 \in \mathcal{M}_0^{d'}(\gamma', \gamma'')$, $f_2 \in \mathcal{M}_0^{d''}(\gamma'', \gamma')$, where the orbit γ'' is one of the bad orbits which we excluded from $\mathcal{P}$. The orbit γ'' has even multiplicity $2k$, and hence on the boundary of $\mathcal{M}_0^d(\gamma', \gamma)$ there are $2k$ strata which correspond to $2k$ different possible positions of the asymptotic marker at the punctures mapped to γ''. The Poincaré return map of the Reeb flow along the orbit γ'' has an odd number of eigenvalues in the interval $(-1, 0)$, and hence according to Lemma 1.8.8 the coherent orientation will automatically assign to these orbits opposite signs, which means that these strata will not contribute to the sum (10). This explains why the exclusion of bad orbits is *possible*. Remark 1.9.6 below explains why this exclusion is *necessary*.

Now we follow Steps 1–3 in section 1.9.1 above to show the independence of the homology group

$$\oplus H_k(\mathfrak{F}, \partial) = \mathrm{Ker}\partial / \mathrm{Im}\partial,$$

graded by the degree k, of the choice of a nice contact form α and a compatible almost complex structure J.

Suppose now that we have a directed symplectic cobordism $W = \overrightarrow{V^- V^+}$, and J is a compatible almost complex structure on W. Suppose that the inclusions $V^\pm \hookrightarrow W$ induce isomorphisms on 2-dimensional homology. Then we can define a homomorphism $\Phi = \Phi_W : \mathfrak{F}^+ \to \mathfrak{F}^-$ by the formula

$$\Phi(q_\gamma) = \sum_{\gamma', d} \frac{1}{\kappa_{\gamma'}} n_{\gamma, \gamma', d} z^d q_{\gamma'}, \tag{11}$$

where the sum is taken over all trajectories $\gamma' \in \mathcal{P}^-$ and d with $\mathrm{CZ}(\gamma') = \mathrm{CZ}(\gamma) + 2\langle c_1, d\rangle$, and the coefficient $n_{\gamma,\gamma',d}$ counts the algebraic number of points of the compact 0-dimensional moduli space $\mathcal{M}_0^d(\gamma', \gamma; W, J)$. If the condition on the second homology is not satisfied then the above construction gives us only a correspondence, rather than a homomorphism. See section 2.5 for the discussion of a more general case.

PROPOSITION 1.9.3. *Suppose that the contact forms $\alpha^\pm$ associated to the ends satisfy the condition* $\deg_{\mathrm{disk}}(\gamma) \neq 0, 1$ *for any* contractible in W *periodic orbit $\gamma \in \mathcal{P}^\pm$. Then the homomorphism Φ_W commutes with ∂.*

PROPOSITION 1.9.4. *Let $J_t, t \in [0,1]$, be a family of almost complex structures compatible with the directed symplectic cobordism $W = \overrightarrow{V^-V^+}$. Suppose that the forms $\alpha^\pm$ associated to the ends satisfies the condition* $\deg_{\mathrm{disk}}(\gamma) \neq -1, 0, 1$ *for any* contractible in W *periodic orbit $\gamma \in \mathcal{P}^\pm$. Then the homomorphisms $\Phi_0 = \Phi_{W,J_0}$ and $\Phi_1 = \Phi_{W,J_1}$ are chain homotopic, i.e there exists a homomorphism $\Delta : \mathfrak{F}^+ \to \mathfrak{F}^-$ such that $\Phi_1 - \Phi_0 = \partial\Delta + \Delta\partial$.*

PROPOSITION 1.9.5. *Given two cobordisms W_1 and W_2, and a compatible almost complex structure J on the composition $W_1 \circledcirc W_2$, the homomorphism $\Phi_{W_1 \circledcirc W_2}$ is chain-homotopic to $\Phi_{W_1} \circ \Phi_{W_2}$.*

Together with an obvious remark that for the cylindrical cobordism W_0 the homomorphism Φ_{W_0} is the identity, Propositions 1.9.1–1.9.4 imply that if a contact structure ξ on V admits a *nice* contact form, i.e. a form without contractible periodic orbits of index $-1, 0$ and 1, then the *contact homology group*

$$\oplus HC_k(V, \xi) = \oplus H_k(\mathfrak{F}, \partial)$$

is well defined and independent of the choice of a nice contact form and a compatible almost complex structure (however, if $H_2(V) \neq 0$ and/or $H_1(V) \neq 0$ it depends on a choice of spanning surfaces F_γ and the framing of the bundle ξ over basic loops). Similarly to what was explained in the sketch of the proof Proposition 1.9.1 the "niceness" assumptions guarantees that the top codimension strata on the boundary of the involved moduli spaces consist of height 2 cylindrical curves, and thus the proofs of Propositions 1.9.3–1.9.5 may precisely follow the standard scheme of the Floer theory (see [F], [S]).

REMARK 1.9.6. Similarly to what we explained in Remark 1.9.2 the coefficient $n_{\gamma,\gamma'}$ in the definition (11) of Φ equals 0 if at least one of the orbits γ, γ' is "bad". Hence, in the presence of "bad" orbits the homomorphism

Φ could never be equal to the identity, even for the cylindrical cobordism. This explains why the exclusion of "bad" periodic orbits is *necessary.*

Besides the degree (or Conley-Zehnder) grading, the contact homology group is graded by elements of $H_1(V)$, because the boundary operator preserves the homology class of a periodic orbit. We will denote the part of $HC_*(V,\xi)$ which correspond to a class $a \in H_1(V)$ by $HC_*(V,\xi|a)$. One can similarly construct a contact homology group $HC_*^{\text{contr}}(V,\xi)$, generated only by contractible periodic orbits, which is another invariant of the contact manifold (V,ξ).

Contact structures which admit nice contact forms do exist, as it is illustrated by examples in section 1.9.3 below. However, the condition of existence of a nice form is too restrictive. The general case leads to an algebraic formalism developed in sections 2.2–2.5 below.

1.9.3 Examples.

1. *Contact homology of the standard contact sphere* S^{2n-1}. Take the 1-form $\alpha = \frac{1}{2}\sum(x_i dy_i - y_i dx_i)$, which is a primitive of the standard symplectic structure in $\mathbb{R}^{2n}$. Its restriction to a generic ellipsoid

$$S = \left\{ \sum \frac{x_i^2 + y_i^2}{a_i^2} = 1 \right\}$$

is a nice contact form for the standard contact structure ξ on the sphere $S = S^{2n-1}$. The form $\alpha|_S$ has precisely one periodic orbit for each Conley-Zehnder index $n + 2i - 1$ for $i = 1, \dots,$. Hence the contact homology group $HC_*(S,\xi)$ has one generator in each dimension $2i, i \geq n-1$. See also the discussion in section 2.9.2 below.

2. *Contact homology of Brieskorn spheres.* Ilya Ustilovsky computed ([U]) the contact homology of certain Brieskorn spheres.

Let us consider the Brieskorn manifold

$$\Sigma\Big(p, \underbrace{2,\dots,2}_{n}\Big) = \left\{ z_0^p + \sum_1^n z_j^2 = 0 \right\} \cap \left\{ \sum_0^n |z_j|^2 = 1 \right\} \subset \mathbb{C}^{n+1}.$$

$\Sigma(p, \underbrace{2,\dots,2}_{n})$ carries a canonical contact structure as a strictly pseudo-convex hypersurface in a complex manifold.

Suppose that $n = 2m+1$ is odd, and $p \equiv 1 \mod 8$. Under this assumption $\Sigma(p, \underbrace{2,\dots,2}_{n})$ is diffeomorphic to S^{2n-1} (see [Br]). However, the following theorem of Ustilovsky implies that the contact structures on Brieskorn spheres $\Sigma(p,\underbrace{2,\dots,2}_{n})$ and $\Sigma(p',\underbrace{2,\dots,2}_{n})$ are not isomorphic, unless $p = p'$. This result should be confronted with a computation of Morita ([Mo]), which implies that the formal homotopy class (see section 1.1 above) of the contact structure on $\Sigma(p,\underbrace{2,\dots,2}_{n})$ is standard, provided $p \equiv 1 \mod 2(2m!)$. Hence, Ustilovsky's theorem provides infinitely many non-isomorphic contact structures on S^{4m+1} in the standard formal homotopy class.

Theorem 1.9.7. (I. Ustilovsky, [U]) *The contact homology*

$$HC_*\Big(\Sigma(p,\underbrace{2,\dots,2}_{n})\Big)$$

is defined, and the dimension

$$c_k = \dim HC_k\Big(\Sigma(p,\underbrace{2,\dots,2}_{n})\Big)$$

is given by the formula

$$c_k = \begin{cases} 0, & k \text{ is odd or } k < 2n-4, \\ 2, & k = 2\left[\frac{2N}{p}\right] + 2(N+1)(n-2), \text{ for } N \geq 1,\ 2N+1 \notin p\mathbb{Z}, \\ 1, & \text{in all other cases.} \end{cases}$$

3. *Contact homology of boundaries of subcritical Stein manifolds.* A co-oriented contact manifold (V,ξ) is called Stein-fillable if it can be realized as a strictly pseudoconvex boundary of a complex manifold W, whose interior is Stein, and if the co-oriented contact structure ξ coincides with the canonical contact structure of a strictly pseudo-convex hypersurface. We say that (V,ξ) admits a subcritical Stein filling if the corresponding Stein manifold Int$W$ admits an exhausting plurisubharmonic function without critical points of dimension $\dim_{\mathbb{C}}(W)$. If $\dim V > 3$ then one can equivalently require that W deformation retracts to a CW-complex of dimension $< \dim_{\mathbb{C}} W$ (see [E1]).

Mei-Lin Yau studied in her PhD thesis [Y] contact homology of contact manifolds admitting a subcritical Stein filling. Here is her result.

Theorem 1.9.8. (Mei-Lin Yau, [Y]) *Let (V,ξ) be a contact manifold of dimension $2n-1$ which admits a subcritical Stein filling W. Suppose that $c_1(V) = 0$ and $H_1(V) = 0$. Let $c_1,\dots,c_k$ be generators of $H_*(W)$. Then the contact homology $HC_*(V)$ is defined and generated by elements $q_{i,j}$ of degree $\deg q_{i,j} = 2(n+i-2) - \dim c_j$, where $j = 1,\dots,k$, and $i \geq 1$.*

4. *Contact homology of T^3 and its coverings.* Set $\alpha_n = \cos 2\pi nz\,dx + \sin 2\pi nz\,dy$. This contact form descend to the 3-torus $T^3 = \mathbb{R}^3/\mathbb{Z}^3$ and defines there a contact structure ξ_n. The structure ξ_1 is just the canonical contact structure on T^3 as the space of co-oriented contact elements of T^2. The form α_n for $n > 1$ is equal to the pull-back $\pi_n^*(\alpha_1)$, where $\pi_n : T^3 \to T^3$ is the covering $(x, y, z) \mapsto (x, y, nz)$. Notice that all structures ξ_n are homotopic as plane field to the foliation $dz = 0$.

Theorem 1.9.9. *The contact homology group $HC_*(T^3, \xi_n|w)$, where w is the homology class $(p, q, 0) \in H_1(T^3)$, is isomorphic to $\mathbb{Z}^{2n}$.*

In particular we get as a corollary a theorem of E. Giroux:

COROLLARY 1.9.10. (E. Giroux, [Gi]) *The contact structures ξ_n, $n = 1, \ldots,$ are pairwise non-isomorphic.*

The contact manifold (T^3, ξ_1) is foliated by pre-Lagrangian tori $L_{p,q}$, indexed by simple homology classes $(p, q) \in H_1(T^2)$. Each torus $L_{p,q}$ is foliated by the S^1-family of lifts of closed geodesics which represent the class (p, q). Thus for any given $(p, q) \in H_1(T^2)$ (even when (p, q) have common divisors) the set of closed orbits in $\mathcal{P}_{\alpha_1}$ which represent the class $(p, q, 0) \in H_1(T^3)$ is a circle $S_{p,q}$, and for any $n \geq 1$ the set of closed orbits in $\mathcal{P}_{\alpha_n}$ which represent the class $(p, q, 0) \in H_1(T^3)$ consists of n copies $S^1_{p,q}, \ldots, S^n_{p,q}$ of such circles. The forms α_n have no contractible periodic orbits, but of course, they are degenerate. To compute the contact homology groups, one can either work directly with these degenerate forms, as it is explained in section 2.9.2 below, and show that $HC_*(T^3, \xi_n|w) = H_*(\bigcup_1^n S^i_{p,q}) = \mathbb{Z}^{2n}$, or first perturb the form α_1, and respectively all its covering forms $\alpha_n = \pi_n^*(\alpha_1)$, in order to substitute each circle $S^i_{p,q}$ by two non-degenerate periodic orbits, and then show that the orbits from each of these pairs are connected by precisely two holomorphic cylinders, which cancel each other in the formula for the boundary operator ∂.

1.9.4 Relative contact homology and contact non-squeezing theorems. Let us observe that the complex $(\mathfrak{F}, \partial)$ is filtrated by the values of the action functional S, $\mathfrak{F} = \bigcup_{a \in \mathbb{R}} \mathfrak{F}^a$, where the complex $\mathfrak{F}^a$ is generated by variables q_γ with $S(\gamma) \leq a$. The differential ∂ respects this filtration, and hence descends to $\mathfrak{F}^b/\mathfrak{F}^a, a < b$. Hence, one can define the homology $H_*^{(a,b]}(\mathfrak{F}, \partial) = H_*(\mathfrak{F}^b/\mathfrak{F}^a, \partial)$ in the window $(a, b] \subset \mathbb{R}$. Of course, $H_*^{(a,b]}$ depends on a choice of a particular nice form α. If $\alpha > \beta$ then we have a map $\Phi_* : H_*^{(a,b]}(\mathfrak{F}, \partial; \alpha) \to H_*^{(a,b]}(\mathfrak{F}, \partial; \beta)$. We write H^a instead of $H^{(-\infty,a]}$.

Consider now a contact manifold (V, ξ) which is either closed, or satisfies the following *pseudo-convexity* condition at infinity. A contact manifold $(V, \xi = \{\alpha = 0\})$ with a fixed contact form a is called pseudo-convex at infinity if there exists a compatible almost complex structure J on the symplectization $V \times \mathbb{R}$ for which V can be exhausted by compact domains V_i with smooth pseudo-convex boundary. A sufficient condition for pseudo-convexity is existence of an exhaustion $V = \bigcup V_i$, such that for each $i = 1, \dots,$ trajectories of the Reeb field $R_\alpha|_{V_i}$ do not have interior tangency points with ∂V_i. For instance, for the standard contact form $\alpha = dz - \sum y_i dx_i$ on $\mathbb{R}^{2n+1}$ the latter condition is satisfied for an exhaustion of $\mathbb{R}^{2n+1}$ by round balls, and hence the standard contact form on $\mathbb{R}^{2n+1}$ is pseudo-convex at infinity.

Our goal is to define a relative contact homology group $HC_*(V, U, \xi)$ for a relatively compact open subset $U \subset V$, so that this group would be invariant under a contact isotopy of U in V.

Let us fix a contact form α on V which satisfies the above pseudo-convexity condition. Let us denote by $\mathcal{F}_{U,\alpha}$ the set of C^∞-functions $f : V \to [0, \infty)$ which are ≤ 1 on U, and for which the contact form $f\alpha$ is nice and pseudo-convex at infinity.[8] Take a strictly increasing sequence of functions $f_i \in \mathcal{F}_{U,\alpha}$, such that

a) $\max_K f_i \to_{i\to\infty} \infty$ for each compact set $K \subset (V \setminus \overline{U})$;
b) $f_i|_U \to_{i\to\infty} 1$ uniformly on compact sets.

PROPOSITION 1.9.11. *The limit*

$$HC_*(V, U, \xi) = \lim_{a\to+\infty} \lim_{\leftarrow} HC_*^a(V, f_i\alpha) \tag{12}$$

is independent of α, and thus it is an invariant of the contact pair (V, U). A contact isotopy $f_t : V \to V$ induces a family of isomorphisms $(f_t)_* : H_*(V, U) \to H_*(V, f_t(U))$. An inclusion $i : U_1 \mapsto U_2$ induces a homomorphism

$$i_* : HC_*(V, U_1, \xi) \to HC_*(V, U_2, \xi).$$

One of the most celebrated results in Symplectic topology is Gromov's non-squeezing theorem which states that one cannot symplectically embed a $2n$-ball of radius 1 into $D_r^2 \times \mathbb{R}^{2n-2}$ for $r < 1$. Here D_r^2 denotes a 2-disk of radius r and $D_r^2 \times \mathbb{R}^{2n-2}$ is endowed with the product of standard symplectic structures. Because of the conformal character of contact geometry one

---

[8] Of course, the set $\mathcal{F}_{U,\alpha}$ may be empty, because the niceness condition is very restrictive. In this case one needs to employ a more general construction from section 2.2.3.

cannot expect as strong non-squeezing results for contact manifolds: one can embed any domain in the standard $\mathbb{R}^{2n-1}$ in an arbitrary small ball. However, it turns out that it is not always possible to realize a contact squeezing via a contact isotopy inside a manifold with a non-trivial first Betti number.

As an example, consider the 1-jet bundle

$$V = J^1(\mathbb{R}^n, S^1) = T^*(\mathbb{R}^n) \times S^1$$

of S^1-valued functions with its standard contact structure ξ, given by the contact form $\alpha = dz - \sum y_i dx_i$, $(x,y) \in \mathbb{R}^{2n} = T^*(\mathbb{R}^n)$, $z \in S^1 = \mathbb{R}/\mathbb{Z}$. The contact form α satisfies the condition of pseudo-convexity at infinity and it is nice: the Reeb field equals $\frac{\partial}{\partial z}$, and thus it has no contractible periodic orbits. Let us consider the class $\mathcal{N}$ of domains $\Omega \subset V$ which are images of the split domains $U \times S^1 \subset V$ under a contact isotopy of V, where U is connected. For any $\Omega \in \mathcal{N}$ the relative contact homology group $HC_*(V,\Omega)$ is well defined because for any function $f : \mathbb{R}^{2n} \to \mathbb{R}$ the form $f(x,y)\alpha$ is nice.

Let us denote by $\mathcal{E}_r(\Omega)$, $\Omega \in \mathcal{N}$, the space of contact embeddings $D_r \times S^1 \to \Omega \times S^1$, contact isotopic in V to the inclusion

$$D_r \times S^1 \hookrightarrow \mathbb{R}^{2n} \times S^1 = V\,.$$

Notice, that for any two embeddings $f, g \in \mathcal{E}_r(\Omega)$ there exists a positive $\rho \le r$, such that the restrictions $f|_{D_\rho \times S^1}$ and $g|_{D_\rho \times S^1}$ are isotopic via a *contact* isotopy.

Given a contact embedding $f \in \mathcal{E}_r(\Omega)$ there is defined a homomorphism

$$f_* : HC_*(V, D_r \times S^1, \xi) \to HC_*(V, \Omega, \xi)\,.$$

Let us choose a symplectic trivialization of the contact bundle ξ induced by the projection $V \to \mathbb{R}^{2n}$. We will assume that indices of periodic orbits, and hence the grading of contact homology groups, are associated with this trivialization.

For each homology class $k \in \mathbb{Z} = H_1(D_r \times S^1)$ let us consider the maximal $l = l(f,k)$ such that

$$\operatorname{Ker}\big(f_*|_{HC_l(V, D_r \times S^1, \xi|k)}\big) \ne 0,$$

and define an invariant $w_{\text{cont}}(V,\Omega)$, called the relative *contact width* by the formula

$$w_{\text{cont}}(V,\Omega) = \sup_{k,r>0, f \in \mathcal{E}_r(U)} \frac{2k}{l(f,k)}\,. \tag{13}$$

S.-S. Kim has computed $w_{\text{cont}}(V,\Omega)$ for certain domains Ω. In particular, she proved

Proposition 1.9.12.

$$w_{\text{cont}}(V, D_r^{2n} \times S^1) = \pi r^2;$$
$$w_{\text{cont}}(V, D_r^2 \times D_R^{2n-2} \times S^1) = \pi r^2,$$

if $R \geq r$.

The contact width is clearly a monotone invariant, i.e.

$$w_{\text{cont}}(V, U_1 \times S^1) \leq w_{\text{cont}}(V, U_2 \times S^1)$$

if $U_1 \subset U_2$. Hence Proposition 1.9.12 implies

Corollary 1.9.13. *Suppose that* $r < \min(r', R)$. *Then there is no contact isotopy* $f_t : D_{r'}^{2n} \times S^1 \to V$ *such that* f_0 *is the inclusion, and*

$$f_1(D_{r'}^{2n} \times S^1) \subset D_r^2 \times D_R^{2n-2} \times S^1.$$

Problem 1.9.14. *Suppose there exists a contact isotopy* $f_t : V = \mathbb{R}^{2n} \times S^1 \to V$ *with* $f_0 = \text{Id}$ *and* $f_1(U_1 \times S^1) \subset U_2 \times S^1$. *Does there exist a Hamiltonian isotopy* $g_t : \mathbb{R}^{2n} \to \mathbb{R}^{2n}$ *such that* $g_0 = \text{Id}$ *and* $g_1(U_1) \subset U_2$?

Notice that the converse is obviously true.

2 Algebraic Formalism

2.1 Informal introduction. The Floer-theoretic formalism described in section 1.9 is applicable only to a very limited class of contact manifolds. As it follows from Theorem 1.6.2 the boundary of the moduli space of holomorphic cylinders in the symplectization may consist of stable curves, different from broken cylinders; for instance, it may contain height 2 stable curves which consist of a pair of pants on the upper level, and a copy of $\mathbb{C}$ plus a trivial cylinder at the bottom one. Hence the minimal class of holomorphic curves in symplectizations which has the property that the stable curves of height > 1 on the boundary of the corresponding moduli space are made of curves from the same class, must contain all rational curves with one positive and an arbitrary number of negative punctures. The counting of curves with one positive and arbitrary number of negative punctures can still be interpreted as a differential, but this time defined not on the *vector space* generated by periodic trajectories but on the graded *algebra* which they generate. Thus this leads to a straightforward generalization of the Floer type formalism considered in section 1.9 when instead of the additive Floer complex $\mathfrak{F}$ generated by the variables q_γ, one considers a graded commutative algebra $\mathfrak{A}$ generated by these variables, and when instead of the formula (10) the differential ∂q_γ is defined as a polynomial of a higher degree, rather than a linear expression as in the Floer homology case. Namely, we define

$$\partial q_\gamma = \sum \frac{n_{\Gamma,I,d}}{k!\prod_1^k i_j!\kappa_{\gamma_j}^{i_j}} q_{\gamma_1}^{i_1} \dots q_{\gamma_k}^{i_k} z^d, \tag{14}$$

where the sum is taken over all ordered[9] sets of different periodic orbits $\Gamma = \{\gamma_1, \dots, \gamma_k\}$, multi-indices $d = (d_1, \dots, d_N)$ and $I = (i_1, \dots, i_k), i_j \geq 0$, and where the coefficient $n_{\Gamma,I,d}$ counts the algebraic number of elements of the moduli space

$$\mathcal{M}_0^d\Big(\gamma; \underbrace{\gamma_1, \dots, \gamma_1}_{i_1}, \dots, \underbrace{\gamma_k, \dots, \gamma_k}_{i_k}\Big)\Big/\mathbb{R},$$

if this space is 0-dimensional, and equals 0 otherwise. The differential ∂ extends to the algebra $\mathfrak{F}$ according to the graded Leibnitz rule. Roughly speaking, ∂q_γ is a polynomial, whose monomomials $q_{\gamma_1} \dots q_{\gamma_l}$ are in 1-1 correspondence with rigid, up to translation, rational holomorphic curves with one positive cylindrical end over γ and l negative cylindrical ends over trajectories $\gamma_1, \dots, \gamma_l$.

It turns out that the quasi-isomorphism class of the differential algebra $(\mathfrak{A}, \partial)$ is independent of all extra choices (see section 2.2.3 below). In particular, the *contact homology algebra* $H_*(\mathfrak{A}, \partial)$ is an invariant of the contact manifold (M, ξ).

Having included into the picture the moduli spaces of rational curves with one positive and several negative punctures, one may wonder, what is the role of rational curves with an arbitrary number of positive and negative punctures. One can try to interpret the counting of rational holomorphic curves with fixed number of positive and an arbitrary number of negative punctures as a sequence of bracket type operations on the algebra $\mathfrak{A}$. These operations satisfy an infinite system of identities, which remind the formalism of homotopy Lie algebras.

However, there is a more adequate algebraic formalism for this picture. Let us associate with each periodic orbit γ two graded variables p_γ and q_γ of the same parity (but of different integer grading, as we will see in section 2.2.2 below), and consider an algebra $\mathfrak{P}$ of formal power series $\sum f_\Gamma(q)p^\Gamma$, where $f_\Gamma(q)$ are polynomials of $q = \{q_\gamma\}$ with coefficients in (a completion of) the group algebra of $H_2(V)$. It is useful to think about the algebra $\mathfrak{P}$ as the graded Poisson algebra of functions on the infinite-dimensional

[9]The coefficient $\frac{1}{k!}$ is the price we pay for taking *ordered* sets of periodic orbits.

symplectic super-space $\mathbf{V}$ with the even symplectic form

$$\sum_{\gamma\in\mathcal{P}} \kappa_\gamma{}^{-1} dp_\gamma \wedge dq_\gamma,$$

or rather on its formal analog along the 0-section $\{p = \{p_\gamma\} = 0\}$. With each 0-dimensional moduli space $\mathcal{M}_0^d(\Gamma^-,\Gamma^+)/\mathbb{R}$,

$$\Gamma^\pm = \Big\{ \underbrace{\gamma_1^\pm, \dots \gamma_1^\pm}_{i_1^\pm}, \dots, \underbrace{\gamma_{s^\pm}^\pm \dots \gamma_{s^\pm}^\pm}_{i_{s^\pm}^\pm} \Big\},$$

we associate a monomial

$$\frac{n_{\Gamma^-,\Gamma^+,d}}{s^-!s^+!\kappa_{\gamma^-}^{i^-}\kappa_{\gamma^+}^{i^+}(i_1^-)!\dots(i_{s^-}^-)!(i_1^+)!\dots(i_{s^+}^+)!} q_{\gamma^-}^{I^-} p_{\gamma^+}^{I^+} z^d,$$

where $n_{\Gamma^-,\Gamma^+,d}$ is the algebraic number of elements of the moduli space $\mathcal{M}_0^d(\Gamma^-,\Gamma^+)/\mathbb{R}$, $q_{\gamma^-}^{I^-} = (q_{\gamma_1^-})^{i_1^-}\dots(q_{\gamma_{s^-}^-})^{i_{s^-}^-}$, $p_{\gamma^+}^{I^+} = (p_{\gamma_1^+})^{i_1^+}\cdots(p_{\gamma_{s^+}^+})^{i_{s^+}^+}$ and $\kappa_{\gamma^\pm}^{i^\pm} = (\kappa_{\gamma_1^\pm})^{i_1^\pm}\dots(\kappa_{\gamma_{s^\pm}^\pm})^{i_{s^\pm}^\pm}$.

The sum of all these monomials over all 1-dimensional moduli spaces $\mathcal{M}_0^d(\Gamma^-,\Gamma^+)$ for all ordered sets Γ^-,Γ^+ of periodic orbits is an *odd* element $\mathbf{h} \in \mathfrak{P}$. All the operations on the algebra $\mathfrak{A}$ which we mentioned above appear as the expansion terms of $\mathbf{h}$ with respect to p-variables. It turns out that the infinite system of identities for operations on $\mathfrak{A}$, which we mentioned above, and which is defined by counting holomorphic curves with a certain fixed number of positive punctures, can be encoded into a single equation $\{\mathbf{h},\mathbf{h}\} = 0$. Then the differentiation with respect to the Hamiltonian vector field, defined by the Hamiltonian $\mathbf{h}$:

$$d^{\mathbf{h}}(g) = \{\mathbf{h}, g\}, \ g \in \mathfrak{P},$$

defines a differential $d = d^{\mathbf{h}} : \mathfrak{P} \to \mathfrak{P}$, which satisfies the equation $d^2 = 0$. Thus one can define the homology $H_*(\mathfrak{P}, d^{\mathbf{h}})$ which inherits the structure of a graded Poisson algebra.

The identities, like $d^2 = 0$ and $\partial^2 = 0$, encode in algebraic terms information about the structure of the top-dimensional strata on the boundary of compactified moduli spaces of holomorphic curves, as it is described in Proposition 1.7.2 above. For instance, the codimension 1 strata on the boundary of the moduli space $\mathcal{M}_0^d(\Gamma^-,\Gamma^+)/\mathbb{R}$ consists of height two stable rational curves (f_1, f_2). Each floor of this curve may be disconnected, but precisely one of its components differs from the straight cylinder. Each connected component of f_1 can be glued with a component of f_1 only along one of their ends. One can easily see that the combinatorics of such gluing precisely corresponds to the Poisson bracket formalism and that the algebraic sum of the monomials associated to all stable curves of height two equals $\{\mathbf{h},\mathbf{h}\}$. On the other hand, the algebraic number of such height 2 curves equals 0 because they form the boundary of the a compactified 1-dimensional moduli space of holomorphic curves. Hence we get the identity $\{\mathbf{h},\mathbf{h}\}=0$. The identity is not tautological due to the fact that $\mathbf{h}$ is odd. In view of the super-Jacobi identity it is equivalent to the identity $(d^{\mathbf{h}})^2=0$.

One can go further and include into the picture moduli spaces of punctured holomorphic curves of higher genus. Introducing a new variable, denoted $\hbar$, to keep track of the genus, one can associate with each 0-dimensional moduli space $\mathcal{M}_g^d(\Gamma^-,\Gamma^+)/\mathbb{R}$ a monomial

$$\frac{n_{\Gamma^-,\Gamma^+,d,g}}{s^-!s^+!\kappa_{\gamma^-}^{i^-}\kappa_{\gamma^+}^{i^+}(i_1^-)!\dots(i_{s^-}^-)!(i_1^+)!\dots(i_{s^+}^+)!}q_{\gamma^-}^{I^-}p_{\gamma^+}^{I^+}\hbar^{g-1}z^d,$$

and form a generating function $\mathbf{H}=\hbar^{-1}\sum_{g=0}^{\infty}\mathbf{H}_g\hbar^g$ counting all rigid holomorphic curves of arbitrary genus, whose term $\mathbf{H}_0$ coincides with $\mathbf{h}$. Again, the codimension 1 strata of the boundary of the moduli spaces $\mathcal{M}_g(\Gamma^-,\Gamma^+)$ consists of height 2 stable curves, but unlike the case of rational curves, two connected components on different levels can be glued together along an arbitrary number of ends. The combinatorics of such gluing can be described by the formalism of algebra of higher order differential operators. Fig. 4 illustrates how the composition of differential operators can be interpreted via gluing of Riemann surfaces with punctures. A letter p_i in the picture represents a differential operator $\hbar\frac{\partial}{\partial q_i}$, and a surface of genus g with upper punctures $p_{i_1},\dots,p_{i_k}$ and lower punctures $q_{j_1},\dots,q_{j_l}$ represents a differential operator

$$\hbar^{g-1}q_{j_1}\dots q_{j_l}p_{i_1}\dots p_{i_k}=\hbar^{g-1}q_{j_1}\dots q_{j_l}\left(\hbar\frac{\partial}{\partial q_{i_1}}\right)\dots\left(\hbar\frac{\partial}{\partial q_{i_k}}\right).$$

Thus we are led to consider $\mathbf{H}$ as an element of the Weyl super-algebra $\mathfrak{W}$. This algebra should be viewed as a quantization of the Poisson algebra $\mathfrak{P}$, so that the description of the boundary of the moduli spaces is given by the equation $[\mathbf{H},\mathbf{H}]=0$, where $[\ ,\ ]$ denotes the commutator in $\mathfrak{W}$. As in the rational case, this identity is equivalent to the identity $D_{\mathbf{H}}^2=0$ for the differential $D^{\mathbf{H}}(f)=[H,f]$. Hence we can define the homology algebra $H_*(\mathfrak{W},D^{\mathbf{H}})$, which also turns out to be an invariant of the contact manifold (V,ξ). Similarly to the standard Gromov-Witten theory for closed symplectic manifolds one can develop an even more general formalism by encoding in $\mathbf{H}$ information about higher-dimensional moduli spaces of holomorphic curves. This leads to a deformation of the differential algebra $(\mathfrak{W}, D^{\mathbf{H}})$ along the space of closed forms on V. The corresponding family of homology algebras is then parameterized by $H^*(V)$.

After going that far it is natural to make the above algebraic structure for contact manifolds a part of a formalism in the spirit of topological field theory, which we call *Symplectic Field Theory*, and which also includes the theory of Gromov-Witten invariants of closed manifolds. To do that one considers moduli spaces of holomorphic curves with cylindrical ends in directed symplectic cobordisms $W = \overrightarrow{V^-V^+}$. The generating function counting rational holomorphic curves in W can be naturally written as a function $\mathbf{f}(q^-, p^+)$ of p^+-variables associated with the positive end V^+, and q^--variables associated with the negative end V^- of the cobordism W. It turns out that the Lagrangian submanifold in $(-\mathbf{V}^-) \times \mathbf{V}^+$ generated by the function $\mathbf{f}$ defines a *Lagrangian correspondence* $\mathbf{L}_W \subset (-\mathbf{V}^-) \times (\mathbf{V}^+)$ which transforms the Hamiltonian functions $\mathbf{h}^+$ and $\mathbf{h}^-$ to each other, i.e.

$$\left(\mathbf{h}^-(p^-, q^-) - \mathbf{h}^+(p^+, q^+)\right)|_{\mathbf{L}_W} = 0, \tag{15}$$

where

$$L_W = \begin{cases} q^+_{\gamma^+} &= \kappa_{\gamma^+} \dfrac{\partial \mathbf{f}}{\partial p^+_{\gamma^+}}(q^-, p^+); \\ p^-_{\gamma^-} &= \kappa_{\gamma^-} \dfrac{\partial \mathbf{f}}{\partial q^-_{\gamma^-}}(q^-, p^+). \end{cases}$$

We recall that $\kappa_{\gamma^\pm}$ denotes the multiplicity of the orbit $\gamma^\pm$.

The composition of symplectic cobordisms produces the composition of Lagrangian correspondences, so that if one consider a "Heegard splitting" of a closed symplectic manifold W along a contact hypersurface V, then the computation of Gromov-Witten invariants of W can be viewed as a Lagrangian intersection problem in the symplectic super-space $\mathbf{V}$ associated to the contact manifold V.

After what was said it should not come as a surprise that in the quantized picture Lagrangian correspondences are being replaced by Fourier integral operators, and the composition of correspondences by the convolution of the corresponding operators.

We describe below the SFT-formalism with more details. We treat contact manifolds in section 2.2 and symplectic cobordisms in section 2.3. Section 2.4 is devoted to the SFT-version of the chain homotopy statement in Floer homology theory. In section 2.5 we introduce the composition formula for the SFT-invariants of symplectic cobordisms. In section 2.6 we discuss how the introduced algebraic structures of contact manifolds depend on extra choices. Section 2.7 is devoted to a differential equation for the potential $\mathbf{F}$ of a directed symplectic cobordism with a *non-empty*

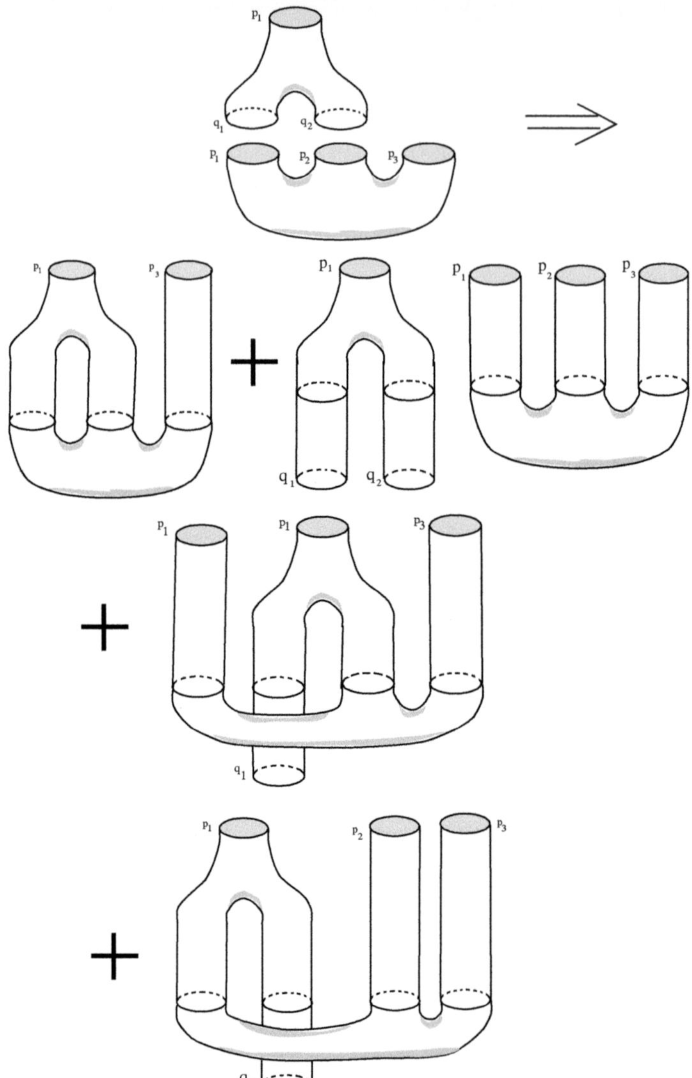

Figure 4: There are four different way to glue the lower and upper surfaces on the picture along their matching ends, i.e. the ends denoted by p's and q's with the same index. These 4 ways correspond to 4 terms in the composition formula for differential operators: $(\hbar^{-1}p_1p_2p_3) \circ (\hbar^{-1}q_1q_2p_1) = p_1p_3 + \hbar^{-2}q_1q_2p_1^2p_2p_3 + \hbar^{-1}q_1p_1^2p_3 + \hbar^{-1}q_2p_1p_2p_3$. We are ignoring here the sign issues and assuming all the boundary components to be simple orbits.

boundary. Together with the gluing formula from section 2.5 this equation provides an effective tool for computing Gromov-Witten invariants. The remainder of the paper has even more sketchy character than the rest of the paper. Section 2.8 is devoted to invariants of Legendrian submanifolds via SFT. Section 2.9 is devoted to various examples and possible generalizations of SFT. In particular, in section 2.9.2 we discuss how one can adapt the theory to include an important for applications, though non-generic, case of contact forms with continuous families of periodic orbits. In section 2.9.3 we describe a new recursive procedure for computing rational Gromov-Witten invariants of $\mathbb{C}P^n$. Finally, in section 2.9.4 we just touch the wealth of other invariants which exist in Symplectic Field Theory.

2.2 Contact manifolds.

2.2.1 Evaluation maps. Let (V,ξ) be a contact manifold with a fixed contact form α, $(W = V \times \mathbb{R}, d(e^t\alpha))$ the symplectization of (V,ξ), and J a compatible almost complex structure. As in section 1.5 we denote by f_V and $f_{\mathbb{R}}$ the V- and $\mathbb{R}$-components of a J-holomorphic curve f in W, and by $\mathcal{M}_{g,r,s^-,s^+}(W,J)$ the disjoint union

$$\bigcup \mathcal{M}^A_{g,r}(\Gamma^-,\Gamma^+),$$

which is taken over all $A \in H_2(V)$, and all sets $\Gamma^-,\Gamma^+ \subset \mathcal{P}_\alpha$ of cardinalities $s^\pm$.

Let us view the set $\mathcal{P} = \mathcal{P}_\alpha$ of periodic orbits of the Reeb fields R_α as a discrete topological space. It naturally splits into the disjoint union

$$\coprod_{k=1}^{\infty} \mathcal{P}_k,$$

of identical subspaces, where $\mathcal{P}_k$ is the space of periodic orbits of multiplicity k.

Consider now three sets of evaluation maps:

$$ev_i^0 : \mathcal{M}_{g,r,s^-,s^+}/\mathbb{R} \to V,\ i = 1,\dots,r,$$
$$ev_j^+ : \mathcal{M}_{g,r,s^-,s^+}/\mathbb{R} \to \mathcal{P},\ j = 1,\dots,s^+,$$

and

$$ev_k^- : \mathcal{M}_{g,r,s^-,s^+}/\mathbb{R} \to \mathcal{P},\ k = 1,\dots,s^-,$$

where ev_i^0 is the evaluation map $f_V(y_i)$ at the i-th marked point y_i, while $ev_j^\pm$ are the evaluation maps at asymptotic markers $\mu_j^{\mathbf{x}^\pm}$. More precise, let

$$\overline{f} = (f, j, \mathbf{x}^-, \mathbf{x}^+, \mathbf{y}, \mu^{\mathbf{x}^-}, \mu^{\mathbf{x}^+}) \in \mathcal{M}_{g,r,s^-,s^+},$$

and f be asymptotically cylindrical over a periodic orbit $\gamma_j^\pm \in \mathcal{P}$ at $\pm\infty$ at the puncture $x_j^\pm$. Then $ev_j^\pm(\overline{f})$ is a point of $\mathcal{P}$ representing the orbit $\gamma_j^\pm$ (comp. Section 2.9.2 below).

All the above evaluation maps can be combined into a map

$$ev : \mathcal{M}_{g,r,s^-,s^+}/\mathbb{R} \to V^r \times (\mathcal{P}^-)^{s^-} \times (\mathcal{P}^+)^{s^+},$$

which extends to the compactified moduli space $\overline{\mathcal{M}_{g,r,s^-,s^+}/\mathbb{R}}$.

2.2.2 Correlators. Now we are ready to define correlators. Given r differential forms $\theta_1, \dots, \theta_r$ on V and $s^\pm$ (0-dimensional) cohomology classes $\alpha_1^\pm, \dots, \alpha_{s^\pm}^\pm \in H^*(\mathcal{P})$ we define the degree -1, or contact *correlator*

$$^{-1}\langle \theta_1, \dots, \theta_r; \alpha_1^-, \dots, \alpha_{s^-}^-; \alpha_1^+, \dots, \alpha_{s^+}^+ \rangle_g^A =$$
$$\int\limits_{\overline{\mathcal{M}^A_{g,r,s^-,s^+}/\mathbb{R}}} ev^*(\theta_1 \otimes \dots \otimes \theta_r \otimes \alpha_1^- \otimes \dots \otimes \alpha_{s^-}^- \otimes \alpha_1^+ \otimes \dots \otimes \alpha_{s^+}^+). \quad (16)$$

Usually we will assume that the forms $\theta_1, \dots, \theta_r$ are closed, but even in this case the above correlator depends on the actual forms, and not just their cohomology classes, because the domain of integration may have a boundary. As we will see below the superscript -1 in $^{-1}\langle \dots \rangle$ corresponds to the grading of the generating function for these correlators. It also refers to the enumerative meaning of the correlators: they count components of 1-dimensional moduli spaces of holomorphic curves. We will consider below also correlators $^0\langle \dots \rangle$ and $^1\langle \dots \rangle$, counting 0-dimensional and -1-dimensional (i.e. appearing in 1-dimensional families) moduli spaces of holomorphic curves.

If we are given K linearly independent differential forms $\Theta_1, \dots, \Theta_K$, then it is convenient to introduce a "general form" $t = \sum_1^K t_i\Theta_i$ from the space $L = L(\Theta_1, \dots, \Theta_K)$ generated by the chosen forms, and view t_i as graded variables with $\deg(t_i) = \deg(\Theta_i) - 2$. In particular, all terms in the sum $\sum_1^K t_i\Theta_i$ have even degrees.

Let us consider two copies $\mathcal{P}^+$ and $\mathcal{P}^-$ of the 0-dimensional space $\mathcal{P}$, one associated with the positive end of W, the other with the negative one. Cohomology classes in $\mathcal{P}_+$ we will denote by p, and in $\mathcal{P}_-$ by q, and will write

$$p = \sum_{\gamma \in \mathcal{P}} \frac{1}{\kappa_\gamma} p_\gamma [\gamma], \quad q = \sum_{\gamma \in \mathcal{P}} \frac{1}{\kappa_\gamma} q_\gamma [\gamma],$$

where κ_γ is the multiplicity of γ, and the cohomology classes $[\gamma]$ form the canonical basis of $H^*(\mathcal{P})$, dual to points in $\mathcal{P}$. Of course, speaking about cohomology classes of a discrete space may sound somewhat ridiculous.

However, this point of view is useful, especially in preparation for a more general case when some periodic orbits may be degenerate and thus the spaces $\mathcal{P}^{\pm}$ need not to be anymore discrete, see section 2.9.2 below. We will also fix a basis $A_1,\dots,A_N$ of $H_2(V)$. The coordinate vector $d=(d_1,\dots,d_N)$ of a class A is called the degree. Here d_j are integers, while we consider t,p,q as graded variables, where the degrees of the variables p,q are defined by the formulas

$$\begin{aligned}\deg(p_\gamma)&=-\mathrm{CZ}(\gamma)+(n-3),\\ \deg(q_\gamma)&=+\mathrm{CZ}(\gamma)+(n-3).\end{aligned}$$

The choice of grading, somewhat strange at the first glance, is explained by Proposition 2.2.1 below.

The correlators

$$\Big\langle \underbrace{t,\dots,t}_{r};\underbrace{q,\dots,q}_{s^-};\underbrace{p,\dots,p}_{s^+}\Big\rangle_g^d$$

with different r,d,g determine all the correlators involving forms from the space L.

2.2.3 Three differential algebras. Similar to the theory of Gromov-Witten invariants of closed symplectic manifolds we will organize all correlators into a generating function, called *Hamiltonian*,

$$\mathbf{H}=\frac{1}{\hbar}\sum_{g=0}^{\infty}\mathbf{H}_g\hbar^g,$$

where,

$$\mathbf{H}_g=\sum_d\sum_{r,s^\pm=0}^{\infty}\frac{1}{r!s^-!s^+!}^{-1}\Big\langle \underbrace{t,\dots,t}_{r};\underbrace{q,\dots,q}_{s^-};\underbrace{p,\dots,p}_{s^+}\Big\rangle_g^d z^d, \tag{17}$$

and $t=\sum_1^K t_i\Theta_i$. We will assume throughout the paper, that all forms $\Theta_1,\dots,\Theta_K$ are closed (see, however, Remarks 2.2.3 and 2.3.4, and section 2.7 below). The variables $\hbar$ and $z=(z_1,\dots,z_N)$ are also considered as graded with $\deg\hbar=2(n-3)$ and $\deg(z_i)=-2c_1(A_i)$, where c_1 is the first Chern class of the almost complex structure J.

Proposition 2.2.1. a) *For each $g=0,\dots$ the series $\mathbf{H}_g$ can be viewed as formal power series in variables p_γ with coefficients which are polynomials of variables q_γ and formal power series of t_i*[10] *with coefficients in the group*

[10] In fact, $\mathbf{H}_g$ depends polynomially on all variables t_i of degree $\neq 0$. The degree 0 variables, i.e. the variables associated with 2-forms, enter into the constant part of

algebra $\mathbb{C}[H_2(V)]$ (which we identify with the algebra of Laurent polynomials of z with complex coefficients);
b) *All terms of* $\mathbf{H}$ *have degree* -1;
c) $\mathbf{H}\big|_{p=0} = \mathbf{H}_{\mathrm{const}}$, *where*

$$\mathbf{H}_{\mathrm{const}} = \hbar^{-1} \sum_{g,r=0}^{\infty} {}^{-1}\Big\langle \underbrace{t,\dots,t}_{r} \Big\rangle_g^0 \frac{\hbar^g}{r!}$$

accounts for the contribution of constant holomorphic curves. In particular, $\mathbf{H}\big|_{p=0}$ *is independent of* q *and* z.

The polynomial dependence of $\mathbf{H}_g$ on variables q_γ and z in a geometric language just means that the union $\overline{\mathcal{M}_g^d(\Gamma^+)}$, $\Gamma^+ = \gamma_1 \dots \gamma_{s^+}$, of the compactified moduli spaces of holomorphic curves of a fixed genus of any degree with prescribed positive ends $\gamma_1^+, \dots, \gamma_{s^+}$ is compact, and in particular that there are only finitely many possibilities for the degrees and the negative ends of these curves. This follows from the fact that for each curve $C \in \mathcal{M}_g(\Gamma^-, \Gamma^+)$ we have

$$0 \leq \int_C d\alpha = \sum_{\gamma_i \in \Gamma^+} \int_{\gamma_i} \alpha - \sum_{\gamma_j \in \Gamma^-} \int_{\gamma_j} \alpha \leq \sum_{\gamma_i \in \Gamma^+} \int_{\gamma_i} \alpha, \tag{18}$$

the fact that there exists a constant $m > 0$ such that $\int_\gamma \alpha > m$ for any periodic orbit $\gamma \in \mathcal{P}_\alpha$ and Theorem 1.6.2 above. Proposition 2.2.1b) follows from the formula (1) for the dimension of the moduli spaces of holomorphic curves, our degree convention and the fact that a correlator ${}^{-1}\langle \theta_1, \dots, \theta_r; \gamma_1^-, \dots, \gamma_{s^-}^-; \gamma_1^+, \dots, \gamma_{s^+}^+ \rangle_g^A$ may be different from 0 only if the total dimension of the forms $\theta_1, \dots \theta_r$ equals the dimension of the moduli space $\mathcal{M}_{g,r}^A(\gamma_1^-, \dots, \gamma_{s^-}^-; \gamma_1^+, \dots, \gamma_{s^+}^+)/\mathbb{R}$. Proposition 2.2.1c) just means that every non-constant holomorphic curves should have at least one positive end, which follows from inequality (18), or alternatively the maximum principle for holomorphic curves.

Let us consider the *Weyl super-algebra* $\mathfrak{W} = \{\sum_{\Gamma,g} f_{\Gamma,g}(q,t) p^\Gamma \hbar^g\}$, where

$$\Gamma = (\gamma_1, \dots, \gamma_a), \gamma_i \in \mathcal{P},\ a = 1, \dots,\ p^\Gamma = p_{\gamma_1} \dots p_{\gamma_a},$$

and $f_{\Gamma,g}(q,t)$ are polynomials of variables q_γ and formal power series of t_i.[11] Proposition 2.2.1a) states that $\mathbf{H} \in \hbar^{-1}\mathfrak{W}$.

$\mathbf{H}_g$ (i.e. the part describing constant holomorphic curves) polynomially, while the non-constant part of $\mathbf{H}_g$ depends polynomially on e^{t_i}. This fact is similar to the standard Gromov-Witten theory and will not discussed in the present paper.

[11] See the previous footnote.

The product operation $F \circ G$ in $\mathfrak{W}$ is associative and satisfies the following commutation relations: all variables are super-commute (i.e. commute or anti-commute according to their grading), except p_γ and q_γ which correspond to the same periodic orbit γ. For these pairs of variables we have the following commutation relation:

$$[p_\gamma, q_\gamma] = p_\gamma \circ q_\gamma - (-1)^{\deg p_\gamma \deg q_\gamma} q_\gamma \circ p_\gamma = \kappa_\gamma \hbar \tag{19}$$

where κ_γ is the multiplicity of the orbit γ. The algebra $\mathfrak{W}$ can be represented as an algebra of formal differential operators with respect to q-variables acting *on the left* on the space of polynomials $f(q, z, \hbar)$, by setting

$$p_\gamma = \kappa_\gamma \hbar \overrightarrow{\frac{\partial}{\partial q_\gamma}}.$$

Alternatively by setting

$$q_\gamma = \kappa_\gamma \hbar \overleftarrow{\frac{\partial}{\partial p_\gamma}}$$

we can represent $\mathfrak{W}$ as an algebra of polynomial differential operators acting *on the right* on the algebra $\{\sum_{\Gamma,g} f_{\Gamma,g}(q,z)\hbar^g p^\Gamma\}$ of formal power series of $\hbar$ and the p-variables.

Notice that the commutator $[F, G]$ of two homogeneous elements $F, G \in \mathfrak{W}$ equals $F \circ G - (-1)^{\deg F \deg G} G \circ F$, and hence if $F \in \mathfrak{W}$ is an odd element (i.e. all its summands are odd) then $[F, F] = 2F \circ F$, and $[F, F] = 0$ if F is even. For any two elements $F, G \in \mathfrak{W}$ the commutator $[F, G]$ belongs to the ideal $\hbar\mathfrak{W}$. According to Proposition 2.2.1 the Hamiltonian $\mathbf{H}$ can be viewed as an element of $\frac{1}{\hbar}\mathfrak{W}$, and hence the above remark shows that for $F \in \mathfrak{W}$ we have $[\mathbf{H}, F] \in \mathfrak{W}$.

Theorem 2.2.2. *The Hamiltonian* $\mathbf{H}$ *satisfies the identity*

$$\mathbf{H} \circ \mathbf{H} = 0. \tag{20}$$

This theorem (as well as Theorems 2.3.3, 2.4.2 and 2.5.3 below) follows from the description of the boundary of the corresponding moduli spaces of holomorphic curves. As it was stated in Proposition 1.7.2 this boundary is tiled by codimension one strata represented by stable curves of height 2, so that the (virtual) fundamental cycles of the boundary of the compactified moduli spaces can be symbolically written as $\partial[\mathcal{M}] = \kappa \sum [\mathcal{M}_-] \times [\mathcal{M}_+]$, where $[\mathcal{M}_\pm]$ are chains represented by the corresponding moduli spaces and where the coefficient κ depends on multiplicities of orbits along which the two levels of the corresponding stable curve are glued. Together with the Stokes formula $\int_{[\mathcal{M}]} d\omega = \int_{\partial[\mathcal{M}]} \omega$, and the fact that the integrand is a closed form, we obtain identity (20).

REMARK 2.2.3. The same argument shows that when the forms Θ_i generating the space L are not necessarily closed we get the following equation

$$d\mathbf{H} + \tfrac{1}{2}[\mathbf{H},\mathbf{H}] = 0, \tag{21}$$

which generalizes (20) and can be interpreted as the zero-curvature equation for the connection $d+[\mathbf{H},\cdot\,]$. We denote here by d the de Rham differential, i.e.

$$\begin{aligned} d\mathbf{H} &= d\Bigg(\sum_{d,g}\sum_{r,s^\pm=0}^{\infty}\frac{1}{r!s^-!s^+!} \\ &\quad \Big\langle \underbrace{\sum_1^K t_i\Theta_i,\ldots,\sum_1^K t_i\Theta_i}_{r};\underbrace{q,\ldots,q}_{s^-};\underbrace{p,\ldots,p}_{s^+}\Big\rangle_g^d z^d\hbar^{g-1}\Bigg) \\ &= \sum_{d,g}\sum_{\substack{s^\pm=0,\\ r=1}}^{\infty}\frac{1}{(r-1)!s^-!s^+!} \\ &\quad \Big\langle \sum_1^K t_i d\Theta_i,\underbrace{\sum_1^K t_i\Theta_i,\ldots,\sum_1^K t_i\Theta_i}_{r-1};\underbrace{q,\ldots,q}_{s^-};\underbrace{p,\ldots,p}_{s^+}\Big\rangle_g^d z^d\hbar^{g-1}\,. \end{aligned} \tag{22}$$

The identity $\mathbf{H}\circ\mathbf{H}=0$ is equivalent to $[\mathbf{H},\mathbf{H}]=0$, because $\mathbf{H}$ is an odd element. Let us define the differential $D=D^{\mathbf{H}}:\mathfrak{W}\to\mathfrak{W}$ by the formula

$$D^{\mathbf{H}}(f) = [\mathbf{H},f] \quad \text{for} \quad f\in\mathfrak{W}. \tag{23}$$

Then Theorem 2.2.2 translates into the identity $D^2=0$. The differential $D^{\mathbf{H}}$ satisfies the Leibnitz rule, and thus $(\mathfrak{W},D)$ is a differential Weyl (super-)algebra. In particular, one can define the homology algebra $H_*(\mathfrak{W},D)$, which inherits its multiplication operation from the Weyl algebra $\mathfrak{W}$. The differential $D^{\mathbf{H}}$ extends in an obvious way to the modules $\hbar^{-k}\mathfrak{W}$, $k=1,\ldots$.

EXAMPLE 2.2.4. Let $V=S^1$. We have in this case

$$\mathbf{H} = \hbar^{-1}\left(\frac{t_1t_0^2}{2} + t_1\sum p_kq_k - \frac{t_1\hbar}{24}\right), \tag{24}$$

where $t=t_0 1+t_1 d\phi$ is a general harmonic differential form on S^1, so that $\deg t_1=-1$, $\deg t_0=-2$, $\deg\hbar=-4$ and $\deg p_k,\deg q_k=-2$, which corresponds to the convention that the Maslov index of any path in the 1-point group $Sp(0)$ equals 0. The term $t_1t_0^2/2=\int_{S^1}t^{\wedge 3}/6$ is the contribution of the moduli space S^1 of constant maps $\mathbb{C}P^1\to\mathbb{R}\times S^1$ with 3 marked points. The term $-\frac{t_1\hbar}{24}$ is accounted for the contribution of constant curves of genus 1 (see [Wi2]), and the term $t_1 p_k q_k$ represents the contribution $t_1 = \int_{S^1} t$ of trivial curves of multiplicity k with one marking. All other curves do not contribute to $\mathbf{H}$ for dimensional reasons and because $t_1^2 = 0$.

Notice that if we organize all variables p_k, q_k into formal Fourier series (comp. [Ge])

$$u(x) = \sum_{k=1}^{\infty} (p_k e^{ixk} + q_k e^{-ixk}), \tag{25}$$

then the term accounting for the contribution of rational curves in the formula (24) takes the form

$$\frac{t_1}{4\pi} \int_0^{2\pi} \left(t_0 + u(x)\right)^2 dx, \tag{26}$$

see further discussion of this u-formalism in section 2.9.2 below.

We will associate now with $(\mathfrak{W}, D)$ two other differential algebras, $(\mathfrak{P}, d)$ and $(\mathfrak{A}, \partial)$, which can be viewed as *semi-classical* and *classical* approximations of the Weyl differential algebra.

Let us denote by

- $\mathfrak{P}$, a graded Poisson algebra of formal power series in variables p_γ with coefficients which are polynomials of q_γ, z_j, z_j^{-1}, and formal power series of t_i,[12] and by
- $\mathfrak{A}$, a graded commutative algebra generated by variables $q = \{q_\gamma\}_{\gamma \in \mathcal{P}}$ with coefficients in the algebra $\mathbb{C}\,[H_2(V)][[t]]$.

The Poisson bracket on $\mathfrak{P}$ is defined by the formula

$$\{h, g\} = \sum_{\gamma} \kappa_\gamma \left(\frac{\partial h}{\partial p_\gamma} \frac{\partial g}{\partial q_\gamma} - (-1)^{\deg h \deg g} \frac{\partial g}{\partial p_\gamma} \frac{\partial h}{\partial q_\gamma} \right), \tag{27}$$

assuming that h and g are *monomials.* When computing partial derivatives, like $\frac{\partial h}{\partial q_\gamma}$, one should remember that we are working in the super-commutative environment, and in particular the operator $\frac{\partial}{\partial q_\gamma}$ has the same parity as the variable q_γ.

REMARK 2.2.5. Notice, that if similarly to Example 2.2.4 above we organize the variables $p_k = p_{\gamma_k}, q_k = q_{\gamma_k}$ corresponding to multiples of each simple periodic orbit $\gamma = \gamma_1$ into a Fourier series

$$u_\gamma = \sum_{k=1}^{\infty} (p_k e^{ixk} + q_k e^{-ixk}),$$

[12] See the first footnote in section 2.2.3.

then the value of the Poisson tensor (27) on covectors $\delta u, \delta v$ takes the form

$$\frac{1}{2\pi i}\int_0^{2\pi}(\delta u)'\delta v dx\,. \tag{28}$$

In order to define differentials on the algebras $\mathfrak{A}$ and $\mathfrak{P}$ let us first make the following observation.

Lemma 2.2.6. *We have*

$$[\mathbf{H},\mathbf{H}] = \frac{1}{\hbar}\{\mathbf{H}_0,\mathbf{H}_0\} + \dots ,$$

and for any $f = \frac{1}{\hbar}\sum f_g \hbar^g \in \frac{1}{\hbar}\mathfrak{W}$ *we have*

$$D^{\mathbf{H}}(f) = \frac{1}{\hbar}\{\mathbf{H}_0, f_0\} + \dots .$$

In particular, $\mathbf{H}_0$ *satisfies the equation* $\{\mathbf{H}_0,\mathbf{H}_0\} = 0$.

To cope with a growing number of indices we will rename $\mathbf{H}_0$ into $\mathbf{h}$. Lemma 2.2.6 allows us to define the differential $d = d^{\mathbf{h}} : \mathfrak{P} \to \mathfrak{P}$ by the formula

$$dg = \{\mathbf{h}, g\} \quad \text{for} \quad g \in \mathfrak{P}. \tag{29}$$

Theorem 2.2.2 then implies

Proposition 2.2.7. *We have* $d^2=0$ *and* $d\{f,g\}=\{df,g\}+(-1)^{\deg f}\{f,dg\}$ *for any homogeneous element* $f \in \mathfrak{P}$. *In other words,* $(\mathfrak{P}, d)$ *is a graded differential Poisson algebra with unit.*

Proposition 2.2.7 enables us to define the homology $H_*(\mathfrak{P}, d)$ which inherits from $\mathfrak{P}$ the structure of a graded Poisson algebra with unit.

Let us recall that according to 2.2.1 $\mathbf{h}|_{p=0} = \mathbf{h}_{const}$, where $\mathbf{h}_{const}$ accounts for constant rational holomorphic curves, and thus it is independent of q-variables. In fact,

$$\mathbf{h}_{\text{const}}(t) = \frac{1}{6}\sum_{i,j,k=1}^{K} c^{ijk} t_i t_j t_k,$$

where $c^{ijk} = \int_V \Theta_i \wedge \Theta_j \wedge \Theta_k$ are the structural coefficients of the cup-product. Hence,

$$\mathbf{h} = \mathbf{h}_{\text{const}} + \sum h_\gamma(q,t,z)p_\gamma + \dots ,$$

where ... denote terms at least quadratic in p_γ. Thus we have

$$\{\mathbf{h},\mathbf{h}\} = 2\sum_{\gamma,\gamma'\in\mathcal{P}} \kappa_{\gamma'} h_{\gamma'}(q,t)\frac{\partial h_\gamma}{\partial q_{\gamma'}}(q,t)p_\gamma + o(p) = 0\,. \tag{30}$$

Therefore,

$$\sum_{\gamma'\in\mathcal{P}} \kappa_{\gamma'} h_{\gamma'}(q,t)\frac{\partial h_\gamma}{\partial q_{\gamma'}}(q,t) = 0 \tag{31}$$

for all t and all $\gamma \in \mathcal{P}$.

Let us define a differential $\partial : \mathfrak{A} \to \mathfrak{A}$ by the formula

$$\partial f = \{\mathbf{h}, f\}|_{\{p=0\}} = \sum_{\gamma\in\mathcal{P}} \kappa_\gamma h_\gamma \frac{\partial f}{\partial q_\gamma}. \tag{32}$$

Then the equation (31) is equivalent to

PROPOSITION 2.2.8. *$\partial^2 = 0$, and hence $(\mathfrak{A}, \partial)$ is a graded commutative differential algebra with unit.*

The homology group $H_*(\mathfrak{A}, \partial)$ inherits the structure of a graded commutative algebra with unit.

As it was already mentioned in section 2.1, it is convenient to view the Poisson algebra $\mathfrak{P}$ as an algebra of functions on an infinite-dimensional symplectic super-space $\mathbf{V}$ with the even symplectic form $\boldsymbol{\omega} = \sum k_\gamma^{-1} dp_\gamma \wedge dq_\gamma$. Then the differential $d^{\mathbf{h}}$ is the Hamiltonian vector field on $\mathbf{V}$ generated by the Hamiltonian function $\mathbf{h}$. One should remember, however, that the p-variables are formal, so all that we have is the infinite jet of the symplectic space $\mathbf{V}$ along the 0-section. The equation $\mathbf{h}|_{p=0} = \mathbf{h}_{const}$ translates into the fact that the vector field $d^{\mathbf{h}}$ is tangent to the 0-section, and the differential ∂ is just the restriction of this vector field to the 0-section. The higher order terms in the expansion of $\mathbf{h}$ with respect to p-variables define a sequence of (co-)homological operations on the algebra $\mathfrak{A}$.

Notice also that the differentials D, d and ∂ do not involve any differentiation with respect to t. Hence the differential algebras $(\mathfrak{W}, D^{\mathbf{H}})$, $(\mathfrak{P}, d^{\mathbf{h}})$ and $(\mathfrak{A}, \partial)$ can be viewed as *families* of differential algebras, parameterized by $t \in H^*(V)$, and in particular, one can compute the homology at any fixed $t \in H^*(V)$. We will sometimes denote the corresponding algebras and their homology groups with the subscript t, i.e. $(\mathfrak{W}, D)_t$, $H_*(\mathfrak{P}, d)_t$, etc., and call them *specialization* at the point $t \in H^*(V)$. We will also use the notation

$$H_*^{\mathrm{SFT}}(V,\xi|\, J,\alpha), \quad H_*^{\mathrm{RSFT}}(V,\xi|\, J,\alpha), \quad \text{and} \quad H_*^{\mathrm{cont}}(V,\xi|\, J,\alpha)$$

instead of $H_*(\mathfrak{W}, \partial)$, $H_*(\mathfrak{P}, \partial)$ and $H_*(\mathfrak{A}, \partial)$, and will usually omit the extra data J, α from the notation: as we will see in section 2.5 below all these homology algebras are independent of J, α and other extra choices, like closed forms representing cohomology classes of V, a coherent orientation of the moduli spaces, etc. The abbreviation RSFT stands here for *Rational Symplectic Field Theory.*

REMARK 2.2.9. It is important to observe that the algebras $\mathfrak{W}, \mathfrak{P}$ and $\mathfrak{A}$ have an additional grading by elements of $H_1(V)$ (comp. Section 1.9 above). This grading is also inherited by the corresponding homology algebras. However, this grading carries a non-trivial information only when we consider homology of algebras, specialized at points $t = \sum_1^K t_i \Theta_i$ with $t_i = 0$ for at least some of the coordinates t_i corresponding 1-dimensional forms. Otherwise all cycles in these algebras are graded by the 0-class from $H_1(V)$.

2.3 Symplectic cobordisms.

2.3.1 Evaluation maps and correlators. Let us now repeat the constructions of the previous section for a general directed symplectic cobordism $W = \overrightarrow{V^- V^+}$ between two contact manifolds V^- and V^+ with fixed contact forms α^- and α^+. As in section 2.2.1 we consider the sets $\mathcal{P}^\pm$ of periodic orbits of the Reeb fields $R^\pm = R_{\alpha^\pm}$ as discrete topological spaces.

We denote by $\mathcal{M}^A_{g,r,s^-,s^+}(W,J)$ the disjoint union

$$\bigcup \mathcal{M}^A_{g,r}(\Gamma^-, \Gamma^+; W, J),$$

where the union is taken over all sets (Γ^-, Γ^+) of cardinality (s^-, s^+), and consider three sets of evaluation maps:

$$ev_i^0 : \mathcal{M}^A_{g,r,s^-,s^+}(W,J) \to W, \; i = 1, \dots, r,$$
$$ev_j^\pm : \mathcal{M}^A_{g,r,s^-,s^+}(W,J) \to \mathcal{P}^\pm, \; j = 1, \dots, s^\pm,$$

where ev_i^0 is the evaluation map $f(y_i)$ of the map f at the i-th marked point y_i, while $ev_j^\pm$ are the evaluation maps at asymptotic markers $\mu_j^{\mathbf{x}^\pm}$, i.e. $ev_j^\pm(f)$ is a point of $\mathcal{P}^\pm$ representing the orbit $\gamma_j^\pm$, which contains the image of the corresponding marker. The evaluation maps ev_i^0 and $ev_j^\pm$ can be combined into a map

$$ev : \mathcal{M}^A_{g,r,s^-,s^+}(W,J) \to W^r \times (\mathcal{P}^+)^{s^+} \times (\mathcal{P}^-)^{s^-}.$$

Now we are ready to define degree 0, or symplectic correlators. We will have to consider on W differential forms, which do not necessarily have compact support, but have, however, *cylindrical ends.* We say that a differential form θ on W is said to have cylindrical ends if it satisfies the following condition:

There exists $C > 0$ such that

$$\theta|_{V^- \times (-\infty, -C)} = (\pi^-)^*(\theta^-) \quad \text{and} \quad \theta|_{V^+ \times (C,\infty)} = (\pi^+)^*(\theta^+),$$

where $\theta^{\pm}$ are forms on $V^{\pm}$, and $\pi^{\pm}$ are the projections of the corresponding ends to $V^{\pm}$. We will denote the forms $\theta^{\pm}$ also by $\mathrm{restr}^{\pm}(\theta)$, or $\theta|_{V^{\pm}}$. In what follows we assume that all considered differential forms on W have cylindrical ends.

Given r differential forms $\theta_1, \dots, \theta_r$ on W and $s^{\pm}$ cohomology classes

$$\alpha_1^{\pm}, \dots, \alpha_{s^{\pm}}^{\pm} \in H^*(\mathcal{P}^{\pm}) = H_0^*(\mathcal{P}^{\pm})$$

we define the degree 0 correlator

$$^0\langle \theta_1, \dots, \theta_r; \alpha_1^-, \dots, \alpha_{s^-}^-; \alpha_1^+, \dots, \alpha_{s^+}^+ \rangle_g^A = \int\limits_{\overline{\mathcal{M}^A_{g,r,s^-,s^+}}} ev^*(\theta_1 \otimes \dots \otimes \theta_r \otimes \alpha_1^- \otimes \dots \otimes \alpha_{s^-}^- \otimes \alpha_1^+ \otimes \dots \otimes \alpha_{s^+}^+). \quad (33)$$

Similar to section 2.2.1 above, we denote the cohomology classes in $H^*(\mathcal{P}^+) = H^0(\mathcal{P}^+)$ (resp. in $H^*(\mathcal{P}^-) = H^0(\mathcal{P}^-)$) by p^+ (resp. q^-), and write

$$p^+ = \sum_{\gamma \in \mathcal{P}^+} k_g^{-1} p_\gamma^+ [\gamma] \quad \left(\text{resp.} \quad q^- = \sum_{\gamma \in \mathcal{P}^-} k_g^{-1} q_\gamma^- [\gamma]\right).$$

We will also fix a basis $A_1, \dots, A_N$ of $H_2(W)$ and denote by $d = (d_1, \dots, d_N)$ the degree of A in this basis.

Let us call a system of linearly independent closed forms $\theta_1, \dots, \theta_m$ on W with cylindrical ends *basic*, if

$$\begin{aligned} &\text{a) the image} \quad \mathrm{restr}^{\pm}\big(L(\theta_1, \dots, \theta_m)\big) \\ &\quad \text{generates} \quad \mathrm{Im}\big(H^*(W) \to H^*(V^{\pm})\big); \\ &\text{b) the homomorphism} \quad \mathrm{Ker}\big((\mathrm{restr}^+ \oplus \mathrm{restr}^-)|_L\big) \to H^*_{\mathrm{comp}}(W) \\ &\quad \text{is bijective.} \end{aligned} \quad (34)$$

Here we denote by $L(\theta_1, \dots, \theta_m)$ the subspace generated by the forms $\theta_1, \dots, \theta_m$, and by $H^*_{\mathrm{comp}}(W)$ the cohomology with compact support. Equivalently, one can say that a basic system of forms consists of a basis of $H^*(W)$ together with a basis of $\mathrm{Ker}(H^*_{\mathrm{comp}}(W) \to H^*(W))$.

A general point $t \in L(\theta_1, \dots, \theta_m)$ we will write in the form $t = \sum_1^m t_i \theta_i$. The grading of the variables t, p^+, q^- is defined as in section 2.2.2:

$$\begin{aligned} \deg(t_i) &= \deg(\theta_i) - 2; \\ \deg(p_\gamma^+) &= -\mathrm{CZ}(\gamma^+) + (n-3), \\ \deg(q_\gamma^-) &= \mathrm{CZ}(\gamma^-) + (n-3) \end{aligned} \quad (35)$$

2.3.2 Potentials of symplectic cobordisms. Let us now organize the correlators into a generating function, called the *potential* of the symplectic cobordism $(W = \overrightarrow{V^-V^+}, J, \alpha^\pm)$

$$\mathbf{F} = \mathbf{F}_{W,J,\alpha^\pm} = \frac{1}{\hbar}\sum_{g=0}^{\infty} \mathbf{F}_g \hbar^g,$$

where (36)

$$\mathbf{F}_g = \sum_d \sum_{r,s^\pm=0} \frac{1}{r!s^+!s^-!} {}^0\Big\langle \underbrace{t,\dots,t}_{r}; \underbrace{q^-,\dots,q^-}_{s^-}; \underbrace{p^+,\dots,p^+}_{s^+} \Big\rangle_g^d z^d .$$

When W is a closed symplectic manifold, then the potential F is just the Gromov-Witten invariant of the symplectic manifold W. However, if W is not closed, then F itself is not an invariant. It depends on particular forms θ_i, rather than their cohomology classes, on J, on a coherent orientation, and several other choices. We will see, however, that the *homotopy class* of F, which we define in section 2.4 below, is independent of most of these choices.

In order to make sense out of the expression for $\mathbf{F}$ let us consider a graded commutative algebra $\mathfrak{D} = \mathfrak{D}(W, \alpha^\pm)$ which consists of power series of the form

$$\sum_{\Gamma,d,g} \varphi_{\Gamma,d,g}(q^-, t) z^d (p^+)^\Gamma \hbar^g, \tag{37}$$

where $\varphi_{\Gamma,d,g}$ are polynomials of q_γ, formal power series of variables t_i,[13] and where Γ and d satisfies the following Novikov type inequality:

$$[\omega](d) = \sum d_i \int_{A_i} \omega > -|\Gamma| = -\sum_{i=1}^{k} |\gamma_i|, \tag{38}$$

where $\Gamma = \{\gamma_1, \dots, \gamma_k\}$, and $|\gamma_i| = \int_{\gamma_i} \alpha^+$ is the period of the periodic orbit $\gamma_i \in \mathcal{P}^+$, or equivalently its action. Recall that $(p^+)^\Gamma = p^+_{\gamma_1} \dots p^+_{\gamma_k}$.

PROPOSITION 2.3.1. *We have*

$$\mathbf{F}_{W,J,\alpha^\pm} \in \frac{1}{\hbar}\mathfrak{D}(W, \alpha^\pm).$$

Let us also consider a bigger algebra $\mathfrak{D}\mathfrak{D}$ which consists of series

$$\sum_{\Gamma,d} \varphi_{\Gamma,d}(q^-, t, \hbar) z^d (p^+)^\Gamma, \tag{39}$$

[13] See the first footnote in section 2.2.3.

where $\varphi_{\Gamma,d}$ are polynomials of q^-_γ, formal power series of t_i and formal *Laurent series* of $\hbar$, while Γ and d still satisfy condition (38). For instance, for any element $f \in \hbar^{-1}\mathfrak{D}$ we have $e^f \in \mathfrak{D}\mathfrak{D}$.

The algebra $\mathfrak{D}\mathfrak{D} = \mathfrak{D}\mathfrak{D}(W, J, \alpha^\pm)$ has a structure of a *left D-module* over the Weyl algebra $\mathfrak{W}^- = \mathfrak{W}(V^-, J, \alpha^-)$, and of a *right D-module* over the Weyl algebra $\mathfrak{W}^+ = \mathfrak{W}(V^+, J, \alpha^+)$. Indeed, we first associate with an element

$$\Delta^- = \sum_{\Gamma=\{\gamma_1,\dots,\gamma_m\},\Gamma',d,g} \delta^-_{\Gamma',\Gamma,d,g}(t)(q^-)^{\Gamma'}(p^-)^{\Gamma} z^d \hbar^g \in \mathfrak{W}^-$$

a differential operator

$$\sum_{\Gamma=\{\gamma_1,\dots,\gamma_m\},\Gamma',d,g} \delta^-_{\Gamma,\Gamma',d,g}(t)(q^-)^{\Gamma'} \hbar^{m+g} \prod_{i=1}^m \kappa_{\gamma_i} \overrightarrow{\frac{\partial}{\partial q^-_{\gamma_i}}} z^d, \tag{40}$$

then change the coefficient ring via the inclusion homomorphism $H_2(V^-) \to H_2(W)$, and finally lift functions $\delta^-_{\Gamma,\Gamma',d,g}(t)$ to the space of forms with cylindrical ends on W via the restriction map $t \mapsto t|_{V^-}$. We will denote the resulting operator by $\overrightarrow{\Delta^-}$. Similarly we associate with $\Delta^+ \in \mathfrak{W}^+$ an operator $\overleftarrow{\Delta^+}$ by first quantizing $q^+_\gamma \Rightarrow \hbar\kappa_\gamma \overleftarrow{\frac{\partial}{\partial p^+_\gamma}}$ and then making an appropriate change of the coefficient ring. It is straightforward to verify that for $f \in \mathfrak{D}$ we have $\overrightarrow{\Delta^-} f,\ f\overleftarrow{\Delta^+} \in \mathfrak{D}$, and for $F \in \mathfrak{D}\mathfrak{D}$ we have $\overrightarrow{\Delta^-} F,\ F\overleftarrow{\Delta^+} \in \mathfrak{D}\mathfrak{D}$.

Let us denote the Hamiltonians (see section 2.2.3 above) in $\mathfrak{W}^\pm$ by $\mathbf{H}^\pm$ and define a map $D_W = D_W : \mathfrak{D}\mathfrak{D} \to \mathfrak{D}\mathfrak{D}$ by the formula

$$D_W(G) = \overrightarrow{\mathbf{H}^-} G - (-1)^{\deg G} G \overleftarrow{\mathbf{H}^+}, \quad G \in \mathfrak{D}\mathfrak{D}, \tag{41}$$

where we assume G dimensionally homogeneous. Clearly, Theorem 2.2.2 implies that $D_W^2 = 0$. However, the differential algebra $(\mathfrak{D}\mathfrak{D}, D_W)$ is too big and instead of considering its homology we will define a differential on the algebra $\mathfrak{D}$, or which is equivalent but more convenient, on the module $\hbar^{-1}\mathfrak{D}$.

For an *even* element $F \in \hbar^{-1}\mathfrak{D}$ let us define a map $D^F = T_F D_W : \hbar^{-1}\mathfrak{D} \to \hbar^{-1}\mathfrak{D}$ by the formula

$$\begin{aligned} D^F(g) &= e^{-F}[D_W, g](e^F) \\ &= e^{-F}\left(D_W(ge^F) - (-1)^{\deg g} g D_W(e^F)\right), \quad g \in \hbar^{-1}\mathfrak{D}. \end{aligned} \tag{42}$$

The map $T_F D_W$ is the linearization of the map $\widetilde{D}_W : \hbar^{-1}\mathfrak{D} \to \hbar^{-1}\mathfrak{D}$, defined by the formula

$$\widetilde{D}_W(F) = e^{-F} D_W(e^F), \quad F \in \hbar^{-1}\mathfrak{D}\,.$$

at the point $F \in \hbar^{-1}\mathfrak{D}$. Notice that if $D_W(e^F) = 0$ then $D^F(g) = e^{-F} D_W(ge^F)$. Let us first state a purely algebraic

PROPOSITION 2.3.2. *Suppose that for $F \in \hbar^{-1}\mathfrak{D}$ we have $D_W(e^F) = 0$. Then*

1. $(D^F)^2 = 0$;
2. *The homology algebra $H_*(\mathfrak{D}, D^F)$ inherits the structure of a left module over the homology algebra $H_*(\mathfrak{W}^-, D^-)$, and the structure of a right module over the homology algebra $H_*(\mathfrak{W}^+, D^+)$;*
3. *The homomorphisms $F^\pm : \mathfrak{W}^\pm \to \mathfrak{D}$, defined by the formulas*
$$\begin{aligned} &f \mapsto e^{-F}\overrightarrow{f}\, e^F,\ f \in \mathfrak{W}^-, \quad \text{and} \\ &f \mapsto e^F \overleftarrow{f}\, e^{-F},\ f \in \mathfrak{W}^+, \end{aligned} \tag{43}$$
commute with the boundary maps of chain complexes, i.e.
$$F^\pm \circ D^\pm = D^F \circ F^\pm,$$
and thus induce homomorphisms of homology
$$F^\pm_* : H_*(\mathfrak{W}^\pm, D^\pm) \to H_*(\mathfrak{D}, D^F),$$
as modules over $H_*(\mathfrak{W}^\pm, D^\pm)$.

Theorem 2.3.3. *The potential $\mathbf{F} \in \hbar^{-1}\mathfrak{D}$ defined above by the formula* (36) *satisfies the equation*
$$D_W e^{\mathbf{F}} = 0, \tag{44}$$
and hence all conclusions of Proposition 2.3.2 *hold for* $\mathbf{F}$.

The appearance of $e^{\mathbf{F}}$ in equation (44) has the following reason. Similar to Theorem 2.2.2 above, equation (44) follows from the description of codimension 1 strata on the boundary of the moduli space of holomorphic curves in the cobordism W, see Proposition 1.7.2 above. Notice that $e^{\mathbf{F}}$ is the generating function counting possibly disconnected holomorphic curves in W. Thus the identity
$$\overrightarrow{\mathbf{H}^-} e^{\mathbf{F}} - e^{\mathbf{F}} \overleftarrow{\mathbf{H}^+} = 0$$
asserts, in agreement with Proposition 1.7.2, that the codimension 1 strata on the boundary of the moduli space $\widetilde{\mathcal{M}}(W)$ of not necessarily connected curves in W correspond to stable curves (f_1, f_2) of height 2, where one of the curves f_1, f_2 belongs to $\widetilde{\mathcal{M}}(W)$, while the second one is contained in the symplectization of $V^\pm$ and has precisely one component different from the straight cylinder over a periodic orbit from $\mathcal{P}^\pm$.

REMARK 2.3.4. (Comp. Remark 2.2.3 above) The potential $\mathbf{F}$, extended to the space of all, not necessarily closed differential forms satisfies the equation
$$d(e^{\mathbf{F}}) = D_W e^{\mathbf{F}}, \tag{45}$$

where d is the de Rham differential. This equation generalizes equation (44).

Following the scheme of section 2.2.3 above we will associate now with the cobordism W two other left-right modules, one over the Poisson algebras $\mathfrak{P}^\pm$, and another over the graded differential algebras $\mathfrak{A}^\pm$.

Consider the graded commutative algebra $\mathfrak{L} = \mathfrak{L}(W, \alpha^\pm)$ of power series of the form

$$\sum_{\Gamma,d} \varphi_{\Gamma,d}(q^-, t) z^d (p^+)^\Gamma, \tag{46}$$

where $\varphi_{\Gamma,d}$ are polynomials of q_γ^- and formal power series of t_i, while Γ and d satisfies the above inequality (38). Let us also consider the larger graded commutative algebra

$$\widehat{\mathfrak{L}} = \left\{ \sum_{\Gamma^+,\Gamma^-,d} \varphi_{\Gamma^+,\Gamma^-,d}(q^-, q^+, t) z^d (p^+)^\Gamma (p^-)^{\Gamma'} \right\}, \tag{47}$$

where the Novikov condition (38) is satisfied for both pairs (d, Γ) and (d, Γ'). The algebra $\widehat{\mathfrak{L}}$ has a natural Poisson bracket so that the homomorphisms $f \mapsto \widehat{f}$, where we denote by $\widehat{f}$ the image in $\widehat{\mathfrak{L}}$ of an element $f \in \mathfrak{P}^\pm$ under the coefficient homomorphism, are Poisson homomorphisms. We set $\widehat{\mathbf{h}} = \widehat{\mathbf{h}^-} - \widehat{\mathbf{h}^+}$, and for any $f \in \mathfrak{L}$ denote by L_f the "Lagrangian variety"

$$\left\{ p_\gamma^- = \kappa_\gamma \frac{\partial \widehat{f}}{\partial q_\gamma^-}, q_\gamma^+ = \kappa_\gamma \frac{\partial \widehat{f}}{\partial p_\gamma^+} \right\}.$$

Strictly speaking L_f is an ideal in the Poisson algebra $\widehat{L}$. However, it is useful to think about L_f as a Lagrangian variety in the symplectic superspace $(\mathbf{V}^-) \oplus \mathbf{V}^+$ with the symplectic form

$$\sum \kappa_{\gamma^-}^{-1} dp_{\gamma^-}^- \wedge dq_{\gamma^-}^- + \kappa_{\gamma^+}^{-1} dq_{\gamma^+}^+ \wedge dp_{\gamma^+}^+ ,$$

and with an appropriate change of the coefficient ring.

For any function $f \in \mathfrak{L}$, which satisfies the Hamilton-Jacobi equation

$$\widehat{\mathbf{h}}|_{L_f} = 0 \tag{48}$$

the Hamiltonian vector field defined by the Hamiltonian $\widehat{\mathbf{h}}$ is tangent to L_f, and hence the differential $d^f : \mathfrak{L} \to \mathfrak{L}$, defined by the formula

$$d^f(g) = \{\widehat{\mathbf{h}}, g\}|_{L_f}$$

has the following meaning: we identify $\mathfrak{L}$ with the space of functions on L_f and differentiate them along the Hamiltonian vector field determined by $\widehat{\mathbf{h}}$.

Here is an analog of Proposition 2.3.5 for the algebra $\mathfrak{L}$.

PROPOSITION 2.3.5. *Suppose that* $\widehat{\mathbf{h}}|_{L_f} = 0$. *Then*

1. $(d^f)^2 = 0$;
2. *The maps* $f^\pm : \mathfrak{P}^\pm \to \mathfrak{L}$, *defined by the formula* $g \mapsto \widehat{g}|_{L_f}$, *are homomorphisms of chain complexes, i.e.*
$$d^f \circ f^\pm = f^\pm \circ d^\pm;$$
3. *If* $g_1, g_2 \in \mathfrak{P}^\pm$ *Poisson commute with* $\mathbf{h}^\pm$ *or, in other words, if* $g_1, g_2 \in \operatorname{Ker} d^\pm$ *then*
$$\{f^\pm(g_1), f^\pm(g_2)\} = f^\pm(\{g_1, g_2\}),$$
where the left-side Poisson bracket is taken in the algebra $\widehat{\mathfrak{L}}$.

Let us recall that $\mathbf{F} = \mathbf{F}_W \in \mathfrak{D}$ has the form $\mathbf{F} = \hbar^{-1}\sum_{g=0}^{\infty}\mathbf{F}_g\hbar^g$. Again, to simplify the notation we will write $\mathbf{f}$ instead of $\mathbf{F}_0$. The following theorem is the reduction of Theorem 2.3.3 to the level of rational Gromov-Witten theory.

Theorem 2.3.6. *The series* $\mathbf{f}(q^-, p^+, t)$ *belongs to the algebra* $\mathfrak{L}$ *and satisfies the equation*
$$\widehat{\mathbf{h}}|_{L_\mathbf{f}} = 0. \tag{49}$$

In particular, all statements of the above Proposition 2.3.5 hold for $\mathbf{f}$, and this allows us to define the homology
$$_{\mathbf{f}}H_*^{\mathrm{RSFT}}(W|J, \alpha^\pm) = H_*(\mathfrak{L}, d^\mathbf{f}).$$
The chain maps $\mathbf{f}^\pm$ induce homomorphism of Poisson homology algebras
$$(\mathbf{f}^\pm)_* : H_*^{\mathrm{RSFT}}(V^\pm|J, \alpha^\pm) = H_*(\mathfrak{P}^\pm, d^\pm) \to {}_{\mathbf{f}}H_*^{\mathrm{RSFT}}(W|J, \alpha^\pm). \tag{50}$$

For the rest of this section we assume that W is a rational homology cobordism, i.e. the restriction maps
$$H^*(V^-; \mathbb{R}) \leftarrow H^*(W; \mathbb{R}) \to H^*(V^+, \mathbb{R})$$
are isomorphisms. Equivalently, this means that the inclusions $V^\pm \to W$ induce isomorphisms of rational homology groups.

The potential $\mathbf{f} \in \mathfrak{L}$ which we defined above can be written in the form
$$\mathbf{f} = \sum_i \sum_{|\Gamma^+|=i} f_\Gamma^i(q^-, t)(p^+)^{\Gamma^+}. \tag{51}$$
Notice that the assumption that W is a homology cobordism implies that $f^0(q^-, t)$ is independent of q^-. Let us now define a homomorphism $\mathbf{\Psi} : \mathfrak{A}^+ \to \mathfrak{A}^-$ by the formula
$$\mathbf{\Psi}(q_\gamma^+) = f_\gamma^1(q^-, t) \in \mathfrak{A}^- \tag{52}$$
on the generators $q_\gamma^+, \gamma \in \mathcal{P}^+$, of the algebra $\mathfrak{A}^+$ and then extend by linearity.

Theorem 2.3.7. *The homomorphism $\mathbf{\Psi}:\mathfrak{A}^+\to\mathfrak{A}^-$ commutes with the boundary operators $\partial^\pm$, i.e. $\partial^-\circ\mathbf{\Psi}=\mathbf{\Psi}\circ\partial^+$, and in particular defines a homomorphism of homology algebras*

$$(\mathbf{\Psi})_*:H_*(\mathfrak{A}^+,\partial^+)\to H_*(\mathfrak{A}^-,\partial^-).$$

Without the assumption that W is a homology cobordism one gets only a correspondence between the algebras $\mathfrak{A}^+$ and $\mathfrak{A}^-$, similar to the "semi-classical" case considered above.

2.4 Chain homotopy. Let $W=\overrightarrow{V^-V^+}$ be a directed symplectic cobordism with fixed contact forms $\alpha^\pm$ on $V^\pm$. We will discuss in this section how the function $\mathbf{F}=\mathbf{F}_{W,J,\alpha^\pm}(p^+,q^-,t)$ and other associated structures change when one replaces J with another compatible almost complex structure J' and replaces t with a form $t'=t+d\theta$ where θ has compact support in W.

Let us begin with some algebraic preliminaries. Two series $F_0,F_1\in\hbar^{-1}\mathfrak{D}$ are called *homotopic*, if they can be included into a family $F_s\in\hbar^{-1}\mathfrak{D}$, $s\in[0,1]$, which satisfies the following differential equation

$$\frac{dF_s}{ds}=e^{-F_s}\left(\overrightarrow{[\mathbf{H}^-,K_s]}e^{F_s}+e^{F_s}\overleftarrow{[K_s,\mathbf{H}^+]}\right),\quad s\in[0,1],\tag{53}$$

for a family $K_s\in\hbar^{-1}\mathfrak{D}$. Here $[\mathbf{H}^-,K_s]$ and $[K_s,\mathbf{H}^+]$ are commutators in the algebra $\hbar^{-1}\widehat{\mathfrak{D}}$, defined similar to $\widehat{\mathfrak{L}}$ in (47) above, i.e.

$$\widehat{\mathfrak{D}}=\left\{\sum_{\Gamma^+,\Gamma^-,d,g}f_{\Gamma^+,\Gamma^-,d}(q^-,q^+,t)z^d\hbar^g(p^+)^{\Gamma^+}(p^-)^{\Gamma^-}\right\},\tag{54}$$

where the Novikov condition (38) is satisfied for both pairs (d,Γ^+) and (d,Γ^-). In other words, we view K_s as an operator on $\hbar^{-1}\mathfrak{D}$, acting by the multiplication by the series K_s, and view $\mathbf{H}^-$ and $\mathbf{H}^+$ as left and right differential operators.

Notice that the family $K_s\in\hbar^{-1}\mathfrak{D}, s\in[0,1]$, defines a flow $\Phi^s=\Phi^s_K:\mathfrak{D}\mathfrak{D}\to\mathfrak{D}\mathfrak{D}$, by a differential equation

$$\frac{d\Phi^s(G)}{ds}=\mathcal{K}_s(\Phi^s(G)),\tag{55}$$

where we set

$$\mathcal{K}_s(G)=\left(\overrightarrow{[\mathbf{H}^-,K_s]}G+G\overleftarrow{[K_s,\mathbf{H}^+]}\right),\quad s\in[0,1].$$

The linear operators Φ^s preserve the "submanifold" $\mathfrak{E}=e^{\hbar^{-1}\mathfrak{D}_{\text{even}}}$, where $\mathfrak{D}_{\text{even}}$ is the even part of $\mathfrak{D}$, and we have

$$\Phi^s(e^{F_0})=e^{F_s},$$

where the family F_s satisfies the equation (53).

The tangent space to $\mathfrak{E}$ at a point e^F, $F \in \hbar^{-1}\mathfrak{D}_{\text{even}}$, consists of series fe^F, $f \in \hbar^{-1}\mathfrak{D}_{\text{even}}$, and thus it is naturally parameterized by $\hbar^{-1}\mathfrak{D}_{\text{even}}$. With respect to this parameterization the differential of the flow $\Phi^s|_{\mathfrak{E}}$ defines a family of maps $T^s_F : \hbar^{-1}\mathfrak{D}_{\text{even}} \to \hbar^{-1}\mathfrak{D}_{\text{even}}$, $F \in \hbar^{-1}\mathfrak{D}_{\text{even}}$, by the formula

$$T^s_F(f) = e^{-F_s}\Phi^s(fe^F), \quad \text{where} \quad F_s = \Phi^s(F). \tag{56}$$

We extend T^s_F to the whole $\hbar^{-1}\mathfrak{D}$ by the same formula (56). Let us list some properties of the flows Φ^s and T^s_F

PROPOSITION 2.4.1. *Suppose that for an element $F \in \hbar^{-1}\mathfrak{D}_{\text{even}}$ we have*

$$D_W(e^F) = \overrightarrow{\mathbf{H}^-}e^F - e^F\overleftarrow{\mathbf{H}^+} = 0.$$

Then

1. *The flow $T^s_F : \hbar^{-1}\mathfrak{D} \to \hbar^{-1}\mathfrak{D}$ satisfies the equation*
$$T^s_F \circ D^F = D^{F_s} \circ T^s_F \tag{57}$$
for all $s \in [0,1]$. In particular, $D_W(F_s) = 0$ for all $s \in [0,1]$, and T^s_F defines a family of isomorphisms $H_*(\mathfrak{D}, D^F) \to H_*(\mathfrak{D}, D^{F_s})$.
2. *The homology class $[e^{F_s}] \in H_*(\mathfrak{D}\mathfrak{D}, D_W)$ is independent of s.*
3. *The diagram*
$$\begin{array}{ccc} \mathfrak{D} & \xrightarrow{T^s_F} & \mathfrak{D} \\ & \nwarrow F^{\pm} \quad \nearrow F^{\pm}_s & \\ & \mathfrak{W}^{\pm} & \end{array}$$
homotopically commutes, i.e. there exist operators $A^{\pm}_s : \mathfrak{W}^{\pm} \to \mathfrak{D}$, such that
$$(T^s_F)^{-1} \circ F^{\pm}_s - F^{\pm} = D^F \circ A^{\pm}_s + A^{\pm}_s \circ D^{\pm}\,. \tag{58}$$
In particular, this diagram commutes on the level of homology algebras.

The proof of this proposition is a straightforward computation by differentiating the corresponding equations. To illustrate the argument, let us verify (58) in 2.4.1.

Take, for instance, $f \in \mathfrak{W}^-$ and set

$$\begin{aligned} G_s(f) &= (T^s_F)^{-1}(e^{-F_s}\overrightarrow{f}e^{F_s})) - e^{-F}\overrightarrow{f}e^F \\ &= e^{-F}(\Phi^s)^{-1}\left(\overrightarrow{f}e^{F_s}\right) - e^{-F}\overrightarrow{f}e^F. \end{aligned}$$

Then we have $G_0(f) = 0$ and

$$\begin{aligned} \frac{dG_s(f)}{ds} &= -e^{-F}\left((\Phi^s)^{-1}(\mathcal{K}_s\overrightarrow{f}e^{F_s})\right) + e^{-F}(\Phi^s)^{-1}\left(\overrightarrow{f}\mathcal{K}_s\left(e^{F_s}\right)\right) \\ &= e^{-F}(\Phi^s)^{-1}\left(\left[\mathcal{K}_s, \overrightarrow{f}\right]e^F\right). \end{aligned}$$

Now remember that $\mathcal{K}_s = [\widehat{\mathbf{H}}, K_s]$, where $\widehat{\mathbf{H}} = \overrightarrow{\mathbf{H}^-} - \overleftarrow{\mathbf{H}^+}$, and using the Jacobi identity we get

$$\begin{aligned}\frac{dG_s(f)}{ds} &= e^{-F}(\Phi^s)^{-1}\left(\left[[\widehat{\mathbf{H}}, K_s], \overrightarrow{f}\right] e^{F_s}\right)\\ &= e^{-F}(\Phi^s)^{-1}\left(\left[[K, \overrightarrow{f}], \widehat{\mathbf{H}}\right] e^{F_s}\right)\\ &\quad + e^{-F}(\Phi^s)^{-1}\left(\left[[\overrightarrow{f}, \widehat{\mathbf{H}}], K_s\right] e^{F_s}\right).\end{aligned}$$

Let us define now a linear operator $B_s : \hbar^{-1}\mathfrak{W}_- \to \hbar^{-1}\mathfrak{D}$ by the formula

$$B_s(g) = e^{-F}(\Phi^s)^{-1}\left(\overrightarrow{[g, K_s]} e^{F_s}\right) \tag{59}$$

for $g \in \hbar^{-1}\mathfrak{W}_-$. Recall that $D_W(e^F) = \widehat{\mathbf{H}}e^F = 0$, and observe that $\widehat{\mathbf{H}}$ commutes with $\mathcal{K}_s = [\widehat{\mathbf{H}}, K_s]$ because $\widehat{\mathbf{H}} \circ \widehat{\mathbf{H}} = 0$. We also have $[\widehat{\mathbf{H}}, f] = D^- f$ for $f \in \hbar^{-1}\mathfrak{W}_-$. Hence $\frac{dG_s(f)}{ds}$ can be rewritten as

$$\frac{dG_s(f)}{ds} = D^F\big(B_s(f)\big) + B_s\big(D^-(f)\big). \tag{60}$$

Finally we integrate B_s into the required linear operator $A^- : \hbar^{-1}\mathfrak{W}^- \to \hbar^{-1}\mathfrak{D}$:

$$A_s^-(g) = \int_0^s B_s(g)ds, \quad \text{for} \quad g \in \hbar^{-1}\mathfrak{W}^-. \tag{61}$$

In view of equation (60) we get

$$(\Phi^s)^{-1} \circ (F^s)^- - (F)^- = D^F \circ A_s^- + A_s^- \circ D^-. \tag{62}$$

□

Let us consider now a generic family J^τ, $\tau \in [0, 1]$, of compatible almost complex structures on W connecting $J^0 = J$ with $J^1 = J'$. We assume that the deformation J^τ is fixed outside of a compact subset of W.

Set

$$\mathcal{M}_{g,r,s^+,s^-}\big(W, \{J^\tau\}\big) = \bigcup_{\tau \in [0,1]} \mathcal{M}_{g,r,s^+,s^-}(W, J^\tau). \tag{63}$$

The evaluation maps defined for each τ can be combined into the evaluation map

$$ev : \mathcal{M}_{g,r,s^+,s^-}\big(W, \{J^\tau\}\big) \to (W \times I)^r \times \big(\mathcal{P}^-\big)^{s^-} \times \big(\mathcal{P}^+\big)^{s^+}. \tag{64}$$

Consider closed forms $\widehat{\theta}_1, \dots \widehat{\theta}_r$ on $W \times I$, such that $\widehat{\theta}_i = \widetilde{\theta}_i + d\beta_i$, $i = 1, \dots, r$, where $\widetilde{\theta}_i$ is the pull-back of a form θ_i on W with cylindrical ends, and β_i has compact support in $W \times \mathrm{Int}I$.

Similarly to correlators of degree -1 and 0 (see 2.2.2 and 2.3.1) we can define correlators of degree 1, or 1-parametric correlators by the formula

$$^1\langle\widehat{\theta}_1,\dots\widehat{\theta}_r;\theta_1^-,\dots,\theta_{s^-}^-;\theta_1^+,\dots,\theta_{s^+}^+\rangle_g^A = \int\limits_{\overline{\mathcal{M}_{g,r,s^+,s^-}(W,\{J^\tau\})}} ev^*(\widehat{\theta}_1\otimes\dots\otimes\widehat{\theta}_r\otimes\theta_1^-\otimes\dots\otimes\theta_{s^-}^-\otimes\theta_1^+\otimes\dots\otimes\theta_{s^+}^+), \quad (65)$$

for a homology class $A\in H_2(W)$, and cohomology classes

$$\theta_i^\pm\in H^*(\mathcal{P}^\pm),\ i=1,\dots,s^\pm.$$

Consider a closed form $T=\widetilde{t}+d\beta$, where the notation $\widetilde{t}$ and β have the same meaning as above, i.e. $\widetilde{t}$ is the pull-back of a form t on W with cylindrical ends, and β has compact support in $W\times \mathrm{Int}I$. We can organize the correlators

$$^1\langle T,\dots,T;q^-,\dots,q^-,p^+,\dots,p^+\rangle_g^A$$

into a generating function $\mathbf{K}=\frac{1}{\hbar}\sum_{g=0}^\infty \mathbf{K}_g\hbar^g\in\frac{1}{\hbar}\mathfrak{D}$, defined by the formula

$$\mathbf{K}=\sum_d\sum_{g,r,s^\pm=0}^\infty \frac{1}{r!s^+!s^-!}\,{}^1\Big\langle \underbrace{T,\dots,T}_{r};\underbrace{q^-,\dots,q^-}_{s^-};\underbrace{p^+,\dots,p^+}_{s^+}\Big\rangle_g^d \hbar^g z^d. \quad (66)$$

Let us define an operator $\mathcal{K}:\mathfrak{D}\mathfrak{D}\to\mathfrak{D}\mathfrak{D}$ by the formula

$$\mathcal{K}(G)=\overrightarrow{[\mathbf{H}^+,\mathbf{K}]}G+G\overleftarrow{[\mathbf{K},\mathbf{H}^-]},\ G\in\mathfrak{D}\mathfrak{D}. \quad (67)$$

Similar to Theorems 2.2.2 and 2.3.3 the next theorem can be viewed as an algebraic description of the boundary of the compactified moduli space $\overline{\mathcal{M}_{g,r,s^+,s^-}}(W,\{J^\tau\})$.

Theorem 2.4.2. *For a generic family $J_\tau,\tau\in[0,1]$, of compatible almost complex structures on W we have*

$$e^{\mathbf{F}^1}=e^{\mathcal{K}}(e^{\mathbf{F}^0}), \quad (68)$$

where $\mathbf{F}^0=\mathbf{F}_{W,J_0}(T|_{W\times 0})$, $\mathbf{F}^1=\mathbf{F}_{W,J_1}(T|_{W\times 1})$, and $\mathcal{K}=\mathcal{K}(T)$.

Notice that if we define $\mathbf{F}^s$, $s\in[0,1]$, by the formula

$$e^{\mathbf{F}^s}=e^{s\mathcal{K}}(e^{\mathbf{F}^0}), \quad (69)$$

then the flow $\Phi^s(F)=F^s$ satisfies the differential equation (55) with $K(s)\equiv\mathbf{K}$. Hence $\mathbf{\Phi}^0$ and $\mathbf{\Phi}^1$ are homotopic, and therefore Theorem 2.4.2 and Proposition 2.4.1 imply

Corollary 2.4.3. 1. *The homology class $[e^{\mathbf{F}}]\in H_*(\mathfrak{D}\mathfrak{D},D_W)$ is independent of the choice of a compatible almost complex structure J on W and of the choice of $t \bmod (d\,(\Omega_{\mathrm{comp}}(W)))$, where $\Omega_{\mathrm{comp}}(W)$ is the space of forms with compact support.*
2. *For a generic compatible deformation J^τ, $\tau\in[0,1]$, the isomorphism $T:\mathfrak{D}\to\mathfrak{D}$ defined by the formula*

$$T(f)=e^{-\mathbf{F}_1}e^{\mathcal{K}}(fe^{\mathbf{F}^0}) \quad (70)$$

satisfies the equation

$$T \circ D^{\mathbf{F}^0} = D^{\mathbf{F}^1} \circ T, \tag{71}$$

and thus defines an isomorphism $H_*(\mathfrak{D}, D^{\mathbf{F}^0}) \to H_*(\mathfrak{D}, D^{\mathbf{F}^1})$.
3. *The diagram*

$$\begin{array}{ccc} \mathfrak{D} & \xrightarrow{T} & \mathfrak{D} \\ \nwarrow (\mathbf{F}^0)^{\pm} & & \nearrow (\mathbf{F}^1)^{\pm} \\ & \mathfrak{W}^{\pm} & \end{array}$$

homotopically commutes, i.e. there exist operators $A^{\pm} : \mathfrak{W}^{\pm} \to \mathfrak{D}$, *such that*

$$T^{-1} \circ (\mathbf{F}^1)^{\pm} - (\mathbf{F}^0)^{\pm} = D^{\mathbf{F}^0} \circ A^{\pm} + A^{\pm} \circ D^{\pm}. \tag{72}$$

Consider now the equivalence relation for rational potentials.

Two series $f_0, f_1 \in \mathfrak{L}$ are called *homotopic* if there exist families $f_s, k_s \in \mathfrak{L}$, $s \in [0,1]$, so that the family f_s connects f_0 and f_1 and satisfies the following Hamilton-Jacobi differential equation

$$\frac{\partial f_s(p^+, q^-, t)}{\partial s} = \mathbf{G}\left(p^+, \frac{\partial f_s(p^+, q^-, t)}{\partial p^+}, \frac{\partial f_s(p^+, q^-, t)}{\partial q^-}, q^-\right), \tag{73}$$

where

$$\begin{aligned} \mathbf{G}(p^+, q^+, p^-, q^-, t) &= \{\mathbf{h}^+ - \mathbf{h}^-, k_s\} \\ = \sum_{\gamma^+ \in \mathcal{P}^+, \gamma^- \in \mathcal{P}^-} & \kappa_{\gamma^-} \frac{\partial \mathbf{h}^-(p^-, q^-, t)}{\partial p^-_{\gamma^-}} \frac{\partial k_s(p^+, q^-, t)}{\partial q^-_{\gamma^-}} \\ &+ \kappa_{\gamma^+} \frac{\partial k_s(p^+, q^-, t)}{\partial p^+_{\gamma^+}} \frac{\partial \mathbf{h}^+(p^+, q^+, t)}{\partial q^+_{\gamma^+}}. \end{aligned} \tag{74}$$

Here κ_γ denotes, as usual, the multiplicity of γ.

We can view the correspondence

$$f(p^+, q^-, t) \mapsto f_s(p^+, q^-, t),$$

where f_s is the solution of the above equation (73) with the initial data $f_0 = f$, as a non-linear operator $S^s : \mathfrak{L} \to \mathfrak{L}$. Let us denote by T^s_f the linearization of S^s at a point f. The next proposition is a rational version of Proposition 2.4.1. It can be either deduced from 2.4.1, or similarly proven by differentiation with respect to the parameter s. Denote by $\mathcal{S}$ the subspace of $\mathfrak{L}$ which consists of solutions of the Hamilton-Jacobi equation (48), i.e.

$$\widehat{\mathbf{h}}|_{L_f} = 0, \quad \text{where} \quad L_f = \left\{p^-_\gamma = \kappa_\gamma \frac{\partial f}{\partial q^-_\gamma}, q^+_\gamma = \kappa_\gamma \frac{\partial f}{\partial p^+_\gamma}\right\}.$$

PROPOSITION 2.4.4. 1. *The subspace $\mathcal{S} \subset \mathfrak{L}$ is invariant under the flow S^s.*
2. *For $f \in \mathcal{S}$ the isomorphism $T^s_f : \mathfrak{L} \to \mathfrak{L}$ satisfies the equation*

$$T^s_f \circ d^f = d^{S^s(f)} \circ T^s, \tag{75}$$

and thus defines an isomorphism $H_*(\mathfrak{L}, d^f) \to H^*(\mathfrak{L}, d^{S^s(f)})$.
3. *For $f \in \mathcal{S}$ the diagram*

$$\begin{array}{ccc} \mathfrak{L} & \xrightarrow{T^s_f} & \mathfrak{L} \\ \nwarrow (f)^{\pm} & & \nearrow (S^s(f))^{\pm} \\ & \mathfrak{P}^{\pm} & \end{array}$$

homotopically commutes.

Theorem 2.4.2 reduces on the level of rational curves to the following

Theorem 2.4.5. *Let $\mathbf{F}^0$, $\mathbf{F}^1$ and $\mathbf{K}$ be as in Theorem 2.4.2. Set $\mathbf{f}^0 = \mathbf{F}^0_0$, $\mathbf{f}^1 = \mathbf{F}^1_0$, $\mathbf{k} = \mathbf{K}_0$. Then $\mathbf{f}_0$ and $\mathbf{f}_1$ are homotopic, i.e. they can be included into a family $\mathbf{f}_s$, $s \in [0,1]$, such that the Hamilton-Jacobi equation (73) holds with $f_s = \mathbf{f}_s$ and $k_s \equiv \mathbf{k}$.*

Hence, Proposition 2.4.4 implies

COROLLARY 2.4.6. *For a generic compatible deformation J_s, $s \in [0,1]$, we have*

1. *The operator $S^1 : \mathfrak{L} \to \mathfrak{L}$ defines an automorphism of the space of solutions of (48);*
2. *The linearization $T^1 = T^1_{\mathbf{f}^0}$ of S^1 at the point $\mathbf{f}_0$ satisfies the equation*

$$T^1 \circ d^{\mathbf{f}^0} = d^{\mathbf{f}^1} \circ T^1, \tag{76}$$

and thus defines an isomorphism $H_*(\mathfrak{L}, d^{\mathbf{f}^0}) \to H^*(\mathfrak{L}, d^{\mathbf{f}^1})$.

3. *The diagram*

$$\begin{array}{ccc} \mathfrak{L} & \xrightarrow{T^1} & \mathfrak{L} \\ \nwarrow (\mathbf{f}^0)^{\pm} & & \nearrow (\mathbf{f}^1)^{\pm} \\ & \mathfrak{P}^{\pm} & \end{array}$$

homotopically commutes.

To formulate the "classical level" corollary of Theorem 2.4.5 we assume, as usual, that W is a homology cobordism.

Theorem 2.4.7. *The homomorphisms $\Psi^1_{J_0}, \Psi^1_{J_1} : \mathfrak{A}^+ \to \mathfrak{A}^-$ associated to two compatible almost complex structures J_0 and J_1 are homotopic, i.e. there exists a map $\Delta : \mathfrak{A}^+ \to \mathfrak{A}^-$ such that*

$$\Psi^1_{J_1} - \Psi^1_{J_1} = \partial^- \circ \Delta + \Delta \circ \partial^+.$$

The map Δ can be expressed through the function $\mathbf{k} \in \mathfrak{L}$. However, unlike the case of usual Floer homology theory, Δ and $\mathbf{k}$ are related via a first order non-linear PDE (which can be deduced from the equation (73)), and thus one cannot write a general explicit formula relating Δ and $\mathbf{k}$.

2.5 Composition of cobordisms. In this section we study the behavior of potentials and associated algebraic structures under the operation of composition of directed symplectic cobordisms.

Let us recall (see section 1.3) that given a dividing contact type hypersurface V in a directed symplectic cobordism $W = \overrightarrow{V^-V^+}$ one can split W into a composition $W = W_- \circledcirc W_+$ of cobordisms $W_- = \overrightarrow{V^-V}$ and $W_+ = \overrightarrow{VV^+}$. From the point of view of an almost complex structure the process of splitting consists of deforming an original almost complex structure $J = J^0$ to an almost complex structure J^∞, such that the restrictions $J_\pm = J^\infty|_{W_\pm}$ are compatible with the structure of (completed) directed symplectic cobordisms $W_\pm$.

Conversely, directed symplectic cobordisms $W_- = \overrightarrow{V^-V}$ and $W_+ = \overrightarrow{VV^+}$ with matching data on the common boundary can be glued into a cobordism $W = \overrightarrow{V^-V^+}$ in the following sense: there exists a family J^τ of almost complex structures on W which in the limit splits W into the composition of cobordisms $W_- = \overrightarrow{V^-V}$ and $W_+ = \overrightarrow{VV^+}$.

In order to write the formula relating the potentials of W and $W_\pm$ we first need to make more explicit the relation between 2-dimensional homology classes realized by holomorphic curves in $W_\pm$ and W. We will keep assuming that there are no torsion elements in H_1. Let us recall (see sections 1.2 and 1.5 above) that we have chosen curves $C^i_- \subset W_-, i = 1,\dots,m_-$, $C^j_+ \subset W_+, j = 1,\dots,m_+$, and $C^k \subset W, k = 1,\dots m$, which represent bases of first homology of the respective cobordisms. We also have chosen for every periodic orbit $\gamma \in \mathcal{P}_\alpha$ of the Reeb field R_α on V a surface $F^\gamma_\pm$ realizing homology in $W_\pm$ between γ and a linear combination of basic curves $C^i_\pm$. For our current purposes we have to make one extra choice: for each curve $C^i_\pm$ we choose a surface $S^i_\pm$ which realizes homology in W between $C^i_\pm$ and the corresponding linear combination of the curves $C^1,\dots,C^m$. All the choices enable us to associate with every orbit $\gamma \in \mathcal{P}_\alpha$ a homology class C^γ which is realized by the chain

$$F^\gamma_+ - F^\gamma_- + \sum_1^{m_+} n^+_j S^j_+ - \sum_1^{m_-} n^-_i S^i_-\,,$$

where $\partial F^\gamma_\pm = [\gamma] - \sum_1^{m_\pm} n_j^\pm C^j_\pm$. We will denote by d^γ the degree of C^γ, i.e. the string of its coordinates in the chosen basis $A_1, \dots, A_N \in H_2(W)$.

Let us define an operation

$$\star : \mathfrak{DD}_- \otimes \mathfrak{DD}_+ \to \mathfrak{DD}, \tag{77}$$

where $\mathfrak{DD}_\pm = \mathfrak{DD}_{W_\pm}$ and $\mathfrak{DD} = \mathfrak{DD}_W$. For $F = \sum_\Gamma f_\Gamma(t^-, q^-, \hbar, z^-)p^\Gamma \in \mathfrak{DD}_-$ and $G = \sum_{\Gamma^+} g_{\Gamma^+}(t^+, q, \hbar, z^+)(p^+)^{\Gamma^+} \in \mathfrak{DD}_+$ we set

$$F{\star}G(t, q^-, p^+, \hbar, z) =$$

$$\left(\sum_\Gamma \widetilde{f}_\Gamma(t, q^-, \hbar, z)\hbar^s \prod_{i=1}^s \kappa_{\gamma_i} z^{d^{\gamma_i}} \overrightarrow{\frac{\partial}{\partial q_{\gamma_i}}} \sum_{\Gamma^+} \widetilde{g}_{\Gamma^+}(t, q, \hbar, z)(p^+)^{\Gamma^+}\right)\Bigg|_{q=0}. \tag{78}$$

Here we denote by $\widetilde{f}_\Gamma$ and $\widetilde{g}_{\Gamma^+}$ the images of f_Γ and g_{Γ^+} under the coefficient homomorphisms $H_2(W_\pm) \to H_2(W)$. Let us explain what happens with the variables t and z in more details. Let $A_1^\pm, \dots, A^\pm_{N^\pm}$, and $A_1, \dots, A_N$, be the chosen bases in $H_2(W_\pm)$ and $H_2(W)$. Then we have

$$i_*^\pm(A_k^\pm) = \sum_{j=1}^{N^\pm} n^\pm_{kj} A_j,$$

where $k = 1, \dots, N$, $(n^\pm_{kj})$ are integer matrices, and $i^\pm : W_\pm \to W$ the inclusion maps.

We have

$$f_\Gamma(t^-, q^-, \hbar, z^-) = \sum_{d=(d_1,\dots,d_{N^-})} f_{\Gamma,d}(t^-, q^-, \hbar)(z_1^-)^{d_1} \dots (z^-_{N^-})^{d_{N^-}},$$

$$g_{\Gamma^+}(t^+, q, \hbar, z^+) = \sum_{d=(d_1,\dots,d_{N^+})} g_{\Gamma^+,d}(t^+, q, \hbar)(z_1^+)^{d_1} \dots (z^+_{N^+})^{d_{N^+}},$$

where we denote by $z_\pm$ the "z-variables" in $W_\pm$. Then

$$\widetilde{f}_\Gamma(t, q^-, \hbar, z) = \sum_{d=(d_1,\dots,d_{N\_})} \widetilde{f}_{\Gamma,d}(t|_{W_-}, q^-, \hbar) z_1^{M_1^-} \dots z_N^{M_N^-},$$

$$\widetilde{g}_{\Gamma^+}(t, q, \hbar, z) = \sum_{d=(d_1,\dots,d_{N_+})} \widetilde{g}_{\Gamma^+,d}(t|_{W_+}, q^-, \hbar) z_1^{M_1^+} \dots z_N^{M_N^+},$$

where $M_j^\pm = \sum\limits_{k=1}^{N^\pm} n^\pm_{kj} d_k$, $j = 1, \dots, N$.

Let us observe

LEMMA 2.5.1. 1. *The operation $\star$ is associative.*
2. *For $F \in \hbar^{-1}\mathfrak{D}_-, G \in \hbar^{-1}\mathfrak{D}_+$ there exists a unique function $H \in \hbar^{-1}\mathfrak{D}$, such that*

$$e^H = e^F \star e^G.$$

We will denote this H by $F\Diamond G$, so that we have $e^{F\Diamond G} = e^F \star e^G$. We will also consider the maps

$$\Diamond G : \hbar^{-1}\mathfrak{D}_- \to \hbar^{-1}\mathfrak{D}, \quad \Diamond G(F) = F\Diamond G,\ F \in \mathfrak{D}_-,$$

and

$$F\Diamond : \hbar^{-1}\mathfrak{D}_+ \to \hbar^{-1}\mathfrak{D}, \quad F\Diamond(G) = F\Diamond G,\ G \in \mathfrak{D}_+,$$

and for even F, G their linearizations:

$$T_F(\Diamond G) : \hbar^{-1}\mathfrak{D}_- \to \hbar^{-1}\mathfrak{D},$$

$$T_F(\Diamond G)(f) = \frac{d(F+\varepsilon f)\Diamond G)}{d\varepsilon}\Big|_{\varepsilon=0} = e^{-F\Diamond G}((fe^F)\star e^G), \quad f \in \hbar^{-1}\mathfrak{D}_-,$$

and

$$T_G(F\Diamond) : \hbar^{-1}\mathfrak{D}_+ \to \hbar^{-1}\mathfrak{D},$$

$$T_G(F\Diamond)(g) = \frac{d(F\Diamond(G+\varepsilon g))}{d\varepsilon}\Big|_{\varepsilon=0} = e^{-F\Diamond G}(e^F \star (ge^G)), \quad g \in \hbar^{-1}\mathfrak{D}_+.$$

Let us first formulate an algebraic

PROPOSITION 2.5.2. *Suppose that $F \in \hbar^{-1}\mathfrak{D}_-$ and $G \in \hbar^{-1}\mathfrak{D}_+$ are even elements, which satisfy the equations*

$$D_{W_-}(e^F) = \overrightarrow{\mathbf{H}^-}e^F - e^F\overleftarrow{\mathbf{H}} = 0$$

and

$$D_{W_+}(e^G) = \overrightarrow{\mathbf{H}}e^F - e^F\overleftarrow{\mathbf{H}^+} = 0,$$

where $\mathbf{H}^\pm = \mathbf{H}_{V^\pm}$, $\mathbf{H} = \mathbf{H}_V$. Then we have

1. $D_W(e^{F\Diamond G}) = \overrightarrow{\mathbf{H}^-}\, e^{F\Diamond G} - e^{F\Diamond G}\,\overleftarrow{\mathbf{H}^+} = 0.$
2. *The homomorphisms*

$$T_G(F\Diamond) : \hbar^{-1}\mathfrak{D}_+ \to \hbar^{-1}\mathfrak{D}$$

and

$$T_F(\Diamond G) : \hbar^{-1}\mathfrak{D}_- \to \hbar^{-1}\mathfrak{D}$$

satisfy the equations

$$T_G(F\Diamond) \circ D^G = D^{F\Diamond G} \circ T_G(F\Diamond)$$

and

$$T_F(\Diamond G) \circ D^F = D^{F\Diamond G} \circ T_F(\Diamond G),$$

and in particular they define homomorphisms of the corresponding homology algebras:

$$\big(T_G(F\Diamond)\big)_* : H_*(\mathfrak{D}_+, D^G) \to H_*(\mathfrak{D}, D^{F\Diamond G})$$

and

$$\big(T_F(\Diamond G)\big)_* : H_*(\mathfrak{D}_-, D^G) \to H_*(\mathfrak{D}, D^{F\Diamond G}).$$

3.

$$T_F(\Diamond G)\circ F^- = (F\Diamond G)^-$$

and

$$T_G(F\Diamond)\circ G^+ = (F\Diamond G)^+.$$

4. *Suppose we are given three cobordisms* W_1, W_2, W_3 *with matching ends so that we can form the composition* $W_{123} = W_1 \circledcirc W_2 \circledcirc W_3$, *and series* $F_i \in \hbar^{-1}\mathfrak{D}_i = \hbar^{-1}\mathfrak{D}_{W_i}$, $i = 1,2,3$, *such that* $D_{W_i}e^{F_i} = 0$, $i = 1,2,3$. *Then*

$$T_{F_1\Diamond F_2}(\Diamond F_3)\circ T_{F_1}(\Diamond F_2) = T_{F_1}\big(\Diamond(F_2\Diamond F_3)\big). \tag{79}$$

The proof of this proposition is immediate from the definition of the corresponding operations. For instance, to prove 2.5.2.1 we write

$$\begin{aligned}\overrightarrow{\mathbf{H}^-}\left(e^{F\Diamond G}\right) - \left(e^{F\Diamond G}\right)\overleftarrow{\mathbf{H}^+} &= \overrightarrow{\mathbf{H}^-}\, e^F \star e^G - e^F \star e^G\overleftarrow{\mathbf{H}^+}\\ &= \left(\overrightarrow{\mathbf{H}^-}\, e^F\right)\star e^G - e^F\star\left(e^G\overleftarrow{\mathbf{H}^+}\right)\\ &= \left(e^F\,\overleftarrow{\mathbf{H}}\right)\star e^G - e^F\star\left(\overrightarrow{\mathbf{H}}e^G\right).\end{aligned}$$

To finish the argument let us consider a cylindrical cobordism $W_0 = V\times\mathbb{R}$, take the function $I = \sum \kappa_\gamma^{-1}p_\gamma q_\gamma$. Taking into account associativity of $\star$ (see 2.5.1) we have

$$\overrightarrow{f}\, e^G = (fe^I)\star e^G \quad\text{and}\quad e^F\overleftarrow{f} = e^F\star(fe^I). \tag{80}$$

Hence, we have

$$\left(e^F\,\overleftarrow{\mathbf{H}}\right)\star e^G - e^F\star\left(\overrightarrow{\mathbf{H}}e^G\right) = e^F\star(\mathbf{H}e^I)\star e^G - e^F\star(\mathbf{H}e^I)\star e^G = 0.$$

Any cohomology class from $H^*(W)$ can be represented by a form t which splits into the sum of forms $t_\pm$ with cylindrical ends on $W_\pm$, so that we have $t_\pm|_V = t_V$. Let us define now a series $\mathbf{F}^\infty \in \hbar^{-1}\mathfrak{D}$ by the formula

$$\mathbf{F}^\infty(q^-,p^+,t) = \mathbf{F}_-(q^-,p,t_-)\Diamond\mathbf{F}_+(q,p^+,t_+), \tag{81}$$

where p,q are variable associated to the space $H^*(\mathcal{P})$ of periodic orbits of the Reeb vector field R_α of the contact form α on V.

The following theorem is the main claim of this section, and similar to Theorems 2.2.2 and 2.3.3 and 2.4.2 it is a statement about the boundary of an appropriate moduli space of holomorphic curves. This time we deal with limits of J^s-holomorphic curves in W when $s\to\infty$, i.e. when the family J^s realizes the splitting of W into the composition $W_-\circledcirc W_+$, see Theorem 1.6.3 above.

Theorem 2.5.3. *The element* $\mathbf{F}^\infty$ *is homotopic to the potential* $\mathbf{F} = \mathbf{F}_{W,J,\alpha^\pm}$ *for any generic compatible almost complex structure* J *on* W.

Let us now describe the above results on the level of rational potentials. Let $W_- = \overrightarrow{V^-V}$, $W_+ = \overrightarrow{VV^+}$ and $W = W_- \circledcirc W_+ = \overrightarrow{V^-V^+}$ be as above. Set $\mathbf{h}^\pm = \mathbf{h}_{V^\pm}, \mathbf{h} = \mathbf{h}_V$, $\widehat{\mathbf{h}}_- = \mathbf{h}^- - \mathbf{h}$, $\widehat{\mathbf{h}}_+ = \mathbf{h} - \mathbf{h}^+$, $\widehat{\mathbf{h}} = \mathbf{h}^+ - \mathbf{h}^-$, $\mathfrak{L}_\pm = \mathfrak{L}_{W_\pm}$ and $\mathfrak{L} = \mathfrak{L}_W$.

The operation $\lozenge : \hbar^{-1}\mathfrak{D}_- \times \hbar^{-1}\mathfrak{D}_+ \to \hbar^{-1}\mathfrak{D}$ defined above reduces on the rational level to the operation

$$\sharp : \mathfrak{L}_- \times \mathfrak{L}_+ \to \mathfrak{L},$$

defined as follows. For $f_\pm \in \mathfrak{L}_\pm$ we set

$$f_-\sharp f_+(q^-,p^+) = \left(f_-(q^-,p) + f_+(q,p^+) - \sum_{\gamma\in\mathcal{P}} \kappa_\gamma^{-1} z^{-d_\gamma} q_\gamma p_\gamma\right)\Bigg|_L, \tag{82}$$

where

$$L = \begin{cases} q_\gamma = & \kappa_\gamma z^{d_\gamma} \frac{\partial f_-}{\partial p_\gamma}; \\ p_\gamma = & \kappa_\gamma z^{d_\gamma} \frac{\partial f_+}{\partial q_\gamma}. \end{cases}$$

Notice that given series

$$F_- = \hbar^{-1}\sum_0^\infty (F_-)_g \hbar^g \in \hbar^{-1}\mathfrak{D}^- \quad \text{and} \quad F_+ = \hbar^{-1}\sum_0^\infty (F_+)_g \hbar^g \in \hbar^{-1}\mathfrak{D}^+$$

with $F_-\lozenge F_+ = \hbar^{-1}\sum_0^\infty (F_-\lozenge F_+)_g \hbar^g \in \hbar^{-1}\mathfrak{D}$ then

$$(F_-\lozenge F_+)_0 = (F_-)_0 \sharp (F_+)_0.$$

We will also consider the operations

$$\sharp f_+ : \mathfrak{L}_- \to \mathfrak{L},\ \ \sharp f_+(f_-) = f_-\sharp f_+ \ \text{ and } \ \sharp f_- : \mathfrak{L}_+ \to \mathfrak{L},\ \ \sharp f_-(f_+) = f_-\sharp f_+,$$

and their linearizations

$$T_{f_-}(\sharp f_+) : \mathfrak{L}_- \to \mathfrak{L},\ \ T_{f_-}(\sharp f_+)(g) = (g|_{L_+})\sharp f_+$$

and

$$T_{f_+}(\sharp f_-) : \mathfrak{L}_+ \to \mathfrak{L},\ \ T_{f_+}(\sharp f_-)(g) = f_-\sharp (g|_{L_-}).$$

Here we view $g|_{L_+}$ (resp. $g|_{L_-}$) as an element of $\mathfrak{L}_+$ which depends on variables q^- as parameters (resp. an element of $\mathfrak{L}_-$, which depends on p^+ as parameters). We have the following rational version of Theorem 2.5.2.

PROPOSITION 2.5.4. *Suppose that even elements $f_\pm \in \mathfrak{L}_\pm = \mathfrak{L}_{W_\pm}$ satisfy equation (48), i.e.*

$$\widehat{\mathbf{h}}_\pm|_{L_{f_\pm}} = 0,$$

where

$$L_{f_-} = \begin{cases} q_\gamma & = k_\gamma \frac{\partial f_-}{\partial p_\gamma};\ \ \gamma \in \mathcal{P} \\ p^-_{\gamma^-} & = \kappa_{\gamma^-} \frac{\partial f_-}{\partial q^-_{\gamma^-}},\ \ \gamma^- \in \mathcal{P}^-, \end{cases}$$

and

$$L_{f_+} = \begin{cases} p_\gamma &= k_\gamma \frac{\partial f_+}{\partial q_\gamma}; \ \gamma \in \mathcal{P} \\ q^+_{\gamma^+} &= \kappa_{\gamma^+} \frac{\partial f_+}{\partial p^+_{\gamma^+}}, \ \gamma^+ \in \mathcal{P}^+. \end{cases}$$

Then

1. *The function $f_- \sharp f_+$ satisfies the Hamilton-Jacobi equation*
$$\widehat{\mathbf{h}}_{L|_{f_-\sharp f_+}} = 0;$$
2. *The homomorphisms $T_{f_-}(\sharp f_+) : \mathfrak{L}_- \to \mathfrak{L}$ and $T_{f_+}(f_-\sharp) : \mathfrak{L}_+ \to \mathfrak{L}$ satisfy the equations*
$$T_{f_-}(\sharp f_+) \circ d^{f_-} = d^{f_-\sharp f_+} \circ T_{f_-}(\sharp f_+),$$
$$T_{f_+}(f_-\sharp) \circ d^{f_+} = d^{f_-\sharp f_+} \circ T_{f_+}(\sharp f_-),$$
and hence define homomorphisms of the corresponding homology algebras:
$$\big(T_{f_-}(\sharp f_+)\big)_* : H_*(\mathfrak{L}_-, d^{f_-}) \to H_*(\mathfrak{L}, d^{f_-\sharp f_+}),$$
$$\big(T_{f_+}(f_-\sharp)\big)_* : H_*(\mathfrak{L}_-, d^{f_-}) \to H_*(\mathfrak{L}, d^{f_-\sharp f_+});$$
3.
$$T_{f_-}(\sharp f_+) \circ (f_-)^- = (f_-\sharp f_+)^-,$$
$$T_{f_+}(\sharp f_-) \circ (f_+)^- = (f_-\sharp f_+)^+;$$
4. *Suppose we are given three cobordisms W_1, W_2, W_3 with matching ends so that we can form the composition $W_{123} = W_1 \circledcirc W_2 \circledcirc W_3$, and series $f_i \in \hbar^{-1}\mathfrak{L}_i = \mathfrak{L}_{W_i}$ which satisfy Hamilton-Jacobi equations $\widehat{\mathbf{h}}_{W_i}|_{L_{f_i}} = 0$, $i = 1, 2, 3$. Then*
$$T_{f_2}(\sharp f_3) \circ T_{f_1}(\sharp f_2) = T_{f_1}\big(\sharp(f_2 \sharp f_3)\big).$$

Let t be a closed form on W which is split into two forms $t_\pm$ on $W_\pm$ with cylindrical ends. Set $\mathbf{f}_\pm = \mathbf{f}_{W_\pm}$ and

$$\mathbf{f}^\infty(q^-, p^+, t) = \mathbf{f}_-(q^-, p, t_-)\sharp\mathbf{f}_+(q, p^+, t_+). \tag{83}$$

Alternatively $\mathbf{f}^\infty$ can be defined as the first term in the expansion $\mathbf{F}^\infty = \hbar^{-1}\sum_0^\infty \mathbf{F}^\infty_g \hbar^g$. The following theorem is a rational analog, and a direct corollary of Theorem 2.5.3.

Theorem 2.5.5. *The series $\mathbf{f}^\infty(q^-, p^+, t)$ and $\mathbf{f}_{W,J}(q, p, t)$ are homotopic for any generic compatible almost complex structure J on W.*

Coming down to the "classical" level, let us assume that W, W_- and W_+ are homology cobordisms (see 2.2.3 above). Thus there are defined the homomorphisms $\mathbf{\Psi} : \mathfrak{A}^+ \to \mathfrak{A}^-$, $\mathbf{\Psi}_+ : \mathfrak{A}^+ \to \mathfrak{A}$ and $\mathbf{\Psi}_- : \mathfrak{A} \to \mathfrak{A}^-$, see section 2.3.2 above. Set $\mathbf{\Psi}_\infty = \mathbf{\Psi}_+ \circ \mathbf{\Psi}_-$. Then we have

Theorem 2.5.6. *For any generic compatible almost complex structure J on W homomorphisms $\mathbf{\Psi}_1 = \mathbf{\Psi}_J, \mathbf{\Psi}_\infty : \mathfrak{A}^+ \to \mathfrak{A}^-$ are chain homotopic.*

2.6 Invariants of contact manifolds. Theorem 2.5.3 allows us to define SFT-invariants of contact manifolds. Let (V, ξ) be a contact manifold, and α^+ and α^- two contact forms for ξ, such that $\alpha^+ > \alpha^-$, i.e. $\alpha^+ = f\alpha^-$, for a function $f > 1$. Then for an appropriately chosen function $\zeta : V \times \mathbb{R} \to (0, \infty)$ the form $\omega = d(\zeta\alpha^-)$ on $W = V \times \mathbb{R}$ is symplectic, and (W, ω) is a directed symplectic cobordism between (V, α^-) and (V, α^+). Let $t^\pm$ be two cohomologous forms on V, and t be a closed form on W with cylindrical ends which restricts to $t^\pm$ on $V^\pm$. Suppose we are also given almost complex structures $J^\pm$ on V, compatible with $\alpha^\pm$, which are extended to a compatible almost complex structure J on W. We will call a directed symplectic cobordism (W, J, t), chosen in the above way, a *concordance* between the data on its boundary. Notice, that concordance becomes an equivalence relation if we identify contact forms proportional with a constant factor. A concordance (W, J, t) is called *trivial* if $W = V \times \mathbb{R}$, the almost complex structure J is translationally invariant, and t is the pullback of a form t_+ under the projection $W \to V$.

Let us denote, as usual, by $(\mathfrak{W}^\pm, D^\pm)$ the differential Weyl algebras associated to the data at the ends of the cobordism W, by $(\mathfrak{D}, D_W)$ the D-module $\mathfrak{D}(W, J, \alpha^\pm)$, by $\mathbf{F} \in \mathfrak{W}$ the potential of the cobordism W, and by $\mathbf{F}^\pm : (\mathfrak{W}^\pm, D^\pm) \to (\mathfrak{D}, D_W)$ the corresponding homomorphisms of differential algebras defined in (43).

Theorem 2.6.1. *For any concordance (W, J, t) the homomorphisms*

$$\mathbf{F}^\pm : (\mathfrak{W}^\pm, d^\pm) \to (\mathfrak{D}, D_W)$$

are quasi-isomorphisms of differential algebras. In particular, the homology algebras $H_*(\mathfrak{W}^-, D^-)$ and $H_*(\mathfrak{W}^+, D^+)$ are isomorphic.

Proof. We will prove 2.6.1 in three steps.
Step 1. Let us begin with the trivial concordance (W, J, t). In this case $\mathfrak{D}$ can be identified with $\mathfrak{W}^\pm$ and we have $\mathbf{F}(q, p, t) = \hbar^{-1} \sum \kappa_\gamma^{-1} q_\gamma p_\gamma$. Hence

$$\mathbf{F}^-(f) = e^{-\mathbf{F}} \overrightarrow{f} e^{\mathbf{F}} = f = e^{-\mathbf{F}} \left(e^{\mathbf{F}} \overleftarrow{f} \right) = \mathbf{F}^+(f).$$

Step 2. If we add now to t a form $d\theta$, where θ has a compact support, and change J in a compact part of W then according to Theorem 2.4.2 the potential $\mathbf{F}_{W,J}(t + d\theta)$ remains the same up to homotopy, and hence Corollary 2.4.3 implies that the homomorphisms $\mathbf{F}_*^\pm$ induced on homology remain unchanged.

Step 3. Now assume that $(W,J,t)=(W^1,J^1,t^1)$ is a general concordance. Consider the reversed concordance (W^2,J^2,t^2), so that the compositions

$$(W^{12}=W^1\circledcirc W^2, J^{12}=J^1\circledcirc J^2, t^{12}=t_1\circledcirc t_2)$$

and

$$(W^{21}=W^2\circledcirc W^1, J^{21}=J^2\circledcirc J^1, t^{21}=t_2\circledcirc t_1)$$

of concordances (W^1,J^1,t^1) and (W^2,J^2,t^2) are as in Step 2. Then according to Theorem 2.5.3 $\mathbf{F}_{W^{12},J^{12}}(t^{12})$ is homotopic to

$$\mathbf{F}_{W^1,J^1}(t^1)\Diamond\mathbf{F}_{W^2,J^2}(t^2)=\mathbf{F}^1\Diamond\mathbf{F}^2$$

and $\mathbf{F}^{21}=\mathbf{F}_{W^{21},J^{21}}(t^{21})$ is homotopic to

$$\mathbf{F}_{W^2,J^2}(t^2)\Diamond\mathbf{F}_{W^1,J^1}(t^1)=\mathbf{F}^2\Diamond\mathbf{F}^1.$$

Hence Proposition 2.5.2 implies

$$\mathrm{Id}=(\mathbf{F}_{W^1\circledcirc W^2})^-_*=\left(T_{\mathbf{F}^1}(\Diamond\mathbf{F}^2)\right)_*\circ(\mathbf{F}^1)^-_*$$

and

$$\mathrm{Id}=\Big(T_{\mathbf{F}^1}\left(\Diamond(\mathbf{F}^2\Diamond\mathbf{F}^1)\right)\Big)_*=\left(T_{\mathbf{F}^1\Diamond\mathbf{F}^2}(\Diamond\mathbf{F}^1)\right)_*\circ\left(T_{\mathbf{F}^1}(\Diamond\mathbf{F}^2)\right)_*.$$

Hence $\left(T_{\mathbf{F}^1}(\Diamond\mathbf{F}^2)\right)_*$, $(\mathbf{F}^1)^-_*$, and similarly $(\mathbf{F}_1)^+_*$ are isomorphisms. □

The following rational and classical analogs of Theorem 2.6.1 can be either deduced directly from Theorem 2.6.1, or alternatively can be proven similarly to 2.6.1 using 2.4.5 (resp. 2.4.7) and 2.5.5 (resp. 2.5.6).

Theorem 2.6.2. *For any concordance (W,J,t) the homomorphisms*

$$\mathbf{f}^\pm:(\mathfrak{P}^\pm,D^\pm)\to(\mathfrak{L},D_W)$$

are quasi-isomorphisms of differential algebras. In particular, Poisson homology algebras $H_*(\mathfrak{P}^\pm,d^\pm)$ are isomorphic.

Theorem 2.6.3. *For any concordance (W,J,t) the homomorphism*

$$\mathbf{\Psi}:(\mathfrak{A}^+,\partial^+)\to(\mathfrak{A}^-,\partial^-)$$

is a quasi-isomorphism of differential algebras.

The definition of the differential algebras $(\mathfrak{W},D)$, $(\mathfrak{P},d)$ and $(\mathfrak{A},\partial)$ depends on the choice of a coherent orientation (see section 1.8), and of spanning surfaces and framings of periodic orbits (see section 1.2). As it is stated in Theorem 1.8.7 a coherent orientation is determined by a choice of asymptotic operators associated with each periodic orbit γ. Let $\mathbf{H}'$ be the new Hamiltonian which one gets by changing the orientation of the asymptotic operator associated with a fixed periodic orbit γ. One can then check that the change of variables $(p_\gamma,q_\gamma)\mapsto(-p_\gamma,-q_\gamma)$ is an isomorphism between the differential algebras $(\mathfrak{W},D^{\mathbf{H}})$ and $(\mathfrak{W},D^{\mathbf{H}'})$. Different choices for spanning surfaces and framings of periodic orbits do not affect mod 2 grading but change the integer grading of the differential algebras.

REMARK 2.6.4. An accurate introduction of virtual cycle techniques would reveal that even more choices should be made. However, an independence of all these extra choices can be also established following the scheme of this section.

2.7 A differential equation for potentials of symplectic cobordisms. In this section we describe differential equations for the potentials $\mathbf{F}_W$ and $\mathbf{f}_W$ of a directed symplectic cobordism with a non-empty boundary. These equations completely determine the potentials, and in combination with gluing Theorems 2.5.3 and 2.5.5 they provide in many cases an effective recursive procedure for computing potentials $\mathbf{F}_W$ and $\mathbf{f}_W$, and even Gromov-Witten invariants of closed symplectic manifolds W (see some examples in section 2.9.3 below).

Let us assume for simplicity that W has only a positive end $E^+ = V \times (0,\infty)$, and choose a basic system $\Delta_1,\dots,\Delta_k$, $\Theta_1,\dots,\Theta_m$ of closed forms so that the following conditions are satisfied:

a) $\Delta_1,\dots,\Delta_k$ form a basis of $H^*(W)$, and the restrictions $\delta_i = \Delta_i|_V$, $i=1,\dots,l$ for $l \le k$ form a basis of $\mathrm{Im}\big(H^*(W) \to H^*(V)\big)$;
b) $\Theta_1,\dots,\Theta_m$ are compactly supported and represent a basis of $\mathrm{Ker}\big(H^*_{\mathrm{comp}}(W) \to H^*(W)\big)$,
c) there exist forms $\theta_1,\dots,\theta_m$ on V and a compactly supported 1-form ρ on $(0,+\infty)$, such that $\Theta_j = e_*\big(\rho \wedge \pi^*(\theta_j)\big)$, $j = 1,\dots,m$, where π is the projection $E = V\times(0,\infty) \to V$ and $e : E \hookrightarrow W$ is the inclusion. In other words, Θ_j equals $\rho\wedge\pi^*(\theta_j)$ viewed as a form on W.

Theorem 2.7.1. *Let $\mathbf{H} = \mathbf{H}_{V,\alpha,J}$ be the Hamiltonian associated with the contact manifold V. Set*

$$\mathbf{H}^j(t_1,\dots,t_l,q,p) = \left(\frac{\partial \mathbf{H}}{\partial s_j}\Big(\sum_{i=1}^{l} t_i\delta_i + s_j\theta_j, q, p\Big)\right)\Bigg|_{s_j=0}, \quad j = 1,\dots m,$$

$$\mathbf{F}^0(t_1,\dots,t_k,p) = \mathbf{F}_{W,J}\Big(\sum t_i\Delta_i, p\Big),$$

and define $\mathbf{F}(t_1,\dots,t_k,\tau_1,\dots,\tau_m,p)$ by the formula

$$e^{\mathbf{F}(t_1,\dots,t_k,\tau_1,\dots,\tau_m,p)} = e^{\mathbf{F}^0(t_1,\dots,t_k,p)}\overleftarrow{\mathbf{G}}(t_1,\dots,t_l,\tau_1,\dots,\tau_m,p), \tag{84}$$

where we denote by $\overleftarrow{\mathbf{G}}$ the operator obtained from

$$\mathbf{G}(t_1,\dots,t_l,\tau_1,\dots,\tau_m,q,p) = e^{\tau_m\mathbf{H}^m(t_1,\dots,t_l,q,p)}\dots e^{\tau_1\mathbf{H}^1(t_1,\dots,t_l,q,p)} \tag{85}$$

by quantizing $q_\gamma = \kappa_\gamma \hbar \overleftarrow{\frac{\partial}{\partial p_\gamma}}$. *Then* $\mathbf{F}(t_1,\dots,t_k,\tau_1,\dots,\tau_m,p)$ *is homotopic to*

$$\mathbf{F}_{W,J}\Big(\sum_{i=1}^{k} t_i\Delta_i + \sum_{r=1}^{m}\tau_r\Theta_r, p\Big).$$

Proof. Set

$$T^j(s) = \mathbf{F}_{W,J}\Big(\sum t_i\Delta_i + \sum_{r=1}^{j-1}\tau_r\Theta_r + s\tau_j\Theta_j, p\Big).$$

We have

$$T^j(1) = T^{j+1}(0) \quad \text{for } j = 1,\dots,m-1,$$

$$T^m(1) = \mathbf{F}_{W,J}\Big(\sum_{i=1}^{k} t_i\Delta_i + \sum_{r=1}^{m}\tau_r\Theta_r, p\Big),$$

and

$$T^1(0) = \mathbf{F}_{W,J}\Big(\sum t_i\Delta_i, p\Big) = \mathbf{F}^0(t_1,\dots,t_k,p).$$

Let $S^j \in \hbar^{-1}\mathfrak{D}$ be defined from the equation

$$e^{S^j} = e^{T^j(0)} e^{\tau_j \overleftarrow{\mathbf{H}^j(t_1,\dots,t_l,q,p)}}.$$

It is enough to prove that $T^j(1)$ is homotopic to S^j for $j = 1,\dots,m$. We have

$$\begin{aligned}\frac{\partial e^{T^j(s)}}{\partial s} &= \frac{\partial T^j(s)}{\partial s} e^{T^j(s)} \\ &= e^{T^j(s)} \sum_d \sum_{g,u,v=0}^{\infty} \frac{1}{u!v!} \\ &\quad {}^0\Big\langle \tau_j\Theta_j, \underbrace{\sum_{i=1}^{k} t_i\Delta_i + \sum_{r=1}^{j-1}\tau_r\Theta_r + s\tau_j\Theta_j}_{u}; \underbrace{p,\dots,p}_{v}\Big\rangle_g^d z^d\hbar^{g-1}.\end{aligned}$$

The compactly supported form Θ_j is exact in W,

$$\Theta_j = d\widetilde{\Theta}_j,$$

where $\widetilde{\Theta}_j$ is closed at infinity, has a cylindrical end, and $\widetilde{\Theta}_j|_V = \theta_j$. Hence,

$$\begin{aligned}\frac{\partial T^j(s)}{\partial s} &= \sum_d \sum_{g,u,v=0}^{\infty} \frac{1}{u!v!} \\ &\quad {}^0\Big\langle \tau_j d\widetilde{\Theta}_j, \underbrace{\sum_{i=1}^{k} t_i\Delta_i + \sum_{r=1}^{j-1}\tau_r\Theta_r + s\Theta_j}_{u}; \underbrace{p,\dots,p}_{v}\Big\rangle_g^d z^d\hbar^{g-1}\end{aligned}$$

$$= d\Big(\sum_d \sum_{g,u,v=0}^{\infty} \frac{1}{u!v!}$$

$$_0\Big\langle \underbrace{\tau_j\widetilde{\Theta}_j, \sum_{i=1}^{k} t_i\Delta_i + \sum_{r=1}^{j-1}\tau_r\Theta_r + s\tau_j\Theta_j}_{u}; \underbrace{p,\dots,p}_{v}\Big\rangle_g^d z^d\hbar^{g-1}\Big)$$

$$= \frac{\partial}{\partial u}\Big(d\Big(\sum_d \sum_{g,u,v=0} \frac{1}{u!v!}$$

$$_0\Big\langle \underbrace{\sum_{i=1}^{k} t_i\Delta_i + \sum_{r=1}^{j-1}\tau_r\Theta_r + s\tau_j\Theta_j + u\tau_j\widetilde{\Theta}_j}_{u}; \underbrace{p,\dots,p}_{v}\Big\rangle_g^d z^d\hbar^{g-1}\Big)\Big)\Big|_{u=0}$$

$$= \frac{\partial}{\partial u}\Big(d\Big(\mathbf{F}_{W,J}\Big(\sum_{i=1}^{k} t_i\Delta_i + \sum_{r=1}^{j-1}\tau_r\Theta_r + s\tau_j\Theta_j + u\tau_j\widetilde{\Theta}_j, p\Big)\Big)\Big)\Big|_{u=0},$$

where d denotes the de Rham differential. Using equation (45) we get

$$d\Big(\mathbf{F}_{W,J}\Big(\sum_{i=1}^{k} t_i\Delta_i + \sum_{r=1}^{j-1}\tau_r\Theta_r + s\tau_j\Theta_j + u\tau_j\widetilde{\Theta}_j, p\Big)\Big)$$

$$= e^{-\mathbf{F}_{W,J}}\Big(e^{\mathbf{F}_{W,J}}\overleftarrow{\mathbf{H}}\Big(\sum_{i=1}^{l} t_i\Delta_i + u\tau_j\theta_j, p\Big)\Big),$$

and hence

$$\frac{\partial T^j(s)}{\partial s}$$

$$= \frac{\partial}{\partial u}\Big(e^{-\mathbf{F}_{W,J}}\Big(e^{T^j(s)}\overleftarrow{\mathbf{H}}\Big(\sum_{i=1}^{l} t_i\Delta_i + u\tau_j\theta_j, q, p\Big)\Big)\Big)\Big|_{u=0}$$

$$= \tau_j e^{-T^j(s)}\Big(-\mathbf{F}^j(t_1,\dots,t_k,\tau_1,\dots,\tau_j,s,p)\Big(e^{T^j(s)}\overleftarrow{\mathbf{H}}\Big(\sum_{i=1}^{l} t_i\Delta_i, q, p\Big)\Big)$$

$$+ (e^{T^j(s)}\mathbf{F}^j(t_1,\dots,t_k,\tau_1,\dots,\tau_j,s,p))\overleftarrow{\mathbf{H}}\Big(\sum_{i=1}^{l} t_i\Delta_i, q, p\Big)$$

$$+ e^{T^j(s)}\overleftarrow{\mathbf{H}}^j(t_1,\dots,t_l,q,p)\Big) = \tau_j e^{-T^j(s)}\Big(\big(e^{T^j(s)}[\mathbf{F}^j, \overleftarrow{\mathbf{H}}] + e^{T^j(s)}\overleftarrow{\mathbf{H}}^j\Big),$$

where

$$\mathbf{F}^j(t_1,\dots,t_k,\tau_1,\dots,\tau_j,s,p)$$

$$= \frac{\partial}{\partial u}\Big(\mathbf{F}\Big(\sum t_i\Delta_i + \sum_{r=1}^{j-1}\tau_r\Theta_r + s\tau_j\Theta_j + u\widetilde{\Theta}_j, p\Big)\Big)\Big|_{u=0}$$

and

$$\mathbf{H}^j(t_1,\dots,t_l,q,p) = \frac{\partial}{\partial u}\Big(\mathbf{H}\Big(\sum t_i\Delta_i + u\theta_j, q, p\Big)\Big)\Big|_{u=0}.$$

Therefore,

$$\frac{\partial e^{T^j(s)}}{\partial s} = e^{T^j(s)}\frac{\partial T^j(s)}{\partial s} = \tau_j e^{T^j(s)}\big([\mathbf{F}^j,\overleftarrow{\mathbf{H}}] + \overleftarrow{\mathbf{H}}^j\big).$$

Let us define now a family $U^j(s) \in \hbar^{-1}\mathfrak{D}$, $s \in [0,1]$ by the formula

$$e^{U^j(s)} = e^{T^j(s)}e^{(1-s)\tau_j\overleftarrow{\mathbf{H}^j}(t_1,\dots,t_l,q,p)}. \tag{86}$$

Then $U^j(s)$ is a homotopy between S^j and $T^j(1)$. Indeed, we have $U^j(0) = S^j$ and $U^j(1) = T^j(1)$. On the other hand we get an equation

$$\frac{\partial e^{U^j(s)}}{\partial s} = \tau_j e^{U^j(s)}\big(-\overleftarrow{\mathbf{H}}^j + [\mathbf{F}^j,\overleftarrow{\mathbf{H}}] + \overleftarrow{\mathbf{H}}^j\big) = e^{U^j(s)}[\tau_j\mathbf{F}^j,\overleftarrow{\mathbf{H}}],$$

which is the definition of homotopy (see section 2.4 above). □

We formulate now a version of Theorem 2.7.1 for rational potentials. Set

$$\mathbf{h}^j(t_1,\dots,t_l,q,p) = \frac{\partial \mathbf{h}}{\partial s_j}\Big(\sum t_i\delta_i + s_j\theta_j, q, p\Big),\ j = 1,\dots m,$$

and for any $\mathbf{g} \in \mathfrak{L}$ we denote by $L_{\mathbf{g}}$ the Lagrangian variety of $\mathbf{V}$, defined by the equation

$$L_{\mathbf{g}} = \Big\{q_\gamma = \kappa_\gamma\frac{\partial \mathbf{g}}{\partial p_\gamma}\Big\}. \tag{87}$$

Theorem 2.7.2. *Let $\mathbf{f}(t_1,\dots,t_k,\tau_1,\dots,\tau_m,p)$ be the solution of the Hamilton-Jacobi equation*

$$\frac{\partial \mathbf{f}}{\partial \tau_j}(t_1,\dots,t_k,\tau_1,\dots,\tau_m) = \mathbf{h}^j(t_0,\dots,t_l,q,p)|_{L_{\mathbf{f}}} \tag{88}$$

with the initial condition

$$\mathbf{f}|_{\tau_j=0} = \mathbf{f}_{W,J}\Big(\sum t_i\Delta_i + \sum_{r\neq j}\tau_r\Theta_r, p\Big).$$

Then

$$\mathbf{f}(t_1,\dots,t_k,\tau_1,\dots,\tau_m,p)$$

is homotopic to

$$\mathbf{f}_{W,J}\Big(\sum t_i\Delta_i + \sum_{r=1}^{m}\tau_r\Theta_r, p\Big).$$

2.8 Invariants of Legendrian knots. Symplectic Field Theory can be extended to include Gromov-Witten invariants of pairs (W, L), where L is a Lagrangian submanifold of a symplectic manifold W. The corresponding relative object is a pair (W, L), where $W = \overrightarrow{V^-V^+}$ is a directed symplectic cobordism between contact manifolds $(V^\pm, \alpha^\pm)$, and L is a Lagrangian cobordism between Legendrian submanifolds $\Lambda^\pm \subset V^\pm$. More precisely, we assume that Lagrangian submanifold L is cylindrical at infinity over $\Lambda^\pm$, i.e. there exists $C > 0$, such that $L \cap V^- \times (-\infty, -C] = \Lambda^- \times (-\infty, -C]$ and $L \cap V^+ \times (C, \infty] = \Lambda^+ \times (C, \infty]$. In other words, we require L to coincide at infinity with symplectizations of Legendrian submanifolds $\Lambda^\pm$.

The moduli space of holomorphic curves to be considered in this case consists of equivalence classes of holomorphic curves with boundary which can have punctures of two types, interior and at the boundary. The boundaries of holomorphic curves are required to be mapped to the Lagrangian submanifold L, the holomorphic curves should be cylindrical over periodic orbits from $\mathcal{P}^\pm$ at interior punctures, while at boundary punctures we require them to be asymptotically cylindrical over Reeb chords connecting points of the Legendrian submanifolds $\Lambda^\pm \subset V^\pm$. A more precise definition is given below. The algebraic structure arising from the stratification of boundaries of these moduli spaces is more complicated than in the closed case. First of all, unlike the interior punctures the punctures at the boundary are cyclically ordered, which leads to associative, rather than graded commutative algebras. Second, the "usual" cusp degenerations of curves with boundary at boundary points (see [Gro1]) has in this case codimension 1, rather than 2 as in the closed case, and hence the combinatorics of such degenerations should also be a part of the algebraic formalism.

We will sketch in this paper only the simplest of three cases of SFT, namely the "classical case", which corresponds to the theory of moduli spaces of holomorphic disks with only 1 positive puncture at the boundary.

Let $(V, \xi = \{\alpha = 0\})$ be a contact manifold with a fixed contact form α, $W = V \times \mathbb{R}$ its symplectization with a compatible almost complex structure J, $\Lambda \subset V$ a compact Legendrian submanifold, and $L = \Lambda \times \mathbb{R} \subset W$ the symplectization of Λ, i.e. the corresponding Lagrangian cylinder in W. We assume that all periodic orbits of the Reeb vector field R_α are non-degenerate and fix a marker on every periodic orbit. We also consider the set $\mathcal{C}$ of Reeb chords connecting points on Λ, and impose an extra non-degeneracy condition along the chords from $\mathcal{C}$ by requiring that the linearized flow of R_α along a chord $c \in \mathcal{C}$ connecting points $a, b \in \Lambda$ sends the tangent space $T_a(\Lambda)$ to a space transversal to $T_b(\Lambda)$. We also require that none of the chords from $\mathcal{C}$ be a part of an orbit from $\mathcal{P}$. Under these assumptions, the set $\mathcal{C}$ is finite: $\mathcal{C} = \{c_1, \dots, c_m\}$.

We will restrict the consideration to the case when

$$\pi_1(V) = 0,\ \pi_2(V, \Lambda) = 0, \quad \text{and} \quad w_2(\Lambda) = 0. \tag{89}$$

First two assumptions are technical and can be removed (comp. Section 1.2 above). However, the third one is essential for orientability of the involved moduli spaces of holomorphic curves. Moreover, the invariants we define depends on a particular choice of a spin-structure on Λ.[14]

As in section 1.2 we choose capping surfaces F_γ for $\gamma \in \mathcal{P}$, and for each chord $c \in \mathcal{C}$ we also choose a surface G_c which is bounded by a curve $c \cup \delta_c$, where $\delta_c \subset \Lambda$. The choice of surfaces F_γ, $\gamma \in \mathcal{P}$, allows us to define Conley-Zehnder indices of periodic orbits (see section 1.2 above). Similarly, surfaces G_c enable us to define *Maslov indices* $\mu(c)$, $c \in \mathcal{C}$. Namely, let us consider a Lagrangian sub-bundle of $\xi|_{\partial G_c}$, which consists of the Lagrangian sub-bundle $T\Lambda|_{\delta_c} \subset \xi|_{\delta_c}$ over δ_c, together with the family of Lagrangian planes $T_u \subset \xi_u, u \in c$, which are images of $T_a(\Lambda)$ under the linearized flow of the Reeb field R_α. Choose a symplectic trivialization of $\xi|_{\partial G_c}$ which extends to G_c. With respect to this trivialization the above sub-bundle can be viewed as a path of Lagrangian planes in a symplectic vector space. The Maslov index of such path is defined as in [RS].

Consider a unit disk $D \subset \mathbb{C}$ with punctures

$$\left(\{z^+, z_1^-, \dots, z_\sigma^-\} \cup \{x_1^-, \dots, x_s^-\}\right),$$

where $\mathbf{z} = \{z^+, z_1^-, \dots, z_\sigma^-\}$, $0 \le \sigma \le m$, is a counter-clockwise ordered set of punctures on ∂D, and $\mathbf{x} = \{x_1^-, \dots, x_s^-\}$ is an ordered set of interior punctures. As usual we provide interior punctures with asymptotic markers.

Let us denote by $\mathcal{M}^A(\{c_{i_1}, \dots, c_{i_\sigma}\}, \{\gamma_1, \dots, \gamma_s\}, c_i; W, \Lambda, J)$ the moduli space of J-holomorphic maps

$$(D \setminus (\mathbf{z} \cup \mathbf{x})), \partial(D \setminus (\mathbf{z} \cup \mathbf{x})) \to (W, L),$$

which are asymptotically cylindrical at the negative end over the periodic orbit γ_k^- at the puncture $x_{i_k}^-$, and over the chord c_{i_k} at the puncture z_k, asymptotically cylindrical at the positive end over the chord c_i at the puncture z^+, and which send asymptotic markers of interior punctures to the markers on the corresponding periodic orbits.

[14]We thank K. Fukaya for pointing this out.

Two maps are called equivalent if they differ by a conformal map $D \to D$ which preserves all punctures, marked points and asymptotic markers. Each moduli space $\mathcal{M}^A(c_i, \{c_{i_1}, \dots, c_{i_\sigma}\}, \{\gamma_1, \dots, \gamma_s\}; W, \Lambda, J)$ is invariant under translations $V \times \mathbb{R} \to V \times \mathbb{R}$ along the factor $\mathbb{R}$, and we denote the corresponding quotient moduli space by

$$\mathcal{M}^A\big(c_i, \{c_{i_1}, \dots, c_{i_\sigma}\}, \{\gamma_1, \dots, \gamma_s\}; W, \Lambda, J\big)/\mathbb{R}\,.$$

Let $(\mathfrak{A}, \partial) = (\mathfrak{A}(V, \alpha), \partial_J)$ be the graded commutative differential algebra defined in section 2.2.3 above, or rather its specialization at the point 0. Consider a graded associative algebra $\mathfrak{K} = \mathfrak{K}(V, \Lambda, \alpha)$ generated by elements $c_i \in \mathcal{C}$ with coefficients in the algebra $\mathfrak{A}$. We define a differential $\partial_\Lambda = \partial_{\Lambda,J}$ on $\mathfrak{K}$ first on the generators c_i by the formula

$$\partial_\Lambda(c_i) = \sum \frac{n_{\Gamma,I,d}}{k! \prod_1^k \kappa_{\gamma_j}^{i_j} i_j!} c_{j_1} \dots c_{j_\sigma} q_{\gamma_1}^{i_1} \dots q_{\gamma_k}^{i_k} z^d, \tag{90}$$

where the sum is taken over all $d \in H_2(V)$, all ordered sets of different periodic orbits $\Gamma = \{\gamma_1, \dots, \gamma_k\}$, all multi-indices $J = (j_1, \dots, j_\sigma)$, $1 \le j_i \le m$, and $I = (i_1, \dots, i_k)$, $i_j \ge 0$, and where the coefficient $n_{\Gamma,I,d}$ counts the algebraic number of elements of the moduli space

$$\mathcal{M}^d\Big(c_i, \{c_{j_1}, \dots, c_{j_\sigma}\}, \{\underbrace{\gamma_1, \dots, \gamma_1}_{i_1}, \dots, \underbrace{\gamma_k, \dots, \gamma_k}_{i_k}\}\Big)\Big/\mathbb{R},$$

if this space is 0-dimensional, and equals 0 otherwise. The differential extends to the whole algebra $\mathfrak{K}$ by the graded Leibnitz rule. However, it does not treat coefficients as constants: we have $\partial_\Lambda(q_\gamma) = \partial(q_\gamma)$, where ∂ is the differential defined on the algebra $\mathfrak{A}$.

Then we have

Proposition 2.8.1.

$$\partial_\Lambda^2 = 0.$$

Given a family of contact forms $\Lambda_\tau, \alpha_\tau, J_\tau$ $\tau \in [0,1]$ of Legendrian submanifolds, contact forms, and compatible almost complex structures one can define, similar to the case of closed contact manifolds (see sections 2.3.2 and section 2.4 above) a homomorphism of differential algebras

$$\Psi_S : \mathfrak{K}(V, \Lambda_0, \alpha_0) \to \mathfrak{K}(V, \Lambda_1, \alpha_1),$$

which is independent up to homotopy of the choice of a connecting homotopy. Composition of homotopies generates composition of homomorphisms, and hence one conclude

Proposition 2.8.2. *The quasi-isomorphism type of the differential algebra*

$$\big(\mathfrak{K}(V, \Lambda, \alpha), \partial_{\Lambda,J}\big)$$

depends only on the contact structure ξ *and the Legendrian isotopy class of* Λ. *The* Legendrian contact homology *algebra* $H_*(\mathfrak{K}, \partial_\Lambda)$ *has a structure of a module over the contact homology algebra* $H_*^{\mathrm{cont}}(V,\xi) = H_*(\mathfrak{A}, d)$, *and it is an invariant of the Legendrian knot (or link)* Λ.

The theory looks especially simple when the contact structure ξ on V admits a contact form α such that the Reeb vector field R_α has no closed periodic orbits. If, in addition the space of trajectories is a manifold M (e.g. when $V = J^1(N) = T^*(N) \times \mathbb{R}$ with a contact form $dz + pdq$), then W is automatically symplectic, and the projection $\pi : W \to V$ sends the Legendrian submanifold $\Lambda \subset V$ to an immersed *Lagrangian* submanifold $L \looparrowright M$ with transverse self-intersection points. These points correspond to Reeb chords c_i connecting points on Λ. Hence, the algebra $\mathfrak{K}$ in this case is just a free associative algebra, generated over $\mathbb{C}$ (or $\mathbb{C}$) by the self-intersection points of L. It is possible to choose a compatible almost complex structures J on the symplectization $W = V \times \mathbb{R}$ and J_M on M to make the projection $W \to M$ holomorphic (comp. section 2.9.2 below). Then punctured holomorphic disks in W from moduli spaces $\mathcal{M}^A(c_i, \{c_{i_1}, \dots, c_{i_\sigma}\}; W, \Lambda, J)$ project to J_M-holomorphic disks in M with boundary in the immersed Lagrangian manifold L. Conversely, one can check that each such disk lifts to a disk from the corresponding moduli space $\mathcal{M}^A(c_i, \{c_{i_1}, \dots, c_{i_\sigma}\}; W, \Lambda, J)$, uniquely, up to translation along the $\mathbb{R}$-factor in $W = V \times \mathbb{R}$. This is especially useful when $\dim M = 2$. In this case L is an immersed curve, and the holomorphic disks are precisely immersed, or branched disks with their boundaries in L. Moreover, branched disks are never rigid, because the branching point may vary. Hence, the differential $\partial : \mathfrak{K} \to \mathfrak{K}$ can be defined in this case in a pure combinatorial way, just summing the terms corresponding to all appropriate immersed disks whose boundary consists of arcs of L, and which are locally convex near their corner.

Yu. Chekanov independently realized (see [C]) this program for Legendrian links in the standard contact $\mathbb{R}^3$. He was also motivated by the hypothetical description of the compactification of the moduli spaces of holomorphic discs, but has chosen to prove the invariance of the quasi-isomorphism type of the differential algebra $(\mathfrak{K}, \partial)$ in a pure combinatorial way. In fact, he proved a potentially stronger form of equivalence of differential algebras of isotopic Legendrian links, which he called stable tamed isomorphism. Stable tame isomorphism implies quasi-isomorphism, but we do not know whether it is indeed stronger. Let also note that Chekanov considered a $\mathbb{Z}_2$-version of the theory. In some examples it works better the $\mathbb{Q}$-version, which is provided by our formalism. J. Etnyre–J. Sabloff ([EtS]) and L. Ng ([N]) worked out the combinatorial meaning of signs dictated by the coherent orientation theory (see section 1.8 above), and proved invariance of the stable tame type of the differential algebra $(\mathfrak{K}, \partial)$ *defined over* $\mathbb{Z}$. Note that Chekanov's paper [C] also contains examples which show that the stable tame $\mathbb{Z}_2$-isomorphism type do distinguish some Legendrian knots, which could not be previously distinguished.

Similar to the absolute case of SFT, one can define further invariants of Legendrian submanifolds by including in the formalism higher-dimensional moduli spaces. For instance, by introducing marked points on the boundary of the disk one gets a non-commutative deformation of Legendrian contact homology along the homology of Legendrian manifolds. This is useful, in particular, to define invariants of Legendrian links with ordered components. However, the full-scale generalization of Symplectic Field Theory to directed symplectic-Lagrangian cobordisms between pairs of contact manifolds and their Legendrian submanifolds, which would formalize information about moduli spaces of holomorphic curves of arbitrary genus and arbitrary number of positive and negative punctures, is not straightforward due to existence of different type of codimension 1 components on the boundary of the corresponding moduli spaces. We will discuss this theory in one of our future papers.

2.9 Remarks, examples, and further algebraic constructions in SFT.

2.9.1 Dealing with torsion elements in H_1. Let us discuss grading issues for a contact manifold $(V, \xi = \{\alpha = 0\})$ in the case when the torsion part of $H_1(V)$ is non-trivial. As we will see it is impossible to assign in a coherent way an integer grading to torsion elements and still keep the property that the Hamiltonian $\mathbf{H}$ has total grading -1. We will deal with this problem by assigning to some elements fractional degrees, and thus obtain a rational grading, incompatible with the canonical $\mathbb{Z}_2$-grading. In fact the term "grading" is misleading in this case, and more appropriately one should talk about an Euler vector field with rational coefficients.

Let us split $H_1(V)$ as $T \oplus F$, where T and F are the torsion and free parts, respectively. As in section 1.2 above let us choose curves $C_1, \dots, C_k$ representing a basis of F, fix a trivialization of the bundles ξ_{C_i}, for any periodic orbit $\gamma \in \mathcal{P}_\alpha$ with $[\gamma] \in F$ choose a surface F_γ which realizes the homology between $[\gamma]$ and a linear combinations $\sum n_i[C_i]$, and trivialize the bundle $\xi|_\gamma$ accordingly. For any other periodic orbit γ let γ_l be its smallest multiple which belong to F. In particular, the bundle ξ_{γ_l} is already trivialized by a framing f. The problem is that in general there is no framing over γ which would produce f over γ_l. Choose then an arbitrary framing g over γ and denote by g_l the resulting framing over γ_l. Let $2m(g_l, f) \in \pi_1(Sp(2n-2, \mathbb{R})) = \mathbb{Z}$ be the Maslov class of the framing g_l with respect to f. The Conley-Zehnder indices of γ_l with respect to these two gradings are then related by the formula

$$\mathrm{CZ}(\gamma|f) = \mathrm{CZ}(\gamma|g_l) + 2m(g_l, f).$$

We then assign to γ the fractional degree

$$\deg \gamma = \mathrm{CZ}(\gamma|g) - \frac{2m(g_l, f)}{l}. \tag{91}$$

With this modification SFT can be extended to the case of contact manifolds with no restrictions on H_1. However, the price we pay is that this grading, even if integer, may not be compatible with the universal $\mathbb{Z}_2$-grading which determines the sign rules.

2.9.2 Morse-Bott formalism. Our assumption that all periodic orbits from $\mathcal{P}_\alpha$ for the considered contact forms α are non-degenerate, though generic, but is very inconvenient for computations: in many interesting examples periodic orbits come in continuous families. Sometimes the Reeb flow is periodic, and it sounds quite stupid to destroy this beautiful symmetry.

In fact the above formalism can be adapted to this "Morse-Bott" case. We sketch below how this could be done for the periodic Reeb flow of an S^1-invariant form of a pre-quantization space. We consider below only the "semi-classical" case which concerns moduli spaces of rational holomorphic curves.

Let (M, ω) be a symplectic manifold of dimension $2n-2$ with an integral cohomology class $[\omega] \in H^2(M)$. We will assume for simplicity that $H_1(M) = 0$. The pre-quantization space V is a circle bundle over M with first Chern class equal to $[\omega]$. The fibration $\pi : V \to M$ admits a S^1-connection form α whose curvature is ω. It defines a S^1-invariant contact structure ξ on V, transversal to the fibers of the fibration. The Reeb flow of R_α is periodic, so all its trajectories are closed and coincide with the fibers of the fibration π, or their multiples.

The fiber of the fibration V is a torsion element in $H_1(V)$, and if l is the greatest divisor of the class $[\omega]$ then the l-multiple of the fiber is homological to 0.

Consider the cylindrical cobordism (the symplectization) $W = V \times \mathbb{R}$

with an almost complex structure J compatible with α and denote by $\mathcal{M}_{0,r}(s|W,J,\alpha)$ the moduli space of rational holomorphic curves in W with s punctures and r marked points. Near punctures the curves are required to be asymptotically cylindrical over some fibers of V, or their multiples. However, we do not specify to which particular fiber they are being asymptotic, or whether this fiber is considered on the positive, or negative end of W. We do not equip curves from $\mathcal{M}_{0,r}(s|W,J,\alpha)$ with asymptotic markers of punctures, because we cannot fix in a continuous way points on each simple periodic orbit, as we did in the non-degenerate case.

As it was already mentioned in section 1.4 above, W can be viewed as the total space of the complex line bundle L associated with the S^1-fibration $V \to M$, with the zero-section removed, and the almost complex structure J can be chosen compatible with the structure of this bundle, so that the projection $W \to M$ becomes holomorphic with respect to a certain almost complex structure J_M on M compatible with ω. Then automatically the bundle induced over any holomorphic curve in the base has a structure of a holomorphic line bundle. With this choice of J each holomorphic curve $f \in \mathcal{M}_{0,r}(s|W,J,\alpha)$ projects to a J_M-holomorphic sphere $\overline{f} : \mathbb{C}P^1 \to M$, and can be viewed as a meromorphic section of the induced bundle $(\overline{f})^*L$ over $\mathbb{C}P^1$. This bundle is ample, and hence poles of its sections correspond to the *negative* ends of f, while zeroes correspond to the positive ones. Notice that although the moduli spaces $\mathcal{M}_{0,r}(s|W,J,\alpha)$ can be identified with the moduli spaces of closed holomorphic curves in a $\mathbb{C}P^1$-bundle over M with prescribed tangencies to two divisors, their compactifications are different, and in particular the compactification of the first moduli space may have codimension one strata on its boundary.

The correspondence $f \mapsto \overline{f}$ define a fibration

$$\mathrm{pr} : \mathcal{M}_{0,r}(s|W,J,\alpha)/\mathbb{R} \to \mathcal{M}_{0,r+s}(M,J_M).$$

The fiber $\mathrm{pr}^{-1}(\overline{f})$ is the union of (an infinite number of) disjoint circles, which are indexed by sequences of integers $(k_1,\dots,k_{s+r})$ with $\sum k_i = d_0 = \int_A \omega$, where $A \in H_2(M)$ is the homology class realized by $\overline{f}$, and where there are precisely s non-zero coefficients k_i.

Let us consider two copies $\mathcal{P}^\pm$ of the space $\mathcal{P} = \mathcal{P}_\alpha$ of periodic orbits, as we need to differentiate between positive and negative ends of holomorphic curves. We will write $\ddot{\mathcal{P}} = \mathcal{P}^+ \cup \mathcal{P}^-$ and define the evaluation maps:

$$ev_0 : \mathcal{M}_{0,r}(s|W,J,\alpha)/\mathbb{R} \to V^{\times r} \quad \text{and} \quad ev^+_- : \mathcal{M}_{0,r}(s|W,J,\alpha/\mathbb{R}) \to \ddot{\mathcal{P}}^{\times s}\,. \tag{92}$$

Here ev^+_- associates with each puncture the corresponding point of $\ddot{\mathcal{P}}$.

The space $\mathcal{P}^{\pm}$ can be presented as $\coprod_{k=1}^{\infty}\mathcal{P}_k^{\pm}$, where each $\mathcal{P}_k^{\pm}$ is a copy of M, associated with k-multiple orbits.

We will denote forms on $\mathcal{P}^+$ by p, on $\mathcal{P}^-$ by q, denote by p_k, q_k their restrictions to $\mathcal{P}_k^{\pm}$, and organize them into Fourier series $u=\sum_{k=1}^{\infty}(p_k e^{ikx}+q_k e^{-ikx})$. If we are given a basis of $H^*(M)$ represented by forms $\Delta_1,\dots,\Delta_a$ we will consider only forms from the space generated by this basis, and write $p_k=\sum_{i=1}^{a}p_{k,i}\Delta_i$, $q_k=\sum_{i=1}^{a}q_{k,i}\Delta_i$ and denote by u_i the Δ_i-component of u, i.e.

$$u_i=\sum_{k=1}^{\infty}\left(p_{k,i}e^{ikx}+q_{k,i}e^{-ikx}\right)\quad\text{and}\quad u=\sum_{1}^{a}u_i\Delta_i.$$

Given a closed form t on V and a class $A\in H_2(V)$ we define the correlator

$$_{-1}\langle\underbrace{t,\dots,t}_{r};\underbrace{u,\dots,u}_{s}\rangle_0^A=$$

$$\int\limits_{\mathcal{M}^A_{0,r}(s|W,J,\alpha)/\mathbb{R}} ev_0^*(t\otimes\cdots\otimes t)\wedge(ev_-^+)^*(u\otimes\cdots\otimes u)\Big|_{x=0}. \tag{93}$$

Let us choose a basis $A_0,\dots,A_N$ in $H_2(M)$ in such a way that $\int_{A_0}\omega=l>0$ and $\int_{A_i}\omega=0$ for $i=1,\dots,N$. Then the classes $A_i, i\geq 1$, lift to classes $\widetilde{A}_i\in H_2(V)$ which under the assumption $H_1(M)=0$ form a basis of $H_2(V)$. The degree $d=(d_1,\dots,d_N)$ of a class $A\in H_2(V)$ is a vector of its coordinates in this basis.

To associate an absolute homology class with a holomorphic curve we pick the l-multiple (recall that l denotes the greatest divisor of ω) of the fiber γ over a point $x\in M$ and choose a lift of the surface representing the class A_0 with $\int_{A_0}\omega=l$ as a spanning surface F_γ. Any other m-multiple of γ we will cap with the chain $\frac{m}{l}[F_\gamma]$. However, to fix a spanning surface for a fiber over any other point $y\in M$ or its multiples, one needs to make some extra choices, for instance fix a path connecting x and y. The condition $H_1(M)=0$ guarantees independence of the homology class of the resulting surface of the choice of this connecting path. Notice that with this choice, the degree of $f\in\mathcal{M}_{0,r}(s|W,J,\alpha)/\mathbb{R}$ equals $(d_1,\dots,d_N)$, if the degree of its projection $\mathrm{pr}(f)\in\mathcal{M}_{0,r+s}(M,J_m)$ is equal to $(d_0,d_1,\dots,d_N)$.

In this notation the rational Hamiltonian $\mathbf{h}=\mathbf{h}_{V,J,\alpha}$ is defined by the formula

$$\mathbf{h}(t,u)=\sum_d\sum_{r,s=0}^{\infty}\frac{1}{r!s!}\,{}_{-1}\Big\langle\underbrace{t,\dots,t}_{r};\underbrace{u,\dots,u}_{s}\Big\rangle_0^d z^d. \tag{94}$$

Suppose that a basis of $H^*(M)$, represented by closed forms $\Delta_1,\dots,\Delta_a$, is chosen in such a way that for $b\le a$ the system forms $\widetilde{\Delta}_j=\pi^*(\Delta_j)$, $j=1,\dots,b$, generate the image $\pi^*(H^*(M))\subset H^*(V)$, and the forms $\widetilde{\Theta}_1,\dots,\widetilde{\Theta}_c$, complete it to a basis of $H^*(V)$. We will denote (graded) coordinates in the space generated by the forms $\widetilde{\Delta}_j$, $j=1,\dots,b$ and $\widetilde{\Theta}_1,\dots,\widetilde{\Theta}_c$ by $(t,\tau)=(t_1,\dots,t_b,\tau_1,\dots,\tau_c)$.

As usual, the Hamiltonian $\mathbf{h}$ is viewed as an element of a graded commutative Poisson algebra $\mathfrak{P}$, which consists of formal power series of coordinates of vectors p_k and $T=(t,\tau)=(t_1,\dots,t_b,\tau_1,\dots,\tau_c)$ with coefficients which are polynomials of coordinates of vectors $q_k=(q_{k,1},\dots,q_{k,a})$. The coefficients of these polynomials belong to a certain completion (see condition (38) above) of the group algebra of $H_2(V)$. All the variables $p_{k,i},q_{k,i}$ have in this case the same parity as forms Δ_i, the parity of variables t_i and τ_j is the same as the degree of the corresponding forms $\widetilde{\Delta}_i$ and $\widetilde{\Theta}_j$. If $l=1$, i.e. when $H_1(V)=0$, then the integer grading of variables which corresponds to the choice of capping surfaces described above is defined as follows:

$$\begin{aligned}\deg t_i &= \deg\widetilde{\Delta}_i-2;\\ \deg\tau_i &= \deg\widetilde{\Theta}_i-2;\\ \deg q_{k,i} &= \deg\Delta_i-2+2ck;\\ \deg p_{k,i} &= \deg\Delta_i-2-2ck;\\ \deg z_i &= -2c_1(A_i)\end{aligned}\tag{95}$$

where $c=c_1(A_0)$. As it was explained in section 2.9.1 if $l>1$ one can only define fractional degrees, given by the above formulas (95) with $c=\frac{c_1(A_0)}{l}$.

The following proposition is useful for applications (see below the discussion of Biran-Cieliebak conjecture about subcritical symplectic manifolds). It follows from the fact that all the moduli spaces $\mathcal{M}_{g,r}(s|W,J,\alpha)$ which we defined above are even-dimensional.

PROPOSITION 2.9.1. *Let (V,ξ) be the contact pre-quantization space for a symplectic manifold (M,ω). Then all contact homology algebras*

$$H_*^{\mathrm{SFT}}(V,\xi)\big|_{t=0},\,H_*^{\mathrm{RSFT}}(V,\xi)\big|_{t=0},\,H_*^{\mathrm{cont}}(V,\xi)\big|_{t=0}$$

specialized at $0\in H^*(V)$ are free graded, respectively Weyl, Poisson, or commutative algebras, generated by elements

$$p_{k,i},q_{k,i},\quad i=1,\dots,a,\ k=1,\dots.$$

In particular, the parts of all these homology algebras graded by the homology class $w\in H_1(V)$ (see 2.2.9 above) *are non-trivial.*

The Poisson tensor on $\mathfrak{P}$ is determined in the "u-notation" by the following generalization of the formula (28):

$$\frac{1}{2\pi i}\int_0^{2\pi}\langle(\delta u)',\delta v\rangle dx, \tag{96}$$

where $\langle\ ,\ \rangle$ denotes Poincaré pairing on cohomology $H^*(M)$, which is given in the basis $\Delta_1,\dots,\Delta_a$ by the matrix

$$\eta_{ij}=\langle\Delta_i,\Delta_j\rangle=\int_M \Delta_i\wedge\Delta_j.$$

The Poisson tensor can be written in (p,q)-coordinates as

$$\sum_{k=1}^{\infty}k\sum_{i,j=1}^{a}\eta_{ij}\frac{\partial}{\partial p_{k,i}}\wedge\frac{\partial}{\partial q_{k,j}}.$$

It can be shown that the above Hamiltonian $\mathbf{h}$ satisfies the identity $\{\mathbf{h},\mathbf{h}\}=0$, and that the differential Poisson algebra $(\mathfrak{P},d^{\mathbf{h}})$ is quasi-isomorphic to the corresponding differential Poisson algebra defined in section 2.2.3 for any non-degenerate contact form for the same contact structure ξ on V.

The following formula (97), which sometimes allows to compute the Hamiltonian $\mathbf{h}$ of V in terms of the Gromov-Witten invariant $\mathbf{f}=\mathbf{f}_{M,J_M}$ of M, emerged in a discussion of the authors with T. Coates and F. Bourgeois.

Proposition 2.9.2. *Set*

$$\mathbf{h}^j_{W,J}(t,q,p,z)=\frac{\partial\mathbf{h}}{\partial\tau_j}\left(\sum_1^b t_i\widetilde{\Delta}_i+\tau_j\widetilde{\Theta}_j,q,p,z\right)\Bigg|_{\tau_j=0},$$

and

$$\widehat{\mathbf{f}}^j(t,z)=\frac{\partial\mathbf{f}}{\partial s}\left(\sum_1^a t_i\Delta_i+s\pi_*\widetilde{\Theta}_j,z\right)\Bigg|_{s=0},$$

for $j=1,\dots,c$ (comp. Theorem 2.7.2). Then we have

$$\mathbf{h}^j_{W,J}(t_1,\dots,t_b,q,p,z)$$
$$=\frac{1}{2\pi}\int_0^{2\pi}\widehat{\mathbf{f}}^j(t_1+u_1(x),\dots,t_b+u_b(x),u_{b+1}(x),\dots,u_a(x),\tilde{z})dx\ , \tag{97}$$

where $z=(z_1,\dots,z_N)$, $\tilde{z}=(e^{-ilx},z_1,\dots,z_N)$ and l is the greatest divisor of ω.

To prove (97) one just observes that the correlator

$${}^{-1}\langle\widetilde{\Delta}_{j_1},\dots,\widetilde{\Delta}_{j_r},\widetilde{\Theta}_j;u_{i_1}\Delta_{i_1},\dots,u_{i_v}\Delta_v\rangle_0^d$$

equals the Fourier coefficient with e^{ilx} of the correlator

$${}^{0}\langle\Delta_{j_1},\dots,\Delta_{j_r},\pi_*\widetilde{\Theta}_j,u_{i_1}\Delta_{i_1},\dots,u_{i_v}\Delta_v\rangle_0^{\tilde{d}}.$$

Notice that if $\widetilde{\Theta}_j$ is an odd form, then

$$\mathbf{h}\left(\sum_1^b t_i\widetilde{\Delta}_i + \tau_j\widetilde{\Theta}_j, q, p, \tilde{z}\right) = \tau_j \mathbf{h}^j(t, q, p, z),$$

because all terms of $\mathbf{h}$ must contain τ_j. In particular, for $M = \mathbb{C}P^{n-1}$ the manifold V is a rational homology sphere, and thus a volume form Θ on S^{2n-1} generates the odd part of $H^*(V;\mathbb{R})$. Hence, the formula (97) completely determines $\mathbf{h}$. Namely, let $\mathbf{f}(t,z)$ be the Gromov-Witten invariant of $\mathbb{C}P^{n-1}$, and let $\Delta_{2i}, i = 0,\dots,n-1$, be closed forms generating $H^*(\mathbb{C}P^{n-1})$, so that $\Delta_{2n-2} = \pi_*(\widetilde{\Theta})$. Set $\widetilde{\Delta}_0 = \pi^*(\Delta_0)$ and

$$\widehat{\mathbf{f}}(t,z) = \frac{\partial \mathbf{f}(t,z)}{\partial t_{2n-2}}.$$

Then we have

$$\mathbf{h}(t_0\widetilde{\Delta}_0 + \tau\Theta, q, p) = \frac{\tau}{2\pi}\int_0^{2\pi} \widehat{\mathbf{f}}(t_0\Delta_0 + u, e^{-ix})dx. \tag{98}$$

Let us consider some applications of the formula (98).

Contact homology of the standard contact 3-sphere. Let $\pi : V = S^3 \to M = \mathbb{C}P^1$ be the Hopf fibration. V is the pre-quantization space for (S^2,ω) with $\int_{S^2}\omega = 1$. The 0-form $\Delta_0 = 1$ and the symplectic 2-form $\Delta_2 = \omega$ generate $H^*(M)$, the 0-form $\widetilde{\Delta}_0 = \pi^*(\Delta_0)$ and the volume form $\widetilde{\Theta}_3$ with $\pi_*(\widetilde{\Theta}_3) = \Delta_2$ on S^3 generate $H^*(S^3)$. Thus we have functional variables

$$u_j(x) = \sum_{k=1}^{\infty}\left(p_{k,j}e^{ikx} + q_{k,j}e^{-ikx}\right),$$

associated to the classes Δ_j, $j = 0, 2$, and variables t_0 and τ associated to $\widetilde{\Delta}_0$ and $\widetilde{\Theta}_3$. According to (95) we have

$$\deg q_{k,0} = -2 + 4k, \quad \deg q_{k,2} = 4k, \quad \deg p_{k,0} = -2 - 4k,$$
$$\deg p_{k,2} = -4k, \quad \deg t_0 = -2, \quad \deg\tau = 1.$$

The potential $\mathbf{f}$ for $M = \mathbb{C}P^1$ can be written, as it well known (see also section 2.9.3 below), as

$$\mathbf{f} = \frac{t_0^2 t_2}{2} + e^{t_2}z, \tag{99}$$

and hence

$$\widehat{\mathbf{f}} = \frac{t_0^2}{2} + e^{t_2}z.$$

Thus applying (97) we get the following expression for the rational Hamiltonian $\mathbf{h}$ for S^3:

$$\mathbf{h} = \frac{\tau}{2\pi}\int_0^{2\pi}\left(\frac{(t_0+u_0)^2}{2}+e^{u_2-ix}\right)dx = \tau\left(\frac{t_0^2}{2}+\sum_{k\geq 1} q_{k,0}p_{k,0}\right.$$

$$\left. + \sum_{t,s\geq 0}\ \sum_{\substack{i_1,\dots,i_s\geq 0\\ j_1,\dots j_t\geq 0\\ \sum_1^s li_l-\sum_1^t mj_m=1}} \frac{q_{1,2}^{i_1}\dots q_{s,2}^{i_s}p_{1,2}^{j_1}\dots p_{t,2}^{i_t}}{i_1!\dots i_s!j_1!\dots j_t!}\right). \tag{100}$$

Let us use (100) to compute the contact homology algebra

$$H_*^{\mathrm{cont}}(S^3,\xi_0) = H_*(\mathfrak{A}(S^3,J,\alpha),\partial).$$

The part of $\mathbf{h}$ linear in the p-variables has the form

$$\tau\sum_1^\infty\left(p_{k,0}q_{k,0}+p_{k,2}h_k(q_{1,2},\dots,q_{k-1,2})\right),$$

so that the differential $\partial:\mathfrak{A}\to\mathfrak{A}$ is given by the formulas

$$\partial q_{k,2} = k\tau q_{k,0},\quad \partial q_{k,0} = k\tau h_k(q_{1,2},\dots,q_{k-1,2}).$$

Here are few first polynomials h_k:

$$\begin{aligned} h_1 &= 1,\\ h_2 &= q_{1,2},\\ h_3 &= q_{2,2}+\tfrac{1}{2}q_{1,2}^2,\\ h_4 &= q_{3,2}+q_{2,2}q_{1,2}+\tfrac{1}{6}q_{1,2}^3. \end{aligned}$$

Notice that $\mathrm{Im}\partial$ coincides with the ideal $I(\tau)$ generated by τ. Hence, the homology algebra $H_*(\mathfrak{A},\partial)$ specialized over a point $t=(t_0,0)$ is a free graded commutative algebra $\mathfrak{A}_0$ generated by variables $q_{k,0},q_{k,2}$, $k=1,\dots,$ and in particular it has one generator in each even dimension. On the other hand, over any point $t=(t_0,\tau)$ with $\tau\neq 0$ the algebra $H_*(\mathfrak{A},\partial)$ is isomorphic to a proper subalgebra $\mathfrak{A}_1$ of $\mathfrak{A}_0$. It has its first non-trivial generator $g_1 = q_{1,2}-\frac{1}{2}q_{1,0}^2$ in dimension 4.

REMARK 2.9.3. The contact homology of the Lens space $V=L(l,1)$ which is the pre-quantization space for (S^2,ω) with $\int_{S^2}\omega = l$ can be computed similarly. The variables $p_{k,0},q_{k,0},p_{k,2},q_{k,2},t_0$ and τ have the same $\mathbb{Z}_2$-grading, as in the case of S^3, i.e. all of them, except τ are even. However, the grading assigned by the Euler field to $p_{k,0},q_{k,0},p_{k,2},q_{k,2}$ is fractional in this case and given by formulas

$$\deg q_{k,0} = -2+\tfrac{4k}{l},\quad \deg q_{k,2}=\tfrac{4k}{l},\quad \deg p_{k,0}=-2-\tfrac{4k}{l},\quad \deg p_{k,2} = -\tfrac{4k}{l}.$$

The formula for $\mathbf{h}$ takes the form

$$\mathbf{h} = \frac{\tau}{2\pi}\int_0^{2\pi}\left(\frac{(t_0+u_0)^2}{2} + e^{u_2 - ilx}\right)dx. \tag{101}$$

We will not carry on here the computation of the contact homology of the Lens space $L(l,1)$, and only note, that as in the case of S^3 the homology algebra $H_*(\mathfrak{A},\partial)$ specialized over a point $t=0$ is a free graded commutative algebra $\mathfrak{A}_0$ generated by variables $q_{k,0}, q_{k,2}$, $k=1,\dots,$ and over any point $t\neq 0$ the algebra $H_*(\mathfrak{A},\partial)$ is isomorphic to a proper subalgebra $\mathfrak{A}_1$ of $\mathfrak{A}_0$. In particular, *over any point the contact homology algebra $H_*(\mathfrak{A},\partial)$ has no odd elements.*

Distinguishing contact structures on pre-quantizations spaces. Formula (97) can be used for distinguishing contact structures on pre-quantization spaces of certain symplectic manifolds, which have different Gromov-Witten invariants. Here is an example.

PROPOSITION 2.9.4. *Let (M_1,ω_1) and (M_2,ω_2,J_2) be two symplectic 4-manifolds with integral cohomology classes of their symplectic forms. Suppose that for compatible almost complex structures J_1 on M_1 and J_2 on M_2 there are no non-constant J_1-holomorphic spheres in M_1, while M_2 contains an embedded J_2-holomorphic (-1)-sphere S. Then the pre-quantization spaces (V_1,ξ_1) and (V_2,ξ_2) are not contactomorphic.*[15]

REMARK 2.9.5. Even when the manifolds M_1 and M_2 are homeomorphic, the prequantization spaces V_1 and V_2 are not diffeomorphic (even not homotopy equivalent!) for most choices of symplectic forms ω_1 and ω_2, and hence the statement of the theorem is trivial in these cases. However, one can show that for homeomorphic M_1 and M_2 the symplectic forms can always be deformed in the class of symplectic forms compatible with the chosen almost complex structures J_1 and J_2, in order to make V_1 and V_2 diffeomorphic.

To prove Proposition 2.9.4 we will show that the "classical" contact homology algebras $H_*^{\mathrm{cont}}(V_1,\xi_1)$ and $H_*^{\mathrm{cont}}(V_2,\xi_2)$ are not isomorphic.

Let $S_0=S,S_1,\dots,S_m$ be the exceptional J_2-holomorphic spheres in M_2. Then the cohomology classes $S_0^*,\dots,S_m^*,[\omega]$ are linearly independent, where we denote by S_i^* the cohomology class Poincaré-dual to $[S_i]\in$

[15]Yongbin Ruan proved in [Ru2] that under the assumptions of Proposition 2.9.4 the symplectic manifolds $(M_1,\omega_1)\times\mathbb{C}P^1$ and $(M_2,\omega_2)\times\mathbb{C}P^1$ are not symplectomorphic (and not even deformationally equivalent), despite the fact that for a certain choice of M_1 and M_2 (e.g. M_1 is the Barlow surface and $M_2=\mathbb{C}P^2\#8\overline{\mathbb{C}P^2}$), and for appropriate symplectic forms ω_1 and ω_2 the underlying manifolds V_1 and V_2 are diffeomorphic.

$H_2(M_2)$. Hence there exists a class $X \in H^2(M_2)$, such that

$$XS_0^* = 1,\ X[\omega] = 0,\ \text{ and } XS_i^* = 0 \text{ for } 1 \geq i \geq m. \tag{102}$$

Then the potential $\mathbf{f}_{M_2,J_2}(tX)$ coincides with

$$\mathbf{f}_{S,J_2|_S}(tX|_S) = e^t z.$$

Let us choose a basis of closed forms $\Delta_i, i = 0,\dots,a$, generating generating $H^*(M_2)$, so that one of the forms, say Δ_1 represents the class X. The form Δ_1 then lifts to a form $\widetilde{\Delta}_1$ such that $\pi_*\widetilde{\Delta}_1 = \Delta_1$. According to the formula (98) we have

$$\mathbf{h}_V(\tau_1\widetilde{\Delta}_1, u) = \frac{\tau_1}{2\pi}\int_0^{2\pi} e^{u_1 - ilx}dx, \tag{103}$$

where $u = \sum_0^a u_j\Delta_j$, $u_j = \sum_1^\infty (p_{k,j}e^{ikx} + q_{k,j}e^{-ikx})$, $l = \int_S \omega$.

Hence, the contact homology algebra $H_*^{\text{cont}}(M_2,\xi_2)$, specialized at the point $\tau\widetilde{\Delta}$ for $\tau \neq 0$ is isomorphic to the contact homology algebra of the standard contact Lens space $L(l,1) = \pi^{-1}(S) \subset V$, specialized at the volume form $\tau\widetilde{\Delta}|_{L(l,1)}$. It follows then from Remark 2.9.3 that $H_*^{\text{cont}}(V_2,\xi_2)$, specialized at any point has no odd elements. On the other hand, for any 2-dimensional cohomology class $t \in H_2(M_1)$ we have $\mathbf{f}_{M_1,J_1}(t) = 0$, and hence for any 3-form $\widetilde{\Delta}$ on V_1, the formula (98) implies that the Hamiltonian $\mathbf{h}_{V_1}(\widetilde{\Delta}, u)$ vanishes as well, and therefore the contact homology algebra $H_*^{\text{cont}}(M_1,\xi_1)$, specialized at the point $\tau\widetilde{\Delta}, \tau \neq 0$ is a free graded commutative algebra generated by the odd variable τ, and even variables $p'_{k,j}, q'_{k,j}$, which correspond to even dimensional generators of $H^*(M_1)$. Hence the contact manifolds (M_1,ξ_1) and (M_2,ξ_2) are not isomorphic. □

Subcritical symplectic manifolds. The content of this example is a result of our discussion with P. Biran and K. Cieliebak.

In [D2] S. Donaldson generalized the Kodaira embedding theorem by proving that for any closed symplectic manifold (W,ω) with an integral cohomology class of the symplectic form there exists an integer $l > 0$ such that the homology class dual to $l[\omega]$ can be represented by an embedded symplectic hypersurface W_0. In fact, S. Donaldson proved a stronger result, which together with an improvement by Biran-Cieliebak asserts that for a sufficiently large l the hypersurface W_0 can be chosen in such a way that in the complement $W \setminus W_0$ there exists a vector field X with the following properties:

- $L_X\omega = -\omega$, where L_X denotes the Lie derivative along X; in other words, X is conformally symplectic and contracting;
- X is forward integrable;

– X is gradient-like for a Morse function $\phi : W \setminus W_0 \to \mathbb{R}_+$, which coincides with $-\log \mathrm{dist}^2$ near W, where $\mathrm{dist}(x)$ is the distance function from a point x to W_0 with respect to some Riemannian metric.

The vector field X retracts $W \setminus W_0$ to the Morse complex K of the function ϕ, which is automatically isotropic for the symplectic form ω (see [EG]), and, in particular, $\dim K \leq n$. Biran-Cieliebak call the pair (W, W_0) *subcritical* if $\dim K < n$. They constructed in [BiC] several interesting examples of subcritical pairs, and conjectured that *if* (W, W_0) *is subcritical, then* $l = 1$. We sketch below the proof of this conjecture.

First, let us observe that the contact structure ξ, defined by the contact form $\alpha = X \lrcorner \omega$ on the boundary V of a small tubular neighborhood of W_0, is equivalent to the contact structure which is defined on V as the pre-quantization space of the symplectic manifold $(W_0, l\omega)$. On the other hand, the condition, that the pair (W, W_0) is sub-critical implies that the contact manifold (V, ξ) is itself subcritical in the sense of Example 1.9.3.4 above, i.e. it is isomorphic to the strictly pseudo-convex boundary of a subcritical Stein (or Weinstein) manifold with its canonical complex structure. Let us recall (see 2.2.9 above) that all SFT-objects, in particular Floer contact homology $HC_*(V, \xi)$ and the contact homology algebra $H_*^{\mathrm{cont}}(V, \xi)$ are graded by elements of $H_1(V)$. Using arguments as in the theorem of Mei-Lin Yau (see 1.9.8 above) one can show that all non-trivial elements in the contact homology algebra $H_*^{\mathrm{cont}}(V, \xi)$ of a subcritical contact manifold (V, ξ) may correspond only to $0 \in H_1(V)$. On the other hand, it follows from Proposition 2.9.1 above that $H_*^{\mathrm{cont}}(V, \xi)$ specialized at $0 \in H^*(V)$ has non-trivial elements which correspond to the homology class of the fiber in $H_1(V)$. Therefore, $l = 1$.

2.9.3 Computing rational Gromov-Witten invariants of $\mathbb{C}P^n$. We will show in this section how SFT can be used for computing rational Gromov-Witten invariants of $\mathbb{C}P^n$. Our method differs from traditional ways (see [Ko1], [FulP], [GrP], [V], [RuT]) for this computation. We will be simultaneously computing the rational potential of $\mathbb{C}^n$ and the rational Gromov-Witten invariant of $\mathbb{C}P^n$ by a recursion using Theorem 2.7.2

Let us choose basic forms in $\mathbb{C}^n$ as in the previous section, i.e $\Delta = 1$, and Θ is a volume form with compact support in $C^n \setminus 0 = S^{2n-1} \times (0, \infty)$ with $\int_{\mathbb{C}^n} \Theta = 1$. We denote by δ the restriction of Δ to S^{2n-1}. We also assume that Θ splits into a product $\hat{\theta} \wedge \rho$, where $\hat{\theta}$ is pull-back of a unit volume form θ on S^{2n-1}, and ρ is a compactly supported form in $(0, \infty)$.

Set

$$\begin{aligned}\mathbf{h}^1(t_0,q,p)) &= \frac{\partial \mathbf{h}}{\partial \tau}(t_0\delta+\tau\theta,q,p)|_{\tau=0}\\ &= \frac{1}{2\pi}\int_0^{2\pi} \widehat{\mathbf{f}}_{\mathbb{C}P^{n-1}}\big(t_0+u_0(x),u_2(x),\dots,u_{2n-2}(x),e^{-ix}\big)\,dx,\end{aligned} \tag{104}$$

where

$$\widehat{\mathbf{f}}_{\mathbb{C}P^{n-1}}(t_0,\dots,t_{2n-2},z) = \frac{\partial \mathbf{f}_{\mathbb{C}P^{n-1}}}{\partial t_{2n-2}}(t_0,\dots,t_{2n-2},z),$$

$\mathbf{f}_{\mathbb{C}P^{n-1}}(t_0,\dots,t_{2n-2},z)$ is the rational Gromov-Witten invariant of $\mathbb{C}P^{n-1}$, and

$$u_{2j}(x) = \sum_1^\infty p_{k,2j}\,e^{ikx} + q_{k,2j}\,e^{-ikx}.$$

Then the equation (88), which determines $\mathbf{f}(t_0,t_{2n},p) = \mathbf{f}_{\mathbb{C}^n}(t_0\Delta+t_{2n}\Theta,p)$ takes the form

$$\begin{aligned}&\frac{\partial \mathbf{f}}{\partial t_{2n}}(t_0,t_{2n},p) =\\ &\qquad \frac{1}{2\pi}\int_0^{2\pi} \widehat{\mathbf{f}}_{\mathbb{C}P^{n-1}}\big(t_0+u_0(x),u_2(x),\dots,u_{2n-2}(x),e^{-ix}\big)\,dx\big|_{L_{\mathbf{f}}},\end{aligned} \tag{105}$$

where

$$L_{\mathbf{f}} = \left\{q_{k,2j} = k\frac{\partial \mathbf{f}}{\partial p_{k,2n-2j-2}}(t_0,t_{2n},p)\right\}.$$

Together with the initial data

$$\mathbf{f}(t_0,0,p) = \begin{cases} p_{1,0}, & \text{if } n=1;\\ 0, & \text{otherwise}\end{cases} \tag{106}$$

the equation (105) provides a recursive procedure for computing coefficients $f_j(t_0,p)$ of the expansion

$$\mathbf{f}(t_0,t_{2n},p) = \sum_0^\infty f_j(t_0,p)t_{2n}^j.$$

For instance for $n=1$ we have (see Example 2.2.4) $\mathbf{h}^1 = \frac{t_0^2}{2}+\sum_1^\infty p_k q_k$, where we write p_k,q_k instead of $p_{k,0},q_{k,0}$, and hence the equation (105) takes the form

$$\frac{\partial \mathbf{f}}{\partial t_2}(t_0,t_2,p) = \frac{t_0^2}{2} + \sum_0^\infty k p_k \frac{\partial \mathbf{f}}{\partial p_k}(t_0,t_2,p) \tag{107}$$

with the initial data $\mathbf{f}(t_0, 0, p) = p_1$. This linear first order PDE is straightforward to solve, and we get

$$\mathbf{f}(t_0, t_2, p) = \frac{t_2 t_0^2}{2} + e^{t_2} p_1.$$

For $n = 2$ the Hamiltonian $\mathbf{h}$ is given by the formula (100), and we have

$$\mathbf{h}(t_0, \tau, p) = \tau \mathbf{h}^1(t_0, p).$$

Thus the equation for the potential of $\mathbb{C}^2$ has the form

$$\begin{aligned} \frac{\partial \mathbf{f}}{\partial t_4}(t_0, t_4, p) &= \frac{t_0^2}{2} + \sum_{k \geq 1} k \frac{\partial \mathbf{f}}{\partial p_{k,2}} p_{k,0} \\ &+ \sum_{t,s \geq 0} \sum_{\substack{i_1,\dots,i_s \geq 0 \\ j_1,\dots j_t \geq 0 \\ \sum_1^s l i_l - \sum_1^t m j_m = 1}} \frac{\left(\frac{\partial \mathbf{f}}{\partial p_{1,0}}\right)^{i_1} \dots \left(s \frac{\partial \mathbf{f}}{\partial p_{s,0}}\right)^{i_s} p_{1,2}^{j_1} \dots p_{t,2}^{i_t}}{i_1! \dots i_s! j_1! \dots j_t!}; \qquad (108) \\ \mathbf{f}(t_0, 0, p) &= 0. \end{aligned}$$

Hence, we get

$$\begin{aligned} \mathbf{f}(t_0, t_4, p) = t_4 \left(\frac{t_0^2}{2} + p_{1,2}\right) + \frac{t_4^2}{2!} p_{1,0} + \frac{t_4^3}{3!}\left(p_{2,2} + \frac{1}{2} p_{1,2}^2\right) \\ + \frac{t_4^4}{4!}(2 p_{2,0} + p_{1,2} p_{1,0}) + \dots . \end{aligned} \qquad (109)$$

To compute $\mathbf{f}_{\mathbb{C}^n}$ for $n > 2$ we need to know $\mathbf{f}_{\mathbb{C}P^{n-1}}$. So to complete the recursion we will explain now how to express the rational Gromov-Witten invariant $\mathbf{f}_{\mathbb{C}P^{n-1}}$ through $\mathbf{f}_{\mathbb{C}^n}$.

First of all we split, as it is described in Example 1.3.2 above, $\mathbb{C}P^n$ along the boundary of a tubular neighborhood of $\mathbb{C}P^{n-1} \subset \mathbb{C}P^n$ into two completed symplectic cobordism $W_1 = \mathbb{C}^n$ and $W_2 = \mathbb{C}P^n \setminus x$, where we introduce on W_2 a complex structure of the holomorphic line bundle over $\mathbb{C}P^{n-1}$ determined by the hyperplane section $\mathbb{C}P^{n-2} \subset \mathbb{C}P^{n-1}$. We will denote by $\mathbf{f}_1$ and $\mathbf{f}_2$ the potentials for W_1 and W_2, respectively.

Let $\Delta_0, \dots, \Delta_{2n-2}$ be closed forms representing the standard basis of $H^*(\mathbb{C}P^{n-1})$. We will keep the same notation for the pull-backs of these forms to W_2. Let Δ_{2n} be a closed $2n$-form with a compact support, which generates

$$\operatorname{Ker}\big(H^*_{\text{comp}}(W_2) \to H^*(W_2)\big).$$

We are interested in the potential $\mathbf{f}_2(t_0, \dots, t_{2n}, q) = \mathbf{f}_2(\sum_{i=1}^n t_{2i}\Delta_{2i}, q)$. First of all notice that by dimensional reasons the moduli spaces of holomorphic curves which project to non-constant curves in $\mathbb{C}P^{n-1}$ do not contribute to the potential

$$\mathbf{f}_2(t_0,\ldots,t_{2n-2},0,q),$$

and hence we have

$$\mathbf{f}_2\left(\sum_{i=1}^{n-1} t_{2i}\Delta_{2i},q\right) = z\sum_{i=0}^{n-1} q_{1,2i} \sum_{\substack{(s_1,\ldots,s_{n-1})\\ \sum s_j(j-1)=n-i-1}} \prod_{j=1}^{n-1} \frac{t_{2j}^{s_j}}{s_j!}\,. \tag{110}$$

In particular, for $n=2$ we get

$$\mathbf{f}_2(t_0\Delta_0+t_2\Delta_2,q) = ze_2^t q_{1,2}.$$

One can recover $\mathbf{f}_2(t_0,\ldots,t_{2n},q)$ for $t_{2n}\neq 0$ using the equation (88), as we did it above for $W_1=\mathbb{C}^n$. However, for the purpose of our computation of Gromov-Witten invariant $\mathbf{f}_{\mathbb{C}P^n}$ this is not necessary, as we can proceed as follows.

Notice that the above chosen forms Δ_2,Δ_{2n-2} extend to CP^n. On the other hand, we will choose a volume form Δ_{2n} on $\mathbb{C}P^n$ to be supported in the affine part, so that the restriction $\Delta_{2n}|_{\mathbb{C}^n}$ coincides with the form Θ introduced above. Then Theorem (2.5.5) implies that

$$\mathbf{f}_{\mathbb{C}P^n}\left(\sum_{i=1}^{n} t_{2i}\Delta_{2i}\right) =$$

$$\left(\mathbf{f}_1(t_0,t_{2n},p)+\mathbf{f}_2(t_0,\ldots,t_{2n-2},0,q) - \sum_{i+j=n-1}\sum_1^\infty \frac{1}{k} p_{k,2i}q_{k,2j}\right)\Bigg|_L\,, \tag{111}$$

where

$$L = \begin{cases} p_{1,2i} = & z \displaystyle\sum_{\substack{(s_1,\ldots,s_{n-1})\\ \sum s_j(j-1)=i}} \prod_{j=1}^{n-1} \frac{t_{2j}^{s_j}}{s_j!}\,; \\ p_{k,2i} = & 0, \text{ if } k>1\,; \\ q_{k,2i} = & k\dfrac{\partial \mathbf{f}_1}{\partial p_{k,2(n-i-1)}}(t_0,t_{2n},p)\,. \end{cases} \tag{112}$$

Plugging expressions from (112) into equation (111) we get

$$\mathbf{f}_{\mathbb{C}P^n}(t_0,\ldots,t_{2n}) = \mathbf{f}_{\mathbb{C}^n}(t_0,t_{2n},p)\big|_{L_1}\,, \tag{113}$$

where

$$L_1 = \begin{cases} p_{1,2i} & = z \displaystyle\sum_{\substack{(s_1,\ldots,s_{n-1})\\ \sum s_j(j-1)=i}} \prod_{j=1}^{n-1} \frac{t_{2j}^{s_j}}{s_j!}\,; \\ p_{k,2i} & = 0 \text{ if } k>1\,. \end{cases}$$

Indeed, two last terms in the formula (111) cancel each other (as it always happens when $\mathbf{f}_2$ is linear with respect to q-variables). For instance, for $n=1$ we get

$$\mathbf{f}_{\mathbb{C}P^1}(t_0,t_2) = \left(\frac{t_0^2 t_2}{2} + e^{t_2}p_1\right)\Big|_{p_1=z} = \frac{t_0^2 t_2}{2} + e^{t_2}z\,. \tag{114}$$

For $n=2$ we have

$$L_1 = \begin{cases} p_{1,0} & = ze^{t_2}\,; \\ p_{k,i} & = 0, \quad \text{for all other} \quad k,i\,, \end{cases} \tag{115}$$

and hence

$$\mathbf{f}_{\mathbb{C}P^2}(t_0,t_2,t_4) = \mathbf{f}_{\mathbb{C}^2}(t_0,t_4,ze^{t_2},0,\dots)\,. \tag{116}$$

REMARK 2.9.6. The method which we used above for computing of the rational potential of $\mathbb{C}P^n$, when applied to an arbitrary symplectic manifold W, allows us to express $\mathbf{f}_W$ through the potential of the affine part $W\setminus M$. The latter computation seems tractable when the Weinstein manifold $W\setminus M$ is subcritical (see section 1.3 above), i.e. when its isotropic skeleton does not have Lagrangian cells. On the other hand, when Lagrangian cells are present this problem is related to central questions of Symplectic topology.

2.9.4 Satellites. Let $(V,\xi=\{\alpha=0\})$ be a contact manifold, $(W=V\times\mathbb{R}, d(e^t\alpha))$ its symplectization, and J a compatible translation-invariant almost complex structure on W. In this section we will show that the homological Poisson super-algebra $H_*(\mathfrak{P},d^{\mathbf{h}})$ comes equipped with some additional structures, rather unfamiliar in abstract Poisson geometry. Namely, the counting of genus g curves with a fixed complex structure and with a fixed configuration of n points gives rise to an odd n-linear totally symmetric poly-form $\mathbf{h}^{g,n}$ on the Poisson super-space $\mathbf{V}$ underlying $\mathfrak{P}$. The poly-form descends well to the homology and thus yields another invariant of the contact structure which we call the genus-g n-point *satellite* of the Poisson structure.

Let us denote by $\overline{\mathcal{M}_{g,m}(V)/\mathbb{R}}$ the compactified moduli space of stable connected J-holomorphic curves in W which are characterized by the arithmetical genus g and by the total number m of punctures and marked points numbered somehow by the indices $1,\dots,m$ (see section 1.6 above). We emphasize that the moduli space in question contains equivalence classes of all such curves, and in particular, may have infinitely many connected components corresponding to different homotopy types of curves in W and different numbering of the m markings. Let $\overline{\mathcal{M}}_{g,n}$ be the Deligne-Mumford compactification of the moduli space of genus g Riemann surfaces with n marked points. For any g, n with $2g-2+n>0$ and $l \geq 0$ there is a natural *contraction map* $\mathrm{ct}: \overline{\mathcal{M}_{g,n+l}(V)/\mathbb{R}} \to \overline{\mathcal{M}}_{g,n}$ defined by forgetting the map to W and the last l markings followed by the contraction of those components of the curve which have become unstable. Given a differential form τ on $\overline{\mathcal{M}}_{g,n}$ we will denote by $\mathrm{ct}^*\tau$ its pull-back to $\overline{\mathcal{M}_{g,m}(V)/\mathbb{R}}$.

Let $u=(p,q,t)$ denote a point in $\mathbf{V}$, that is p,q and t are (closed) differential forms on $\mathcal{P}^-, \mathcal{P}^+$ and V respectively. We will denote $\mathrm{ev}_i^*\, u$, $i=1,\dots,m$, the pull-back by the evaluation map

$$\mathrm{ev}_i : \overline{\mathcal{M}_{g,m}(V)/\mathbb{R}} \to (\mathcal{P}^- \cup \mathcal{P}^+ \cup V)$$

at the i-th marking. Let us emphasize the point that we are treating here the marked points and punctures on equal footing.

Let $\delta u \in \mathbf{V}$ be a tangent vector to $\mathbf{V}$ at a point $u \in \mathbf{V}$. We introduce the formal function

$$\mathbf{h}_\tau^{g,n} := \frac{1}{n!}\sum_{l=0}^{\infty}\frac{1}{l!}\int\limits_{\overline{\mathcal{M}_{g,n+l}(V)/\mathbb{R}}} \mathrm{ct}^*\tau \wedge \mathrm{ev}_1^*\,\delta u \wedge \dots \wedge \mathrm{ev}_n^*\,\delta u \wedge \mathrm{ev}_{n+1}^*\,u \wedge \dots \wedge \mathrm{ev}_{n+l}^*\,u. \tag{117}$$

It is a super-symmetric n-linear form in δu with coefficients depending on the application point u.

Let $d^{\mathbf{h}}(f)$ denote the Lie derivative of a tensor field f along the odd Hamiltonian vector field $d^{\mathbf{h}}$ on $\mathbf{V}$ with the Hamilton function $\mathbf{h}$.

PROPOSITION 2.9.7. *Let τ be a top degree form on $\overline{\mathcal{M}}_{g,n}$. Then $d^{\mathbf{h}}(\mathbf{h}_\tau^{g,n}) = 0$. If the top degree form $\tau = d\alpha$ is exact then $\mathbf{h}^\tau_{g,n} = d^{\mathbf{h}}(\mathbf{h}^\alpha_{g,n})$. In particular, the tensor field $\mathbf{h}_\tau^{g,n}$ descends to the homology algebra $H_*(\mathfrak{P},\partial)$ into a satellite which depends only on the total volume of τ.*

This follows from the Stokes formula applied to $\mathbf{h}^{d\tau}_{g,n} = 0$ and respectively to $\mathbf{h}^{d\alpha}_{g,n}$. Codimension 1 boundary strata of the moduli space $\mathcal{M}_{g,m}(V)/\mathbb{R}$ are formed by stable curves of height 2. Most of the strata do not contribute to the Stokes formula since they are mapped by the contraction map to complex codimension 1 strata of the Deligne-Mumford space $\overline{\mathcal{M}}_{g,n}$, where τ and α restrict to 0 for dimensional reasons. Exceptions to this rule occur only if one of the two curves which form the stable curve is to be contracted. It is therefore a sphere with glued to the other level of the stable curve along at precisely one end, and which have at most one marked points or ends with the index $\leq n$, and with any number of ends or marked points with indices $> n$. Contributions of such curves to the Stokes formula is expressed bi-linearly via the 1-st or 2-nd derivatives of the Hamilton function $\mathbf{h}$ and the satellite. It is easy to see that the whole expression is interpreted correctly as the Lie derivative of the tensor field $\mathbf{h}^{\tau}_{g,n}$ along the Hamiltonian vector field $d^{\mathbf{h}}$. □

We will assume further on that τ is normalized to the total volume 1 and will often drop it from the notation for the satellite $\mathbf{h}^{g,n}$.

Let us consider now a directed symplectic cobordism $W = \overrightarrow{V_-V_+}$ between two contact boundaries $V_{\pm}$. Then we have the Hamilton function $\widehat{\mathbf{h}} = \mathbf{h}^+ - \mathbf{h}^-$ and the satellites $\widehat{\mathbf{h}}^{g,n} = (\mathbf{h}^{g,n})^+ - (\mathbf{h}^{g,n})^-$ defined as elements of the algebra $\widehat{\mathfrak{L}}$, which in the case when the cobordism is a concordance just equal to the tensor product of the Poisson algebras $\mathfrak{P}_{\pm}$. Also, we have the potential $\mathbf{f}(p_-, q_+, t)$ counting rational J-holomorphic curves in W which defines a Lagrangian correspondence between $\mathfrak{P}^{\pm}$ invariant under the vector field $d^{\widehat{\mathbf{h}}}$ with the Hamilton function $\widehat{\mathbf{h}}$. Finally, using the moduli spaces $\mathcal{M}_{g,m}(W)$ of J-holomorphic curves in the cobordism, we can introduce the satellites $\mathbf{f}^{g,n}_{\tau}$ as symmetric n-forms on the space (p_-, q_+, t)-space parameterizing the Lagrangian correspondence. Then the arguments similar to the above proof of the proposition, but applied this time to $\mathbf{f}^{g,n}_{d\tau} = 0$, show that *the restriction of* $\widehat{\mathbf{h}}^{g,n}_{\tau}$ *to the Lagrangian correspondence defined by* $\mathbf{f}$ *coincides with the Lie derivative of* $\mathbf{f}^{g,n}_{\tau}$ *along the vector field* $d^{\widehat{\mathbf{h}}}$ *restricted to the Lagrangian correspondence* (comp. Theorem 2.3.6 above). In this sense the Lagrangian correspondences defined by symplectic cobordisms preserve the satellite structures defined by $\left(\mathbf{h}^{g,n}\right)^{\pm}$ on the homology $H_*(\mathfrak{P}_{\pm}, d^{\mathbf{h}^{\pm}})$. In particular, *the satellite structures of a contact manifold* V *on the homology* $H_*(\mathfrak{P}, d^{\mathbf{h}})$ *depend only on the contact structure.*

The following discussion is the first steps in the study of the geometric structure defined by the satellites.

Let $\mathbf{h}^{g,n}_{\alpha_1,\dots,\alpha_n}$ denote components of the satellite tensors on $\mathfrak{P}$. Using the Poisson tensor $\pi^{\mu\nu}$ we can couple two satellites with respect to some indices:

$$\mathbf{h}^{g',n'+1}_{\dots\mu}\pi^{\mu\nu}\mathbf{h}^{g'',n''+1}_{\nu\dots}.$$

Similarly, we can couple two indices in $\mathbf{h}^{g-1,n+2}$ with two indices in the 2-nd differential $\delta^2\mathbf{h}$ of the Hamilton function $\mathbf{h}$.

PROPOSITION 2.9.8. *If* $g = g' + g'' > 0$ *then the coupling of* $\mathbf{h}^{g',n'}$ *and* $\mathbf{h}^{g'',n''}$ *is a Lie derivative along* ∂ *and thus vanishes in the homology* $H_*(\mathfrak{P}, \partial)$. *Similarly, the coupling of* $\mathbf{h}^{g-1,n+2}$ *with* $\delta^2\mathbf{h}$ *vanishes in the homology* $H_*(\mathfrak{P}, \partial)$.

The proof is based on some famous but non-trivial property of the spaces $\overline{\mathcal{M}}_{g,n}$ with $g > 0$ to have complex codimension one strata homologically independent. Such strata correspond to different ways of cutting a (g,n)-surface along one circle and can be identified either with $\overline{\mathcal{M}}_{g',n'+1} \times \overline{\mathcal{M}}_{g'',n''+1}$ where $g' + g'' = g, n' + n'' = n$ or with $\overline{\mathcal{M}}_{g-1,n+2}$. The independence property implies that a volume form τ on the stratum, say $\tau' \otimes \tau''$ in the first case, can be obtained as the restriction of a closed codegree two form ω on $\overline{\mathcal{M}}_{g,n}$ which have exact (or even zero, for suitable choices of τ) restrictions to all other codimension-1 strata in $\overline{\mathcal{M}}_{g,n}$. Applying the Stokes formula to $0 = \mathbf{h}^{g,n}_{d\omega}$ we find that the coupling of $\mathbf{h}^{g',n'+1}_{\tau'}$ and $\mathbf{h}^{\tau'',n''+1}_{\tau''}$ (or — in the second case — of $\mathbf{h}^{g-1,n+2}_{\tau}$ and $\delta^2\mathbf{h}$) equals $d^{\mathbf{h}}(\mathbf{h}^{g,n}_{\omega})$.

REMARK 2.9.9. To the contrary, coupling $\mathbf{h}^{0,3}$ with itself via one index is not, in general, a $d^{\mathbf{h}}$-derivative, but instead the following triple sum is:

$$\mathbf{h}^{0,3}_{\alpha\beta\mu}\pi^{\mu\nu}\mathbf{h}^{0,3}_{\nu\gamma\delta} + (-1)^{(\deg\alpha+\deg\beta)\deg\gamma}\mathbf{h}^{0,3}_{\gamma\alpha\mu}\pi^{\mu\nu}\mathbf{h}^{0,3}_{\nu\beta\delta} + (-1)^{\deg\alpha(\deg\beta+\deg\gamma)}\mathbf{h}^{0,3}_{\beta\gamma\mu}\pi^{\mu\nu}\mathbf{h}^{0,3}_{\nu\alpha\delta} \equiv 0\,.$$

This follows from the property of the 3 boundary strata in $\overline{\mathcal{M}}_{0,4}$ to represent the same homology class (use the Stokes formula for $\omega = 1$). In fact $\mathbf{h}^{0,3}$ coincides with the 3-rd differential $\delta^3\mathbf{h}/6$ of the Hamilton function, and the above Jacobi-like identity can be derived by 4 differentiations of $\{\mathbf{h},\mathbf{h}\} = 0$ in $\alpha,\beta,\gamma,\delta$. One can interpret the integrability property $(d^{\mathbf{h}})^2 = 0$ of the odd vector field $d^{\mathbf{h}}$ on $\mathbf{V}$ as a homotopy Lie super-algebra structure on $\Pi\mathbf{V}^*$, the dual space with changed parity. The identity in question corresponds to the Jacobi identity for the remnant Lie super-algebra structure in homology.

It is sometimes convenient to extend the definition of genus 0 satellites to unstable values of n by $\mathbf{h}^{0,n} = \delta^n\mathbf{h}/n!$ for $n = 0,1,2$. Also, one can define the function $\mathbf{h}^{1,0}$ at least locally as a potential for $\mathbf{h}^{1,1}$, using the following

PROPOSITION 2.9.10. *The differential* 1*-form* $\mathbf{h}^{1,1}$ *is closed.*

Indeed, the partial derivatives $\delta_\mu\mathbf{h}^{1,1}_\nu$ and $\delta_\nu\mathbf{h}^{1,1}_\mu$ are identified with the satellites $(\mathbf{h}^{1,2}_\omega)_{\mu\nu}$ corresponding to the 2-form ω on $\overline{\mathcal{M}}_{1,2}$ pulled-back from $\overline{\mathcal{M}}_{1,1}$ by forgetting the 1-st and respectively the 2-nd marked point. But the two maps $\overline{\mathcal{M}}_{1,2} \to \overline{\mathcal{M}}_{1,1}$ coincide.

It would be interesting to study other general properties of satellites which may depend on more sophisticated geometry of Deligne-Mumford compactifications. For instance, what can be said about Poisson brackets among the functions $\mathbf{h}^{g,0}$?

We complete the section with the computation of the satellites in the example $V = S^1$. Let $t = t_0 1 + t_1 d\phi$ denote the general harmonic form on

S^1, $\delta t = \tau_0 1 + \delta t_1 \delta\phi$, $u(x) = t_0 + \sum p_k e^{ikx} + q_k e^{-ikx}$, $\delta u = \delta t_0 + \sum \delta p_k e^{ikx} + \delta q_k e^{-ikx}$.

PROPOSITION 2.9.11. *For $2g - 2 + n \geq 0$ we have*

$$\mathbf{h}^{g,n+1} = \frac{\delta t_1}{2\pi\ n!} \int_0^{2\pi} (u_{xx})^g\ (\delta u)^n\ dx.$$

Let us begin with the remark that the formula does not (and cannot) contain $t = t_0 + t_1\phi$ because $\deg t < 2$, and therefore pushing forward from $\mathcal{M}_{g,m+1}(V)/\mathbb{R} \to \mathcal{M}_{g,m}(V)/\mathbb{R}$ by forgetting the corresponding marked point would send t to 0. Exceptions to this rule could occur only if $\mathcal{M}_{g,m}(V)$ were ill-defined, that is only in the case of constant maps with "unstable" indices, $2g-2+m \leq 0$, which has no effect on the satellites with "stable" indices. On the other hand the factor δt_1 is (and must be) present in the formula since the dimension of the moduli spaces is odd. With this information in mind, the enumerative question equivalent to computation of the satellites can be described as follows. On a Riemann surface Σ of genus g, we are given a divisor D of n distinct points with (possibly zero) multiplicities $m_1, ..., m_n$. We have to count the divisors $l_1 P_1 + ... + l_g P_g$ which in the sum with D form the divisor of a rational function. (In particular, the degree $\sum m_i + \sum l_j$ of the total divisor must vanish.) The answer to this question is equal to the degree of the Abel-Jacobi map $\Sigma^g \to J_\Sigma$ defined by integration of holomorphic differentials $\vec{\omega} = (\omega_1, ..., \omega_g)$ on Σ as

$$(P_1, ..., P_g) \mapsto l_1 \int^{P_1} \vec{\omega} + ... + l_g \int^{P_g} \vec{\omega}.$$

When the multiplicities $(l_1, ..., l_g) = (1, ..., 1)$, the degree equals $g!$ (it is well-known that $S^g\Sigma^g \to J_\Sigma$ is a bi-rational isomorphism). For arbitrary $(l_1, ... l_g)$ the Abel-Jacobi map has the Jacobi matrix $[l_j \omega_i(P_j)]$. Thus the degree equals $l_1^2 ... l_g^2 g!$. Taking these answers as the coefficients in the generating function on the variables t_0, p_l, q_{-l} corresponding to $l = 0, l > 0$ and $l < 0$ we arrive at the factor u_{xx}^g. The other factor $(\delta u)^n/n!$ is similarly accountable for all possible choices of multiplicities $m_1, ..., m_n$ in the divisor D. The contour integration of the product couples the choices with $m_1 + ... + m_n + l_1 + ... + l_g = 0$.

References

[AH] C. ABBAS, H. HOFER, Holomorphic curves and global questions in contact geometry, to appear in Birkhäuser.

[Ar1] V.I. ARNOLD, Sur une propriété topologique des applications globalement canoniques de la méchanique classique, C. R. Acad. Paris 261 (1965), 3719–3722.

[Ar2] V.I. ARNOLD, On a characteristic class entering in quantization conditions, Funct. Anal. and Applic. 1 (1967), 1–14.

[Ar3] V.I. ARNOLD, First steps in symplectic topology, Russian Math. Surveys 41 (1986), 1–21.

[B] D. BENNEQUIN, Entrelacements et équations de Pfaff, Astérisque (1983), 106–107.

[BiC] P. BIRAN, K. CIELIEBAK, Symplectic topology on subcritical manifolds, preprint, 2000.

[Br] E. BRIESKORN, Beispiele zur differentialtopologie von singularitäten, Invent. Math. 2 (1966), 1–14.

[C] YU. CHEKANOV, Differential algebra of a Legendrian link, preprint, 1997.

[CoZ] C. CONLEY, E. ZEHNDER, The Birkhoff–Lewis fixed point theorem and a conjecture of V.I. Arnold, Invent. Math. 73 (1983), 33–49.

[D1] S.K. DONALDSON, Polynomial invariants for smooth four-manifolds, Topology 29 (1990), 257–315.

[D2] S.K. DONALDSON, Symplectic submanifolds and almost-complex geometry, J. Diff. Geom. 44 (1996), 666–705.

[E1] Y. ELIASHBERG, Topological characterization of Stein manifolds of complex dimension > 2, Int. J. of Math. 1 (1991), 29–46.

[E2] Y. ELIASHBERG, Invariants in contact topology, Proc. of ICM-98, Berlin, Doc. Math. (1998), 327–338.

[EG] Y. ELIASHBERG, M. GROMOV, Convex symplectic manifolds, Proc. of Symp. in Pure Math. 52:2 (1991), 135 162.

[EH] Y. ELIASHBERG, H. HOFER, A Hamiltonian characterization of the three-ball, Differential Integral Equations 7 (1994), 1303–1324.

[EHS2] Y. ELIASHBERG, H. HOFER, S. SALAMON, Lagrangian intersections in contact geometry, Geom. and Funct. Anal. 5 (1995), 244–269.

[EtS] J. ETNYRE, J. SABLOFF, Coherent orientations and invariants of Legendrian knots, preprint, 2000.

[F] A. FLOER, The unregularised gradient flow of the symplectic action, Comm. Pure Appl. Math. 41 (1988), 775–813.

[FH] A. FLOER, H. HOFER, Coherent orientations for periodic orbit problems in symplectic geometry, Math. Z. 212 (1993), 13–38.

[FuO1] K. FUKAYA, K. ONO, Arnold conjecture and Gromov-Witten invariants, preprint, 1996.

[FuO2] K. FUKAYA, K. ONO, Arnold conjecture and Gromov-Witten invariant for general symplectic manifolds, in "The Arnoldfest" (Toronto, ON, 1997), Fields Inst. Commun., 24, Amer. Math. Soc., Providence, RI (1999), 173–190.

[FuOOO] K. FUKAYA, K. ONO, Y.-G. OH, H. OHTA, Lagrangian intersection Floer theory. Anomaly and obstruction, preprint, 2000.

[FulP] W. FULTON, R. PANDHARIPANDE, Notes on stable maps and quantum cohomology, Proc. of Symp. in Pure Math. 62:2 (1995), 45–96

[G] A. GATHMANN, Absolute and relative Gromov-Witten invariants of very ample hypersurfaces, preprint, 1999.

[Ge] E. GETZLER, Topological recursion relations in genus 2, in "Integrable Systems and Algebraic Geometry" (Kobe/Kyoto, 1997), World Science Publishing (1998), 73–106.

[Gi] E. GIROUX, Une structure de contact, même tendue est plus ou moins tordue, Ann. Scient. Ec. Norm. Sup. 27 (1994), 697–705.

[Giv1] A. GIVENTAL, Nonlinear generalization of the Maslov index, Adv. Soviet Math. 1 (1990), 71–103.

[Giv2] A. GIVENTAL, A symplectic fixed point theorem for toric manifolds, in "The Floer Memorial Volume", Progr. Math., 133, Birkhäuser, Basel (1995), 445–481.

[Giv3] A. GIVENTAL, Homological geometry and mirror symmetry, Proc. Int. Congress of Math., Zürich-1994, Birkhäuser, 1 (1995), 472–480.

[Giv4] A. GIVENTAL, Homological geometry I: Projective hypersurfaces, Selecta Math. 1:2 (1995), 325–345.

[Giv5] A. GIVENTAL, Equivariant Gromov–Witten invariants, Intern. Math. Res. Notices 13 (1996), 613–663.

[GivK] A. GIVENTAL, B. KIM, Quantum cohomology of flag manifolds and Toda lattices, Commun. Math. Phys. 168:3 (1995), 609–641.

[GrP] T. GRABER, R. PANDHARIPANDE, Localization of virtual classes, Invent. Math. 135 (1999), 487–518.

[Gra] J.W. GRAY, Some global properties of contact structures, Annals of Math. 69 (1959), 421–450.

[Gro1] M. GROMOV, Pseudo-holomorphic curves in symplectic manifolds, Invent. Math. 82 (1985), 307–347.

[Gro2] M. GROMOV, Partial Differential Relations, Springer-Verlag, 1986.

[H] H. HOFER, Pseudo-holomorphic curves and Weinstein conjecture in dimension three, Invent. Math. 114 (1993), 515–563.

[HWZ1] H. HOFER, K. WYSOCKI, E. ZEHNDER, Properties of pseudo-holomorphic curves in symplectisations. I. Asymptotics, Ann. Inst. H. Poincaré, Anal. Non Linéaire 13 (1996), 337–379.

[HWZ2] H. HOFER, K. WYSOCKI, E. ZEHNDER, The dynamics on a strictly convex energy surface in $\mathbf{R}^4$, Annals of Math. 148 (1998), 197–289.

[HWZ3] H. HOFER, K. WYSOCKI, E. ZEHNDER, Finite energy foliations of tight three-spheres and Hamiltonian dynamics, preprint, 1999.

[I] E.-N. IONEL, Topological recursive relations in $H^{2g}(M_{g,n})$, preprint, 1999.

[IP1] E.-N. IONEL, T.H. PARKER, Gromov-Witten invariants of symplectic sums, Math. Res. Lett. 5 (1998), 563–576.

[IP2] E.-N. IONEL, T.H. PARKER, Relative Gromov-Witten invariants, preprint, 1999.

[K] J. KOLLAR, Rational curves on algebraic varieties, Springer-Verlag, 1996.

[Ko1] M. KONTSEVICH, Enumeration of rational curves via torus action, in "The Moduli Space of Curves" (R. Dijgraaf, C. Faber and G. van der Geer, eds.), Birkhauser (1995), 335–368.

[Ko2] M. KONTSEVICH, Deformation quantization of Poisson manifolds, I, preprint, 1997.

[KoM] M. KONTSEVICH, YU. MANIN, Gromov-Witten classes, quantum cohomology, and enumerative geometry, Commun.Math.Phys. 164 (1994), 525-562.

[LT] J. LI, G. TIAN, Virtual moduli cycles and Gromov-Witten invariants of general symplectic manifolds, in "Topics in Symplectic 4-Manifolds" (Irvine, CA, 1996), First Int. Press Lect. Ser., I, Internat. Press, Cambridge, MA (1998), 47–83.

[LiT] G. LIU, G. TIAN, Floer homology and Arnold conjecture, J. Diff. Geom. 49 (1998), 1–74.

[Lu] R. LUTZ, Structures de contact sur les fibrés principaux en cercles de dimension 3, Ann. Inst. Fourier, XXVII, 3 (1977), 1–15.

[M] J. MARTINET, Formes de contact sur les variétés de dimension 3, Lecture Notes in Math. 209 (1971), 142–163.

[Mc] D. MCDUFF, The virtual moduli cycle, in "Northern California Symplectic Geometry Seminar", Amer. Math. Soc. Transl. Ser. 2, 196, Amer. Math. Soc., Providence, RI (1999), 73–102.

[Mo] S. MORITA, A topological classification of complex structures on $S^1 \times \Sigma^{2n-1}$, Topology 14 (1975), 13–22.

[MyM] Y. MYAYOKA, S. MORI, A numerical criterion for uniruleness, Annals of Math 124 (1986), 65–69.

[N] L. NG, On invariants of Legendrian knots, preprint, 2000.

[RS] J. ROBBIN, D. SALAMON, The Maslov index for paths, Topology 32 (1993), 827–844.

[Ru1] Y. RUAN, Topological sigma model and Donaldson-type invariants in Gromov theory, Duke Math. J. 83 (1996), 461–500.

[Ru2] Y. RUAN, Virtual neighborhoods and pseudo-holomorphic curves, Proceedings of 6th Gökova Geometry-Topology Conference, Turkish J. Math. 23 (1999), 161–231.

[RuL] Y. RUAN, A.-M. LI, Symplectic surgery and Gromov-Witten invariants of Calabi-Yau 3-folds I, preprint, 1999.

[RuT] Y. RUAN, G. TIAN, A mathematical theory of quantum cohomology, J. Diff. Geom. 42 (1995), 259–367.

[S] D. SALAMON, Lectures on Floer homology, in "Symplectic Geometry and Topology", IAS/Park City Mathematics Series, vol. 7, AMS/IAS (1999),

144–229.

[Si] B. SIEBERT, Gromov-Witten invariants of general symplectic manifolds, preprint, 1997.

[U] I. USTILOVSKY, Infinitely many contact structures on S^{4m+1}, Int. Math. Res. Notices 14 (1999), 781–791.

[V] R. VAKIL, The enumerative geometry of rational and elliptic curves in projective space, preprint, 1997.

[Vi] C. VITERBO, in preparation.

[W] A. WEINSTEIN, On the hypotheses of Rabinowitz's periodic orbits theorems, J. Diff. Eq. 33 (1979), 353–358.

[Wi1] E. WITTEN, Supersymmetry and Morse theory, J. Diff. Geom. 17 (1982), 661–692.

[Wi2] E. WITTEN, Two-dimensional gravity and intersection theory on moduli space, Surveys in Diff. Geom. 1 (1991), 243–310.

[Y] M.-L. YAU, Contact homology of subcritical Stein manifolds, thesis, Stanford University, 1999.

YAKOV ELIASHBERG, Department of Mathematics, Stanford University, Stanford, CA 94305, USA eliash@math.stanford.edu

ALEXANDER GIVENTAL, Department of Mathematics, University of California, Berkeley, CA 94720, USA and Department of Mathematics, CALTECH, Pasadena, CA 91125, USA givental@math.berkeley.edu

HELMUT HOFER, Courant Institute of Mathematical Sciences, New York University, 251 Mercer Street, New York, NY 10012, USA hofer@cims.nyu.edu

GAFA, Geom. funct. anal.
Special Volume – GAFA2000, 674 – 704
1016-443X/00/S20674-31 $ 1.50+0.20/0

GAFA Geometric And Functional Analysis

HOLOMORPHIC CURVES AND REAL THREE-DIMENSIONAL DYNAMICS

H. Hofer

1 A Relationship Between Certain Vector Fields and a Holomorphic Curve Theory

In this paper we describe new tools for studying smooth dynamical systems on three-manifolds. We also mention some interesting open problems.

Assume that M^3 is a closed, connected, orientable three-manifold and let X be a nowhere vanishing vector field. We are interested in describing the orbit structure of the dynamical system

$$\dot{x} = X(x). \tag{1.1}$$

This, in general, is a very hard problem. As it turns out, there is, however, a very interesting class of vector fields for which a general theory can be developed (at least for certain manifolds).

A classical idea, going back at least to Poincaré and successfully used by Birkhoff, is to try to reduce the study of (1.1) to a 2-dimensional problem by finding a global surface of section.

Definition 1.1. *A global surface of section for (M, X) consists of an embedded compact surface $\Sigma \subset M$ having the following three properties:*

1. *If Σ has a boundary $\partial\Sigma$ its boundary components are periodic orbits for X.*
2. *The interior $\dot{\Sigma} := \Sigma \setminus \partial\Sigma$ of the surface Σ is transversal to X.*
3. *The orbit through a point not lying on Σ hits $\dot{\Sigma}$ in forward and backward time.*

If a global surface of section exists, the problem of understanding the global flow in three dimensions, is reduced to the understanding of the discrete dynamics of a self map of a surface. Indeed, starting on $\dot{\Sigma} = \Sigma \setminus \partial\Sigma$ and following the orbit we can define a global return map

$$\Psi : \dot{\Sigma} \to \dot{\Sigma}.$$

The dynamics of (1.1) is entirely encoded in Ψ. It is a particularly nice case, when we can take Σ to be closed. The reader should note, however, that global surfaces of section need not to exist in general. For example assume that X is a nowhere vanishing vector field on S^3 without periodic orbits. Such vector fields exist by recent work of K. Kuperberg, [K1]. Hence, if we have a global surface of section, it must be closed. This implies immediately that S^3 fibers over S^1. This is impossible since the fundamental group of S^3 is trivial.

By the previous comment we have to restrict ourselves to a smaller class of vector fields. But even volume preserving vector fields are still too general. In fact, G. Kuperberg, see [K2], showed that there are nowhere vanishing C^1-vector fields without periodic orbits on S^3 (in fact on every M). It is an open problem, if there exists a smooth volume-preserving non-singular vector field without periodic orbits.

QUESTION 1. *Does there exist a smooth, nowhere vanishing, volume-preserving vector field on S^3 without periodic orbits?*

For completeness, though it will not be relevant to us, we mention the following even harder question.

QUESTION 2. *Does there exist in the standard symplectic vector space $(\mathbb{R}^4, \omega_0)$ a compact smooth energy surface diffeomorphic to S^3 without periodic orbits?*

One knows that the higher dimensional vector spaces admit smooth energy surfaces diffeomorphic to the sphere without periodic orbits. Such examples are due to V. Ginzburg and M. Herman, [G1], [G2], [H].

At this point it might be useful to focus for a while on the question of the existence of periodic orbits. More than 20 years ago P. Rabinowitz, [R1], proved in a landmark result that an autonomous Hamiltonian system in $\mathbb{R}^{2n}$ has a periodic orbit on every energy surface bounding a star-shaped domain. Simultaneously, A. Weinstein, [W1], proved the weaker result for the case that the energy surface bounds a convex domain. The results of Rabinowitz, [R1], [R2], prompted A. Weinstein, [W2], to make a very influential conjecture concerning the existence of periodic orbits for vector fields:

CONJECTURE 1.2 (Weinstein, 78). *Let X be a non-singular smooth vector field on a closed odd-dimensional manifold M. Assume that there exists a 1-form λ such that $d\lambda$ has maximal rank and*

$$\lambda(X) > 0 \quad \text{and} \quad d\lambda(X, .) = 0\,.$$

Then X admits a periodic orbit.

We call a vector field X for which such a one-form exists Reeb-like. The assumption on the vector field is also quite often called the "contact-type hypothesis". Note that the one-form λ occurring in the above conjecture has the property that $\lambda \wedge d\lambda^n$ is a volume form, where $\dim(M) = 2n+1$. Such a one-form is called contact form.

In order to understand the geometric meaning of the Weinstein conjecture let us call two vector fields, say X_j on M_j, $j = 1,2$, conjugated, if there exists a diffeomorphism $M_1 \to M_2$ mapping flow lines of X_1 to flow lines of X_2.

The hypothesis of the Weinstein conjecture is in some sense a stability assumption. Namely, it requires that the flow is conjugated to an autonomous Hamiltonian flow on some energy surface E_0 having close by energy surfaces with the property that their flow is symplectically conformal to that on E_0. We refer the reader to Weinstein's seminal note, [W2], and to [HoZ] for more explanations and consequences of the contact-type hypothesis.

The Weinstein conjecture, in the generality stated, is still open. In fact, only a few manifolds are known for which every Reeb vector field admits a periodic orbit. All these examples are in three dimensions, [Ho1]:

Theorem 1.3. *Every Reeb-like vector field on S^3 has a periodic orbit. The same is true for every closed oriented three-manifold M with $\pi_2(M) \neq 0$.*

One of the very exiting features of the Weinstein conjecture is the fact that it is very closely connected to a holomorphic curve theory for contact manifolds. In the remaining part of this section we describe this relationship in more detail.

Assume that M is a connected, orientable, closed $(2n+1)$–dimensional manifold, equipped with a 1-form λ satisfying $\lambda \wedge d\lambda^n \neq 0$ (pointwise). Such a 1-form λ defines (uniquely) a vector field X, called the Reeb vector field, by

$$\lambda(X) = 1 \quad \text{and} \quad d\lambda(X,.) = 0\,.$$

The 1-form λ is called a contact form. The contact form also defines a contact structure $\xi = \mathrm{kern}(\lambda)$, which is a hyperplane bundle in TM. We note that $(\xi, d\lambda) \to M$ is a symplectic vector bundle. Moreover X and ξ are transversal.

Now we can explain the main point in the proof of Theorem 1.3. It is the observation that the Weinstein conjecture is completely equivalent to an existence result for an associated pseudoholomorphic curve equation.

Namely, pick any complex multiplication $J : \xi \to \xi$ compatible with $d\lambda$, so that $(h,k) \to d\lambda(h, J(m)k)$, for $(h,k) \in \xi_m \oplus \xi_m$, defines a positive metric for the contact structure. Such complex structures exist in abundance. In fact, it is well-known, that the set of all such structures equipped with the obvious C^∞-topology is a contractible space.

Given J we extend it to an $\mathbb{R}$-invariant almost complex structure $\tilde{J}$ on $\mathbb{R} \times M$ by mapping $1 \in T\mathbb{R}$ to X.

Using $\tilde{J}$ we can formulate a nonlinear partial differential equation of Cauchy-Riemann type:

Find a closed Riemann surface (S,j), a finite subset Γ of S and a proper(!) and non-constant(!) map $\tilde{u} = (a,u) : S \setminus \Gamma \to \mathbb{R} \times M$ satisfying

$$\begin{aligned} &\tilde{J} \circ T\tilde{u} = T\tilde{u} \circ j \\ &\int_{S\setminus\Gamma} u^* d\lambda < \infty \,. \end{aligned} \tag{1.2}$$

We call a solution of (1.2) a finite energy map. We refer to the points in Γ as the punctures. A puncture $p \in \Gamma$ is said to be positive (negative) if the $\mathbb{R}$-component a near p satisfies $a(z) \to \infty$ as $z \to p$ ($a(z) \to -\infty$ as $z \to p$). We note that for a solution $\tilde{u} = (a,u)$ of the Cauchy Riemann type equation in (1.2) the integrand $u^* d\lambda$ is non-negative.

Let us stress here that the properness assumption is really important since it excludes certain solutions which are not interesting (see below). Using Stokes' theorem one shows that

LEMMA 1.4. *The set of punctures Γ is necessarily non-empty by the assumption that $\tilde{u}$ is proper and not constant.*

Given a periodic orbit for the Reeb vector field X it is easy to construct solutions for (1.2). Indeed, if $x : \mathbb{R} \to M$ is a T-periodic orbit, define $\tilde{u} : \mathbb{C}/iT\mathbb{Z} \to \mathbb{R} \times M$ by $\tilde{u}([s+it]) = (s, x(t))$. We may view $\tilde{u}$ as a map defined on the Riemann sphere punctured at 0 and ∞. It is clearly proper and non-constant. Moreover, it satisfies (1.2).

In order to see that the properness assumption excludes certain solutions let $x : \mathbb{R} \to M$ be an orbit of the Reeb vector field. Then $\tilde{u} : \mathbb{C} \to \mathbb{R} \times M$ defined by

$$\tilde{u}(s,t) = \big(s, x(t)\big)$$

is non-constant, it satisfies (1.2), with the exception that it is not proper. So the properness assumption excludes such solutions.

As we will see shortly, periodic orbits for the Reeb vector field play an important role for the behaviour of a finite energy map near a puncture.

Assume that $\tilde{u}$ solves (1.2). Let us introduce the notion of holomorphic polar coordinates near the puncture $p \in S$. Take a closed disk-like neighbourhood $\mathcal{D}$ around p and fix a biholomorphic map $\phi : D \to \mathcal{D}$ between the closed unit-disk D and $\mathcal{D}$ which maps 0 to p. Then define polar coordinates around p by $\sigma(s,t) = \phi(e^{-2\pi(s+it)})$ for $(s,t) \in \mathbb{R}^+ \times (\mathbb{R}/\mathbb{Z})$.

For a finite energy map $\tilde{u}$ consider $\tilde{v}(s,t) := \tilde{u} \circ \sigma(s,t)$, for some holomorphic polar coordinates σ around p. It can be shown that the limit

$$m := \lim_{s \to \infty} \int_{S^1} v(s,.)^* \lambda$$

exists and is nonzero. Moreover, for every sequence $s_k \to \infty$ there is a subsequence, so that

$$x(mt) := \lim_{j \to \infty} v(s_{k_j}, t) \tag{1.3}$$

exists (in C^∞) and defines a $|m|$-periodic orbit $(x, |m|)$ for X, see [Ho1], [HoWZ3]. This limit exists for $s \to \infty$ if the limiting periodic orbit is non-degenerate. The behavior near a puncture is shown in Figure 1.

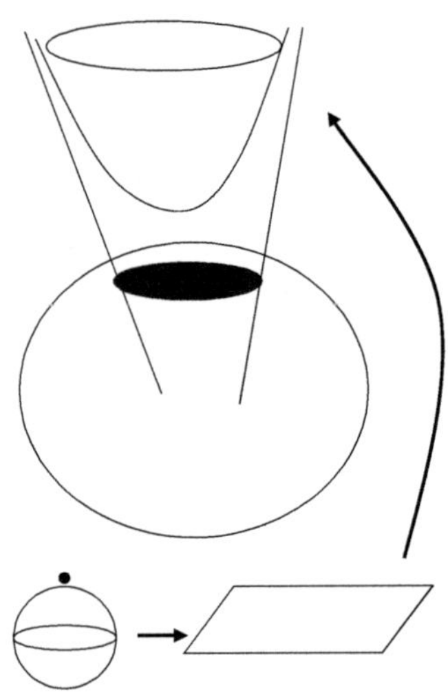

Figure 1: The figure shows the behavior of a finite energy map near a puncture. The domain is the one-punctured Riemann sphere.

Hence we can state the following, [Ho1]:

Theorem 1.5. *Let M be a closed, orientable, connected $(2n+1)$-dimensional manifold equipped with a contact form λ. Then the Reeb* *vector field X associated to λ has a periodic orbit iff for some admissible J the problem* (1.2) *has a solution.*

For details we refer the reader to [Ho1], [AH], [HoK].

Theorem 1.5 shows that the Weinstein conjecture is equivalent to an existence result for (pseudo-)holomorphic curves associated to an adapted almost complex structure $\tilde{J}$.

There is, of course, a legitimate question which should be addressed. Why does one replace a question concerned with the behavior of ordinary differential equations (i.e. finding periodic orbits) by the seemingly much more complicated question about the existence of certain solutions for a nonlinear first order elliptic equation? The deeper reason for this is the following. If there exist periodic orbits for a Reeb vector field the reason has to be global and connected with the topology of the underlying manifold and the fact that the vector field is a Reeb vector field. As it turns out the Reeb property makes the problem variational in a quite controlled way. First of all there exists a functional on the loop space of the underlying manifold whose critical points on non-zero level are related to periodic orbits. Such functionals, however, can be also found (in dimension three) for certain divergence free vector fields which are not Reeb. In the Reeb case it is an important additional fact that the value of the functional controls the length of the periodic orbit. To make this precise denote by $\mathcal{L}$ the free loop space of M, i.e. it consists of smooth maps $x : S^1 \to M$. We associate to the contact form λ the functional $\Phi : \mathcal{L} \to \mathbb{R}$ via

$$\Phi(x) = \int_{S^1} x^*\lambda\,.$$

If x is a critical point of Φ, then for every smooth section h of x^*TM we have

$$\int_{S^1} d\lambda\big(x(t)\big)\big(h(t), \dot{x}(t)\big)dt = 0\,.$$

This implies that for some smooth function on S^1, say f, we have

$$\dot{x}(t) = f(t)X\big(x(t)\big),$$

where X is the Reeb vector field. Clearly $\Phi(x) = \int_{S^1} f(t)dt$. Assume that $\Phi(x) \neq 0$. Let A be the orbit of X through $x(0)$. We must have $x(S^1) \subset A$. However $x : S^1 \to A$ cannot be contractible because otherwise $\Phi(x) = 0$ by Stokes' theorem. This implies that A is a periodic orbit and x some parameterization with nontrivial degree. If we fix a complex multiplication J for the contact structure ξ we obtain a Riemann metric g_J for M via

$$g_J(k, k') = \lambda(k)\lambda(k') + d\lambda(k, Jk')\,.$$

The length of A with respect to g_J can be estimated as follows:

$$\ell_J(A) = \int_A \lambda = \left| \int_A \lambda \right| \leq \left| \int_{S^1} x^* \lambda \right| = |\Phi(x)| .$$

Hence the value of the functional Φ allows to control the length of the underlying periodic orbit. In summary we see that the periodic orbits are related to critical points of some functional Φ, which allows some "apriori estimate". However, this relationship is more subtle and we return to it after the following remarks about "classical critical point theory".

How to find critical points? Even in finite dimensions one needs some machinery for finding critical points. A single critical point does not carry too much information. Indeed, one can always create by a small C^0-perturbation a pair of critical points with consecutive Morse indices (the so-called death-birth mechanism). On the other hand by the same reasoning two critical points may vanish simultaneously if we perturb the functional. In order to find critical points one needs more information. Already Morse observed that one has to take the gradient flow lines into consideration. Using the gradient flow he obtained his famous Morse relations which relate the local information of all(!) critical points with the global data of the underlying manifold. If one is only interested in finding critical points it is a well-known fact that the easiest way to find them is to establish the existence of global flow lines for the gradient flow on which the functional is bounded. Given a sufficient amount of compactness such flow lines have to converge in forward and backward time to critical points. These critical points can be identical in the case that the flow-line is constant.

In our context of Reeb vector fields a deeper study reveals the following nonobvious fact, which is slightly different from our previous heuristics. Finite unions of periodic orbits of the Reeb vector field can be viewed as critical points and the finite energy surfaces can be viewed as gradient lines. To be more precise, given a finite energy surface $\tilde{u} = (a, u) : S \setminus \Gamma \to \mathbb{R} \times M$ we know that the surface $\tilde{u}(S \setminus \Gamma)$ has cylindrical ends asymptotic to cylinders over periodic orbits. Let us assume that the surface is embedded with simply covered asymptotic limits. Sliding the surface via the $\mathbb{R}$-action up or down the part of the surface staying in a bounded region $[-R, R] \times M$ will converge to a union of cylinders over periodic orbits. In fact sliding it down it converges in C^∞_{loc} to a union of cylinders over the periodic orbits associated to positive punctures and in the other case the negative periodic orbits. The reparameterization of a flow line $x : \mathbb{R} \to M$ in the classical case by replacing it by x_c, defined by $x_c(t) = x(t + c)$ (which has the same image) defines a $\mathbb{R}$-action on the (parameterized) flow-lines. In the Reeb case it turns out that this $\mathbb{R}$-action corresponds to the obvious (additive) $\mathbb{R}$-action on the $\mathbb{R}$-factor replacing $\tilde{u} = (a, u)$ by $\tilde{u}_c = (a + c, u)$. As a side remark let us mention that the objects of interest are, in fact, not the finite energy maps, but equivalence classes of such maps. Here we call $((S, j), \Gamma, \tilde{u})$ and $((S', j'), \Gamma', \tilde{u}')$ equivalent if there exists a biholomorphic map $\phi : (S, j) \to (S', j')$ mapping Γ onto Γ', so that $\tilde{u} = \tilde{u}' \circ \phi$. Note that the $\mathbb{R}$-action described above is compatible with the equivalence relation.

The previous remarks somewhat motivate why it makes sense to study the pseudoholomorphic curve problem even if one only wants to find periodic orbits. Indeed if our problem of studying the Reeb vector field behaves like a classical variational problem we should expect that the topology of the underlying situation is represented by all gradient lines or (what suffices) bounded gradient lines.

Even knowing that finding a periodic orbit for a Reeb vector field is a variational problem, the pseudoholomorphic curve problem exhibits an enormous amount of structure. This wealth of structure is astounding. Using the moduli spaces of finite energy maps it is possible to construct a symplectic field theory. There one associates to contact manifolds symplectic (super-)vector spaces and to symplectic cobordisms between contact manifolds Lagrangian relations. This theory is described in a joint paper with Y. Eliashberg and A. Givental, [EGH], contained in this volume. Offspring of such a theory is a Floer-like homology theory for contact manifolds called contact homology. Here the connecting orbits are, however, not finite energy cylinders, but in fact punctured Riemann spheres. The nontriviality of the contact homology implies the Weinstein conjecture for the Reeb vector fields associated to contact forms inducing the given contact structure. It is also possible to relate the Weinstein conjecture with Gromov-Witten invariants and Quantum-cohomology, see [C] and [LT]. That this is possible has its origins in results by A. Floer, C. Viterbo and the author, [FHV], [HoV], who used moduli spaces of rational curves in symplectic manifolds in order to prove the Weinstein conjecture in certain cases.

2 The Behavior of a Finite Energy Map Near a Puncture

In this paragraph we discuss the precise asymptotic approach to a periodic orbit in the case that the periodic orbit is non-degenerate. The knowledge of this asymptotic behavior is very important in a Fredholm theory for finite energy maps as well as the study of their geometric properties in dimension three. We outline some of the results in [HoWZ6] and [HoWZ3].

Let M be a closed and connected three-manifold equipped with the contact form λ determining the contact bundle ξ and the Reeb vector field X. Choosing a complex multiplication J on ξ we denote by $\widetilde{J}$ the associated distinguished $\mathbb{R}$-invariant almost complex structure on $\mathbb{R} \times M$ and consider the finite energy surface

$$\tilde{u} = (a, u) : S \setminus \Gamma \to \mathbb{R} \times M$$

with the non empty finite set of punctures Γ. Near the puncture $z_0 \in \Gamma$ we take holomorphic polar coordinates σ. Then $z_0 = \lim_{s\to\infty} \sigma(s,t)$. In these coordinates, $\tilde{u}$ becomes near z_0 the positive half cylinder

$$\tilde{v} = (b, v) = \tilde{u} \circ \sigma : [0, \infty) \times S^1 \to \mathbb{R} \times M\,.$$

The map $\tilde{v}$ solves the Cauchy-Riemann equation

$$\tilde{v}_s + \tilde{J}(\tilde{v})\tilde{v}_t = 0$$

and has bounded energy $E(\tilde{v}) \leq E(\tilde{u}) < \infty$. Here the energy is defined as follows. Denote by Σ the collection of all smooth maps $\varphi : \mathbb{R} \to [0,1]$ with non-negative derivative. Given $\varphi \in \Sigma$ we can define a one-form λ_φ on $\mathbb{R} \times M$ by

$$\lambda(a, m)(h, k) = \varphi(a)\lambda(m)(k)\,.$$

It is easy to verify that for a finite energy map $\tilde{u}$ the two-form $\tilde{u}^* d\lambda_\varphi$ is a non-negative integrand. The energy $E(\tilde{u})$ is now defined by taking the supremum of the numbers $\int_{S\setminus\Gamma} \tilde{u}^* d\lambda_\varphi$, where φ varies over Σ.

One can prove the following proposition.

PROPOSITION 2.1. *Let $\tilde{u} = (a, u) : S \setminus \Gamma \to \mathbb{R} \times M$ be a solution of*

$$T\tilde{u} \circ j = \tilde{J} \circ T\tilde{u}\,,$$

where (S, j) is a closed Riemann surface and Γ is a finite set. Assume in addition that

$$E(\tilde{u}) < \infty\,.$$

Then $\tilde{u}$ can be smoothly extended over every point $p \in \Gamma$, where the $\mathbb{R}$-component a is bounded. After removing these singularities $\tilde{u}$ is proper in the neighbourhood of the remaining punctures. Alternatively if $\tilde{u}$ is a finite energy surface in the original sense the energy $E(\tilde{u})$ is finite.

Because of the energy bound the following limit exists:

$$m(\tilde{u}, z_0) = \lim_{s\to\infty} \int_{S^1} v(s, \cdot)^* \lambda\,.$$

Indeed, denoting by $|h|_J$ the norm associated to $g_J(h,h)$ and by $\pi : TM \to \xi$ the projection along X, Stokes' theorem implies that:

$$\begin{aligned}\int_{S^1} v(s,\cdot)^*\lambda &= \int_{S^1} v(0,\cdot)^*\lambda + \int_{[0,s]\times S^1} v^* d\lambda \\ &= c_0 + \frac{1}{2}\int_{[0,s]\times S^1} \left[|\pi v_s|_J^2 + |\pi v_t|_J^2\right] ds\, dt\end{aligned}$$

so that the map $s \to \int_{S^1} v(s,\cdot)^*\lambda$ is monotonic and bounded. The real number $m = m(\tilde{u}, z_0)$ is called the charge of the puncture z_0. This number is positive if z_0 is a positive puncture and negative for a negative puncture. If $m = 0$ the puncture is removable. The behaviour of the surface near z_0 is determined by periodic solutions of the Reeb vector field having periods $T = |m(\tilde{u}, z_0)|$. Namely, every sequence $s_k \to \infty$ possesses a subsequence, still denoted by s_k, such that

$$v(s_k, t) \to x(mt) \quad \text{in} \quad C^\infty(S^1) \tag{2.4}$$

for an orbit $x(t)$ of the Reeb vector field $\dot{x}(t) = X(x(t))$. Here m is the charge of z_0. If $m \neq 0$ the solution is necessarily a periodic orbit of X having the period $T = |m|$.

Call a periodic orbit (x, T) non-degenerate if the linearization of a local Poincaré section for the periodic orbit, as well as all its iterates do not have 1 in the spectrum. Clearly, a non-degenerate periodic orbit is isolated among all periodic orbits having periods close to $|m|$. It is an important fact that the non-degeneracy of the periodic orbit x in (2.4) implies that

$$\lim_{s\to\infty} v(s,t) = x(Tm) \quad \text{in} \quad C^\infty(S^1)$$

and

$$\lim_{s\to\infty} \frac{b(s,t)}{s} = m \quad \text{in} \quad C^\infty(S^1)\,.$$

In this case we call the uniquely determined periodic solution (x, T) with period $T = |m|$ the asymptotic limit of the puncture z_0.

In the non-degenerate case, the finite energy surface $\tilde{v}$ approaches the orbit cylinder $\tilde{v}_\infty(s,t) = (sm + d, x(mt))$ in $\mathbb{R} \times M$ in an exponential way for a suitable constant $d \in \mathbb{R}$, as $s \to \infty$. In order to describe this in detail we represent the contact structure λ in a tubular neighbourhood of the asymptotic limit (x, T) in a normal form. In the next lemma we denote by

$$\lambda_0 = d\vartheta + x dy$$

the (standard) contact form on $S^1 \times \mathbb{R}^2$ with coordinates (ϑ, x, y).

LEMMA 2.2. *Let M be a three-dimensional manifold equipped with a contact form λ and let (x,T) be a periodic solution of the Reeb vector field X. Denote by τ the minimal period of x so that $T = k\tau$ for an integer k. Then there exist open neighbourhoods U of $S^1 \times \{0\} \subset S^1 \times \mathbb{R}^2$ and $V \subset M$ of $P = x(\mathbb{R}) \subset M$, and a diffeomorphism $\varphi : U \to V$ mapping $S^1 \times \{0\}$ onto P and satisfying*

$$\varphi^* \lambda = f \lambda_0 .$$

The smooth function $f : U \to (0,\infty)$ has the properties $f(\vartheta,0,0) = \tau$ and $df(\vartheta,0,0) = 0$ for all $\vartheta \in S^1$.

Working in the covering space $\mathbb{R}$ of S^1, the curve $\tilde{v}$ is, in the coordinates of the lemma, represented as a map

$$\begin{aligned} \tilde{v} &= (b,v) : [0,\infty) \times \mathbb{R} \to \mathbb{R}^4 \\ \tilde{v}(s,t) &= \big(b(s,t), \vartheta(s,t), z(s,t)\big) , \end{aligned} \tag{2.5}$$

where the functions $b : [0,\infty) \times \mathbb{R} \to \mathbb{R}$ and $z : [0,\infty) \times \mathbb{R} \to \mathbb{R}^2$ are 1-periodic in t, while $\vartheta : [0,\infty) \times \mathbb{R} \to \mathbb{R}$ satisfies $\vartheta(s,t+1) = \vartheta(s,t) + k$. The last factor $\mathbb{R}^2$ in (2.5) is the contact plane along the asymptotic limit in these coordinates.

Theorem 2.3. *Let $z_0 \in \Gamma$ be a non-removable puncture of a finite energy surface $\tilde{u} : S \setminus \Gamma \to \mathbb{R} \times M$ whose charge is $m(\tilde{u}, z_0) = m$ and whose non-degenerate asymptotic limit is (x,T), where $T = |m| = k\tau$ with the minimal period τ. Introduce near z_0 the cylindrical coordinates $[0,\infty) \times S^1$ and near the asymptotic limit the normal form coordinates of the lemma. Then*

Either: there exists a constant c such that

$$\tilde{w}(s,t) = \Big(ms + c, \frac{m}{\tau} t, 0\Big) \quad \text{for} \quad (s,t) \in [0,\infty) \times \mathbb{R}$$

or, $\tilde{w}$ is of the form:

$$\begin{aligned} a(s,t) &= ms + c + \hat{a}(s,t) \\ \vartheta(s,t) &= \frac{m}{\tau} t + d + \hat{\vartheta}(s,t) \\ z(s,t) &= e^{\int_0^s \mu(\tau) d\tau} \big[e(t) + \hat{r}(s,t)\big] \end{aligned}$$

for a constant $d \in \mathbb{R}$, where

$$\partial^\alpha \hat{r}(s,t) \to 0 \quad \text{as} \quad s \to \infty$$

uniformly in $t \in \mathbb{R}$ and for all multi-indices $\alpha = (\alpha_1, \alpha_2)$. In addition, there are constants $M_\alpha > 0$ and $\beta > 0$ such that

$$\big|\partial^\alpha \hat{a}(s,t)\big| , \quad \big|\partial^\alpha \hat{\vartheta}(s,t)\big| \le M_\alpha e^{-\beta s}$$

for $s \geq 0$ and all multi-indices α. Moreover, $\gamma : [0, \infty) \to \mathbb{R}$ is a smooth function converging, $\gamma(s) \to \mu < 0$ as $s \to \infty$, to a negative eigenvalue μ of a self-adjoint operatorA_∞ in $L^2(S^1, \mathbb{R}^2)$, defined by linearization of the Reeb vector field X along the asymptotic limit $(x, |m|)$. Finally, the function $e(t) = e(t+1) \neq 0$ represents an eigenvector of A_∞ belonging to the eigenvalue μ.

The proofs of these statements can be found in [AH], [Ho2], [Ho1], [HoWZ3]. We would like to add that the above asymptotic formula is used in the Fredholm theory [HoWZ7] for embedded finite energy surfaces. It also plays an important role in the geometric description of finite energy surfaces in [HoWZ6], [HoWZ1], [HoWZ2], [HoWZ5].

We summarize the previous discussion: Given a finite energy surface $\tilde{u} = (a, u)$ and a puncture p we have near p a precise directional convergence to the periodic orbit . Namely, for suitable polar coordinates σ around p (assuming p is a positive puncture), and identifying the neighbourhood of the asymptotic limit with an open neighbourhood of the zero section in the contact bundle along the periodic orbit, the map $u \circ \sigma(s, t)$, for large s, has (modulo lower terms) in special local coordinates the form $e^{\mu s} e(t)$, where $\mu < 0$ and $e(t) \in \xi_{x(Tt)} \setminus \{0\}$. Moreover, e is the eigenvector of some operator associated to the negative eigenvalue μ. If the puncture is negative, one should introduce negative polar coordinates $\mathbb{R}^- \times S^1 \to \mathcal{D}$, and the asymptotic is similar for $s \to -\infty$, except that this time $\mu > 0$.

In both cases we can associate with the puncture a winding number measuring the winding of e provided, of course, we have a framing of ξ along the periodic orbit. We will discuss this later.

As a remark, the reader should note that if the asymptotic limit is multiple covered, i.e. its period is not the minimal period, then the higher order terms which we neglected in the previous argument are geometrically very important.

3 The Conley–Zehnder Index

In this section we introduce the important Conley–Zehnder index and some other indices relevant for our constructions.

We begin with the definition of a non-degenerate contact form.

DEFINITION 3.1. *Call a contact form on M^3 non-degenerate if for every periodic orbit (x, T) the linearized return map for a transversal section at $x(0)$ and all its iterates do not have 1 in the spectrum. A non-degenerate* *periodic orbit is called even if the eigenvalues of the linearized return map are positive real numbers* $(1/\beta < 1 < \beta)$*. Otherwise it is called odd.*

The Conley–Zehnder index is an integer measure for the behaviour of the flow in a neighbourhood.

Assume now that we are given a non-degenerate contact form λ. In that case, if $\tilde{u}$ is a solution of (1.2), we can extend its M-part to the circle compactification (see next paragraph) $\overline{S}$ of $S \setminus \Gamma$. This is due to the fact $u \circ \sigma(s,t)$, where σ are holomorphic polar coordinates around a puncture, converges to a reparameterized periodic orbit as $s \to \infty$ as described in (1.3). The limit exists for $s \to \infty$ rather than $s_k \to \infty$ since the periodic orbit is assumed to be non-degenerate.

The circle compactification is obtained by taking holomorphic polar coordinates near a puncture. The punctured neighbourhood looks like $[0,\infty) \times S^1$. Then add a circle $\{\infty\} \times S^1$. So the extension $\bar{u} : \bar{S} \to M$ maps the boundary of $\bar{S}$ to some periodic orbits. Note that $\bar{S}$ has a natural orientation inducing orientations on its boundary components. We call an asymptotic limit positive if the corresponding boundary component is mapped orientation-preservingly onto the periodic orbit (x,T) (which is oriented by X) and negative otherwise.

The Conley–Zehnder index of a periodic orbit now -roughly speaking- measures how the neighbouring orbits wind around it with respect to some natural framing.

Consider the pullback $\bar{u}^*\xi$ of the contact structure. This is a trivial symplectic vector bundle and we can take a nowhere vanishing section A. It determines some framing along the asymptotic periodic orbits. Different nowhere vanishing sections A may lead to different framings. However, the following construction is independent of the choices involved.

Fix an asymptotic limit (x,T). Then consider the flow η_t associated to X. For $v \in \xi_{x(0)}$ we observe that $T\eta_t(x(0))v \in \xi_{x(t)}$ for all t. This follows easily from the fact that by the definition of the Reeb vector field

$$\frac{d}{dt}\eta_t^*\lambda = \eta_t^*\big([i_X d + d i_X]\lambda\big) = \eta_t^*\big[i_X d\lambda + d(1)\big] = 0\,.$$

Hence, the tangent map to the flow $T\eta_t(x)$ maps ξ_x onto $\xi_{\eta_t(x)}$.

Using the complex multiplication J on ξ we can write

$$T\eta_t\big(x(0)\big)v = f(t)\cdot A\big(x(t)\big)\,.$$

For non-zero $v \in \xi_{x(0)}$ we write $\delta(x,T,v)$ for the change of argument of f along $[0,T]$. The collection $\{\frac{1}{2\pi}\delta(x,T,v) | v \in \xi_{x(0)} \setminus \{0\}\}$ we call the winding interval of (x,T) with respect to (J,A) and denote it by $\Delta(J,A,(x,T))$.

The length of the winding interval $\Delta \subset \mathbb{R}$ is always less than 1/2. Hence it follows that either the winding interval is contained in $(k, k+1)$ or contains k, where k is a uniquely determined integer. We define the Conley–Zehnder index $\mu_{(J,A)}(x,T)$ of (x,T) with respect to (J,A) by $2k+1$ in the first case and $2k$ in the second case. It can be shown that the Conley–Zehnder index of a periodic orbit is even (odd) if and only if the orbit is even (odd). This is true independent of the choices which have been made.

We define the total Conley–Zehnder index as the integer

$$\mu(\tilde{u}) = \sum_{(x,T)\text{ positive}} \mu_{(J,A)}(x,T) - \sum_{(x,T)\text{ negative}} \mu_{(J,A)}(x,T)\,.$$

The definition of the total Conley–Zehnder index does not depend on the choices involved.

Since our main results will be concerned with the three-sphere it makes sense to elaborate a little bit further. In S^3 we can define the Conley–Zehnder index for every periodic orbit (x,T) by spanning in a disk map $u : D \to S^3$ so that $u(e^{2\pi it}) = x(Tt)$. Then we take a nowhere vanishing section A of $u^*\xi$ and proceed as before. We obtain a number $\mu(x,T)$. This number does not depend on our choice and we call it the Conley–Zehnder index of (x,T).

This construction is related to the previous one as follows. For a finite energy surface it can be shown that the total Conley–Zehnder index $\mu(\tilde{u})$ is the difference of the sums of the Conley–Zehnder indices of positive and negative asymptotic limits, respectively. Let us note that the section A defines a framing of ξ along the periodic orbit.

In the following, working on S^3, periodic orbits (x,T), with minimal period T and with Conley–Zehnder indices 1, 2 and 3 will be important. Let A be the section of ξ along a non-degenerate periodic orbit (x,T), where T is the minimal period. Let $v \in \xi_{x(0)}$ be a non-zero vector. Then we can write

$$T\eta_t\big(x(0)\big)v = f(t) \cdot A\big(x(t)\big)\,.$$

If the Conley–Zehnder index is 1 the change of argument of f along a full period is strictly positive, but less than 2π. If the Conley–Zehnder index is 2 and v is an eigenvector for the linearized Poincaré map it is precisely 2π. Otherwise it is strictly between π and 3π. If the Conley–Zehnder index is 3 it is more than 2π but less than 4π.

If $\tilde{u}$ is a finite energy map and $p \in \Gamma$ a puncture, we know from the previous section that we have a precise directional convergence towards the asymptotic periodic orbit.

Assume for example p is positive and the asymptotic limit has Conley–Zehnder index 2 or 3. Then one can show, see [HoWZ7], [HoWZ9], that the eigenfunction e describing the approach towards the periodic orbit has winding number at most 1 with respect to A. This in particular means that the flow turns faster (more than 2π) around the periodic orbit than the approaching surface if the Conley–Zehnder index is 3. If the Conley–Zehnder index is 2 the situation is a little bit more subtle. Assume that the vector e describing the directional convergence has winding number 1. The linearized stable and unstable manifold of the periodic orbit have winding number precisely 1 (they are generated by nowhere vanishing sections whose winding can be measured relatively to A) and intersect a transversal section in a pair of transversal lines creating four quadrants. The trace of e on this transversal section will lie in one of these quadrants. See Figure 2 for more details.

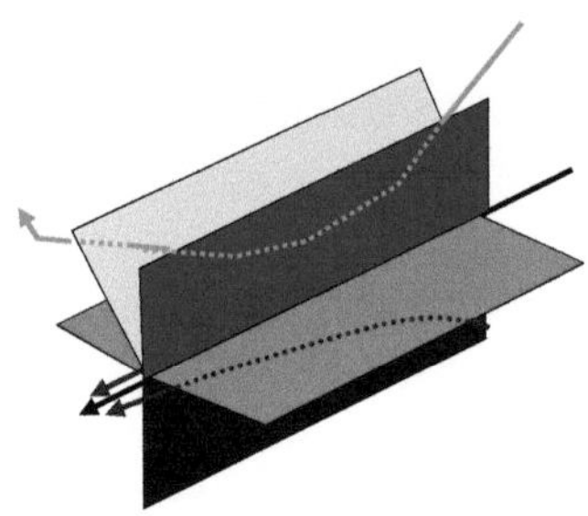

Figure 2: The figure shows the stable and unstable manifold as well as the surface asymptotic to the periodic orbit.

If p is negative (so that for $s \to -\infty$ the loops $u \circ \sigma(s, \cdot)$ are mapped orientation reversingly to the reparametrized periodic orbit) one can show if the asymptotic limit has Conley–Zehnder index 1 or 2 that the winding number of the eigenfunction describing the asymptotic approach is at least 1. This time the surface turns faster than the flow around the orbit if the Conley–Zehnder index is 1. The behavior, in case the Conley–Zehnder index is 2, is similar as above.

Figure 3 shows the M-part of a cylindrical finite energy cylinder.

With the help of the Conley–Zehnder index we can define the index of

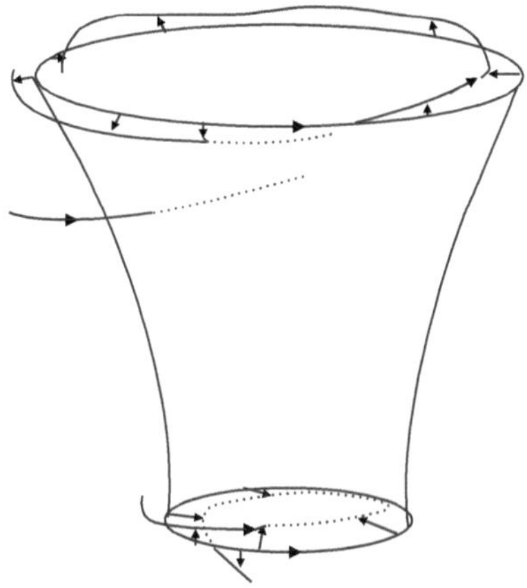

Figure 3: The flow near periodic orbits of Conley–Zehnder index 1 and 3 relative to a finite energy cylinder. The Reeb vector field will be transversal to the surface. One orbit intersecting the surface is shown.

$\tilde{u} : S \setminus \Gamma \to \mathbb{R} \times M$ by

$$\text{ind}(\tilde{u}) := \mu(\tilde{u}) - \chi(S) + \sharp\Gamma .$$

This number is, in fact, the Fredholm index of the linearization of some non-linear Cauchy-Riemann type operator describing all finite energy surfaces near a given one, allowing the punctures as well as the complex structure to vary in Teichmueller space. We will need this quantity in the next section.

4 Holomorphic Curves and More Dynamics

In the following parts we will restrict ourselves to the case of Reeb-like vector fields on three-dimensional manifolds.

It is instructive to look at some special case. Consider $S^3 \subset \mathbb{C}^2$ equipped with the contact form

$$\lambda_0 := \left(\frac{1}{2}\sum_{j=1}^{2} q_j dp_j - p_j dq_j\right)\Big| S^3 ,$$

where $q+ip$ are the coordinates in $\mathbb{C}^2$. Then the Reeb vector field X_0 generates the Hopf fibration and the associated contact structure ξ_0 is the bundle of complex lines in TS^3. Now we observe that the map $z \to (1/2 \ln |z|, z/|z|)$

defines a biholomorphic map Φ between $\mathbb{C}^2 \setminus \{0\}$ and $\mathbb{R} \times S^3$. Here the domain is equipped with i and the target space with $\tilde{J}$, where we take $J = i|\xi_0$. The important fact is that the images of solutions of (1.2) correspond to the 1-dimensional affine algebraic sets (take away 0), see [HoK]. So, in some sense, studying (1.2) is a (non-integrable) deformation of studying affine algebraic sets. In fact, the solutions of (1.2) have many of the properties which one might expect and are familiar within the algebraic context.

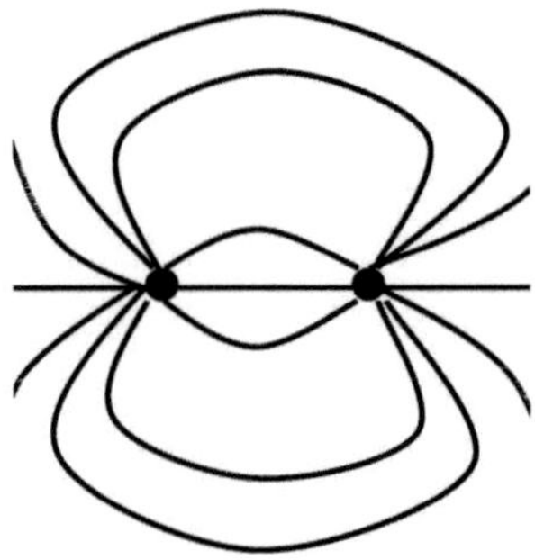

Figure 4: The picture shows the trace of the open book decomposition.

To get more insight, take again λ_0 on S^3. Then consider the family of affine algebraic sets $\mathbb{C} \times \{c\}$, $c \neq 0$, and add to it the punctured plane $(\mathbb{C} \setminus \{0\}) \times \{0\}$. We note that this establishes a foliation of $\mathbb{C}^2 \setminus \{0\}$. Under the map Φ we obtain a foliation $\mathcal{F}$ of $\mathbb{R} \times S^3$ by finite energy surfaces. With the exception of the finite energy cylinder $Z := \mathbb{R} \times (S^1 \times \{0\})$ all surfaces are finite energy planes asymptotic to Z. Moreover $\mathcal{F}$ is $\mathbb{R}$-invariant. In fact if F is a leaf so is $a + F$, where we use the obvious $\mathbb{R}$-action on $\mathbb{R} \times S^3$. If F is not a fixed point for the $\mathbb{R}$-action then the projection of the surface into S^3 is an embedded disk. However, the surface Z projects to $P := S^1 \times \{0\}$. The correct interpretation of the circle P is that of a periodic orbit for the Reeb vector field. This loop may be viewed as the boundary for all the other disks. We have here an open book decomposition of S^3 with disk-like page. Figure 4 gives a sketch of the situation. There we view S^3 as $\mathbb{R}^3 \cup \{\infty\}$ and take the trace of the foliation on a suitable two-dimensional plane. So the trace of the periodic orbit P will be two points and surfaces will be represented by lines connecting these two points.

This observation brings us to a new concept.

DEFINITION 4.1. *Let λ be a contact form on the closed orientable and connected three-manifold M. Let J be an admissible complex multiplication for the underlying contact structure ξ. Then a finite energy foliation $\mathcal{F}$ for (M, λ, J) is a foliation of $\mathbb{R} \times M$ by surfaces, which are embedded solutions of* (1.2), *so that with F also $a + F$ belongs to $\mathcal{F}$.*

It is not difficult to show that the fixed points for the $\mathbb{R}$-action on $\mathcal{F}$ are precisely the cylinders over periodic orbits which belong to $\mathcal{F}$. If F is not a fixed point it is also easy to show that $a + F \neq F$ for all $a \neq 0$, This immediately implies that the projection into M is injective. So, in some sense the collection of elements of $\mathcal{F}$ projected into M defines in the complement of periodic orbits coming from the fixed point set a foliation. We will make this more precise later.

We have the following result, [HoWZ9]:

Theorem 4.2 (Hofer-Wysocki-Zehnder). *Let $\lambda = f\lambda_0$ be a non-degenerate contact form on S^3. Then there exists a Baire set Ξ of admissible complex multiplications J so that the following holds. For every $J \in \Xi$ there exists a finite energy foliation $\mathcal{F}$ of (S^3, λ, J) having the following additional properties: The leaves are images of punctured Riemann spheres having precisely one positive puncture but an arbitrary number of negative punctures. The asymptotic limits have minimal period and Conley–Zehnder indices in $\{1, 2, 3\}$. For every leaf F which is not a fixed point for the $\mathbb{R}$-action we have* $\operatorname{ind}(F) \in \{1, 2\}$. *Moreover the associated projected surfaces are transversal to X_λ and their $d\lambda$-area is uniformly bounded. Further there always exists a finite energy plane.*

One can also show that the periodic orbits which span the surfaces have self-linking number -1.

If we project the elements of $\mathcal{F}$ into S^3 we obtain a finite number of periodic orbit coming from $\mathcal{F}_{fix}$ (the fixed point set in $\mathcal{F}$ under the $\mathbb{R}$-action) and a foliation of their complement in S^3 by the projected other surfaces. This might be viewed as the natural generalization of a global surface of section: after removing a finite number of periodic orbits the complement is foliated by properly embedded transversal surfaces. Each of these surfaces can be compactified by adding a suitable number of the previously removed periodic orbits. One can also describe the behavior of the foliation near a spanning orbit. If the Conley–Zehnder index is 2 this is shown in Figure 5.

In Figure 6 we show the situation near an orbit of index 1 or 3. Such an orbit is either elliptic or it is hyperbolic with negative eigenvalues.

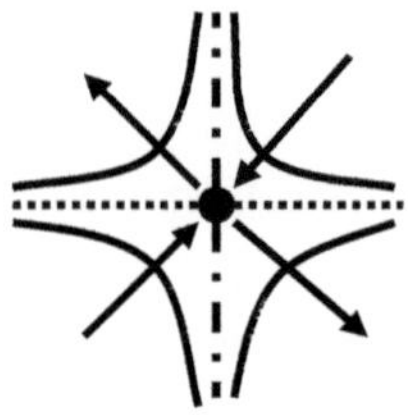

Figure 5: The figure shows the behavior of the foliation near a spanning orbit. We assume that the periodic orbit is perpendicular to the page and we draw only the trace of the foliation on our page. The big dot is the trace of the periodic orbit. The periodic orbit shown has Conley–Zehnder index two. The inward pointing arrows show the trace of the stable manifold, whereas the outward pointing arrows show the unstable manifold of the periodic orbit. The black lines are the traces of surfaces in one-dimensional geometric families which split up into to rigid surfaces while approaching the periodic orbit. One of these two surfaces is shown by a dotted line and the other by a dotted dashed line.

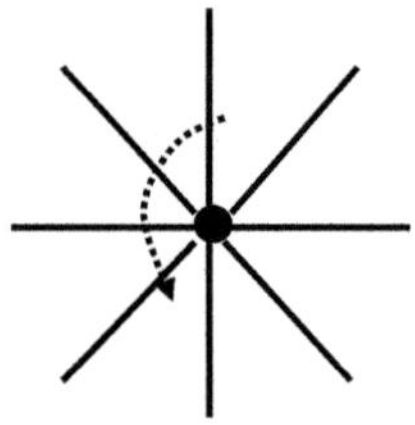

Figure 6: The figure shows the behavior of the foliation near a spanning orbit of index 1 or 3. Such orbits are either elliptic or hyperbolic with negative eigenvalues. In the figure it is assumed that the orbit is elliptic. The foliation would be the same in the hyperbolic case. However, then the stable and unstable manifolds would be Möbius bands cutting transversally through the leaves of the projected foliation. The square dotted arrow shows a flow line.

Using the positivity of intersection results for holomorphic curves, see [Gr] and [Mc], and a certain form of the implicit function theorem, see [HoWZ7], it is possible to show the following. If $F \in \mathcal{F}$ and $\mathrm{ind}(F) = 2$, then there is a 2-dimensional family of finite energy surfaces near F which are given by an implicit function theorem. Moreover all these surfaces belong to $\mathcal{F}$. One of the dimensions accounts for the $\mathbb{R}$-action. The other dimension accounts for the fact that the projection of F in S^3 lies in a 1-dimensional family of mutually disjoint projected finite energy surfaces.

If F satisfies $\mathrm{ind}(F) = 1$ then it belongs to an obvious 1-dimensional family produced through the $\mathbb{R}$-action. The projection of such a surface is rigid and does not belong to any (nontrivial) family.

Denote by $\mathcal{S}$ the collection of all sets obtained from $\mathcal{F}$ by projecting them into S^3. This set decomposes into three parts:

$$\mathcal{S} = \mathcal{S}_0 \cup \mathcal{S}_1 \cup \mathcal{S}_2.$$

Here $\mathcal{S}_0$ consists of periodic orbits obtained by projecting the surfaces in $\mathcal{F}_{fix}$, $\mathcal{S}_1$ consists of the rigid surfaces and $\mathcal{S}_2$ of those occurring in 1-dimensional families. The 1-dimensional families are parameterized either by S^1 or an open interval $I := (0,1)$. If we have an I-family F_τ and $\tau \to 0$ or 1 the surface starts decomposing along a Conley–Zehnder index 2 orbit. This is illustrated in Figure 7.

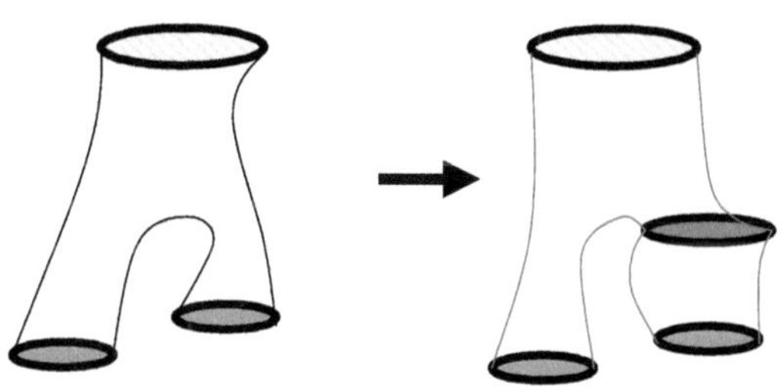

Figure 7: This figure shows how a surface belonging to a 1-dimensional family (of projected surfaces) decomposes.

5 Finite Energy Foliations and Dynamics

The existence of the finite energy foliation has certain corollaries. Call $f : S^3 \to (0,\infty)$ generic if $\lambda = f\lambda_0$ is non-degenerate and the stable and unstable manifolds of hyperbolic orbits intersect transversally. The set of generic f's is known to be a Baire set. In [HoWZ9] the following theorem is proved.

Theorem 5.1. *For a generic f either the associated Reeb flow admits a disk-like global surface of section or there exists a Bernoulli shift in the system. In any case there are either precisely two or infinitely many geometrically distinct periodic orbits.*

This result prompts the following conjecture

CONJECTURE 5.2. *The Reeb flow associated to $\lambda = f\lambda_0$ for any smooth map $f : S^3 \to (0,\infty)$ has either precisely two or infinitely many geometrically distinct periodic orbits.*

This conjecture is known to be true if the set $\{f(z)^{1/2}z \mid z \in S^3\}$ bounds a strictly convex domain in $\mathbb{C}^2$, see [HoWZ2]. Hence the conjecture raises the question if in fact conditions are needed so that there are either precisely two or infinitely many periodic orbits.

Observe that for every Riemannian metric g on S^3 the geodesic flow restricted to the unit sphere bundle is a Reeb flow. The unit sphere bundle T_1S^2 has as a double covering S^3. It can be shown that the double-covered geodesic flow is conjugated to a Reeb flow of $\lambda = f\lambda_0$. By a classical result of Liusternik and Shnirelman we know that there are at least three geometrically distinct geodesics. If the conjecture is true it would imply immediately infinitely many geometrically distinct periodic orbits and hence infinitely many prime geodesic. The latter is known to be true by results of Bangert and Franks, [B] and [Fr2] .

Let us indicate how Theorem 4.2 implies Theorem 5.1. Assume first that there is no spanning orbit with Conley–Zehnder index 2. Then we must have a global disk-like surface of section. Indeed, $\mathcal{F}$ contains a finite energy plane F. It is known that its asymptotic limit has Conley–Zehnder index at least 2, see [HoWZ1]. Since we have just excluded 2 we must have 3 (Recall that the existence theorem for finite energy foliations guarantees indices 1, 2 and 3.) The projected F, say Q must lie in a 1-dimensional family. So it is parameterized either by I or S^1. The I-case is impossible, since the surfaces would degenerate towards the ends implying the existence of index 2 spanning orbits. Consequently, the family of projected surfaces gives an open book decomposition of S^3 with disk-like pages. Any of these surfaces can act as a global surface of section. The return map is area-preserving for the form $d\lambda$ restricted to the disk and the total area equals the period of the spanning orbit by Stokes' theorem. Using Brouwer's translation theorem it must have a fixed point. Removing the fixed point we obtain an open annulus invariant by the (area-preserving) return map. By a result of Franks, [Fr2], such a map has infinitely many periodic points if it has at least one periodic point. This implies that we have precisely either two or infinitely many geometrically distinct periodic orbits provided we have a global disk-like surface of section which is implied by the non-existence of an index 2-spanning orbit.

Let us assume next that we have a spanning orbit H with Conley–Zehnder index 2. Fix a component of the unstable manifold. Take a leaf F_0 of our foliation (on S^3) lying in a one-dimensional family close by to the spanning orbit so that the unstable manifold cuts out an embedded S^1, say P from F_0. The λ-integral over P equals the period of H, see Figure 8.

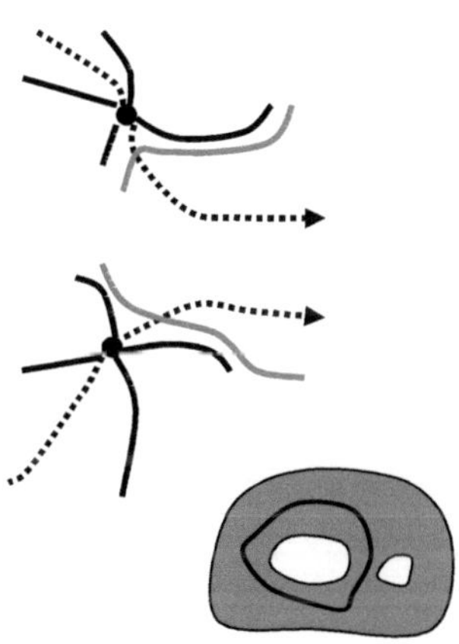

Figure 8: The unstable manifold cuts out a loop of a projected finite energy surface lying in a 1-dimensional family. The underlying finite energy surface has one positive puncture and two negative puncture and is modeled on a Riemann sphere.

We take now the 1-dimensional family containing F_0 and move away from H. The unstable manifold cuts out loops. We claim that some part of the moving loops will hit in forward time the unstable manifold of some spanning orbit. Indeed if this does not happen we can conclude the following. Our one-dimensional family will decompose into two rigid surfaces and the loop cut out by the unstable manifold has entirely to lie on one of these surfaces. On the other side of this surface our foliation continues by a one-dimensional family and we can argue as before. Since there are only a finite number of rigid surfaces and a finite number of one-dimensional families we find a rigid surface R which is hit infinitely often in forward time. By the uniqueness of the initial value problem the loops cut out by the unstable manifold on this surface have to be mutually disjoint and the λ integral over each such loop is the same. This gives immediately a contradiction since $d\lambda$ induces a volume form on W with finite total volume. This argument shows that there exists a heteroclinic orbit between two different hyperbolic spanning orbits or a homoclinic orbit to the Conley–Zehnder index 2 orbit we started with. In the first case we continue with the unstable manifold of the newly obtained hyperbolic spanning orbit. After a finite number of steps we obtain a heteroclinic loop since there are only a finite number of spanning orbits. Since stable and unstable orbits intersect transversally by assumption we can construct a homoclinic orbit. So in both cases we obtain a homoclinic orbit and ultimately a Bernoulli shift in our system. In particular we have infinitely many geometrically distinct periodic orbits.

6 About Possible Generalizations to Other Manifolds

One might raise the question how these results can be generalized to other manifolds?

The definition of a finite energy foliation makes sense on every three-manifold M. However, as already seen in the S^3-case, it makes sense to impose additional conditions. This prompts the following definition.

DEFINITION 6.1. *A good finite energy foliation $\mathcal{F}$ is a finite energy foliation with the following additional properties. Every leaf F is the image of an embedded finite energy surface $\tilde{u}$ with index satisfying* $\mathrm{ind}(F) \in \{1,2\}$ *provided $\pi \circ Tu$ does not vanish identically. Moreover, among the occurring asymptotic limits there is at most one which has an even Conley–Zehnder index*[1].

[1] One should note the following. Even if the Conley–Zehnder index is not well-defined due to the topology of M, it nevertheless makes sense to talk about an even or odd Conley–Zehnder index.

How good is this definition? It is useful to rewrite the equations of finite energy surfaces, see (1.2). In fact, they can be written as:

$$\begin{gathered}\text{The map } \tilde{u} \text{ is proper and not constant.} \qquad (6.6)\\ \pi \circ Tu \circ j = J \circ \pi \circ Tu\\ (u^*\lambda) \circ j = da\\ \int_{S\setminus\Gamma} u^* d\lambda < \infty\,.\end{gathered}$$

Consequently, $(u^*\lambda) \circ j$ defines the trivial cohomology class in $H^1(S \setminus \Gamma, \mathbb{R})$. This observation will be useful later when we start modifying our definitions.

There is an interesting relationship between the number of zeros of $\pi \circ Tu$ and $\text{ind}(u)$. Observe that for $z \in S \setminus \Gamma$ the map

$$\pi \circ Tu(z) : T_z S \to \xi_{u(z)}$$

is a complex linear map between complex 1-dimensional spaces. This follows immediately from the differential equation. Indeed, $h := \pi \circ Tu$ satisfies

$$h \circ j = J \circ h. \qquad (6.7)$$

Observe that h can be viewed as a smooth section of the complex one-dimensional bundle

$$\text{Hom}_{\mathbb{C}}(T\dot{S}, u^*\xi) \to \dot{S}$$

where $\dot{S} = S \setminus \Gamma$. One can show that h satisfies a first order partial differential equation of Cauchy-Riemann type by basically differentiating (6.7). This fact is, however, no entirely trivial since $J = J(u)$ and $\pi = \pi_u$. It is crucial that π projects onto a contact structure implying a magical cancellation. We refer the reader to [HoWZ1] for details. As a consequence of these facts the zeros of $\pi \circ Tu$ are isolated unless the map is identically 0. Moreover the isolated zeros have a positive local index. The asymptotic behavior near a puncture implies that there are no zeros near a puncture for $\pi \circ Tu$ unless it vanishes identically.

Define the global index $\text{wind}_\pi(u)$ as the sum of the local indices. This number is always non-negative. Let Γ_{even} and Γ_{odd} be the subset of Γ consisting of those punctures having an asymptotic limit with even or odd Conley–Zehnder index, respectively.

The following estimate is proved in [HoWZ1]:

Theorem 6.2. *For a finite energy surface $\tilde{u}$ the following estimate holds provided $\pi \circ Tu$ does not vanish identically:*

$$2 \cdot \text{wind}_\pi(u) \leq \mu(\tilde{u}) - 2\chi(S) + \sharp\Gamma + \sharp\Gamma_{\text{even}}\,.$$

The proof uses the asymptotic behavior of a finite energy surface near a puncture. What is crucial is the understanding of the asymptotic winding numbers and their relationship to the Conley–Zehnder index.

Since the Fredholm index is given by $\mathrm{ind}(\tilde{u}) = \mu(\tilde{u}) - \chi + \sharp\Gamma$, we obtain, combining this with the formula of the previous theorem:

$$2 \cdot \mathrm{wind}_\pi(u) \leq \mathrm{ind}(\tilde{u}) + 2g - 2 + \sharp\Gamma_{\mathrm{even}} \,. \tag{6.8}$$

For example, if we only allow finite energy spheres, i.e. $g = 0$, with at most one asymptotic limit of even Conley–Zehnder index and the Fredholm index belonging to $\{1, 2\}$ the right hand side can be estimated from above by 1. This implies that $\pi \circ Tu$ does not have any zero. In particular the image of u is transversal to the Reeb vector field.

Assume we have a finite energy foliation, where we allow surfaces of genus g with at most one asymptotic limit of even Conley–Zehnder index and Fredholm index at most 2. Then we obtain as an upper bound for the right-hand side of (6.8) the number $1 + 2g$. So there can be zeros of $\pi \circ Tu$, which in some sense implies that u should not be injective, contradicting the fact that $\tilde{u}$ parameterizes one of the surface in an $\mathbb{R}$-invariant foliation. Hence, in the case that the surfaces are allowed to have genus this suggests strongly that having a Fredholm index 1 or 2 does not imply that the projected surfaces have to be embedded. This in turn seems to suggest that one should not expect $\mathbb{R}$-invariant foliations with surfaces of higher genus. It could be, that allowing intersections might be useful for topological applications.

There is, however, an interesting modification of the equations, which in the case of genus 0 gives the old equation and where the intersections can be controlled as before. Let us study (6.6) in more detail. We required in (6.6) that $(u^*\lambda) \circ j = da$. This implies that $(u^*\lambda) \circ j$ defines the trivial cohomology class. The idea is now to require that $(u^*\lambda) \circ j$ defines a cohomology class, which, however, needs not to be trivial. For example we could replace the part $(u^*\lambda) \circ j = da$ of (6.6) by the requirement

$$d\big((u^*\lambda) \circ j\big) = 0 \,.$$

However, it turns out that it is useful to supplement this by an additional requirement. Indeed, if we insist on keeping the characteristic behavior of the solutions of the original PDE near the punctures we have to require that the cohomology class is trivial on a punctured neighborhood of each puncture. Then, on such neighborhood we can still write $(u^*\lambda) \circ j = da$ (We call (a, u) a local lift.) In particular we can introduce the notion of properness near a puncture for u, since a is unique up to an additive constant. This additional cohomology condition can be formulated as follows. Denoting by $\tau : S \setminus \Gamma \to S$ the inclusion we require that

$$\big[(u^*\lambda) \circ j\big] \in \tau^* H^1(S, \mathbb{R}).$$

Now, combining our alternative equation with the remaining part of (6.6) we obtain the desired modification:

$$\begin{gathered}\text{The map } u \text{ is not constant.}\\ \pi \circ Tu \circ j = J \circ \pi \circ Tu\\ d\big((u^*\lambda) \circ j\big) = 0\\ [u^*\lambda \circ j] \in \tau^* H^1(S, \mathbb{R})\\ \text{Near every puncture a local lift } (a,u) \text{ is proper}\\ \textstyle\int_{S\setminus\Gamma} u^* d\lambda < \infty .\end{gathered} \tag{6.9}$$

If S is the Riemann sphere this equation is equivalent to the old one. Observe that we basically got rid of the $\mathbb{R}$-component. The new partial differential equation has first order and second order parts. However, as far as the analysis is concerned, we can locally always introduce a first order elliptic system having precisely the same solutions (locally). We allow the punctures and j to vary. The object of interest is as before the equivalence class of u. We say u and u' are equivalent if there exists a biholomorphic map $\phi : S \to S'$ mapping punctures to punctures so that $u' \circ \phi = u$. The following index computation refers to the (virtual) dimension of the space of $[\tilde{u}]$ close-by. For this new problem the Fredholm index ind* is now

$$\operatorname{ind}^*(u) = \mu(u) - 2\chi(S) + \sharp\Gamma + 2.$$

The reader should note that the new index is the old index plus two times the genus (Two times the genus is the dimension of $\tau^* H^1(S, \mathbb{R})$).

PROPOSITION 6.3. *Let u be a solution of* (6.9) *for which $\pi \circ Tu$ is not identically* 0. *Then the following inequality holds*

$$2 \cdot \operatorname{wind}_\pi(u) \le \operatorname{ind}^*(u) + \sharp\Gamma_{\text{even}} - 2.$$

Here $\sharp\Gamma_{\text{even}}$ is the number of punctures with even asymptotic limit.

The first part of the proof is identical to that of Theorem 6.2 since it only uses the ξ-part of the partial differential equation. Then one replaces appropriate combination of terms in the formula by ind$^*(\tilde{u})$.

Note if there is only one even asymptotic limit and the index ind$^*(\tilde{u})$ is at most 2 we see that $\operatorname{wind}_\pi(u) = 0$ since it is a nonnegative integer. So the projected surfaces are again transversal.

We could generalize the notion of finite energy foliation by requiring the following

DEFINITION 6.4. *Let λ be a non-degenerate contact form on the closed three-manifold M^3 and J an admissible complex multiplication for ξ. Then a finite energy foliation for (M, λ, J) consists of a foliation of the complement of a finite number of periodic orbits by images of solutions u of (6.9), so that each surface has at most one asymptotic limit of even Conley–Zehnder index. If $\pi \circ Tu \neq 0$ we require that* $\text{ind}^*(u) \in \{1, 2\}$.

Our previous existence result claims that such foliations exist for every non-degenerate positive tight contact form on S^3 for a generic choice of J. It would be of interest to see a finite energy foliation with surfaces having a nontrivial genus.

It is not clear for which manifolds such foliations exist. In any case the following question is interesting:

QUESTION 3. *Let λ and λ_1 be contact forms inducing the same contact structure and co-orientation. Assume that J and J_1 are generic complex multiplications. Is the following true: If (M, λ, J) possesses a finite energy foliation so does (M, λ_1, J_1).*

The answer to the question might, of course, depend on the precise definition of finite energy foliation. One could speculate that a proof of this assertion should be based on some (higher-dimensional homotopy) invariance property, of the kind generalizing the one we are familiar with in Floer theory. Of course, it is not clear if our proposed definition (6.4) is the right one. However, the question has an affirmative answer for S^3 with the tight contact structure.

If the answer to Question 3 is affirmative, having a finite energy foliation would be a property of the contact structure rather than the contact form.

Every oriented, closed and connected three-manifold admits a contact form so that the Reeb flow is transversal to an open book decomposition with one binding orbit which is a periodic orbit for the associated Reeb vector field. This is not too difficult to prove. It seems plausible that the pages can be deformed into finite energy surfaces for a suitable structure using the Definition 6.4.

This prompts the following conjecture

CONJECTURE 6.5. *Every closed oriented and connected three-manifold admits for suitable data a finite energy foliation (as in Definition* 6.4*).*

Having such an existence theorem together with a deformation result (see the previous question) would give a perhaps powerful tool in studying three-manifolds and would definitely have its dynamical ramifications.

As already remarked studying the moduli spaces of finite energy surfaces leads to a symplectic field theory. For contact three-manifolds the study of finite energy foliations (whatever the definition) could perhaps be seen as a geometrization of the more algebraic symplectic field theory. The next decade hopefully clarifies this picture. At least for the moment it looks tempting to study real three manifolds with (pseudo-) holomorphic curve methods. So far this new approach produced a new understanding of important classes of dynamical systems, rather than new topological results. The subtle interplay between dynamical and topological phenomena looks intriguing. It has produced new results in the theory of dynamical systems and it has to be seen if new topological insight can be obtained as well.

Acknowledgement. I would like to thank L. Polterovich and K. Wysocki for helpful comments.

References

[AH] C. Abbas and H. Hofer, Holomorphic curves and global questions in contact geometry, to appear in Birkhäuser.

[Ar] V.I. Arnold, First steps in symplectic topology, Russian Mathematical Surveys 41 (1986), 1–21.

[B] V. Bangert, On the existence of closed geodesics on two-spheres, Internat. J. Math. 4:1 (1993), 1–10.

[Be] D. Bennequin, Entrelacements et équations de Pfaff, Astérisque 107-108 (1983), 83–161.

[C] W. Chen, Pseudo-holomorphic curves and the Weinstein conjecture, Comm. Anal. Geom. 8:1 (2000), 115–131.

[CoZ] C. Conley, E. Zehnder, The Birkhoff-Lewis fixed point theorem and a conjecture by V.I. Arnold, Inv. Math. 73 (1983), 33–49.

[E1] Y. Eliashberg, Classification of overtwisted contact structures on three manifolds, Inv. Math. (1989), 623–637.

[E2] Y. Eliashberg, Filling by holomorphic discs and its applications, London Math. Society Lecture Notes, Series 151 (1991), 45–67.

[E3] Y. Eliashberg, Contact 3-manifolds, twenty year since J. Martinet's work, Ann. Inst. Fourier 42 (1992), 165–192.

[E4] Y. Eliashberg, Legendrian and transversal knots in tight contact manifolds, in "Topological Methods in Modern Mathematics", Publish or Perish, 1993.

[E5] Y. Eliashberg, Classification of contact structures on $\mathbf{R}^3$, Inter. Math.

Res. Notices 3 (1993), 87–91.

[E6] Y. ELIASHBERG, Invariants in contact topology, Proceedings of the International Congress of Mathematicians, Vol. II (Berlin, 1998), Doc. Math., 327–338.

[EGH] Y. ELIASHBERG, A. GIVENTAL, H. HOFER, in this volume.

[EH] Y. ELIASHBERG, H. HOFER, A Hamiltonian characterization of the three-ball, P. Hess Memorial Volume, Differential and Integral Equations 7 (1994), 1303–1324.

[FHV] A. FLOER, H. HOFER, C. VITERBO, The Weinstein conjecture in $P \times \mathbb{C}^l$, Math. Zeit. 203 (1989), 335–378.

[Fr1] J. FRANKS, A new proof of the Brouwer plane translation theorem, Ergodic Theory and Dynamical Systems 12 (1992), 217–226.

[Fr2] J. FRANKS, Geodesics on S^2 and periodic points of annulus homeomorphisms, Invent. Math. 108 (1992), 403–418.

[Fr3] J. FRANKS, Area preserving homeomorphisms of open surfaces of genus zero, preprint, 1995.

[G1] V.L. GINZBURG, An embedding $S^{2n-1} \to \mathbf{R}^{2n}$, $2n-1 \geq 7$, whose Hamiltonian flow has no periodic trajectories, Inter. Math. Research Notices 2 (1995), 83–97.

[G2] V.L. GINZBURG, A smooth counterexample to the Hamiltonian Seifert conjecture in $\mathbf{R}^6$, Inter. Math. Research Notices 13 (1997), 641-650.

[Gi] E. GIROUX, Convexité en topologie de contact, Comm. Math. Helvetici 66 (1991), 637–677.

[Gr] M. GROMOV, Pseudoholomorphic curves in symplectic manifolds, Invent. Math. 82 (1985), 307–347.

[H] M. HERMAN, Examples of compact hypersurfaces in $\mathbf{R}^{2p}$, $2p \geq 6$ with no periodic orbits, manuscript, 1994.

[Ho1] H. HOFER, Pseudoholomorphic curves in symplectisations with application to the Weinstein conjecture in dimension three, Invent. Math. 114 (1993), 515–563.

[Ho2] H. HOFER, Holomorphic curves and dynamics in dimension three, Lecture Notes for the Proceedings of the IAS/Park City Institute, 7 (1999), 37–101.

[HoK] H. HOFER, M. KRIENER, Holomorphic curves in contact geometry, in "Proceedings of Symposia in Pure Mathematics, Differential Equations; La Pietra 1996, Conference on Differential Equations", marking the 70th birthday of Peter Lax and Louis Nirenberg, 65 (1996), 77-132.

[HoV] H. HOFER, C. VITERBO, The Weinstein conjecture in the presence of holomorphic spheres, Comm. Pure Appl. Math. 45:5 (1992), 583–622.

[HoWZ1] H. HOFER, K. WYSOCKI, E. ZEHNDER, Properties of pseudoholomorphic curves in symplectisations II: Embedding controls and algebraic invariants, GAFA 5 (1995), 270–328.

[HoWZ2] H. HOFER, K. WYSOCKI, E. ZEHNDER, A characterisation of the tight three-sphere, Duke Math. J. 81:1 (1995), 159–226.

[HoWZ3] H. HOFER, K. WYSOCKI, E. ZEHNDER, Properties of pseudoholomorphic curves in symplectisations I: Asymptotics, Ann. Inst. Henri Poincaré 13 (1996), 337–379.

[HoWZ4] H. HOFER, K. WYSOCKI, E. ZEHNDER, Properties of pseudoholomorphic curves in symplectisations IV: Asymptotics with degeneracies, in "Contact and Symplectic Geometry" (C. Thomas, ed.), Cambridge University Press (1996), 78–117.

[HoWZ5] H. HOFER, K. WYSOCKI, E. ZEHNDER, Unknotted periodic orbits for Reeb flows on the three-sphere, Top. Methods in Nonlinear Analysis 7:2 (1996), 219–244.

[HoWZ6] H. HOFER, K. WYSOCKI, E. ZEHNDER, The dynamics on a strictly convex energy surface in $\mathbf{R}^4$, Ann. of Math. (2) 148:1 (1998), 197–289.

[HoWZ7] H. HOFER, K. WYSOCKI, E. ZEHNDER, Properties of pseudoholomorphic curves in symplectisations III: Fredholm theory, in "Topics in Nonlinear Analysis, Progress in Nonlinear Differential Equations and Their Applications", 35 (1999), 381–476.

[HoWZ8] H. HOFER, K. WYSOCKI, E. ZEHNDER, A characterization of the tight three-sphere II, Comm. Pure and Appl. Math. LII (1999), 1139–1177.

[HoWZ9] H. HOFER, K. WYSOCKI, E. ZEHNDER, Finite energy foliations of tight three-spheres and Hamiltonian dynamics, preprint.

[HoZ] H. HOFER, E. ZEHNDER. Hamiltonian Dynamics and Symplectic Invariants, Birkhäuser, 1994.

[K1] K. KUPERBERG, A smooth counter example to the Seifert conjecture in dimension three, Annals of Mathematics 140 (1994), 723–732.

[K2] G. KUPERBERG, A volume-preserving counterexample to the Seifert conjecture, Comm. Math. Helvetici 71:1 (1996), 70–97.

[LT] G. LIU, G. TIAN, Weinstein conjecture and GW invariants, preprint, 1999.

[M] J. MARTINET, Formes de contact sur les variétés de dimension 3, Springer LNM 209 (1971), 142–163.

[Mc] D. MCDUFF, The local behaviour of J–holomorphic curves in almost complex 4–manifolds, J. Diff. Geom. 34 (1991), 143–164.

[McS] D. MCDUFF, D. SALAMON, Introduction to Symplectic Topology, Oxford University Press, 1998.

[MiW] M. MICALLEF, B. WHITE, The structure of branch points in minimal surfaces and in pseudoholomorphic curves, Ann. of Math. 141 (1994), 35–85.

[R1] P. RABINOWITZ, Periodic solutions of Hamiltonian systems, Comm. Pure Appl. Math. 31 (1978), 157–184.

[R2] P. RABINOWITZ, Periodic solutions of Hamiltonian systems on a prescribed energy surface, J. Diff. Equ. 33 (1979), 336-352.

[Ro] C. ROBINSON, A global approximation theorem for Hamiltonian systems, Proc. Symposium in Pure Math. XIV (1970), 233–244.

[W1] A. WEINSTEIN, Periodic orbits for convex Hamiltonian systems, Ann. Math. 108 (1978), 507–518.

[W2] A. WEINSTEIN, On the hypothesis of Rabinowitz's periodic orbit theorems, J. Diff. Equ. 33 (1979), 353–358.

HELMUT HOFER, Courant Institute of Mathematical Sciences, New York University, 251 Mercer Street, New York, NY 10012, USA hofer@cims.nyu.edu

GAFA, Geom. funct. anal.
Special Volume – GAFA2000, 705 – 741
1016-443X/00/S20705-37 $ 1.50+0.20/0

GAFA Geometric And Functional Analysis

PERSPECTIVES ON THE ANALYTIC THEORY OF L-FUNCTIONS

H. Iwaniec and P. Sarnak

Contents:

1 Introduction and Background

To the general mathematician L-functions might appear to be an esoteric and special topic in number theory. We hope that the discussion below will convince the reader otherwise. Time and again it has turned out that the crux of a problem lies in the theory of these functions. At some level it is not entirely clear to us why L-functions should enter decisively, though in hindsight one can give reasons. Our plan is to introduce L-functions and describe the central problems connected with them. We give a sample (this is certainly not meant to be a survey) of results towards these conjectures as well as some problems that can be resolved by finessing these conjectures. We also mention briefly some of the successful present-day tools and the role they might play in the big picture.

An L-function is a type of generating function formed out of local data associated with either an arithmetic-geometric object (such as an abelian variety defined over a number field) or with an automorphic form (it is expected that the latter set contains the former one, Shimura-Taniyama for

Supported in part by NSF Grants DMS-98-01642 and DMS-99-70386 as well as by grants to the Institute for Advanced Study from the Ambrose Monell Foundation, the Hansmann Membership and the Veblen Fund.

special cases and Langlands in general). Fix a number field K (i.e. a finite algebraic extension of $\mathbb{Q}$), the reader will not lose too much by restricting to $\mathbb{Q}$. An L-function takes the form of a product of degree $m \geq 1$ over all primes p of K

$$L(s) = \prod_p L_p(s) , \tag{1}$$

where the local factors are

$$L_p(s) = \prod_{j=1}^{m} \left(1 - \alpha_j(p)\,(Np)^{-s}\right)^{-1} \tag{2}$$

for suitable complex numbers $\alpha_j(p)$ and where Np is the norm of p. As a function of s this product converges absolutely for $\Re(s) > 1$ (see below) and we can multiply out to get the series

$$L(s) = \sum_{a \neq 0} c(a) N(a)^{-s} , \tag{3}$$

the sum being over integral ideals.

We give some concrete examples all being for $K = \mathbb{Q}$.

(1) The Riemann zeta function ($m = 1$)

$$\zeta(s) = \prod_p \left(1 - p^{-s}\right)^{-1} = \sum_{n=1}^{\infty} n^{-s} . \tag{4}$$

(2) Dirichlet L-functions ($m = 1$)

$$L(s, \chi) = \prod_p \left(1 - \chi(p) p^{-s}\right)^{-1} = \sum_{n=1}^{\infty} \chi(n) n^{-s} , \tag{5}$$

where χ is a character of the group of primitive residue classes $a (\operatorname{mod} q)$ (more precisely a multiplicative function on $\mathbb{Z}$ which is periodic of period q). The minimal period q is called the conductor of χ.

(3) For $m = 2$ we give the example of the L-functions of elliptic curves defined over $\mathbb{Q}$. Let E be such a nonsingular curve given by the equation

$$E : \quad y^2 = x^3 + ax + b , \tag{6}$$

$a, b \in \mathbb{Q}$. For a prime p at which reducing E modulo p yields a nonsingular curve over $\mathbb{F}_p$ (the field with p-elements), one defines the local factor $L_p(s, E)$ as follows: Let $N_E(p)$ be the number of solutions of (6) with x, y in $\mathbb{F}_p$ (not counting the point at infinity) and let $a_E(p) = p - N_E(p)$. Define

$$L_p(s, E) = \left(1 - \frac{a_E(p)}{\sqrt{p}} p^{-s} + p^{-2s}\right)^{-1} . \tag{7}$$

Note that this is not the standard algebraists normalization but it is very convenient for analytic purposes. The L-function $L(s,E)$ is defined by

$$L(s,E) = \prod_p L_p(s,E) , \tag{8}$$

where at primes p for which E has singular reduction (there being finitely many of these) special care must be taken in defining the local factor $L_p(s,E)$.

(4) Again we take $m=2$ and give an example of a holomorphic modular form and its L-function. Let $\mathbb{H}$ be the upper half plane. For $z \in \mathbb{H}$ and $m \geq 1$ set

$$F(z) = \sum_{\mu \in \mathbb{Z}\left[\sqrt{-1}\right]} \mu^{4m} \, e^{2\pi i \, N(\mu) z} , \tag{9}$$

where $N(\mu) = \mu\bar{\mu}$. It turns out that $F(z)$ is a holomorphic modular form of weight $k = 4m+1$ for the subgroup

$$\Gamma_0(4) = \left\{ \begin{pmatrix} a\, b \\ c\, d \end{pmatrix} \in SL_2(\mathbb{Z}); 4|c \right\}$$

of the modular group $SL_2(\mathbb{Z})$. That is to say it transforms appropriately under $z \to \frac{az+b}{cz+d}$ for $\gamma = \left(\begin{smallmatrix} ab \\ cd \end{smallmatrix}\right) \in \Gamma_0(4)$, with nebentypus character $\left(\frac{-4}{d}\right)$. Its L-function is

$$L(s,F) = \sum_{t=1}^{\infty} \Bigg(\sum_{\substack{\mu=\mu_1+i\mu_2 \\ \mu_1 \geq 0, \mu_2 > 0 \\ N(\mu)=t}} \left(\frac{\mu}{|\mu|}\right)^{4m} \Bigg) t^{-s} = \prod_p L_p(s,F) . \tag{10}$$

For $p \equiv 3(4)$

$$L_p(s,F) = (1-p^{-2s})^{-1} \tag{11}$$

while for $p \equiv 1(4)$

$$L_p(s,F) = (1 - c(p)p^{-s} + p^{-2s})^{-1} , \tag{12}$$

where

$$c(p) = \frac{1}{4} \sum_{N(\mu)=p} \left(\frac{\mu}{|\mu|}\right)^{4m} . \tag{13}$$

Alternatively we have

$$L(s,F) = \prod_p \left(1 - \lambda(p)(Np)^{-s}\right)^{-1} , \tag{14}$$

where p runs over the prime ideals of $\mathbb{Q}\left(\sqrt{-1}\right)$ and λ is the "Grossencharacter" given by $\lambda((\alpha)) = (\alpha/|\alpha|)^{4m}$. The form F is special among the modular forms for $\Gamma_0(4)$ in having an expression (14) in terms of a quadratic extension. It is an example of a "CM" modular form.

(5) An example of a non-CM holomorphic modular form is the popular

$$\Delta(z) = q\prod_{n=0}^{\infty}(1-q^n)^{24} := \sum_{n=1}^{\infty}\tau(n)q^n\ , \tag{15}$$

where $q=e^{2\pi iz}$ is a holomorphic cusp form of weight 12 for $SL(2,\mathbb{Z})$ (that is $\Delta\left(\frac{az+b}{cz+d}\right) = (cz+d)^{12}\Delta(z)$ for $\binom{ab}{cd}\in SL(2,\mathbb{Z})$). Its L-function is

$$L(s,\Delta) = \sum_{n=1}^{\infty}\frac{\tau(n)}{n^{11/2}}n^{-s} = \prod_p\left(1-\frac{\tau(p)}{p^{11/2}}p^{-s}+p^{-2s}\right)^{-1}. \tag{16}$$

That it factors into an "Euler product" as indicated is a consequence of Δ being a Hecke eigenform (see below).

(6) Our last example is a Maass cusp form for $GL(2,\mathbb{Z})$ on $\mathbb{H}$. That is a real-analytic function $\phi(z)$ on $\mathbb{H}$ satisfying
   (a) $\phi(\gamma z)=\phi(z)$ for $\gamma\in SL(2,\mathbb{Z})=\Gamma$
   (b) $\phi(-\bar{z})=\phi(z)$
   (c) $\Delta\phi+\lambda\phi=0,\ \ \lambda>\frac{1}{4}$
   (Δ being the Laplacian for the hyperbolic metric)
   (d) ϕ is square integrable on $\Gamma\backslash\mathbb{H}$ (in this particular example this property is equivalent to being a *cusp* form).

These ϕ's are far less tangible than the previous examples. Indeed that there are any such ϕ's is far from obvious. The only proof of their existence is through the trace formula (indeed this demonstration was part of Selberg's original motivation for developing the trace formula). Our present understanding is that these elusive forms exist in abundance only for Γ's as above, such as congruence subgroups of $SL_2(\mathbb{Z})$ [PhS], [Wo]. The Hecke operators T_n defined by

$$(T_n\phi)(z) = \frac{1}{\sqrt{n}}\sum_{ad=n}\sum_{b(\mathrm{mod} d)}\phi\left(\frac{az+b}{d}\right)$$

act on these eigenspaces, they commute with each other and are self-adjoint on $L^2(\Gamma\backslash\mathbb{H})$. We may therefore simultaneously diagonalize and assume that

$$T_n\phi = \lambda_\phi(n)\phi,\ \text{for all } n\geq 1\ . \tag{17}$$

For such Maass eigenforms we have corresponding degree two L-functions:

$$L(s,\phi) = \sum_{n=1}^{\infty}\lambda_\phi(n)n^{-s} = \prod_p L_p(s,\phi)\ , \tag{18}$$

where

$$L_p(s, \phi) = (1 - \alpha_{1,\phi}(p)p^{-s})^{-1} (1 - \alpha_{2,\phi}(p)p^{-s})^{-1} \qquad (19)$$

and

$$\alpha_{1,\phi}(p)\, \alpha_{2,\phi}(p) = 1, \qquad \alpha_{1,\phi}(p) + \alpha_{2,\phi}(p) = \lambda_\phi(p)\,. \qquad (20)$$

That completes our list of examples of L-functions.

We now dive in with the general modern definition of an automorphic cusp form and its L-function. We consider only the group GL_m since it is expected from general conjectures of Langlands [La] that all L-functions are products of these standard L-functions (we emphasize that other groups and even the exceptional groups play an important role in understanding these L-functions (see below)). Let $\mathbb{A}_K$ be the adele ring of K, that is the restricted product, $\Pi_v K_v$ over the completions of K. An automorphic cusp form F on $GL_m(\mathbb{A}_K)$ is an irreducible representation of $GL_m(\mathbb{A}_K)$ occurring in $L_0^2(GL_m(K)\backslash GL_m(\mathbb{A}_K))$ under the right regular representation of $GL_m(\mathbb{A}_K)$. Here we are assuming that all functions transform under the action of the center by a unitary central (idele class) character and the subscript zero refers to the cuspidal subspace [GelP]. The relation to the classical description of modular forms is that there are special functions in such an irreducible representation which are classical modular forms. Now an F as above is of the form $\otimes_v F_v$ where v ranges over all places of K (finite and archimedean) and F_v is a unitary (in fact generic [JS]) representation of $GL_m(K_v)$. Using suitable parameters of the local representation of F_v (Satake parameters [Sa2] if F_v is unramified, which is the case for all but a finite number of parameters, and Langlands parameters in general) one defines the numbers $\alpha_{F,j}(v)$, $j = 1, \dots, m$ in (2) above. In particular, this yields the definition of the local factors $L_v(s, F) = L(s, F_v)$ for v finite. At the archimedean places there are similar parameters for the local representations of $GL_m(\mathbb{R})$ or $GL_m(\mathbb{C})$. The local L-factors take the form

$$L(s, F_v) = \prod_{j=1}^{m} \Gamma_v\big(s - \mu_{j,F}(v)\big)\,, \qquad (21)$$

where

$$\Gamma_v(s) = \begin{cases} \pi^{-s/2}\,\Gamma\left(\frac{s}{2}\right), & \text{if } K_v \sim \mathbb{R} \\ (2\pi)^{-s}\,\Gamma(s), & \text{if } K_v \sim \mathbb{C}\,. \end{cases} \qquad (22)$$

The standard global L-function of F is then

$$L(s, F) = \prod_{v \text{ finite}} L(s, F_v). \qquad (23)$$

The product converges absolutely for $\Re(s) > 1$. Moreover, the analogue of Riemann's analytic continuation and functional equation are known in this generality (Hecke [Hec], Godement-Jacquet [GoJ] and see also Tamagawa [T]). Define the completed function

$$\Lambda(s,F) = \Big(\prod_{v \text{ archim.}} L(s,F_v)\Big) \cdot L(s,F) \, . \tag{24}$$

Then $\Lambda(s,F)$ extends to an entire function (except in the case $m=1$ and F is the trivial representation when Λ has poles at $s=0$ and $s=1$) and satisfies the functional equation

$$\Lambda(1-s,\widetilde{F}) = \bar{\epsilon}_F \, N_F^{s-\frac{1}{2}} \Lambda(s,F) \, . \tag{25}$$

Here $N_F \geq 1$ is an integer called the conductor of F, ϵ_F is the root number (which has modulus 1) and $\tilde{F}$ is the representation contragredient to F.

The cuspidal spectrum of $L^2(GL_m(K)\backslash GL_m(\mathbb{A}_K))$ is discrete so that the set of standard L-functions is countable. In the form that we have described them, these L-functions are not unrelated to each other (see for example (10) and (14) above). It is known (see [AC] that each $L(s,F)$ for K is a product of $L(s,F')$ for $\mathbb{Q}$ (with $m'=md$, $d=\deg(K/\mathbb{Q})$). For many purposes it is convenient to think of $L(s,F)$ over K rather than of larger degree over $\mathbb{Q}$. The cuspidal standard $L(s,F)$ over $\mathbb{Q}$ are all independent of each other and they form the basic building blocks for all L-functions.

One can form more general L-functions from these basic ones, that is the tensor powers. Very special cases of these are known to have analytic continuations and functional equations. There are at present two methods to attack this problem of continuation both depend on the analytic properties of Eisenstein series. These are the Rankin-Selberg method, see [Bu], and the Langlands-Shahidi method [Sh1]. For example given F and F' automorphic cuspidal on GL_m and $GL_{m'}$ respectively, then $L(s,F\otimes F')$ is an L-function of degree mm' whose local factor at a place v of K at which both F and F' are unramified is:

$$L(s,F_v \otimes F'_v) = \prod_{j=1}^{m}\prod_{k=1}^{m'} \big(1-\alpha_{F,j}\,\alpha_{F',k}\,N(v)^{-s}\big)^{-1} \, . \tag{26}$$

The precise analytic continuations and functional equations for these are known [JPS]. The function $L(s,F\otimes\tilde{F})$ has non-negative coefficients in its expansion (3). This together with its analytic properties (i.e. pole at $s=1$) imply that the product (23) converges absolutely in $\Re(s)>1$.

Another special case that is known [G] is the degree 8 triple product of GL_2 forms. Let F,G and H be three cusp forms on GL_2/K. At a place v

where F, G and H are unramified define the local L-function of degree 8 by

$$L(s, F_v \otimes G_v \otimes H_v) = \prod_{\substack{\epsilon_j \in \{1,2\} \\ j=1,2,3}} \left(1 - \alpha_{F,\epsilon_1}(v)\,\alpha_{G,\epsilon_2}(v)\,\alpha_{H,\epsilon_3}(v) N(v)^{-s}\right)^{-1}. \quad (27)$$

Set

$$L(s, F \otimes G \otimes H) = \prod_v L(s, F_v \otimes G_v \otimes H_v). \quad (28)$$

Then $L(s, F \otimes G \otimes H)$ has an analytic continuation and functional equation $s \to 1 - s$.

Some special cases of the symmetric power L-functions of GL_2 forms are known to have analytic continuations and functional equations. For $n \geq 1$ and F on GL_2/K define the local factor of the n-th symmetric power (at an unramified place) by

$$L(s, \mathrm{sym}^n F_v) = \prod_{j=0}^{n} \left(1 - (\alpha_{F,1}(v))^j \, (\alpha_{F,2}(v))^{n-j} \, N(v)^{-s}\right)^{-1}. \quad (29)$$

The global n-th symmetric power L-function $L(s, \mathrm{sym}^n F)$ is defined to be the product of these local factors. For $n = 1$ this is just the standard L-function. For $n = 2$ the analytic properties were established by Shimura [Shi2]. Recently, Kim and Shahidi [KiSh] established the expected analytic properties for $n = 3$.[1] Their proof uses at one point the unitary dual of the exceptional group G_2!

The above discussion has indicated why L-functions of automorphic forms enjoy certain analytic properties. For the examples (3) of L-functions of elliptic curves over $\mathbb{Q}$ this follows from the spectacular progress by Wiles [Wi] and [TaW] which asserts that "elliptic curves over $\mathbb{Q}$ are modular." This implies that $L(s, E)$ is an $L(s, F)$ for a suitable holomorphic weight 2 cusp form F on a congruence subgroup of $SL_2(\mathbb{Z})$. Indeed, the construction of automorphic forms (and hence of L-functions) from arithmetic-geometric settings is one of the major thrusts of modern number theory. Our interest here is beyond this and at the same time much older. That is, we are given an automorphic form and its L-function and we investigate its properties (beyond just analyticity) and their applications.

[1] Added in proof: Recently they have also established this for $n = 4$.

2 Fundamental Conjectures

We turn to some of the basic problems which until their resolution are expected to be a focus of the subject. The first is a well known generalization of Riemann's Conjecture.

A) **Grand Riemann Hypothesis** (GRH)
The zeros ρ_F of any $\Lambda(s,F)$ have real part equal to $\frac{1}{2}$.

Comments:

(A1) Crisp, falsifiable and far reaching this conjecture is the epitome of what a good conjecture should be. Moreover, it has many striking consequences (some described below). One of its powers lies in that it ensures uniform (up to square root of the number of terms - like random numbers) cancellations in sums over $c_F(a)$ or $c_F(p)$ (as in (2) and (3)). It is in this form that one often uses it in applications to problems in which the local data in $c_F(p)$ is being used to analyze something global and *visa versa*. In practice, GRH is often used as a working hypothesis (and an apparently very reliable one at that) in that one proceeds by using it, and in this way many results are established under GRH. However, there have been sufficiently powerful advances in the theory that in a number of cases one can dispense with GRH and the desired result is established unconditionally.

(A2) The true strength of GRH lies in the statement for the general $L(s,F)$, or at least for some infinite family of L-functions such as Dirichlet L-functions $L(s,\chi)$. For example, the case of $\zeta(s)$ itself has few consequences (it is of course directly equivalent to the size of the remainder term in the Prime Number Theorem). For a recent description and discussion of RH, see Bombieri [B2]. There is no $L(s,F)$ for which GRH is known. For families such as $L(s,\chi)$, there are results, "density theorems", which assert that almost all their zeros lie near $\Re(s)=\frac{1}{2}$. These can often be used as a substitute for GRH (see section 4 below).

(A3) For $\zeta(s)$, $L(s,\chi)$ and some $GL_2/\mathbb{Q}$ L-functions extensive numerical experimentations have confirmed GRH in impressive ranges. This is important supporting evidence for the truth of GRH. The function field analogues (see section 3) are known to be true and this is further strong evidence in favor of GRH.

The direct results that have been established towards GRH are modest. The method of Hadamard and de la Vallée Poussin for $\zeta(s)$ (in their proof of the Prime Number Theorem) can be used together with the analytic properties of $L(s,F)$ and $L(s,F\otimes\tilde{F})$ to show that $L(1+it,F)\neq 0$ for $t\in\mathbb{R}$, cf. [R]. The lower bounds for $|L(1+it,F)|$ that one obtains this way are all roughly of the same quality[2] except for one major lacuna. That is the case of $L(1,\chi)$, χ quadratic over $\mathbb{Q}$ (this being the first instance of nonvanishing of L-functions and is due to Dirichlet in his proof of the infinitude of primes in arithmetic progressions). For this case instead of the lower bound of $(\log q)^{-1}$ for $L(1,\chi_q)$ the best known *effective* lower bound is $L(1,\chi_q)\gg\frac{\log q}{\sqrt{q}}$, (if q is prime and slightly weaker in general [Gol], [GrZ]). The last is an excellent example of the use of GL_2 L-functions (in particular L-functions of elliptic curves of high rank) to give information about GL_1 L-functions. *Ineffectively* Siegel [Si2], following Landau, established the lower bound; given $\epsilon>0$ there is $C_\epsilon>0$ such that for any $q>1$

$$L(1,\chi_q)\geq C_\epsilon q^{-\epsilon}\,. \tag{30}$$

GRH implies the so-called Lindelöff Hypothesis which, if true, is a very useful bound for L-functions on the critical line.

Precisely for the purpose of estimating $L(\frac{1}{2}+it,F)$ we introduce the quantity (the "analytic conductor")

$$C(F,t)=N_F\prod_{j=1}^{m}\prod_{v\text{ archim.}}\left(1+|\mu_{j,F}(v)+it|^{d(v)}\right)\,, \tag{31}$$

where for v archimedean

$$d(v)=1\quad\text{if}\quad K_v=\mathbb{R}\quad\text{and}\quad d(v)=2\quad\text{if}\quad K_v=\mathbb{C}\,. \tag{32}$$

Fix m and K. Let $d=\deg(K/\mathbb{Q})$. The Lindelöff Hypothesis asserts that for any $\epsilon>0$,

$$L(\tfrac{1}{2}+it,F)\underset{\epsilon}{\ll}\left(C(F,t)\right)^{\epsilon}\,. \tag{33}$$

It follows from the functional equation for $\Lambda(s,F)$ and the convexity bounds of Phragmen-Lindelöff that for $\epsilon>0$

$$L\left(\tfrac{1}{2}+it,F\right)\underset{\epsilon}{\ll}\left(C(F,t)\right)^{\frac{1}{4}+\epsilon}\,. \tag{34}$$

Because of its many applications we single out the following problem as a basic one.

B) **Subconvexity Problem**

For m and K fixed to show there is $\delta>0$ such that

$$L\left(\tfrac{1}{2}+it,F\right)\ll\left(C(F,t)\right)^{\frac{1}{4}-\delta}.$$

[2]In the special case of $\zeta(s)$ some improvements have been given using far reaching methods of I.M. Vinogradov.

Actually in applications we usually have some subfamily (i.e. only one of the parameters t, N_F or $\| F \|_{\text{archim.}}$ varies) and we seek subconvexity estimates uniformly for the subfamily. This problem (B) is solved in a number of cases and we discuss this and some of their applications in section 6.

Next we discuss the generalized Ramanujan Conjecture. It is the local analogue of GRH and is a spectral problem concerning the local representations F_v of $GL_m(K_v)$ of the global automorphic cuspidal representation F. It asserts that for a place v at which F_v is unramified, F_v should be tempered (see [Sa1]). Equivalently this can be stated in terms of $L(s, F_v)$ as follows:

C) **Generalized Ramanujan Conjecture** (GRC)

Let F be an automorphic cuspidal representation of $GL_m(\mathbb{A}_k)$ which is unramified at a place v. Then for v finite $|\alpha_{j,F}(v)| = 1$ while for v archimedean, $\Re(\mu_{j,F}(v)) = 0$.

Comments:

(C1) Again this is a clean and far reaching conjecture. It is in the background in many applications of the spectral theory of automorphic forms to problems in analytic number theory.

(C2) The original problem of Ramanujan was concerned with the case $F = \Delta(z)$ (see (15) above). In this case GRC is equivalent to the original Ramanujan Conjecture:

$$|\tau(p)| \le 2p^{\frac{11}{2}} . \tag{35}$$

For this case and more generally the case of holomorphic cusp forms of even integral weight for congruence subgroups of $SL(2,\mathbb{Z})$ the conjecture was established by Deligne [D2].

(C3) For $K = \mathbb{Q}$ and $\mathbb{Q}_v = \mathbb{R}$, GRC is equivalent to the Selberg Eigenvalue Conjecture; that for any N

$$\lambda_1\big(\Gamma(N)\backslash\mathbb{H}\big) \ge \tfrac{1}{4} . \tag{36}$$

Here $\Gamma(N)$ is the principal congruence subgroup of $\Gamma(1) = SL(2,\mathbb{Z})$, that is $\Gamma(N) = \{\gamma \in \Gamma(1); \gamma \equiv \left(\begin{smallmatrix}1&0\\0&1\end{smallmatrix}\right) \bmod \mathrm{N}\}$, and λ_1 is the smallest eigenvalue of the Laplacian on the cuspidal space $L_0^2(\Gamma(N)\backslash\mathbb{H})$.

There are nontrivial and very useful general bounds towards GRC. Firstly, there are purely local bounds which use only that F_v is unitary and generic [JS] (these properties of F_v follow from F being a cusp form). These bounds are

$$\Big|\log_{N(v)} |\alpha_{j,F}(v)|\Big| < \tfrac{1}{2}, \text{ for } v \text{ finite} \tag{37}$$

$$\left|\Re\big(\mu_{j;F}(v)\big)\right| < \tfrac{1}{2}, \quad \text{for } v \text{ archimedean} . \tag{38}$$

This can be viewed as the analogue of the convexity bound for L-functions. For this problem a "subconvex" bound is known in general [LuRS]. For F on $GL_m(\mathbb{A}_K)$ cuspidal

$$\left|\log_{N(v)} \alpha_{j,F}(v)\right| \le \tfrac{1}{2} - \tfrac{1}{m^2+1} \ , \text{ for } v \text{ finite} \tag{39}$$

$$\left|\Re\big(\mu_{j,F}(v)\big)\right| \le \tfrac{1}{2} - \tfrac{1}{m^2+1} \ , \text{ for } v \text{ archimedean} . \tag{40}$$

The proof of this is global relying on the analytic properties of Rankin-Selberg L-functions as well as a technique of persistence of zeros for families of L-functions (in this case twists by ray class characters) and also a positivity argument. This theme of families will recur often in what follows. Combining the above bounds for $m = 3$ together with the Gelbart-Jacquet [GeJ] symmetric square lift from GL_2 to GL_3 yields an improved bound[3] for GL_2:

$$\left|\log_{N(v)} \alpha_{j,F}(v)\right| \le \tfrac{1}{5} \ , \text{ for } v \text{ finite} \tag{41}$$

$$\left|\Re\big(\mu_{j,F}(v)\big)\right| \le \tfrac{1}{5} \ , \text{ for } v \text{ archimedean} . \tag{42}$$

Remarkably the last at finite places can be proven by a quite different method, see Shahidi [Sh2] who uses exceptional groups. For $K = \mathbb{Q}$ and $Q_\infty \cong \mathbb{R}$ (42) yields $\lambda_1(\Gamma(N)\backslash\mathbb{H}) \ge \frac{21}{100}$ for the Selberg problem (36) above. This being greater than $\frac{3}{16}$ has significant corollaries (see section 7). The use of the family of twists of L-functions by Dirichlet characters for the purpose of obtaining estimates towards the GRC for Maass forms for $GL_2/\mathbb{Q}$, at finite places p, was introduced in [DuI1]. In that case it was used to exploit the extra functional equations afforded by the family as well as to overcome the lack of positivity for the coefficients of the symmetric square L-function. Their method may be used to give a slight improvement of (41) in the case $K = \mathbb{Q}$.

It was observed early on (Tate, Langlands, Serre) that the expected analytic properties (i.e. meromorphic continuation and location of poles) of the symmetric power L-functions imply GRC as well as conjectures about the distribution of the "angles" $\{\alpha_{F,1}(v), \dots, \alpha_{F,m}(\nu)\}$ as $N(v) \to \infty$ (the so-called Sato-Tate Conjectures). For F on $GL_2/\mathbb{Q}$ and $n \ge 1$ consider

$$G_n(s) = L\big(s, \mathrm{sym}^n F \otimes \widetilde{\mathrm{sym}^n F}\big) \ , \tag{43}$$

[3] We have just learned that for this case, Kim and Shahidi have established the improved bound replacing $\frac{1}{5}$ by $\frac{5}{34}$ in (41) and (42). Added in proof: Their methods when combined with the method of twisting leads for the case of $K = \mathbb{Q}$ to bounds with $\frac{1}{5}$ replaced by $\frac{7}{64}$ (see [KiS])

where $L(s, \mathrm{sym}^n F)$ is given following (29). The series for $G_n(s)$ is of the form

$$G_n(s) = \sum_{m=1}^{\infty} b(m) m^{-s} \tag{44}$$

with $b(m) \geq 0$. If as expected $G_n(s)$ is analytic for $\Re(s) > 1$ (it certainly has a pole at $s = 1$) then the positivity of the coefficients easily implies that

$$b(m) \ll_\epsilon m^{1+\epsilon}, \text{ for any } \quad \epsilon > 0 \ . \tag{45}$$

Now examining the coefficients of (43) we have for p a prime at which F in unramified and $e \geq 1$

$$\left| \sum_{j=0}^{n} \left[\left(\alpha_{F,1}(p)\right)^j \left(\alpha_{F,2}(p)\right)^{n-j} \right]^e \right|^2 \leq e\, b(p^e) \ . \tag{46}$$

Hence, combining (45) and (46), the fact that $|\alpha_{F,1}(p)\alpha_{F_2}(p)| = 1$ and letting $e \to \infty$ and $\epsilon \to 0$ we conclude that

$$\max \left\{ \left|\alpha_{F,1}(p)\right|, \left|\alpha_{F,2}(p)\right| \right\} \leq p^{\frac{1}{2n}} \ . \tag{47}$$

Thus the knowledge that $G_n(s)$ is analytic for $\Re(s) > 1$ for all n yields GRC for F. The GRC for $p = \infty$, i.e. the Selberg Conjecture would also follow from similar considerations. The behavior of the distribution of the angles $\{\alpha_{F,1}(p), \alpha_{F,2}(p)\}$ requires a little more, that is the analytic properties of $G_n(s)$ up to and including $\Re(s) = 1$.

The last fundamental conjecture that we mention is the Birch and Swinnerton-Dyer Conjecture. This conjecture was discovered experimentally (i.e. through numerical experimentation) in looking for elliptic curve analogues of the Siegel Mass Formula (see section 6) for quadratic forms.

D) **Birch and Swinnerton-Dyer Conjecture** (BSC)

Let $E/\mathbb{Q}$ be an elliptic curve and $L(s, E)$ its L-function. Then the order of vanishing of $L(s, E)$ at $s = \frac{1}{2}$ is equal to the rank of the group of $\mathbb{Q}$-rational points on E.

Comments:

(D1) Again this qualifies as an excellent and perhaps somewhat unexpected conjecture at the time. It contains highly nontrivial local to global information. Recall that the L-function $L(s, E)$ is defined from local data while the rank of the group of rational points is one of the most interesting global geometric invariants of $E(\mathbb{Q})$.

(D2) The point $s = \frac{1}{2}$ is the only explicitly known point at which any $\Lambda(s, F)$ vanishes. Vanishing at $s = \frac{1}{2}$ could happen simply because of the sign of the functional equation (if $F = \tilde{F}$ and $\epsilon_F = -1$), but as in the case of $\Lambda(s, E)$, it could vanish to order greater than 1 for deeper arithmetical reasons.

As with the last conjecture there are substantial results towards the BSC for elliptic curves over $\mathbb{Q}$. The works of Coates-Wiles [CoW] for CM elliptic curves and Kolyvagin-Lugachev [KoL] and Gross-Zagier [GrZ] in general imply essentially that the Conjecture is true if the order of vanishing at $s = \frac{1}{2}$ is at most 1. It should be noted that the only general method to construct rational points on a given E is Heegner's construction [Hee] (we are not asking to find elliptic curves containing a given rational point). It is unclear what role these play when $L(s, E)$ vanishes to order ≥ 2.

3 Function Field Analogues

As was mentioned in section 2 the function field analogue of GRH is known. There is a lot to be learned from this algebro-geometric analogue and it has led to many insights for L-functions over number fields. As in the number field case the Riemann Hypothesis in the function field has striking implications. In particular it yields optimal bounds for exponential and character sums over finite fields and these are a basic tool in many of the results mentioned already as well as ones mentioned below. In fact, the special cases of the GRC that have been established make use of GRH in the function field. So the function field is an important part of our story and we review this analogue briefly.

The starting point is to replace the field K by a finite extension k of the field $\mathbb{F}_q(t)$, $\mathbb{F}_q$ being the field with q elements. We define, following Artin, the zeta function of k

$$\zeta_k(T) := \prod_v \left(1 - T^{\deg(\nu)}\right)^{-1}, \tag{48}$$

the product being over all the places (i.e. primes) of k and $\deg(v)$ is the corresponding local extension degree. The field k may be realized as the field of functions of a nonsingular projective curve C over $\mathbb{F}_q$. This allows one to give an alternate useful expression for $\zeta_k(T)$. If N_n is the number of points on C defined over $\mathbb{F}_{q^n}$, then

$$\zeta_k(T) = \zeta(T, C/\mathbb{F}_q) = \exp\left(\sum_{n=1}^{\infty} \frac{N_n T^n}{n}\right). \tag{49}$$

Using this and the Riemann-Roch theorem for the curve C over $\bar{\mathbb{F}}_q$ one can show the analogue of the analytic continuation and functional equation of $\zeta(s)$, for $\zeta_k(T)$. That is,

$$\zeta_k(T) = \frac{P(T, C/\mathbb{F}_q)}{(1-T)(1-qT)}, \tag{50}$$

where P is an integral polynomial of degree $2g$ with g the genus of C and P satisfies a functional equation relating its values at T to $1/qT$. The analogue of GRH is the statement that all the zeros of P be on the circle $|T| = 1/\sqrt{q}$. This was established by Weil [We1]. There are a number of ideas that go into his proof (he gave two quite different proofs). The numbers N_n can be realized as the number of points on $C(\bar{\mathbb{F}}_q)$ which are fixed by the nth power of the Frobenius morphism (raising coordinates to the power q). This suggests the use of a Lefschetz trace formula to linearize this counting. To achieve this Weil passes to the Jacobian X of C. For ℓ prime to q and $\nu \geq 1$ the corresponding Frobenius endomorphism α acts on the ℓ^ν division points of X, giving rise to an ℓ-adic matrix realization of α. Its eigenvalues are shown to be the inverses of the zeros of $P(T)$. This gives an important spectral interpretation of the zeros. The proof that the zeros are on the circle $1/\sqrt{q}$ requires a further elaborate analysis of α in the endomorphism ring of X and in particular the use of the positivity of Rosati involutions.

The definition of the zeta functions $\zeta(T, V/\mathbb{F}_q)$ for smooth projective varieties V over $\mathbb{F}_q$ was given by Weil. He put forth conjectures about the rationality, functional equations and analogues of the Riemann Hypothesis for these zetas. The first was proven by Dwork. A different proof was given by Grothendieck who also established the other analytic properties (i.e. functional equations, location of poles) by using his ℓ-adic cohomology theory. In particular, Grothendieck gives a spectral interpretation of $\zeta(T, V/\mathbb{F}_q)$ in terms of the characteristic polynomial of the induced linear action of Frobenius on the cohomology groups of the variety.

The proof of the analogue of GRH is due to Deligne [D2]. An important methodological difference in his proof being that the zeros are not shown to have a given absolute value ("purity") in one step and with one variety V. For example, if V is a smooth hypersurface in $\mathbb{P}^{2n}$ then he places $\zeta(T, V)$ in a family V_t, $t \in U$ a parameter space. The arithmetic fundamental group $\pi_1(U)$ has representations via monodromy in the various cohomology groups $H^i(\bar{V}_0, \mathbb{Q}_\ell)$, where V_0 is a fixed base point (in this hypersurface example only the middle dimensional cohomology group contains nontrivial information). In this way one may realize $\zeta(T, V/\mathbb{F}_q)$ as a local factor of an L-function associated with the monodromy representation above. One can furthermore examine various tensor powers of this representation. Also these new L-functions have known analytic properties (or at least one can locate their poles using invariant theory for the representations of the monodromy groups). One is now very much in the position that one is in deriving the local GRC from the global analytic properties of the symmetric power L-functions (see section 2). In fact, similar positivity arguments with arbitrarily high dimensional representations of the monodromy groups yield in the limit that the zeros of $\zeta(T, V/\mathbb{F}_q)$ are all on the circle $|T| = q^{-n+1/2}$. So the family, its symmetry and positivity are the key ingredients in the known proof of the GRH for varieties over finite fields.

The solution by Deligne of these Weil Conjectures allowed him to solve the special cases of GRC mentioned in section 2. The reduction itself is deep and is due to Eichler [E] and Igusa [I] in the special case of weight 2 and Ihara [Ih] and Deligne [D1] in general.

We end this section by mentioning the function field analogue of automorphic forms F on GL_m. Replacing, as we did at the start of this section, K by k we may consider the space $L^2(GL_m(k)\backslash GL_m(\mathbb{A}_k))$ and its cuspidal subspace. In a recent paper Lafforgue [L] has completed the program started by Drinfeld of (amongst other things) establishing the GRC for these automorphic cusp forms. A key ingredient of course is Deligne's proof of the Weil Conjectures above. There are many other crucial ingredients such as the trace formula [A] and the converse theorem [CogP].

4 Dirichlet L-Functions $GL(1)/\mathbb{Q}$

The work of Linnik [Li1] marked the beginning of a series of developments which give in some sense GRH (for Dirichlet L-functions) on average. This is not just an exercise but is a powerful tool which produces results not covered by GRH. In many applications of GRH one has, say sums of sums over primes in different arithmetic progressions, and GRH would give approximations for each sum. Since one is averaging over different progressions it is just as useful in such situations to know that the approximation offered by GRH is correct on average. The many developments during the period 1950-1970 mentioned above are based on a penetrating study of the orthogonality of Dirichlet characters (to different moduli!) and culminated in the Bombieri-Vinogradov Theorem.

For $(a,q)=1$, let

$$\psi(x;q,a) = \sum_{\substack{p\le x\\ p\equiv a(q)}} \log p\ , \tag{51}$$

the sum being over primes. According to GRH

$$\psi(x;q,a) = \frac{x}{\varphi(q)} + O\big(x^{1/2}(\log x)^2\big)\ , \tag{52}$$

where $\varphi(q)$ is the number of residue classes (modulo q) prime to q (this equivalence is essentially due to Riemann).

The Bombieri-Vinogradov Theorem (in a slightly stronger form by Bombieri [B1]) asserts that for $A>0$ there is $B>0$ such that

$$\sum_{q\le Q} \max_{(a,q)=1} \left|\psi(x;q,a) - \frac{x}{\varphi(q)}\right| \ll \frac{x}{(\log x)^A}\ ,$$

where $Q = x^{1/2}/(\log x)^B$. So this comes close to (52) on average in this range. By the way (53) is closely related to statements giving nontrivial bounds for the number of zeros $\rho=\beta+i\gamma$ of $L(s,\chi)$ of conductor $q\le Q$ and with $\beta\ge\sigma$ ($\sigma>\frac12$), $|\gamma|\le T$, known as Density Theorems.

More recent results [BFI] use much more sophisticated tools including bounds for exponential sums over finite fields as well as $GL_2/\mathbb{Q}$ spectral theory in the form connected with sums of Kloosterman sums (see section 7). Their results concern primes in progressions to moduli beyond $\sqrt{x}$ and cannot be derived from GRH. For example it is shown (among stronger, but more complicated results) that for any $a\neq 0$, $A>0$ there is $B>0$ such that

$$\sum_{\substack{(a,q)=1\\ q\le\sqrt{x}(\log x)^A}} \left|\psi(x;q,a) - \frac{\psi(x)}{\varphi(q)}\right| \ll \frac{x}{(\log x)^3}(\log\log x)^B\ . \tag{53}$$

We turn to Problem B of section 2 which concerns the size of $L(s,\chi)$ on the critical line. The first developments are much older. Weyl's method [Wey] of shifting the argument and repeated squaring in estimating sums $\sum_n e(\alpha f(n))$, where $e(z)=e^{2\pi iz}$ and f is a polynomial, led to the subconvexity estimate (the convexity exponent here is $1/4$)

$$\zeta\big(\tfrac12+it\big) \ll \big(|t|+1\big)^{\frac16} \tag{54}$$

for the Riemann zeta function (the same can be done this way for $L(s,\chi)$ in the t-aspect). There have been many improvements of the exponent $\frac16$, but our emphasis here is on subconvexity.

For the case of $L(s,\chi)$ (s fixed with real part equal to one half) in the conductor q of χ aspect, there is the result of Burgess [Bur]. It gives the subconvexity estimates (again the convexity exponent is $\frac{1}{4}$)

$$L(s,\chi) \underset{\epsilon}{\ll} q^{\frac{3}{16}+\epsilon} . \tag{55}$$

Burgess proceeds by estimating the sums

$$S = \sum_{N<n\leq N+H} \chi(n) \tag{56}$$

for N and H of certain sizes. He obtains nontrivial bounds by summing S and its shifts to large (even) powers which allows him to make use of bounds for complete character sums which in turn rely on Weil's GRH in the function field for curves of suitably large genus. Interestingly there are ranges in (56) where Burgess obtains nontrivial bounds and for which the GRH for $L(s,\chi)$ yields nothing nontrivial.

In the next section we discuss a recent improvement of (56) for χ quadratic.

5 Special Values

The question as to whether an L-function $L(s,F)$ vanishes at a special point on the critical line has arisen in various contexts and is apparently a fundamental one (note that such a question is not addressed by GRH). It arises in the problem of examining the instability of the elusive Maass cusp forms (see section 1). For this problem the L-functions in question are $L(s,\phi\otimes Q)$, ϕ a Maass form and Q a holomorphic cusp form of weight 4 (all this for $GL_2/\mathbb{Q}$). The special points being $s=\frac{1}{2}\pm ir$, where the Laplace eigenvalue of ϕ is $\frac{1}{4}+r^2$. The other special point that arises is $s=\frac{1}{2}$ for self-dual forms F (i.e. $F=\tilde{F}$). This point is the central symmetry point for the functional equation of $L(s,F)$. In the case that $L(s,F)$ is the L-function of an elliptic curve (or an abelian variety) then the vanishing at $s=\frac{1}{2}$ is related to rank of the group of rational points, this being presented in D) of section 2. Note that if F is self dual then $L(s,F)$ is real for s real and since $L(s,F)\to 1$ as $s\to\infty$, it follows that if we admit GRH then

$$L\left(\tfrac{1}{2},F\right) \geq 0 . \tag{57}$$

In the simplest case, that is F being a quadratic Dirichlet character (57) is not known (in fact one can show that if $L\left(\frac{1}{2},\chi\right)\geq 0$ for χ quadratic then one can eliminate in part the Landau-Siegel lacuna mentioned in section 2). So it is quite striking that for $PGL(2)/K$ cusp forms F (these are self-dual) one can show that

$$L\left(\tfrac{1}{2},F\otimes\chi\right) \geq 0 \tag{58}$$

for any quadratic ray class character χ. This final version is due to Guo [Gu], it completed a series of developments beginning with Waldspurger [W]. The theta function treatments of (58) (Guo proceeds differently using the relative trace formula) proceed by expressing $L\left(\frac{1}{2}, F\otimes\chi\right)$ as a sum of squares of the χ-th Fourier coefficient of a form of half-integral weight which corresponds to F as in Shimura [Shi1], [Shi3]. We will exploit this in the next section.

An application of (59) for the case $K=\mathbb{Q}$ and ϕ a Maass cusp form was given recently in [ConrI]. By incorporating (for χ quadratic) $\int_{-\infty}^{\infty} |L(\frac{1}{2}+it,\chi)|^6\, e^{-t^2}dt$ as part of a family involving $L^3\left(\frac{1}{2},\phi\otimes\chi\right)$ it is shown that for s fixed with $\Re(s)=\frac{1}{2}$

$$L(s,\chi) \ll q^{\frac{1}{6}+\epsilon}\,. \tag{59}$$

This gives the first improvement over (58) and is another pleasing example of the use of GL_2 theory to understand GL_1 L-functions. It highlights our point of view that L- functions be considered as a whole and especially in families. We add that the proof of the above appeals to bounds for exponential sums in two variables over finite fields and in particular to Deligne's estimates which follow from the general GRH for varieties over finite fields.

Another case where (58) has been established is the following [HK]. Let F_1,F_2,F_3 be three forms on $PGL(2)/K$ and $L(s,F_1\otimes F_2\otimes F_3)$ the L-function (28) which has degree eight. Then

$$L\left(\tfrac{1}{2},F_1\otimes F_2\otimes F_3\right)\geq 0\,. \tag{60}$$

As in the previous example the proof of (60) involves expressing the special value as a sum of squares of "periods" of $F_1F_2F_3$. For analytic applications one needs an entirely explicit relation between these special values and periods. In his thesis [Wa] has proved such an explicit relation for forms over $\mathbb{Q}$. For example, for Maass forms of full level (i.e. for forms on $SL(2,\mathbb{Z})$) as in (6) of section 1 he shows that

$$\frac{\Lambda\left(\frac{1}{2},\phi_1\otimes\phi_2\otimes\phi_3\right)}{\prod_{j=1}^{3}\Lambda(1,\mathrm{sym}^2\phi_j)} = \frac{\pi^4}{216}\left|\int_{SL(2,\mathbb{Z})\backslash\mathbb{H}} \phi_1(z)\phi_2(z)\phi_3(z)\,\frac{dxdy}{y^2}\right|^2, \tag{61}$$

where Λ is the completed L-function

$$\Lambda(s,\phi_1\otimes\pi_2\otimes\phi_3) = \pi^{-4s}\prod_{\epsilon_j=\pm1}\Gamma\left(\frac{s+\epsilon_1r_1+\epsilon_2r_2+\epsilon_3r_3}{2}\right)L(s,\phi_1\otimes\phi_2\otimes\phi_3)$$

and ϕ_1, ϕ_2, ϕ_3 are normalized to have L^2-norm equal to one on $SL(2,\mathbb{Z})\backslash\mathbb{H}$. We will exploit this beautiful formula in the next section.

Of special interest in applications is to know how often a family of L-functions vanish at a special point. The technique of mollification (championed by Selberg in his proof that a positive proportion of the zeros of $\zeta(s)$ lie on $\Re(s) = \frac{1}{2}$) has been successfully developed in the context of special values of GL_2 forms (at least over $\mathbb{Q}$) in [IwS] and [KowMV2]. We mention some results in this direction. Let N be squarefree and fix $k \geq 2$. Let $H_k^*(N)$ denote the set of holomorphic newforms F of weight k for $\Gamma_0(N)$, that is on the modular curve $X_0(N)$. The sign ϵ_F of the functional equation for $L(s,F)$ is ± 1. When N is large (which is our interest here) roughly one half of the forms have each sign, the total number being $|H_k^*(N)| \sim \frac{k-1}{12}\varphi(N)$. If $\epsilon_F = -1$ then $L\left(\frac{1}{2}, F\right) = 0$ and we are interested in $L'\left(\frac{1}{2}, F\right)$. It is shown [IwS], [KowMV2] that

$$\varliminf_{N\to\infty} \frac{\#\{F \in H_k^*(N); \epsilon_F = 1, L\left(\frac{1}{2}, F\right) \geq (\log N)^{-2}\}}{\#\{F \in H_k^*(N); \epsilon_f = 1\}} \geq \frac{1}{2}, \tag{62}$$

$$\varliminf_{N\to\infty} \frac{\#\{F \in H_k^*(N); \epsilon_F = -1, L'\left(\frac{1}{2}, F\right) \neq 0\}}{\#\{F \in H_k^*(N); \epsilon_F = -1\}} \geq \frac{7}{8}, \tag{63}$$

$$\varlimsup_{N\to\infty} \frac{1}{|H_k^*(N)|} \sum_{F\in H_k^*(N)} \operatorname{ord}_{s=\frac{1}{2}} L(s,F) \leq 1.2. \tag{64}$$

We expect that the constants $\frac{1}{2}$ and $\frac{7}{8}$ in (62) and (63) can be replaced by 1 while 1.2 in (64) can be replaced by $\frac{1}{2}$. It is tantalizing that an improvement in (62) of the $\frac{1}{2}$ to any $c > \frac{1}{2}$, would resolve the Landau-Siegel lacuna (section 2). The proof of this implication [IwS] exploits the positivity (58).

Next (62) and (63) together with the results [KoL], [GrZ] towards BSC mentioned in D of section 2 imply results on the ranks of the Mordell-Weil groups of the Jacobian varieties $J_0(N) = JAC(X_0(N))/\mathbb{Q}$. Precisely (62) yields a quotient ("winding quotient" [M]) $M_0(N)$ of $J_0(N)$ over $\mathbb{Q}$, which has only finitely many rational points and has dimension which is asymptotically at least $\frac{1}{4}\dim J_0(N)$. Moreover, (63) implies that for large N the rank of $J_0(N)$ is asymptotically at least $\frac{7}{16}\dim J_0(N)$. Finally, (64) together with BSC imply that rank $J_0(N) \leq 1.2 \dim J_0(N)$, for N large.

The application of nonvanishing to spectral deformation theory also concerns $X_0(N) = \Gamma_0(N)\backslash\mathbb{H}$ (for N fixed). Let ϕ_j (with eigenvalue λ_j) be an orthonormal basis of Maass Hecke cusp forms and let Q be a fixed holomorphic cusp form of weight $k \geq 1$. Recently Luo [Lu] using the mollification methods as above has shown that

$$\varliminf_{\lambda\to\infty} \frac{\#\{\lambda_j \le \lambda; L\left(\frac{1}{2}+ir_j, Q\otimes\phi_j\right)\neq 0\}}{\#\{\lambda_j\le\lambda\}} > 0 . \tag{65}$$

This has striking applications to the question of nonexistence of Maass cusp forms for the general quotient $\Gamma\backslash\mathbb{H}$, Γ in the deformation (Teichmuller) space of $\Gamma_0(N)$, see [PhS] and [Wo].

6 Subconvexity and Equidistribution

Up to now the discussion has centered around L-functions only. In this section we give two examples of applications to problems which at first sight appear to have nothing to do with L-functions. First we describe some results on the subconvexity problem B of section 2 for Euler products of degree at least two. In these cases the coefficients of the L-functions are arithmetical and inexplicit so that the methods of Weyl and Burgess don't apply. Instead sophisticated new methods are needed (see section 7).

For $L(s,F)$ with F a cusp form on $GL_2/\mathbb{Q}$, subconvexity has been established in all the parameter aspects (in s aspect in [Goo] and [Me] while in the other parameters in the series of papers by Duke Friedlander and Iwaniec). We concentrate on the twisting by Dirichlet characters. For a fixed F on $GL(2)/\mathbb{Q}$ a cuspidal eigenform (i.e. a holomorphic or Maass form on $\Gamma_0(N)\backslash\mathbb{H}$) and χ a (primitive) Dirichlet character, the following subconvexity estimate (s is fixed with $\Re(s)=\frac{1}{2}$) was established in [DuFI]:

$$L(s,F\otimes\chi) \ll q^{\frac{5}{12}+\epsilon} . \tag{66}$$

Here q is the conductor of χ and the convexity bound is $q^{1/2}$.

The methods used to deal with F on $GL(2)/\mathbb{Q}$ run into a number of difficulties (not the least of which are the units) for number fields. Recently the authors of [CogPS] have resolved these difficulties. Let K be a totally real extension of $\mathbb{Q}$. Fix a holomorphic Hilbert modular cusp form of even integral weight (i.e. a form on GL_2/K). Let χ range over primitive ray class characters of conductor $\mathcal{Q}$ (we have in mind $N(\mathcal{Q})\to\infty$). Then for s fixed with $\Re(s)=\frac{1}{2}$

$$L(s,F\otimes\chi) \ll N(\mathcal{Q})^{\frac{49}{100}+\epsilon} . \tag{67}$$

Again the conductor of $F\otimes\chi$ is $N(\mathcal{Q})^2$ so that the convexity bound for (67) is $N(\mathcal{Q})^{1/2}$.

Some progress has also been made for Euler products (over $\mathbb{Q}$) of higher degree. Fix a holomorphic or Maass cusp form G for $\Gamma_0(N)\backslash\mathbb{H}$ (so N is fixed) and let F vary over the holomorphic newforms for $\Gamma_0(N)$ of weight k.

Then for s fixed with $\Re(s) = \frac{1}{2}$, the Rankin-Selberg L-functions $L(s, F \otimes G)$ satisfy the subconvexity estimate [S2] (in the k-aspect)

$$L(s, F \otimes G) \ll k^{\frac{27}{28}+\epsilon} \tag{68}$$

(here the "analytic" conductor is k^4 so that the convexity bound is k).

Also, for Rankin-Selberg L-functions, but in the level aspect, [KowMV1] have established a subconvexity estimate. Precisely, fix G and let F vary over holomorphic newforms of the same weight as G, but of level $N \to \infty$. Then for s fixed with $\Re(s) = \frac{1}{2}$,

$$L(s, F \otimes G) \ll N^{\frac{1}{2}-\frac{1}{96}+\epsilon} \tag{69}$$

(the convexity bound being $N^{\frac{1}{2}}$).

We turn to the applications. The first is to Hilbert's 11-th problem: which integers are integrally represented by a given quadratic form over a number field? The case of binary quadratic forms is equivalent to the theory of relative quadratic extensions and their class groups and Hilbert class fields. For forms in four or more variables the situation is quite different and has been understood for some time. The case of three variables has remained open and we describe below the essential part of its resolution.

Fix the number field K. The problem of which integers ν in K are represented by the genus of a given integral quadratic form $f(x_1, x_2, \dots x_n)$ is answered completely by Siegel's mass formula [Si4] (which gives the number of solutions in terms of local data, via the product of local masses). So if there is one class in the genus of f the formula resolves the representation problem for f. If $n \geq 3$ and f is indefinite at an archimedean place v of K then Kneser's [Kn] results on the class numbers and weights of the spinor genus of f show that we are more or less in the one class in the genus situation. So we restrict to the difficult case when $f(x_1, \dots, x_n)$ is definite over a totally real field K. For four or more variables one can proceed either by using analytic methods of Hilbert modular forms and in particular the bounds towards GRC for GL_2/K (see (39) for weight two holomorphic cusp forms) or by using algebraic methods ([HsKK], [C]), to prove the following: There is C_f (depending on f effectively) such that if $\nu \in \mathcal{O}_K$ is (totally) positive and $N(\nu) \geq C_f$, then ν is primitively represented by f iff it is primitively represented locally at every completion v (the local conditions are satisfied for all but finitely many primes and are easily checked).

For f a form in three variables the problem is much more difficult and is resolved (at least for squarefree ν) by the estimates (66) and (67). The connection is as follows: using the relation between the special value $L\left(\frac{1}{2}, F \otimes \chi\right)$, $\chi^2 = 1$ and the "χ-th" Fourier coefficient of half-integral cusp forms ([W], [Shi3]) mentioned in the last section, one finds that the bound (66) for $\mathbb{Q}$ and (67) in general, give nontrivial bounds for the *squarefree* Fourier coefficients of half-integral weight holomorphic cusp forms. Here and in other such problems, the convexity bound for the L-function corresponds exactly to the "trivial" bound for the Fourier coefficients. Moreover, the Lindelöff Hypothesis in the quadratic twisting χ aspect, for $L\left(\frac{1}{2}, F \otimes \chi\right)$, is equivalent (or determines) the *half-integral weight* GRC. Put another way, a nontrivial bound for the squarefree Fourier coefficients of a half-integral weight cusp form is equivalent to a subconvexity bound for $L\left(\frac{1}{2}, F \otimes \chi\right)$ while GRH for $L(s, F \otimes \chi)$ (via Lindelöff) implies the optimal bound for these coefficients. In the case $K = \mathbb{Q}$ a nontrivial bound for the Fourier coefficients of such forms was derived earlier in [Iw1] by a different method. For the case of Hilbert modular forms the passage via (67) gives the first bounds towards the Ramanujan Conjectures of half-integral weight.

We return to the form $f(x_1, x_2, x_3)$. Its theta function $\theta_f(z)$ is a Hilbert modular form of weight $\frac{3}{2}$ whose coefficients give the number of representations by f. Write $\theta = E + C$ where E is an Eisenstein series and C a cusp form. The ν-th coefficient of the Eisenstein series (a linear combination of standard Eisenstein series) depends only on the genus of f and is a product of local masses. It can be estimated from below by $C_\epsilon N(\nu)^{1/2-\epsilon}$ (when ν is represented locally) where $\epsilon > 0$ and C_ϵ an ineffective positive constant depending on ϵ. It is ineffective since the lower bound appeals to Siegel's ineffective lower bound for $L(1, \chi)$ (see (30)). Now the bound (67) leads as above to the ν-th coefficient of C being $O(N(\nu)^{\frac{49}{200}})$. Thus we conclude: if $N(\nu)$ is sufficiently large and squarefree then ν is represented integrally iff it is represented locally. This yields a solution (albeit ineffective and for squarefree ν) of the representation problem for definite ternary forms. Using the results of Schulze-Pillot [Sc] one can extend these results to all ν except perhaps for an explicit finite set of square classes ($\nu = \nu_0 t^2, t > 0$) along which the local to global principle can fail.

Of special interest is the long studied problem of sums of squares in a number field. Over $\mathbb{Q}$ as is well-known all positive numbers ν are sums of four squares (Lagrange) and such a ν is a sum of three squares iff $\nu \neq 4^a(8b+7)$ (Legendre), that is iff there are no local obstructions. That the answers for these are so neat is a consequence of $x_1^2 + x_2^2 + x_3^2$ and $x_1^2 + x_2^2 + x_3^2 + x_4^2$ having one class in their genus. This happens for very few totally real fields. In fact Siegel [Si3] shows that $\mathbb{Q}(\sqrt{5})$ is the only field for which *every* totally positive number is a sum of three squares. In general Siegel [Si1] showed that every sufficiently large (in norm) totally positive ν is a sum of five squares and the result mentioned above settles four squares similarly. For three squares (67) implies that there is an ineffective C_K depending on K such that if $N(\nu) \geq C_K$ and ν is squarefree and totally positive, then ν is a sum of three squares iff it is so locally (the local condition only involves the primes dividing 2 and for many fields K there are no such local obstructions). Note these results are ones of equidistribution. Indeed for $N(\nu)$ large and ν satisfying the local solvability conditions, ν is represented by the genus of f in roughly $N(\nu)^{1/2}$ ways. The subconvexity estimate ensures that each class in the genus represents ν roughly equally often. We mention that Linnik [Li2] gave an interesting ergodic theoretic approach to equidistribution problems associated to ternary quadratic forms which yield some partial results.

We end this part of the discussion with some general comments about integer solutions to Diophantine equations. The problem of establishing the existence of any or many such solutions for equations for which solutions are expected, has proven formidable. Success has been limited to varieties which are homogeneous (such as the case of quadrics which were discussed above) for an action of an algebraic group, or to varieties defined by many variables compared to the degree and number of equations. For the latter the circle method of Hardy and Littlewood can be applied. An example of the last is due to Heath-Brown [He] who showed that any nonsingular cubic form in 10-variables over $\mathbb{Q}$ has infinitely many rational (projective) points (i.e. for $f(x) = 0$). New methods for exhibiting rational points on varieties would be very welcome.

The second equidistribution problem comes from "Quantum Chaos". One of the central problems in this subject concerns the behavior of individual eigenstates of the quantization of classically chaotic systems in the semi-classical limit. This problem cannot at present be addressed with the techniques of analysis or partial differential equations. To gain insight we therefore specialize to systems of classical mechanics defined by the geodesic motion on a hyperbolic manifold (compact or finite volume). These are well known examples of chaotic Hamiltonian dynamics. We even specialize to such manifolds of arithmetic type. For these it turns out that the questions that we have been discussing about general L-functions lie at the heart of the problem. Consider the case of a hyperbolic such surface $X = \Gamma\backslash\mathbb{H}$. A quantization of the geodesic motion on the cotangent space $T^*(X)$, is the Laplacian Δ. Let $dv(x)$ denote the Riemannian measure on X. Perhaps the most fundamental problem concerns the behavior as $\lambda \to \infty$ of the probability measures $\mu_\phi := |\phi(x)|^2 dv(x)$ on X, where $\Delta\phi + \lambda\phi = 0$; $\int_X |\phi(x)|^2\, dv(x) = 1$. These measures have the well-known interpretation of being the probability distribution in configuration space of a particle in eigenstate ϕ. These ϕ's are the familiar Maass forms, especially if we restrict to $X_0(N) = \Gamma_0(N)\backslash\mathbb{H}$, N fixed (one can similarly analyze the compact surfaces arising as quotients of $\mathbb{H}$ by units in quaternion groups). We consider the question of the behavior of these measures μ_ϕ for Maass forms ϕ which are also eigenforms for the Hecke operators. Since the multiplicity of cusp forms with eigenvalue λ is expected to be very small, the latter assumption is probably not necessary, but we will certainly exploit it. Once we are assuming that ϕ is a Hecke eigenform we can also allow holomorphic eigenforms as well. We formulate the problem precisely for these modular forms: Fix N and $X_0(N) = \Gamma_0(N)\backslash\mathbb{H}$. Each of the spaces of holomorphic newforms of weight k with given Hecke eigenvalues, and Maass newforms with given Hecke eigenvalues, are one dimensional. So there are unique normalizations so that the following are probability measures on $X_0(N)$:

$$\begin{aligned} \mu_f &:= y^k \left|f(z)\right|^2 dv(z) \\ \mu_\phi &:= \left|\phi(z)\right|^2 dv(z)\ , \end{aligned} \tag{70}$$

where $dv(z) = y^{-2}dxdy$. We have the following Equidistribution of Mass Conjecture [RudS2]:

$$\lim_{k\longrightarrow\infty} \mu_f = d\tilde{v} \tag{71}$$

$$\lim_{\lambda\longrightarrow\infty} \mu_\phi = d\tilde{v}\ , \tag{72}$$

where $d\tilde{v} = dv/\mathrm{Vol}\,(X_0(N))$.

Comments:

(1) This conjecture looks reasonable, but at the time it was made it went against certain beliefs (i.e. that eigenstates might concentrate on periodic geodesics for example). The conjecture is made more generally for compact manifolds of negative curvature, yet the only theoretical evidence is in the case of arithmetic manifolds (there is some numerical evidence as well [Hej], [AuS]).

(2) In the holomorphic case the condition that f be a Hecke eigenform is essential. For example the masses μ_{Δ^k} of the holomorphic forms $(\Delta(z))^k$ of weight $12k$ on $X_0(1)$ certainly don't become equidistributed.

(3) The Conjecture (71), if true, has the pleasant corollary that the zeros in $X_0(N)$ of such an $f(z)$ (there are about k of them) become equidistributed with respect to $d\tilde{v}$ as $k \to \infty$ [Rud].

The connection between Conjectures (71), (72) and L-functions is (61) and its generalizations. GRH for the triple product L-function $L(s, F \otimes F \otimes G)$ and even subconvexity in the k or λ aspects (with $s = \frac{1}{2}$) already imply these conjectures! To see this, consider the case of $GL_2(\mathbb{Z})\backslash\mathbb{H}$. The equidistribution (72) in this case is equivalent to

$$\mu_{\phi_\lambda}(\phi_0) = \int_{SL_2(\mathbb{Z})\backslash\mathbb{H}} \phi_0(z)|\phi_\lambda(z)|^2 y^{-2}\, dxdy \longrightarrow 0 \tag{73}$$

as $\lambda \to \infty$, for any fixed Maass cusp form ϕ_0 (one needs also to consider the continuous spectrum, that is the unitary Eisenstein series in place of ϕ_0, but these are slightly easier to handle so we ignore them here). Now according to (61) the right hand side of (73) is up to gamma factors equal to $L(\frac{1}{2}, \phi_\lambda \otimes \phi_\lambda \otimes \phi_0)/L^2(1, \mathrm{sym}^2\phi_\lambda)L(1, \mathrm{sym}^2\phi_0)$. There is no problem dealing with $L(1, \mathrm{sym}^2\phi_\lambda)$ since it is bounded below by $\lambda^{-\epsilon}$ and above by λ^ϵ for any $\epsilon > 0$, and effectively so [Iw2], [HoL]. A simple analysis with Stirling formula then shows that a bound of $O(\lambda^{-\delta})$ for the right hand side in (73) is equivalent to the subconvexity estimate in λ;

$$L\left(\tfrac{1}{2}, \phi_\lambda \otimes \phi_\lambda \otimes \phi_0\right) \ll \lambda^{\frac{1}{2}-\delta_0}\,. \tag{74}$$

Unfortunately we have not been able to establish (74) in general. The L-function $L(s, \phi_\lambda \otimes \phi_\lambda \otimes \phi_0)$ factors into $L(s, \mathrm{sym}^2\phi_\lambda \otimes \phi_0)\, L(s, \phi_0)$. So the key case is subconvexity for $L\left(\frac{1}{2}, \mathrm{sym}^2\phi_\lambda \otimes \phi_0\right)$ which is an Euler product of degree six. For the special case of "CM forms" ϕ_λ or f on $\Gamma_0(N)\backslash\mathbb{H}$ (see (9) for example) this Euler product factors further into a Rankin Selberg L-function of degree four times $L(s, \phi_0)$ which is of degree two. So in this case the subconvexity estimate (68) (and its λ analogue) gives a proof of (71) and (72). That is (71) and (72) are true for CM forms.

These two applications show, of course, the power of GRH, however they also show that in certain problems a complete resolution can be achieved by finessing GRH and establishing the more approachable subconvexity estimate.

7 $GL(2)$ Tools

We give some flavor of some of the modern techniques that have been successful in studying L-functions by indicating how subconvexity estimates are proven. Suppose for example that we want to estimate $L\left(\frac{1}{2}, F\right)$, where

F is a self-dual cusp form on $GL_m(\mathbb{A}_K)$. Using the series representation (3) together with standard arguments involving contour shifts and the functional equations, we obtain

$$L\left(\tfrac{1}{2}, F\right) = 2\sum_{a \neq 0} \frac{c_F(a)}{\sqrt{Na}} W\left(\frac{Na}{X}\right), \tag{75}$$

where $W(t)$ is a smooth function which is essentially independent of F and is rapidly decreasing as $t \to \infty$ and $X = \sqrt{C(F)}$, $C(F)$ being the analytic conductor (31) (here $C(F)$ denotes $C(F,0)$ from (31)). The coefficients $c_F(a)$ are known in some cases to satisfy the GRC so that

$$c_F(a) \underset{\epsilon}{\ll} (Na)^\epsilon . \tag{76}$$

In any case for F on $GL_m(\mathbb{A}_k)$ we have (76) on average ([Iw2], [Mo])

$$\sum_{Na \leq Y} |c_F(a)|^2 \underset{\epsilon}{\ll} Y\big(C(F)\big)^\epsilon . \tag{77}$$

From (75) and (77) the "trivial" convexity bound

$$L\left(\tfrac{1}{2}, F\right) \underset{\epsilon}{\ll} \big(C(F)\big)^{\frac{1}{4}+\epsilon} \tag{78}$$

follows. To go beyond (78) one needs to exhibit some cancellation in the sum (75). We can no longer appeal to any functional equations since these have already been exploited in deriving (75) (that is if we "dualize" using the functional equation we arrive back at a similar sum). Also to directly estimate (75) is problematic since one knows very little about $c_F(a)$. We proceed according to the theme of this account and embed F in a family $\mathcal{F}$ (sometimes even fake families!). Finding suitable families is part of the problem. The idea is to consider averages

$$S(\mathcal{F}) = \sum_{F \in \mathcal{F}} \big|L(\tfrac{1}{2}, F)\big|^2 . \tag{79}$$

The conductors $C(F), F \in \mathcal{F}$ are all assumed to be the same (or nearly the same) size. In some cases one might take higher powers of L in (79) (the choice here of high moments rather than something like high tensor powers as in (43) and in section 3 is no doubt a poor one). Now GRH (via Lindelöff) asserts that $L(\frac{1}{2}, F) \ll_\epsilon \big(C(F)\big)^\epsilon$ so we can expect that

$$S(\mathcal{F}) \underset{\epsilon}{\ll} |\mathcal{F}|\big(C(F_0)\big)^\epsilon . \tag{80}$$

Here F_0 is our particular $F \in \mathcal{F}$ which we seek to bound. Using orthogonality and completeness of the family $\mathcal{F}$, (80) can often be established. By positivity (an apparently precious tool) we have from (80).

$$L(\tfrac{1}{2}, F_0) \underset{\epsilon}{\ll} |\mathcal{F}|^{1/2}\big(C(F_0)\big)^\epsilon . \tag{81}$$

So if the family $\mathcal{F}$ is sufficiently small (precisely $|\mathcal{F}| \ll (C(F))^{1/2-\delta}$) and at the same time rich enough to establish (80), then (81) will yield a subconvex bound. In practice when this method succeeds one finds that one can establish (80) with $|\mathcal{F}| \asymp (C(F))^{1/2}$ in a relatively straight-forward analysis involving summing over $\mathcal{F}$ and analyzing only the "diagonal" contribution. This however simply recovers the convexity bound (78) and the heart of the problem is to decrease somewhat the size of $|\mathcal{F}|$. This is done at the expense of off-diagonal terms now appearing in the analysis. Cancellations in these new sums has to be gotten from some new input. In some problems the decrease in size of $|\mathcal{F}|$ can be achieved analytically (e.g. in the case of (68), $F \otimes G$ with G fixed on $GL_2/\mathbb{Q}$, F varying over $\mathcal{F}_K$ the holomorphic forms of fixed level and weight $\frac{K}{2} \leq k \leq 2K$. Shortening here can be done by restricting $K - H \leq k \leq K + H$ with $H = K^{1-\delta}, \delta > 0$). In many interesting examples (e.g. (66) and (67)) shortening cannot be achieved by such a device. One appeals to a technique known as "amplification" which arithmetically shortens $\mathcal{F}$ by introducing weights. The method was introduced in [FI] in connection with estimating $L(s, \chi)$, however its true power has been shown in the more general setting of $L(s, F)$s. Roughly the idea is as follows: Let M be a small parameter ($M = X^\delta, \delta$ small) and let $\alpha(b)$ with $Nb \leq M$, be complex numbers of modulus at most 1. Consider the amplified sums

$$A = \sum_{F \in \mathcal{F}} \left| \sum_{Nb \leq M} \alpha(b)\, c_F(b) \right|^2 \left| L\left(\tfrac{1}{2}, F\right) \right|^2 . \tag{82}$$

This time we shoot for the expected bound

$$A \underset{\epsilon}{\ll} M |\mathcal{F}| X^\epsilon . \tag{83}$$

In establishing (83) one faces off-diagonal terms and if these can be successfully estimated then choosing $\alpha(b) = \overline{c_{F_0}(b)}$, i.e. amplifying F_0, we get

$$\left| L\left(\tfrac{1}{2}, F_0\right) \right|^2 \left| \sum_{Nb \leq M} |c_{F_0}(b)|^2 \right|^2 \underset{\epsilon}{\ll} |\mathcal{F}| M X^\epsilon . \tag{84}$$

This implies a subconvex bound for $L\left(\frac{1}{2}, F_0\right)$.

So the key features are the family and dealing with the off-diagonal sums. For example for (66) the family used is $L(s, F \otimes \chi)$ with χ a Dirichlet character of conductor q. For (67) one cannot use ray class characters of conductor $\mathcal{Q}$ since there may be very few of these (such characters have to be trivial on the units). One proceeds with nonnegative expressions which make sense for all numerical characters of $(\mathcal{O}_K/\mathcal{Q})^*$, but which have no meaning in terms of L-functions ("fake families"). In both (66) and (67) amplification is used. The key off-diagonal sums that need to be treated are of type

$$\sum_{\nu\alpha-\mu\beta=h} c_F(\alpha)\, c_F(\beta)\, W\Big(\frac{N(\alpha)}{X}\Big)\, W\Big(\frac{N(\beta)}{X}\Big)\, G(\alpha,\beta)\ , \tag{85}$$

where ν and μ are fixed small integers in K, $h \neq 0$ and G is a smooth function depending on the arguments of α and β in the embeddings of K into $\mathbb{R}$. Over $\mathbb{Q}$ a reasonably elementary treatment of these sums is given in [DuFI] (it appeals among other things to Weil's bounds on Kloosteman sums - see below). In general, one uses the full Maass form spectral theory for $GL_2(\mathbb{A}_K)$ and a suitable theory of Poincaré series. Crucial ingredients are the GRC bounds (41) (42) as well as the spectral analysis method in [S1].

For the case of $L(s, F \otimes G)$ and the estimate (69), where F varies over holomorphic cusp forms of a fixed weight for $\Gamma_0(N)$ and $N \to \infty$ (all over $\mathbb{Q}$), the averaging is carried out by means of the Petersson Formula [P]. Let $B_k(N)$ be an orthogonal basis for $S_k(N)$ (the space of holomorphic cusp forms of weight k for $\Gamma_0(N)$). Normalize the Fourier coefficients of $a_F(n)$ for $F \in S_k(N)$ by setting

$$\psi_F(n) \ = \ \left(\frac{\Gamma(k-1)}{(4\pi n)^{k-1}}\right)^{1/2} \frac{a_F(n)}{\langle f, f\rangle^{1/2}}\ . \tag{86}$$

Then the formula reads that for $m, n \geq 1$

$$\sum_{F\in B_k(N)} \overline{\psi_F(n)}\, \psi_F(m)$$
$$= \delta(m,n) \ + \ 2\pi i^k \sum_{c\equiv 0(N)} \frac{S(m,n,c)}{c}\, J_{k-1}\left(\frac{4\pi\sqrt{mn}}{c}\right)\ . \tag{87}$$

Here $\delta(m,n)$ is 0 if $m \neq n$ and is 1 if $m = n$, $J_k(x)$ is the Bessel function and $S(m,n,c)$ is the Kloosterman sum

$$S(m,n,c) \ = \sum_{\substack{x \bmod c\\ x\bar{x}\equiv 1(c)}} e\Big(\frac{mx+n\bar{x}}{c}\Big)\ . \tag{88}$$

Thus (87) converts averages of this family to sums of exponential sums over finite fields and allows one to analyze averages over the family of $L(s,F)$, for $F \in S_k(N)$. The Petersson formula (87) and its generalizations due to Kuznetsov [Ku] are important tools in the subject. They lie at the bottom of many of the applications of the $GL_2/\mathbb{Q}$ analytic theory. We mention one other such application, that is the Density Theorem for exceptional eigenvalues. Recall the Selberg Conjecture (36) which is a special case of GRC. In [Iw2] the following Density Theorem is proven; for any r with $0 \leq r \leq \frac{1}{2}$,

$$\#\Big\{\lambda < \tfrac{1}{4} - r^2; \lambda \text{ an eigenvalue of } \Delta \text{ on } L_0^2\big(\Gamma_0(N)\backslash\mathbb{H}\big)\Big\} \underset{\epsilon}{\ll} \Big[\mathrm{Vol}\big(\Gamma_0(N)\backslash\mathbb{H}\big)\Big]^{1-4r+\epsilon} . \tag{89}$$

The proof uses the generalizations of (87) as well as Weil's bound [We2] for Kloosterman sums

$$\big|S(m,n,p)\big| \leq 2\sqrt{p} \tag{90}$$

if p does not divide m or n ((90) is a consequence of the function field RH for curves). A point to note about (89) is the exponent $1-4r$, which in particular implies Selberg's well-known bound

$$\lambda_1\big(\Gamma_0(N)\backslash\mathbb{H}\big) \geq \tfrac{3}{16} . \tag{91}$$

An exponent of $1-2r$ in (89) is easily deduced from the Selberg Trace Formula. Just as with the Density theorems for $GL(1)/\mathbb{Q}$ mentioned in section 4, (89) can sometimes be used as a substitute for (36) in applications.

Formula (87) and its generalizations are also useful when applied in the other direction, that is to capture cancellations in sums of Kloosterman sums (the Linnik-Selberg Conjecture [Li3], [Se]). Kuznetsov [Ku] used his formula and the fact that $\lambda_1(SL(2,\mathbb{Z})\backslash\mathbb{H}) \geq \frac{1}{4}$, to show that for m,n fixed

$$\sum_{c\leq X} \frac{S(m,n,c)}{\sqrt{c}} \underset{\epsilon}{\ll} X^{\frac{2}{3}+\epsilon} . \tag{92}$$

Note that Weil's bound (which is sharp in c for m,n fixed [Mi]) gives $O_\epsilon(X^{1+\epsilon})$ for the sum in (92). So indeed (92) asserts cancellations in the sums of these sums. The development (42) in as much as it goes beyond (91) (i.e. $\lambda_1 \geq \frac{21}{100}$), shows that there is cancellation in the sums (92) for c in an arithmetic progression. Fix m,n,N and a, then

$$\sum_{\substack{c\leq X\\ c\equiv a(q)}} \frac{S(m,n,c)}{\sqrt{c}} \underset{\epsilon}{\ll} X^{\frac{9}{10}+\epsilon} . \tag{93}$$

This concludes our brief discussion of GL_2 tools. Also fundamental is the Selberg Trace Formula which we have mentioned a few times in passing. So at least over $\mathbb{Q}$ and for GL_2, the analytic theory can be considered to be in quite good shape, much like $GL_1/\mathbb{Q}$ (see section 4) was at the end of the 70's.

The trace formula has been successfully extended to GL_n as well as other groups by Arthur [A]. His form is very suitable for comparisons of the geometric sides of the trace formula for different groups. This implies on the spectral side some striking conjectured "liftings" of automorphic forms between various groups. The analytic type of spectral and trace formula that have been developed for GL_n with applications such as those of this section in mind, have met only with mild success. For now this can be viewed as a challenging new direction which might provide important new information on the basic problems. For example we expect that such developments are needed to resolve the basic problem B in general.

8 Symmetry and Attacks on GRH

In the previous sections we described progress made not by climbing the summit (GRH), but by going around it. In this section we discuss some structural phenomenon and insights that might play a role in the accent.

As we have mentioned a number of times, families of L-functions play a central role even (or especially) when examining the deeper aspects of a *given* L-function. One might ask whether something like a monodromy group of a family of L-functions (section 3) exists in the number field setting. One way to detect such symmetry groups for families is to look at the local distribution of zeros of L-functions. For a fixed L-function $L(s,F)$, F cuspidal on $GL_m(\mathbb{A}_{\mathbb{Q}})$ (note here we demand over $\mathbb{Q}$ so that these L-functions do not factor further) one can examine the high zeros $\rho_F = \frac{1}{2} + i\gamma_F$ (for this part of the discussion we assume GRH). One can show that

$$N_F(T) := \#\{0 \le \gamma_F \le T\} \sim \frac{m \log T}{2\pi} T\,. \tag{94}$$

Hence, in studying the local distribution of spacings between the zeros one considers the unfolded numbers $\frac{m \log \gamma_F}{2\pi}\gamma_F$, whose mean spacing is 1. Remarkably these follow the local scaled spacing laws for eigenvalues of large unitary matrices, that is the CUE (Circular Unitary Ensemble) laws from random matrix theory (at least in leading order asymptotics). This was proven analytically in restricted ranges for the distribution of pairs of zeros ("pair correlation") in [Mon] and for higher correlations in [RudS1]. Moreover, extensive numerical experiments [O], [Rum], [Ru] confirm this phenomenon for various $GL_1/\mathbb{Q}$ and $GL_2/\mathbb{Q}$ L-functions. In [KS] an analogue of this phenomenon about local spacings of zeros is proven for the function field zeta functions of section 3. Moreover, the source of this universal behavior is identified. While in this case the spectral interpretation of the zeros is through the eigenvalues of Frobenius on cohomology groups, the local distribution of zeros is governed by the scaling limits of the eigenvalue distributions of the monodromy groups of families of L-functions. The calculations of these scaling limits appeals to methods from random matrix theory and these limits, at least for monodromies which come from the classical groups, are universally the CUE distributions. In the function field case one can also show [KS] that the low-lying zeros (i.e. zeros $\frac{1}{2}+i\gamma_F$ with γ_F near zero) as F runs over a family of L-functions, follow the laws governed by the corresponding scaling limit of monodromy groups of the family. This time the distributions are not universal and depend on the monodromy, or the symmetry of the family. Again, it is remarkable that this phenomenon of distribution of low-lying zeros persists for L-functions $L(s, F)$ for F in suitable families [KS]. This has been confirmed (again in restricted but wider ranges than for the high zeros) analytically for a number of families [IwLS], with different symmetry types. It has also been confirmed numerically in [Ru]. These distributions attached to each family and its symmetry also explain the specific fractions that appear in (62) and (63) (see also [So]). In (62) the symmetry is an orthogonal one SO(even) while for (63) it is SO(odd) (it is worth noting that subfamilies are independent entities and may have different symmetry types). Random matrix theory via these symmetries has recently been used to predict the asymptotics of all moments of L-functions on the line $\frac{1}{2}+it$ and for suitable families $L(\frac{1}{2}, F)$, $F \in \mathcal{F}$ [ConrF], [KeS].

The results above about the distribution of zeros give ample evidence for there being a natural spectral interpretation of the zeros of $L(s, F)$ as well as the existence of a glue that marries different L-functions. However, these insights offer no real clue as to where such a spectral interpretation, or such symmetry groups may be found. There have been some interesting attempts to find nontautological spectral interpretations of the zeros of a given L-function such as $\zeta(s)$. In particular Connes [Con] suggests the singular space $X = K^*\backslash\mathbb{A}_K$ (P. Cohen has also pointed to this space and its intimate connection to the zeros of L-functions). The idele class group $J_K = K^*\backslash\mathbb{A}_K^*$ acts on X by multiplication $x \to yx$, $x \in X, y \in J_K$. He shows that with a suitable interpretation (via regularization) and assuming GRH, the decomposition of this action of multiplication over addition into irreducibles of J_K, yields exactly all the zeros of $\Lambda(s, F)$ where F is a $GL_1(\mathbb{A}_K)$ automorphic form. It turns out that this is closely connected to the explicit formula of Riemann which relates sums over zeros of an L-function to sums over primes and their powers, of the coefficients of the

L-function. Connes analysis gives a group action interpretation of the explicit formula. Weil [We3] had previously pointed to an arrangement of the terms in the explicit formula in attempting to interpret them in suggestive ways (so that they look like various key players in the function field setting). We note that anyway the explicit formula is a basic tool which is used analytically. For example, it is used directly in the analysis above concerning the distribution of the zeros. It is also used indirectly in the zero density theorems mentioned in section 4.

Whether these interpretations of the zeros or the explicit formula can be of any use in further understanding the zeros or attacking RH is unclear. Right now the use of families as a tool to study the zeros has been the most successful. We believe that families and understanding further what quantities to average as well as positivity will continue to play a central role perhaps even in the big ascent. After all, it is this analysis that "puts the zeros on the line $\frac{1}{2}$" in the general case of varieties over finite fields. One can imagine a scenario, a short cut, where GRH is established via families before a suitable spectral interpretation is given (for example fictitious zeros off the line might be ruled out spectrally before the true zeros are spectrally understood). More likely however is that suitable spaces and spectral interpretations of the zeros will be given and their analysis through families lead to the complete understanding (i.e. GRH, distribution of zeros ...). Anyway, all this is wild speculation and this is no doubt a good place to stop.

References

[A] J. Arthur, The trace formula for noncompact quotients, Proceedings of International Congr of Math., Warsaw (1983), 849-859.

[AC] J. Arthur, L. Clozel, Simple algebras, base change and the advanced theory of the trace formula, in "Annals of Math Studies", 120, Princeton University Press (1989).

[AuS] R. Aurick, F. Steiner, Exact theory for the quantum eigenstates of a strongly chaotic system, Physica D48 (1991), 445-470.

[B1] E. Bombieri, On the large sieve, Mathematika 12 (1965), 201-225.

[B2] E. Bombieri, Problems of the millennium: The Riemann hypothesis, in the home page of the Clay Mathematics Institute (webmaster@claymath.org).

[BFI] E. Bombieri, J. Friedlander, H. Iwaniec, Primes in arithmetic progressions to large moduli, Acta Math. 156 (1986), 203-251.

[Bu] D. Bump, The Rankin-Selberg method, in "Number Theory, Trace Formulas and Discrete Groups" (Aubert, Bombieri and Goldfeld, eds.), Academic Press (1988), 49-109.

[Bur] D. BURGESS, On character sums and L-series. II, Proc. London Math. Soc. 3:13 (1963), 24-36.

[C] J. CASSELS, Rational Quadratic Forms, Academic Press, 1978.

[CoW] J. COATES, A. WILES, On the conjecture of Birch and Swinnerton-Dyer, Invent. Math. 39 (1977), 223-251.

[CogP] J. COGDELL, I. PIATETSKI-SHAPIRO, Converse theorems for $GL(n)$, Publ. IHES 79 (1994), 157-214.

[CogPS] J. COGDELL, I. PIATETSKI-SHAPIRO, P. SARNAK, in preparation.

[Con] A. CONNES, Formule de trace en géometrie non-commutative et hypothesése de Riemann, C.R. Acad. Sc. Paris 323, ser I (1996), 1231-1236.

[ConrF] B. CONREY, D. FARMER, Mean values of L-functions and symmetry, preprint, 2000.

[ConrI] B. CONREY, H. IWANIEC, The cubic moment of central values of automorphic L-functions, Annals of Math., to appear.

[D1] P. DELIGNE, Formes madulaires et représentations ℓ-adiques, Sém. Bourbaki, (1968-1969), exposé 355, Lecture Notes in Math. 179, Springer-Verlag (1971), 139-172.

[D2] P. DELIGNE, La Conjecture de Weil I, II, Publ. IHES 48 (1974), 273-308; 52 (1981), 313-428.

[DuFI] W. DUKE, J. FRIEDLANDER, H. IWANIEC, Bounds for automorphic L-functions, Invent. Math. 112 (1993), 1-8.

[DuI1] W. DUKE, H. IWANIEC, Estimates for coefficients of L-functions. I, in "Automorphic Forms and Analytic Number Theory" (1989), CRM, Montreal (1990), 43-47.

[DuI2] W. DUKE, H. IWANIEC, Estimates for coefficients of L-functions. II, Proceedings of the Amalfi Conference on Analytic Number Theory (1989), Universita di Salerno (1992), 71-82.

[E] M. EICHLER, Quaternäre quadratische formen und die Riemannsche vernutung für die kongruenz zetafunktion, Arch. Math. 5 (1954), 355-366, 113-120.

[FI] J. FRIEDLANDER, H. IWANIEC, A mean value theorem for character sums, Michigan Math. J. 39 (1992), 153-159.

[G] P. GARRET, Decomposition of Eisenstein series; Rankin triple products, Annals of Math. 125 (1987), 209-235.

[GeJ] S. GELBART, H. JACQUET, A relation between automorphic representations of $GL(2)$ and $GL(3)$, Ann. Sci. Ecole Norm. Sup 11 (1978), 411–452.

[GelP] I. GELFAND, I. PIATETSKI-SHAPIRO, Automorphic functions and the theory of representations, Trupy Moskov. Mat 12 (1963), 389-412.

[GoJ] R. GODEMENT, H. JACQUET, Zeta functions of simple algebras, L.N.M. 260, Springer (1972).

[Gol] D. GOLDFELD, Gauss class number problem for imaginary quadratic fields, BAMS 13:1 (1985), 23-27.

[Goo] A. GOOD, The square mean of Dirichlet series associated with cusp forms, Mathematika 29 (1982), 2778-295.

[GrZ] B. GROSS, D. ZAGIER, Heegner points and derivatives of L-series, Invent. Math. 84 (1986), 225-320.

[Gu] J. GUO, On the positivity of the central value of automorphic L-functions for $GL(2)$, Duke Math. J. 83 (1996), 1-18.

[HK] M. HARRIS, S. KUDLA, The central value of a triple product L-function, Ann. of Math. 133 (1991), 605-672.

[He] R. HEATH-BROWN, Cubic forms in ten variables, Proc. London Math. Soc. 47 (1983), 225-257.

[Hec] E. HECKE, Mathematische Werke, Göttingin, 1959.

[Hee] K. HEEGNER, Diophantische Analysis und Modulfunktionen, Math. Z. 56 (1952), 227-253.

[Hej] D. HEJHAL, On the topography of Maass wave forms for $PSL(2,\mathbb{Z})$, J. Exp. Math. 1:4 (1992), 275-305.

[HoL] J. HOFFSTEIN, P. LOCKHART, Coefficients of Maass forms and the Siegel zero, Ann. Math. 140 (1994), 177-181.

[HsKK] J. HSIA, Y. KITAOKA, M. KNESER, Representations of positive definite quadratic forms, J. Reine Angew Math. 301 (1978), 132-141.

[I] J. IGUSA, Fibre systems of Jacobian varieties III. Fibre systems of elliptic curves, Amer. J. Math. 81 (1989), 453-476.

[Ih] Y. IHARA, Hecke polynomials as congruence ζ functions in elliptic modular case, Ann. Math. 85 (1967), 267-295.

[IwS] H. IWANIEC, P. SARNAK, The nonvanishing of central values of automorphic L-functions and Landau-Siegel zeros, Israel J. Math., to appear.

[Iw1] H. IWANIEC, Fourier coefficients of modular forms of half-integral weight, Invent. Math. 87 (1987), 385-401.

[Iw2] H. IWANIEC, Small eigenvalues of Laplacian for $\Gamma_0(N)$, Acta Arith. 56 (1990), 65-82.

[IwLS] H. IWANIEC, W. LUO, P. SARNAK, Low lying zeros of families of L-functions, I.H.E.S., to appear.

[JS] H. JACQUET, J. SHALIKA, On Euler products and the classification of automorphic representations I, Amer. J. Math. 103 (1981), 499-558.

[JPS] H. JACQUET, I. PIATETSKI-SHAPIRO, J. SHALIKA, Rankin-Selberg convolutions, Amer. J. Math. 105 (1983), 367-464.

[KS] N. KATZ, P. SARNAK, Zeros of zeta functions and symmetry, BAMS 36 (1999), 1-26.

[KeS] J. KEATING, N. SNAITH, Random matrix theory and L-functions at $s = 1/2$, preprint, 2000.

[KiS] H. KIM, P. SARNAK, Refined estimates towards the Ramanujan and Selberg conjectures, preprint, 2000.

[KiSh] H. KIM, F. SHAHIDI, Symmetric cube L-functions for GL_2 are entire, Annals of Math. 150 (1999), 645-662.

[Kn] M. KNESER, Darstelungsmasse indefiniter quadratische Formen, Math. Zeit. 77 (1961), 188-194.

[KoL] V. KOLYVAGIN, D. LUGACHEV, Finiteness of the Shafarevich-Tate group and the group of rational points for some modular abelian varieties, Leningrad Math. J. 1:5 (1990), 1229-1253.

[KowMV1] E. KOWALSKI, P. MICHEL, J. VANDERKAM, Rankin-Selberg L-functions in the level aspect, preprint, 2000.

[KowMV2] E. KOWALSKI, P. MICHEL, J. VANDERKAM, Nonvanishing of high derivatives of automorphic L-functions at the center of the critical strip, J. Reine Angew Math., to appear.

[Ku] N. KUZNETZOV, Petersson's conjecture for cusp forms of weight zero and Linnik's conjecture, sums of Kloosterman sums, Math. SB 111 (1980), 334-383.

[L] L. LAFFORGUE, preprint, 2000.

[La] R. LANGLANDS, Problems in the theory of automorphic forms, S.L.N. 170, Springer (1970), 18-86.

[Li1] Y.V. LINNIK, Dokl. Akad. Nauk SSSR 30 (1941), 292-294.

[Li2] Y.V. LINNIK, Additive problems and eigenvalues of the modular operators, Proc. ICM (Stockholm) (1962), 270-284.

[Li3] Y.V. LINNIK, Ergodic properties of algebraic fields, Ergebnisse, 45 Springer Verlag (1968).

[Lu] W. LUO, Nonvanishing of L-values and the strong Weyl law, preprint, 2000.

[LuRS] W. LUO, Z. RUDNICK, P. SARNAK, On the generalized Ramanujan conjecture for $GL(n)$, Proc. Symp. Pure Math. 66, part 2, AMS, (1999).

[M] L. MEREL, Bornes pour la torsion de courbes elliptiques sur les corps de nombres, Invent. Math. 124 (1996), 437-449.

[Me] T. MEURMAN, On the order of the Maass L-function on the critical line, Number Theory, Vol. I, Budapest Colloq. Math. Soc., Janos Bolyai, 51 (1990).

[Mi] P. MICHEL, Autour de la conjecture de Sato-Tate pour les sommes de Kloosterman. I, Invent. Math. 121 (1995), 61-68; Minorations de sommes d'exponentielles, Duke Math. J. 95:2 (1998), 227-240.

[Mo] G. MOLTENI, Upper and lower bounds at $s = 1$ for certain Dirichlet series with Euler product, preprint, 2000.

[Mon] H. MONTGOMERY, The pair-correlation of zeros of the zeta function, Proc. Symp. Pure Math., AMS 24 (1973), 181-193.

[O] A. ODLYZKO, The 10^{20}-th zero of the Riemann zeta function and 70 million of its neighbors, preprint, A.T.T., 1989.

[P] H. PETERSSON, Zur analytischeu theorie de grenzkreisgruppen, Math. Z. 44 (1938), 127-155.

[PhS] R. PHILLIPS, P. SARNAK, On cusp forms for cofinite subgroups of $PSL(2,\mathbb{R})$, Invent. Math. 80 (1985), 339-364.

[R] R. RANKIN, Contributions to the theory of Ramanujan's function $\tau(n)$, Proc. Camb. Phil. Soc. 35 (1939), 351-356.

[Ru] M. RUBINSTEIN, Evidence for a spectral interpretation of zeros of L-functions, 1998, thesis, Princeton.

[Rud] Z. RUDNICK, Letter to Peter Sarnak (1999).

[RudS1] Z. RUDNICK, P. SARNAK, Zeros of principle L-functions and random matrix theory, Duke Math. J. 82 (1996), 269-322.

[RudS2] Z. RUDNICK, P. SARNAK, The behavior of eigenstates of arithmetic hyperbolic manifolds, CMP 161 (1991), 195-213.

[Rum] R. RUMELY, Numerical computations concerning ERH, Math. Comp. 61 (1993), 415-440.

[S1] P. SARNAK, Integrals of products of eigenfunctions, IMRN 6 (1994), 251-260.

[S2] P. SARNAK, Subconvexity estimates for Rankin-Selberg L-functions, in preparation.

[Sa1] I. SATAKE, Spherical functions and the Ramanujan conjecture, Proc. Symp., AMS 9 (1971), 258-264.

[Sa2] I. SATAKE, Theory of spherical functions on reductive algebraic groups over p-adic fields, Publ. IHES 18 (1983).

[Sc] R. SCHULZE-PILLOT, Exceptional integers for genera of integral ternary positive definite quadratic forms, Duke Math. J. 102 (2000), 351-357.

[Se] A. SELBERG, On the estimation of Fourier coefficients of modular forms, Proc. Symp. Pure Math. VIII, AMS, Providence (1965), 1-15.

[Sh1] F. SHAHIDI, Automorphic L-functions-a survey, in "Automorphic Forms, Shimura Varieties and L-functions" (L. Clozel and A. Milne, eds.), Academic Press (1988), 49-109.

[Sh2] F. SHAHIDI, On Ramanujan's conjecture and the finiteness of poles of certain L-functions, Ann. of Math. 127 (1988), 547-584.

[Shi1] G. SHIMURA, On modular forms of half integral weight, Annals of Math. 97 (1973), 440-481.

[Shi2] G. SHIMURA, On the holomorphy of certain Dirichlet series, Proc. London Math. Soc. 31 (1975), 79-98.

[Shi3] G. SHIMURA, Hilbert modular forms of half-integral weight, Duke Math. J. 71 (1993), 501-557.

[Si1] C.L. SIEGEL, Additive Theorie der Zahlkörper II, Math. Ann. 88 (1923), 184-210.

[Si2] C.L. SIEGEL, Über die classenzahl quadratischer Zahlkörper, Acta Arith. 1 (1935), 83-86.

[Si3] C.L. Siegel, Sums of m-th powers of algebraic integers, Ann. of Math. 46 (1945), 313-339.

[Si4] C.L. Siegel, Lectures on the Analytic Theory of Quadratic Forms, Robert Peppermüller, Göttingen, 1963.

[So] K. Soundararajan, Nonvanishing of quadratic L-functions at $s = \frac{1}{2}$, Annals of Math., to appear.

[T] T. Tamagawa, On the zeta functions of a division algebra, Annals of Math. 77 (1963), 387-405.

[TaW] R. Taylor, A. Wiles, Ring-theoretic properties of certain Hecke algebras, Annals of Math. 141 (1995), 553-572.

[W] J. Waldspurger, Sur les coefficients de Fourier de formes modulaires de poids demi entier, J. Math. Pures et Appl. 60 (1981), 365-384.

[Wa] T. Watson, Central value of the Rankin triple L-function for unramified Maass cuspforms, in preparation.

[We1] A. Weil, On the Riemann hypothesis in function fields, Proc. Nat. Acad. Sci. 27 (1941), 345-349.

[We2] A. Weil, On some exponential sums, Proc. Nat. Acad. Sci., USA, 34 (1948), 204-207.

[We3] A. Weil, Sur les formules explicites de la théorie des nombres, Izv. Mat. Nauk. 36, 3-18.

[Wey] H. Weyl, Zur abschätzung von $\zeta(1 + it)$, Math. Zeit. 10 (1921), 88-101.

[Wi] A. Wiles, Modular elliptic curves and Fermat's last theorem, Annals of Math. 141 (1995), 443-551.

[Wo] S. Wolpert, Disappearance of cusp forms in special families, Annals of Math. 139 (1994), 239-291.

Henryk Iwaniec, Department of Mathematics, Rutgers University, New Brunswick, NJ 08903-2101, USA iwaniec@math.rutgers.edu

Peter Sarnak, Department of Mathematics, Princeton University, Princeton, NJ 08544, USA sarnak@math.princeton.edu

GAFA, Geom. funct. anal.
Special Volume – GAFA2000, 742 – 791
1016-443X/00/S20742-50 $ 1.50+0.20/0

GAFA Geometric And Functional Analysis

COMBINATORICS WITH A GEOMETRIC FLAVOR

Gil Kalai

Dedicated to the Memory of Rodica Simion

Abstract

In this paper I try to present my field, combinatorics, via five examples of combinatorial studies which have some geometric flavor. The first topic is Tverberg's theorem, a gem in combinatorial geometry, and various of its combinatorial and topological extensions. McMullen's upper bound theorem for the face numbers of convex polytopes and its many extensions is the second topic. Next are general properties of subsets of the vertices of the discrete n-dimensional cube and some relations with questions of extremal and probabilistic combinatorics. Our fourth topic is tree enumeration and random spanning trees, and finally, some combinatorial and geometrical aspects of the simplex method for linear programming are considered.

Introduction

There is a delicate balance in mathematics between examples and general principles, and in this paper I try to present my field, combinatorics, via five examples of combinatorial studies which have some geometric flavor.

In order to make the presentation self-contained, detailed and interesting, the choice of material (even within the individual sections) is subjective and nonuniform. For an unbiased and comprehensive point of view the reader is referred to the many links and references. I have tried to include many open problems and to point out various possible connections, some of which are quite speculative.

Section 1 deals with configurations of points in Euclidean spaces and specifically with Tverberg's theorem which asserts that every set of $(r-1)(d+1)+1$ points in $\mathbb{R}^d$ can be divided into r parts whose convex hulls have nonempty intersection. A principal question is to find conditions which will guarantee the conclusion of Tverberg's theorem for a smaller set of points.

Section 2 is devoted to McMullen's upper bound theorem for convex polytopes which asserts that among all d-polytopes with n vertices the cyclic polytope has the maximal number of faces of any dimension. Sharp and general forms of this theorem and what will take to prove them are discussed.

The topic of section 3 is the cube: the combinatorics of subsets of the vertices of the discrete cube, discrete isoperimetric relations and especially the notion of influence. General facts about Boolean and real valued functions defined on the discrete cube are useful for various problems in extremal combinatorics, probability and mathematical physics.

In section 4, I discuss some recent results concerning random spanning trees and tree enumeration and mention the recent emerging picture of random spanning trees of grids in the plane.

The principal problem in the final section, §5 is to find a polynomial-time version of the simplex algorithm for linear programming. Combinatorial and geometric aspects of the problem are considered.

Although there are relations between the five sections they can be read in any order. The reader can safely skip any place where she or he feels that the mathematics becomes too heavy-going. Probably these places reflect the fact that either the mathematics should be improved or my understanding of it should.

In the (unusual) style of the "Vision in Mathematics" meeting, each section concludes with brief comments of a philosophical nature.

1 Combinatorial Geometry: An Invitation to Tverberg's Theorem

1.1 Radon's theorem and order types (oriented matroids).

Theorem 1.1 (Radon's Theorem). *Every $d+2$ points in $\mathbb{R}^d$ can be partitioned into two parts such that the convex hulls of these parts have nonempty intersection.*

A pair of disjoint subsets of X whose convex hulls intersect are called a *Radon partition.* The points in the intersection of the convex hulls are called *Radon points.*

Radon's theorem follows at once from the fact that $d+2$ points in $\mathbb{R}^d$ are always affinely dependent. It implies at once another basic theorem on convex sets – Helly's theorem: *For every finite family of convex sets, if every $d+1$ of its members have a point in common then all sets in the family have a point in common.* The reader is referred to [14], [20] for much information on Helly type theorems.

Given n points on the line, the (minimal) Radon partitions determine (up to orientation of the line) the ordering of these points. In a similar way we can classify configurations of points in the plane or in $\mathbb{R}^d$ according to their Radon partitions. This leads to the theory of oriented matroids or order types (see [10]).

1.2 Tverberg's theorem.

Theorem 1.2 (Tverberg's Theorem). *Every $(d+1)(r-1)+1$ points in $\mathbb{R}^d$ can be partitioned into r parts such that the convex hulls of these parts have nonempty intersection.*

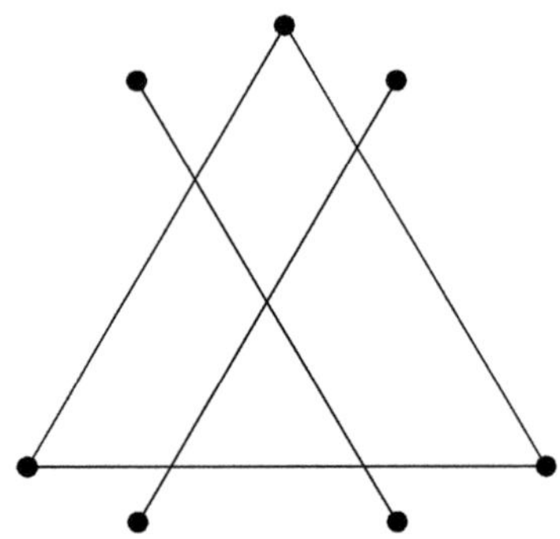

Figure 1: Seven points in the plane and their Tverberg partion.

Proofs of Tverberg's theorem were given by Tverberg ('66) [27], Doignon and Valette ('77), Tverberg ('81), Tverberg and Vrecica ('92), Sarkaria ('92) [25], and Roudneff ('99) [23]. While the original proof was quite difficult, the proofs of Sarkaria and Roudneff are remarkably simple.

Roudneff's recent proof is by minimizing the *sum of squares of the r distances* between a point x and the convex hulls of r pairwise disjoint subsets of the points. It turns out (under mild assumptions of genericity) that if this minimum is positive, it is attained without using one of the points and this extra point can be used to push this minimum down.

While the recent proofs of Tverberg's theorem give an algorithm to find the partition, the computational complexity of finding such a partition is not known.

PROBLEM 1.1. Find a polynomial-time algorithm to obtain a Tverberg partition when Tverberg's theorem applies.

Note that (as will be clear below) deciding for a configuration of points of less than $2d+3$ in $\mathbb{R}^d$ if a Tverberg partition to 3 parts exists is an **NP**-complete problem. However, it is possible that when the number of points is large enough to guarantee a partition then finding such a partition is computationally feasible.

1.3 Topological versions.

CONJECTURE 1.2 (The Topological Tverberg Conjecture). *Let f be a continuous function from the m-dimensional simplex σ^m to $\mathbb{R}^d$. If $m \geq (d+1)(r-1)$ then there are r pairwise disjoint faces of σ^m whose images have a point in common.*

The case $r=2$ was proved by Bajmoczy and Bárány using the Borsuk-Ulam theorem. The case where r is a prime number was proved in a seminal paper of Bárány, Shlosman and Szücs [8]. The prime power case was proved by Ozaydin (unpublished), Volovikov [30] and Sarkaria. For this case the proofs are quite difficult and are based on computations of certain characteristic classes.

If f is a linear function this conjecture reduces to Tverberg's theorem. For a discussion of the topological extensions of Tverberg's theorem in a larger context, see [32]. It turns out that topological methods are crucial for proving various Tverberg type theorems even for linear maps.

1.4 The dimension of Tverberg's points. For a set A, denote by $T_r(A)$ those points in $\mathbb{R}^d$ which belong to the convex hull of r pairwise disjoint subsets of X. We call these points *Tverberg points of order r*.

If we have $(d+1)(r-1)+1+k$ points in $\mathbb{R}^d$, then we expect that the dimension of Tverberg points of order r will be at least k. This is so in the "generic" case. Reay conjectured that it is enough to assume the points are in general position. Various special cases were recently proved by Roudneff [23], [24].

In another direction, I conjectured that failing to have the "right" dimension for the Tverberg points of order r implies the existence of a Tverberg point of order $r+1$.

CONJECTURE 1.3 (Kalai, 1974). *For every $A \subset \mathbb{R}^d$,*

$$\sum_{r=1}^{|A|} \dim T_r(A) \geq 0\,.$$

Note that $\dim \emptyset = -1$. This conjecture includes Tverberg's theorem as a special case: if $|A| = (r-1)(d+1)+1$ $\dim A = d$ and $T_r(A) = \emptyset$, then the sum in question is at most $(r-1)d + (|A|-r+1)(-1) = -1$.

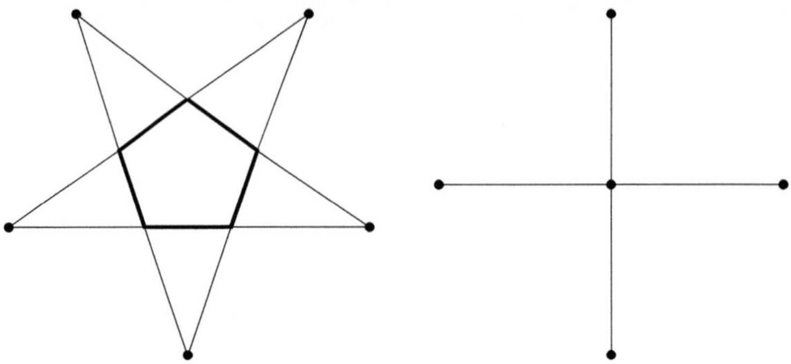

(a) $T_1=2$, $T_2=1$, $T_i=-1$, $i\geq 3$ (b) $T_1=2$, $T_2=T_3=0$, $T_4=T_5=-1$.

Figure 2: Two planar configurations of five points.

It may even be true that Conjecture 1.3 holds if we replace $T_r(A)$ by the minimum of $T_r(A')$ over all configurations A' of the same order type as A.

Kadari proved (around 1980) Conjecture 1.3 for planar configurations. Crucial to his proof is the fact that in the plane (but not in higher dimensions), the convex hull of Tverberg points of order r is precisely the $(r-1)$-core of A: The intersection of all subsets of A of cardinality $|A|-(r-1)$. (Of course, every Tverberg point of order r belongs to the $(r-1)$-core.)

1.5 Conditions for Tverberg partitions and graph colorings.

1.5.1 Conditions for a Tverberg partition into 3 parts. The following problem seems important.

PROBLEM 1.4. Find conditions on the order type for a configuration A of m points ($m < 2d+3$) in $\mathbb{R}^d$ that guarantee the existence of a Tverberg partition into three parts.

Note that deciding the existence of Tverberg partitions into three parts when $m < 2d+3$ is **NP**-complete, as will become evident below, and does not depend only on the order type of the configuration. However, I do expect that there are useful topological sufficient conditions. Conjecture 1.3 gives one such condition: $\dim T_2(A) < |A| - d - 2$.

1.5.2 Point configurations from graphs. For a graph $G = \langle V, E\rangle$ consider the configuration of points in R^V which are the incidence vectors of edges of the graph. Thus, the vector associated to an edge $\{u, v\}$ has the value '1' in the coordinates that correspond to u and v and the value

'0' in all other coordinates.

PROBLEM 1.5. What can be said about affine dependencies and Radon points (and Tverberg points) of such point configurations?

The Radon partitions of such configurations arising from graphs, especially regular graphs, seem to be related to matching theory for graphs [17].

Note that a proper 3-coloring for the edges of a connected cubic graph G is *equivalent* to the existence of a Tverberg partition into 3 parts for the point configuration corresponding to G. Indeed, given a Tverberg partition into 3 parts, color every edge according to the part it belongs to. Every vertex which is incident to one colored edge must be incident to three edges colored with the 3 different colors and therefore the colored edges describe a proper 3-coloring of some cubic subgraph. Since G is connected this must be the entire graph.

1.5.3 The four color theorem. The four color theorem (Appel-Haken, 1977, see [28]) asserts that every planar map is four colorable. An equivalent formulation of the four color theorem is: Every 2-connected cubic planar graph is 3-edge colorable. (A cubic graph or a 3-regular graph is a graph all of whose vertices have degree 3. A graph is 2-connected if it remains connected after deleting every vertex.)

Now, consider a configuration of points P corresponding to a cubic planar graph with n vertices. Note, we have $3n/2$ points in a $(n-1)$-dimensional space. (If G is bipartite, these points are in a $(n-2)$-dimensional subspace.) Finding sufficient conditions for the existence of Tverberg partitions when the number of points is smaller than $2d+3$ may thus be relevant to finding new avenues towards the 4-color theorem (and its many open generalizations).

REMARKS. 1. The idea of trying to relate Tverberg's theorem and the four color theorem (in a different way) goes back to Tverberg himself.

2. There are, of course, 2-connected cubic graphs which are not 3-edge colorable. The most famous example is the Petersen graph (identify pairs of antipodal vertices in the graph of the dodecahedron). It is worth noting that the 2-core of point configurations associated to 2-connected cubic graphs is always nonempty. (The 2-core is the intersection of all convex hulls of all but two of the points.)

3. The Radon partitions of a set A of $d+2+k$ points in $\mathbb{R}^d$ correspond to the faces of a k-dimensional zonotope. Every point in the boundary of this zonotope corresponds to a (normalized) affine dependence of the points

and is mapped to a Radon point of A. This map maps two antipodal points on the zonotope to the same Radon point and thus induces a map from RP^k to $\mathbb{R}^d$ whose image is the Radon points of A.

For generic $3n/2$ points in $\mathbb{R}^{n-1}$ the Radon points form an embedding of the $(n/2-1)$-dimensional real projective space into $\mathbb{R}^{n-1}$. The case of configurations of points arising from graphs is, of course, highly non-generic.

1.6 Other problems and connections.

1.6.1 Halving hyperplanes and colored Tverberg's theorems. An important problem in combinatorial geometry is to determine the maximal number of ways a configuration of $2m$ points in $\mathbb{R}^d$ can be divided into two equal parts by a hyperplane. More generally, to determine the maximum number of ways a configuration of n points in $\mathbb{R}^d$ can be divided by a hyperplane to parts of sizes k and $n-k$ (see [6]). Equivalently, this is the minimal possible number of Radon partitions into two equal parts (or parts of prescribed sizes).

Even in the plane there is a substantial gap between the best known lower bound $C_1 n \cdot \exp(\sqrt{\log n})$ (Toth, [29]) and the upper bound $C_2 n^{4/3}$ (Dey, [13]).

The planar case of the problem is closely related to the following algebraic question: Given a reduced (=minimal) representation of a permutation in S_n as the product of adjacent transpositions, what is the maximum number of appearances of a specific transposition? To see the connection, project the points on a line and slowly rotate the line (see [15]).

For dimension d, it is easy to bound the maximal number of halving hyperplanes between n^{d-1} and n^d. Toth's lower bound extends to a lower bound of $n^{d-1} \cdot \exp(\sqrt{\log n})$ in any dimension. In space, the best known upper bound is $n^{5/2}$ [26]. In higher dimensions, the only known way for finding upper bounds for the halving hyperplane problem is via generalizations of Tverberg's theorem for colored configurations of points. Remarkably, the only proofs of these generalizations are by the topological method [33], [34]. This gives, in every dimension d, an upper bound for the number of halving hyperplanes of the form n^{d-c_d}, for some $c_d > 0$.

1.6.2 Eckhoff's partition conjecture. Ekchoff raised the possibility of finding a purely combinatorial proof of Tverberg's theorem based on Radon's theorem. He considered replacing the operation "taking the convex hull of a set A" by an arbitrary closure operation.

Let X be a set endowed with an abstract closure operation $X \to cl(X)$.

The only requirements from the closure operation are: (1) $cl(cl(X)) = cl(X)$ and (2) $A \subset B$ implies $cl(A) \subset cl(B)$.

Define $t_r(X)$ to be the largest size of a (multi)set in X which cannot be partitioned to r parts whose closures have a point in common. Eckhoff conjectured that always

$$t_r \leq t_2 \cdot (r-1)\,.$$

Thus, if X is the set of subsets of $\mathbb{R}^d$ and $cl(A)$ is the convex hull operation then Radon's theorem asserts that $t_2(X) = d+1$ and Eckhoff's partition conjecture reduces to Tverberg's theorem.

1.7 Some links and references. The reader will find additional references to earlier works and survey papers in the more recent ones. Personal web sites (listed before the references at the end of the paper) will be cited by the name appearing in square brackets. Many of the papers in the references as well as related ones can be found there. The handbooks [3], [2], [1] contain many chapters which are relevant to this paper and we cite only a few.

Helly and Radon type theorems [14], [20]; topological proofs of Radon type theorems [8], [18], [31], [33], [34]; combinatorial geometry [22]; topological methods in combinatorics [9], [32]; oriented matroids [10]; halving lines and hyperplanes [6], [7], [26]; colorings of graphs [16], [5]; developments concerning the four color theorem [28]; matching theory [17]; graph theory [11]; a Radon type theorem of Larman which deserves simple proofs and better understanding [19].

Proofs, more proofs, "proofs from the book" and computer proofs

Science has a dual role: exploring and explaining. In mathematics, unlike other sciences, mathematical proofs are used as the basic tool for both tasks: to explore mathematical facts and to explain them.

The meaning of a mathematical proof is quite stable. It seems unharmed by the "foundation crisis" and the incompleteness results at the beginning of the 20th century, and unaffected by the recent notions of randomized and interactive proofs in theoretical computer science. Still, long and complicated proofs, as well as computerized proofs, raise questions about the nature of mathematical explanations.

Proofs are gradually becoming intolerably difficult. This may suggest that our days of successfully tackling a large percentage of the problems we pose will soon be over. This may also reflect the small incentives to simplify.

Be that as it may, we cannot be satisfied without repeatedly finding new connections and new proofs, and we should not give up hope of finding simple and illuminating proofs that can be presented in the classroom. For some "proofs from the book", see the lovely book by Aigner and Ziegler [4].

Some believe that computer proofs will take over [Zeilberger]. Appel and Haken's proof of the four color theorem was a landmark in this respect. The role of computers in exploring mathematical facts is already significant. As for explaining mathematical facts, it raises, for instance, the question "Explaining to whom? To humans, or to other computers?"

2 Polytopes and Algebraic Combinatorics: How General is the Upper Bound Theorem?

2.1 Cyclic polytopes and the upper bound theorem.

2.1.1 Cyclic polytopes. Consider the *moment curve* $x(t) = (t, t^2, \dots, t^d) \subset \mathbb{R}^d$. The cyclic polytope $C(d, n)$ is the convex hull of n (distinct) points $x(t_1)$, $x(t_2), \dots, x(t_n)$ on the moment curve. The face structure does not depend on the choice of these points.

Cyclic polytopes are $d/2$-*neighborly*, namely the convex hull of every set of k vertices forms a face of the polytope when $k \le d/2$. Thus $f_k(C(d, n))$, the number of k-dimensional faces (in brief, k-faces) of $C(d, n)$ is $\binom{n}{k+1}$, whenever $k < d/2$. Cyclic polytopes were discovered by Carathéodory and were rediscovered by Gale, who described their face-structure.

2.1.2 The upper bound theorem. The upper bound theorem (UBT), conjectured by Motzkin in 1957, asserts that the face numbers of a d-polytope with n vertices are bounded from above by the face numbers of the cyclic d-polytope with n vertices. This conjecture is of special interest in connection with optimization, because it gives the maximum number of vertices that can be possessed by a d-polytope P defined by means of n linear inequality constraints; hence it represents the maximum number of local strict maxima that can be attained by a convex function over P.

The assertion of the upper bound theorem was proved for polytopes (McMullen, 1970 [73]), for simplicial spheres (Stanley, 1975 [79], [82]) and for simplicial manifolds with either vanishing middle homology or the same Euler characteristic as a sphere (Novik, 1998 [74]). It was also been proved when n is large w.r.t. d ($n \ge d^2/4$, will do) for all Eulerian simplicial complexes (Klee, 1964 [65]). (An Eulerian simplicial complex is a pure simplicial complex in which the link of each simplex has the same Euler characteristic as the sphere of the appropriate dimension.)

2.1.3 A stronger form of the UBT. A stronger version of the UBT (referred to, below, as SUBC: strong upper bound conjecture) was proved for simplicial d-polytopes and full dimensional subcomplexes of their boundary complexes by Kalai [62]. It asserts (roughly) that for every k, $0 \leq k < d-1$, if one fixes the number of k-dimensional faces, then the number of $(k+1)$-dimensional faces is maximized by a cyclic d-polytope. (More precisely, it gives a bound on the number of $(k+1)$-faces in terms of the number of k-faces that is similar in form to the Kruskal-Katona theorem, which provides a similar bound for arbitrary simplicial complexes.) I conjecture that the SUBC applies to arbitrary polytopes (and more general complexes considered below).

The SUBC was also motivated by a problem from optimization, namely by an attempt to show expansion properties of graphs of d-polytopes. However, applications in this direction were quite limited.

2.2 Stanley-Reisner rings and their generic initial ideals (algebraic shifting). Stanley's proof of the upper bound theorem for triangulation of spheres relies on the notion of the Stanley-Reisner ring associated to a simplicial complex and on the fact that this ring is Cohen–Macaulay. We will describe below an algebraic statement concerning generic initial ideals of the Stanley-Reisner rings which implies the strong upper bound theorem.

Let me first explain the situation informally. The Stanley-Reisner ring is constructed by associating to each vertex i of a simplicial complex a variable x_i, and considering the ring of monomials which "live" on the complex. Consider next generic linear combinations of these variables y_1, $y_2, \ldots, y_n$ and a Gröbner basis for this ring w.r.t. monomials in the new variables. This construction associates to every simplicial complex K a basis of monomials $GIN(K)$ (in the new variables) which record many topological, combinatorial and algebraic properties of K.

An algebraic statement for a $(d-1)$-dimensional simplicial complex K which immediately implies the UBT, and in fact also the SUBC, is that $GIN(K)$ is a subset of $GIN(C(d,n))$, where n is the number of vertices of K and $C(d,n)$ is the boundary complex of a cyclic d-polytope with n vertices. When K is isomorphic to the boundary complex of a simplicial polytope this relation follows from the Hard Lefschetz Theorem for toric varieties; see [62].

Here is a more accurate description of the Stanley-Reisner ring and $GIN(K)$. Associate to each vertex i of a simplicial complex K a variable

x_i and consider the quotient

$$R(K) = R[x_1, x_2, \ldots, x_n]/I\,,$$

where I is the ideal spanned by monomials $x_{i_1} \cdot x_{i_2} \cdots x_{i_r}$ with $\{i_1, i_2, \ldots, i_r\} \notin K$.

Consider now $y_1, y_2 \ldots, y_n$, which are n generic linear combinations of $x_1, x_2 \ldots x_n$ and construct the Gröbner basis $GIN(K)$ w.r.t. the lexicographic order on the monomials in the y_i's. (Clearly, all monomials in the y_i's span the ring $R(K)$.) Thus, a monomial m belongs to $GIN(K)$ if and only if its image $\tilde{m}$ in $R(K)$ is not a linear combination of (images of) monomials which are lexicographically smaller. (Recall that the lexicographic order is defined as follows: $m_1 <_L m_2$ if the variable with smallest index which divides precisely one of the two monomials divides m_1. Thus $y_1^2 <_L y_1y_2 <_L y_1y_3 <_L \cdots <_L y_1y_n <_L y_2^2 <_L \cdots$.)

2.3 How general is the upper bound theorem?

2.3.1 Witt spaces. Witt spaces [56], [77], [50] are orientable triangulated pseudomanifolds K such that for every K' which is an even-dimensional (proper) link of a face of K, the (middle perversity) intersection homology $IH_{dimK'/2}(K')$ vanishes. For these spaces middle perversity intersection homology is defined and satisfies Poincaré duality. These spaces include all (real) manifolds and (complex, possibly singular) algebraic varieties.

We come now to the main conjecture of this section.

CONJECTURE 2.1. *(i) For every triangulation K of a Witt space with vanishing middle intersection homology*

$$GIN(K) \subset GIN\big(C(d,n)\big)\,. \tag{2.1}$$

(ii) The strong upper bound conjecture holds for arbitrary polyhedral complexes (and even for all regular cell complexes whose face-poset form a lattice) whose underlying space is a Witt space with vanishing middle intersection homology.

What seems to be needed for a proof is an interpretation of intersection homology for simplicial pseudomanifolds in terms of the Stanley-Reisner ring and generic initial ideals. For the polyhedral case, what is needed is a suitable analog of the Stanley-Reisner ring. For this purpose too, intersection homology may play a crucial role. Intersection homology of toric varieties already plays an important role in the combinatorial study of (rational) polytopes [51], [81] (see below).

2.3.2 Embeddability. The upper bound theorem seems closely related to questions concerning embeddability. At the root of things is the assertion that K_5, the complete graph on 5 vertices, cannot be embedded in the plane.

Van Kampen proved that the r-skeleton of σ^{2r+2} (the $(2r+2)$-dimensional simplex) cannot be embedded in $\mathbb{R}^{2r}$. It seems that this property and corresponding local properties (for links of faces) would imply the assertions of the UBT, SUBC and relation (2.1). To understand such a connection it will be useful to know if the Van Kampen theorem holds when $\mathbb{R}^{2r}$ is replaced by any $2r$-dimensional manifold with vanishing middle homology, or even by any Witt space with vanishing middle intersection homology.

2.3.3 An upper bound conjecture for j-sets. Emo Welzl [88] has recently proposed another far-reaching extension for the upper bound theorem. Given a configuration A of n points in general position in $\mathbb{R}^d$ consider the set of all hyperplanes, $\mathcal{H}^j$, which are determined by points in A and have at most j vertices in one of their (open) sides (compare section 1.6.1). For $j=0$ these are supporting hyperplanes for $conv(A)$. Next, let $a_r^j(A)$ be the number of r-dimensional simplices which are determined by point in A and belong to a hyperplane in $\mathcal{H}^j$. (For $j=0$ these are just r-faces of $conv(A)$.) Welzl asked whether $a_r^j(A)$ is maximized for every r and j by n points on the moment curve in $\mathbb{R}^d$. For $j=1$ this is just the UBT and it is also known to be true for every j when $d=2$ (Alon and Gyori) and when $d=3$ (Welzl).

2.4 Duality and h-numbers. I have described very general cases for which I conjecture that the UBT and even the SUBC hold, and a strong property of $GIN(K)$ needed to prove these conjectures. However, these strong algebraic and combinatorial conjectures are known only in very limited cases. For the known cases of the UBT, weaker combinatorial and algebraic statements are sufficient if certain duality relations are also used.

2.4.1 The Dehn–Sommerville relations. For a $(d-1)$-dimensional simplicial manifold K define its h-numbers by the relation:

$$\sum_{k=0}^{d} h_k(K)x^{d-k} = \sum_{k=0}^{d} f_{k-1}(K)(x-1)^{d-k}. \tag{2.2}$$

The Dehn–Sommerville Relations asserts that if K is the boundary complex of a simplicial polytope then

$$h_k(K) = h_{d-k}(K). \tag{2.3}$$

In fact, these relations hold whenever K is *Eulerian* simplicial complex, namely K and all links of faces of K have the same Euler characteristics as a sphere of the same dimension.

2.4.2 The Cohen–Macaulay property. For Stanley's proof of the UBT when K is a simplicial sphere we need to know in addition to the Dehn–Sommerville relations that $R(K)$ is a Cohen–Macaulay ring. When $R(K)$ satisfies the Cohen–Macaulay property, then $h_k(K)$ is the number of monomials of degree k in $GIN(K)$ which use only the variables $y_{d+1}, y_{d+2}, \dots, y_n$. Novik [74] used $GIN(K)$ to prove the UBT for several classes of simplicial manifolds and she relied on the fact that $R(K)$ is still close enough to being a Cohen–Macaulay ring (the technical term is *Buchsbaum* ring). In addition, she needed the analogs of Dehn–Sommerville relations and Poincaré duality. We would like to have a better understanding of these duality relations in terms of $GIN(K)$ and for more general classes of simplicial complexes.

2.4.3 Partial unimodality and the Braden–MacPherson theorem. The face numbers of polytopes are not unimodal. Indeed, the face numbers of the cyclic polytope are highly concentrated near dimension $3d/4$ and therefore, by gluing a cyclic polytope and its dual, you will get two peaks at $d/4$ and at $3d/4$. To get a simplicial example glue a cyclic polytope to the cross polytope (with roughly the same total number of faces). You will get two peaks at $3d/4$ and at $2d/3$.

An appealing application (using an argument of Björner [47]) of the SUBC for general polytopes will be:

Conjecture 2.2. *The face numbers f_i of d-polytopes are nondecreasing for $i \le [(d+3)]/4$ and nonincreasing for $i \ge [3(d-1)/4]$.*

It is possible that this conjecture as well as a suitable (weaker) version of the SUBC will follow in a purely combinatorial way from a recent result by Braden and MacPherson [51] which relates the combinatorics of a polytope with that of faces and quotients.

McMullen's original proof of the upper bound theorem relied on the observation that for a simplicial polytope P and a vertex v,

$$h_k\big(lk(v,P)\big) \le h_k(P)\,. \tag{2.4}$$

Here, $lk(v,P)$ is the link of v in P and h_k is the h-number mentioned above.

The result of Braden and MacPherson is a sharpening as well as a far-reaching generalization of (2.4) for general polytopes. (It is proved, however, only for rational polytopes.) I will now state this result without explaining properly the background and I refer the reader to [51], [81], [63] for more. For a d-polytope P let

$$h_P(x) = \sum_{k=0}^{d} h_i(P)x^k\,, \quad g_P(x) = \sum_{i=0}^{[d/2]} g_k(P)x^k\,.$$

Here $h_k(P) = \dim IH_{2k}(T_P)$, and $g_k(P) = h_k(P) - h_{k-1}(P)$, where T_P is the toric variety associated to P and IH is intersection homology. (The quantities $\dim IH_k(T_P)$ can be described in a purely combinatorial way from the face structure of P and when P is simplicial this is just h_k.) Braden and MacPherson proved that for every rational polytope P and a face F of P,

$$g_P(x) \geq g_F(x)g_{P/F}(x)\,. \tag{2.5}$$

(Namely, every coefficient of the polynomial in the left hand side is at least as large as the corresponding coefficient on the right hand side.)

The Braden–MacPherson inequality has already been used by Bayer [41] to deduce a very sharp form of the UBT for general (rational) polytopes.

2.4.4 Other duality relations. The Dehn–Sommerville duality relations $h_k(P) = h_{d-k}(P)$ applies for arbitrary Eulerian simplicial complexes. For simplicial polytopes this numerical duality manifests Poincaré duality for the associated toric varieties. When we adopt the combinatorial formulas of intersection homology the relations $h_k = h_{d-k}$ extend even to arbitrary Eulerian partially ordered sets. For toric varieties associated to rational polytopes these duality relations manifest Poincaré duality for intersection homology. In commutative algebra this duality relations manifest the Gorenstein property for the Stanley-Reisner ring of homology spheres.

Another important notion of duality is duality between polytopes given by the polar polytope. (Thus, the cube is dual to the octahedron, the dodecahedron is dual to the icosahedron and the tetrahedron is self-dual.) In 1985 I observed some mysterious numerical formulas relating h-numbers of a polytope and those of its dual. The simplest non-trivial relation of this type asserts that for every 4-dimensional polytope $g_2(P) = g_2(P^*)$ [61]. Some extensions were proved by Bayer and Klapper and by Stanley [81] who realized the correct combinatorial context (incidence algebras) for understanding these formulas. Geometric or algebraic understanding of these relations is still missing but for very special cases of toric varieties (which give rise to Calabi–Yau manifolds) it turned out that these numerical relations manifest mirror symmetry [39].

Added in proof. Tom Braden has recently found an algebraic explanation for these duality relations via Koszul's duality.

Yet another important notion of duality is duality of oriented matroids which includes the notions of Gale transform and linear programming duality as special cases (see [10, Chapter 10]). The effect of this duality on the combinatorial notions discussed here (as well as on the algebraic and geometric ones) is yet to be explored.

2.5 Neighborliness.

2.5.1 Neighborly polytopes and spheres. For an extremal combinatorial problem, studying the cases of equality is often as important as proving the inequality. Equality for the upper bound theorem is attained by all neighborly d-polytopes, namely polytopes for which every $[d/2]$ vertices form a face.

Neighborly polytopes form an exciting but mysterious class of polytopes (see [76]). Their face numbers are determined by the number of vertices. It is conjectured that every simplicial polytope is the quotient (link) of an even dimensional neighborly polytope [67]. (The same conjecture can be made for simplicial spheres.) For a generalization of the notion of neighborly polytopes to the nonsimplicial case, see Bayer [40].

2.5.2 Triangulations of manifolds. Triangulations of $2k$-dimensional manifolds can be even $(k+1)$-neighborly. An example is the 6-vertex triangulation of the 2-dimensional projective plane obtained by identifying the opposite faces of the icosahedron. This is quite a fundamental combinatorial object and its dual graph is no other than the Petersen graph.

Heawood, who around 1890 studied colorings of graphs embedded on surfaces (in the context of extending the four color conjecture), conjectured that K_n (the complete graph on n vertices) can be embedded in a surface M (except for the Klein bottle) if and only if

$$n \leq \left(7 + \sqrt{49 - 24\chi(M)}\right).$$

(Here, $\chi(M)$ is the Euler characteristic of M.) Such embeddings giving 2-neighborly triangulations of M were indeed found in all cases by Ringel and Youngs (in some cases with other coauthors). See Ringel's book [75].

However, there are only a handful of examples of $(k+1)$-neighborly $2k$-manifolds, for $k > 1$ (see [68]). Perhaps the most famous example is the remarkable 9-vertex triangulation of the complex 2-dimensional projective space by Kühnel and Lassman (see [70], [69]).

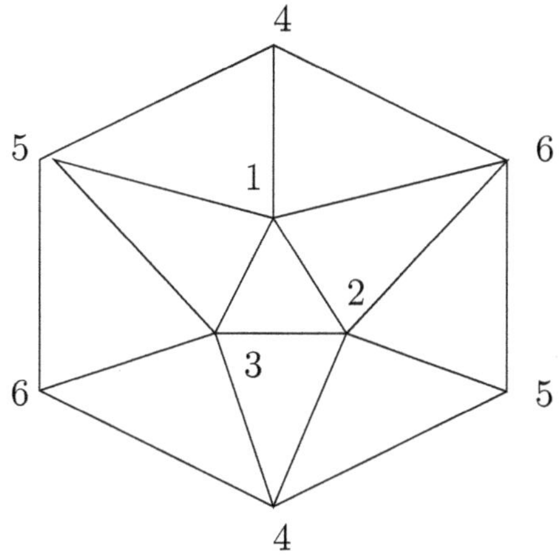

Figure 3: The 6 vertex triangulation of the real projective plane

In Kühnel's own words [68]: "To construct the triangulation we denote the nine vertices by 1,2,3, ..., 9 and take the union of the 4-dimensional simplices 12456 and 12459 under the action of a group of permutations H_{54} generated by: $\alpha = (147)(258)(369)$, $\beta = (123)(465)$ and $\gamma = (12)(45)(78)$. This group is a 2-fold extension of the Heisenberg group over Z_3. γ corresponds to the action of the complex conjugation, in fact its fixed point set is combinatorially isomorphic to the 6-vertex triangulation of the real projective plane."

Novik [74] proved an extension of the upper bound theorem for all triangulations of manifolds and it is plausible that this theorem, and a related conjecture by Kühnel concerning how large the Euler characteristic can be, apply to arbitrary triangulations of Witt spaces.

2.5.3 Neighborly embedded manifolds. The moment curve $x(t) = (t, t^2, \ldots, t^d) \subset \mathbb{R}^d$ is an example of 1-dimensional $[d/2]$-neighborly manifolds in $\mathbb{R}^d$. Namely for every $[d/2]$ points on the curve there is a hyperplane which supports the curve precisely at these points.

While there are many different neighborly polytopes there is only one (in terms of order types) $[d/2]$-neighborly embedding of $\mathbb{R}$ into $\mathbb{R}^d$, for d even. Moreover, for d even, the moment curve is the only order type of an embedding of $\mathbb{R}$ into $\mathbb{R}^d$ where all points are in general position. This indicates that the as yet unexplored area of understanding the "order type" of nondiscrete subsets in $\mathbb{R}^d$ (such as embedded manifolds) may exhibit some simpler phenomena than the discrete (finite) case.

Perles asked: what is the smallest dimension $d(k,n)$ of the ambient space in which a k-neighborly n-dimensional manifold exists? A simple dimension count shows that we must have $d(k,n) \geq (k+1)n$. On the other hand, a straightforward extension of the moment curve gives a bound for $d(k,n)$ which is exponential. Kalai and Wigderson found a simple construction showing a polynomial upper bound on $d(k,n)$, and Vassiliev [86] showed by an intricate topological argument that $d(k,n) \geq 2kn - bin(n)$, where $bin(n)$ is the number of ones in the binary expansion of n.

2.6 Other problems and connections.

2.6.1 Clique complexes and spheres. Start from a graph G and consider its clique complex $K(G)$, a simplicial complex whose faces correspond to the complete subgraphs of G. Understanding the possible face numbers of such complexes is an important problem in extremal combinatorics related to Turan's theorem; see [49], [72]. Suppose that $K(G)$ is a triangulated sphere. What can be said then? Charney and Davis [52] formulated a conjecture concerning the face numbers of such complexes which is closely related to conjectures of Hopf on the Euler characteristic of manifolds M with nonpositive sectional curvature. For some recent developments, see [71].

2.6.2 Cubical upper bound theorems. Cubical complexes seem of equal importance yet quite different from simplicial complexes, and much less is known about them. (A structure of a cubical complex on a manifold seems to tell more on the geometry of the manifold.) Only recently some cubical analogs of the cyclic polytopes were constructed. Joswig and Ziegler [60] constructed d-polytopes with 2^n vertices with $[d/2]$-skeletons of the n-dimensional cube. Previously, Babson, Billara and Chan [38] constructed cubical spheres with this property and found connections between questions on immersions of manifolds and the existence of certain cubical spheres. There are analogs for cyclic polytopes, but the analog of the upper bound theorem is false for spheres and probably also for polytopes. Adin [35] found the right notion of h-numbers, but a construction for a "cubical Stanley-Reisner ring" is yet unknown.

2.7 Some links and references. Polytope theory [89], [58], [59], [42], [66], [46], [78] [Ziegler]; face numbers and h-numbers of polytopes and complexes [43], [48], [45], [80], [63], open problems [85]; cubical spheres and polytopes [38], [60];

a continuous version of the UBT [87]; Kuhnel's CP^2 and other special triangulations [70], [69], [68]; Kruskal-Katona theorem and related results [54]; Turan type theorems [49], [55]; commutative algebra and combinatorics [82], [53] [Herzog], [Bayer]; generic initial ideals and algebraic shifting [48], [74], [37], [57], [53] [Herzog],[Bayer],[Kalai]; intersection homology [56], [50], and some combinatorial applications [81], [51]; h-numbers and polytope duality [81], [63], and mirror symmetry [39]; algebraic combinatorics à la Stanley [Stanley] [44], [82], [83], [84], [85]; Various proofs for the UBT: for Eulerian complexes with many vertices [65], for polytopes using shellability [73], a simple dual form using linear objective functions, [187], for spheres using the Cohen–Macaulay property [79], using shellability and the Cohen–Macaulay property [64], using shellability and a strong form of an extremal theorem of Bollobàs [36], for manifolds, using relations between face numbers and Betti numbers of Buchsbaum rings [74], a strong form for general polytopes using the Braden–MacPherson theorem [73].

Problems and conjectures

The posing of problems and conjectures is part of the process of exploring the factual matters as well as of proposing explanations for them. Is the development of mathematics shaped by problems? And what are good problems? Do they arise naturally like the sphere-packing conjecture, or are they perhaps sporadic and ingenious like Fermat's last theorem and the four color problem? To what an extent are good mathematical problems suggested by other sciences?

Modern combinatorics was greatly shaped by problems posed by Erdős, who was very cautious concerning our ability to predict the future of a problem.

3 Extremal and Probabilistic Combinatorics: the Discrete Cube and Influence of Variables

3.1 Influence of variables on Boolean functions.

3.1.1 The discrete cube. We consider the discrete cube $\Omega_n = \{-1,1\}^n$ and will try to understand real and Boolean functions defined on Ω_n. Boolean functions on Ω_n are of course in 1-1 correspondence with subsets of Ω_n. It turns out that many specific problems in extremal combinatorics, probability, mathematical physics and theoretical computer science can be formulated in terms of Boolean or real functions on Ω_n and that general properties of such functions are very useful.

For $x, y \in \Omega_n$ the Hamming metric $d(x,y)$ is defined by $d(x,y) = |\{i : x_i \neq y_i\}|$. Some related metrics will also be considered.

Denote by $\Omega_n(p)$ the discrete cube endowed with the product probability measure $\mathbf{P}_p$, where $\mathbf{P}_p\{x : x_j = 1\} = p$. Usually, we consider the uniform measure $p = 1/2$. (More general measures like FKG-measures should also be considered, but we will not attempt doing it here.)

Notation: In addition to the standard big O and little o notation we use the following notation: For positive real functions $f(x)$ and $g(x)$, we write $f(x) = \Theta(g(x))$ if, for some positive constants c_1 and c_2, $c_1 g(x) \leq f(x) \leq c_2 g(x)$, as x tends to infinity. We write $f(x) = \Omega(g(x))$ if for some positive constant c, $f(x) \geq cg(x)$.

3.1.2 Influence of variables. Consider an event $A \subset \Omega_n(p)$ and the associated Boolean function $f(x_1, x_2, \dots, x_n) = \chi_A$, the characteristic function of A. The *influence* of the variable k on the Boolean function f, denoted by $I_k^p(f)$ (and also by $I_k^p(A)$), is the probability that flipping the value of x_k will change the value of f. The total influence $I^p(f)$ equals $\sum I_k^p(f)$. We define also $II^p(f) = \sum (I_k^p(f))^2$. (We will not use the superscript p for $p = 1/2$.)

Influence of variables (in a much greater generality) was introduced and studied by Ben-Or and Linial [99] in the context of "collective coin flipping", an important notion in theoretical computer science. The problem they considered is, in short: "Is there a protocol for a society of n processors to produce a random bit immune against a situation where a fraction of the processors is cheating?" Having each processor produce a single random bit, and using a Boolean function to produce the "collective bit" is a simple such protocol. But it turns out (from Theorem 3.3 below) that it can never be immune against $w(n)n/\log n$ cheaters, when $w(n)$ tends to infinity with n. A multistage protocol immune against $\Omega(n)$ cheaters was found by Alon and Naor [92], see also Feige [110].

$I_k(f)$ is essentially identical to the Banzhaff value in game theory. In [99], influence of larger sets of variable is also considered.

A function f is monotone if its value does not decrease when we flip the value of a variable from -1 to 1. Some basic facts on influences are given by:

Theorem 3.1 (Loomis-Whitney, Hart, Harper).

$$\sum I_k(f) \geq \mathbf{P}(A) \log \left(1/\mathbf{P}(A)\right).$$

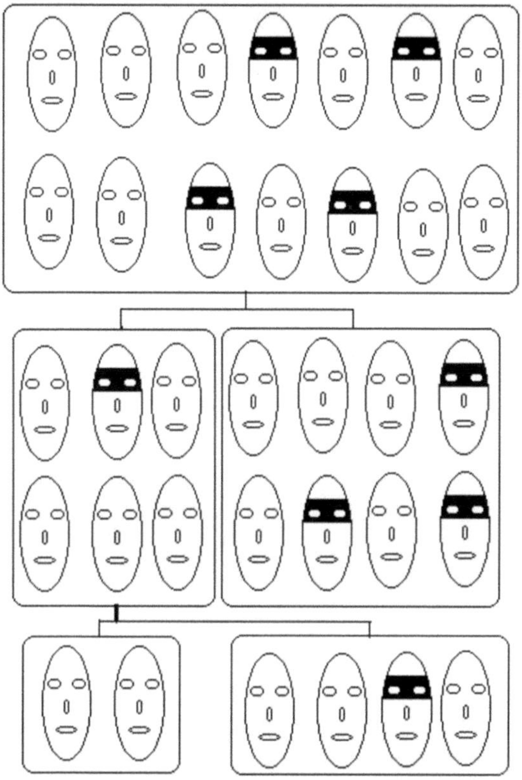

Figure 4: Two steps in Feige's protocol for a collective coin flipping. The agents enter a random room and the process continues with the room with the *least* number of agents.

Theorem 3.2 (Banzhaff). *For monotone Boolean functions* f, $II(f) \leq 1$.

The following result of Kahn, Kalai and Linial (KKL) has a central role in this section.

Theorem 3.3 (Kahn, Kalai and Linial, [119]).

$$\max_k I_k(f) \geq K\mathbf{P}(A) \log n/n\,.$$

Here, K is an absolute positive constant. In fact, $K = 1/2$ will do. Note that this theorem implies that when all individual influences are the same (e.g., when A is invariant under the induced action from a transitive permutation group on $[n]$), then the total influence is larger than $C \log n$.

For the ultimate sharpening of this result,

$$\sum_{k=1}^{n} I_k^p(A)/\big(\log(I_k^p(A)\big) \geq K\mathbf{P}_p(A)\,;$$

see Talagrand [131].

3.1.3 Russo's lemma and threshold intervals. For a monotone event $A \subset \Omega_n$ (i.e., $\mathcal{X}_A$ is a monotone function), let $\mathbf{P}_p(A)$ be the measure of A with respect to the product measure $\mathbf{P}_p$. Note that $\mathbf{P}_p(A)$ is a monotone function of p. Russo's lemma (see [115]) asserts that

$$\tfrac{d\mathbf{P}(A)}{dp} = I^p(f)\,.$$

Given a small real number $\epsilon > 0$, consider the *threshold interval* $[p_1, p_2]$ where $\mathbf{P}_{p_1}(A) = \epsilon$ and $\mathbf{P}_{p_2}(A) = 1 - \epsilon$. Denote by p_C the value so that $\mathbf{P}_{p_C}(A) = 1/2$, and call it the critical probability for the event A. A basic result of Bollobás and Thomasson asserts that the threshold interval is always bounded by a constant times the critical probability. By Russo's lemma, large total influence around the critical probability implies a short threshold interval.

3.1.4 Fourier-Walsh expansion. Consider a Boolean function f on $\Omega_n(1/2)$. Consider the Fourier-Walsh expansion $f = \sum_{S\subset[n]} \widehat{f}(S)u_S$, where $u_S = \prod_{i\in S} x_i$. (For the case of general p, see [131].)

Now, $\|f\|_2^2 = \mathbf{P}(A) = \sum \widehat{f}^2(S)$ (Parseval) and one can show that $I(f) = \sum_{S\subset[n]} \widehat{f}^2(S)|S|$. The result of Kahn, Kalai and Linial (Theorem 3.3) follows using certain hypercontractive estimates of Bonamie and Beckner [95], [101]. In recent years, harmonic analysis on Z_2^n plays an important role in extremal and probabilistic combinatorics and in complexity theory. Bonamie's and related hypercontractive estimates are crucial for the proofs of several of the results discussed in this section.

3.1.5 Noise sensitivity. The effect of random changes in the variables is called the *noise sensitivity* of f. A class of functions is *uniformly noise stable* if for every $\epsilon > 0$ there is $\delta > 0$ such that if you flip the values of δn randomly chosen variables, the correlation of the new value of f with the original value is at least $1 - \epsilon$. It can be shown that this is equivalent to the property that most of the 2-norm of f is concentrated in small Fourier coefficients (i.e. f is well approximated (in ℓ_2) by a small degree polynomial.) A sequence of Boolean functions f_m is (asymptotically) noise sensitive if for every $\delta > 0$ the correlation just defined tends to zero as m tends to infinity. These concepts were introduced by Benjamini, Kalai and

Schramm [97] (we will mention some of their results below) and (in a different language and motivations) by Tsirelson, who described remarkable relations and applications in [140], [128], [139] [Tsirelson].

Start a simple random walk (SRW) from a random point in A. How quickly will you converge to the uniform distribution on Ω_n? If $\mathbf{P}(A)$ is bounded away from 0 and 1, the answer is $O(n)$. For a sequence A_m of such events, the answer is $o(n)$ if and only if they are asymptotically noise sensitive.

3.2 Other general properties of subsets of the discrete cube.

3.2.1 Discrete isoperimetric inequalities. For $A \subset \Omega_n$ and $x \in A$ let $h(x)$ be the number of neighbors of x which are not in A. The vertex boundary of A denoted by $\partial_v(A)$ is the set of $x \in A$ with $h(x) > 0$.

Theorem 3.4. *Consider $A \subset \Omega_n$.*

1. *Given the size of A the vertex boundary of A is minimized for Hamming balls.*
2. *More generally, for every fixed $T > 0$, the number of the points whose distance from A is at least T is maximized when A is a Hamming ball. Therefore,*

$$\mathbf{P}\{x \in \Omega_n : d(x,A) > t\sqrt{n}\} \le \exp(-t^2/2)/\mathbf{P}(A)\,. \tag{3.1}$$

Talagrand [133], [135], [134] found several deep extensions and surprising applications of the isoperimetric inequality. A very useful sharpening is to replace $d(x,A)$ by the Euclidean distance $d_T(x,A)$ from x to the *convex hull* of the points in A (considered as points in $\mathbb{R}^d$).

Theorem 3.5 (Talagrand isoperimetric inequality).

$$\int_{\Omega_n(p)} \exp\left(\frac{1}{4}d_T^2(x,A)\right) \le \frac{1}{\mathbf{P}_p(A)}\,. \tag{3.2}$$

This inequality (in a dual formulation) is extremely useful for proving tail-estimates [133], [135], [130]. Inequalities (3.1), (3.2) manifest the "concentration of measure phenomenon". We use the term "hyperconcentration" in cases where asymptotically stronger inequalities are valid.

For $A \subset \Omega_n(p)$ let $B(A) = \int_{x\in A} \sqrt{h(x)}$. Talagrand proposed $B(A)$ as the "correct" notion of boundary for A and proved it in [132] (sharpening a result by Margulis).

Theorem 3.6 (Margulis-Talagrand [132]). *Let A be an event in $\Omega_n(p)$, $t = \mathbf{P}_p(A)$ and $C_p = \min(p,q)/\sqrt{pq}$, where $q = 1-p$. Then*

$$B(A) \ge KC_p t(1-t)\sqrt{\log(t(1-t))}\,.$$

For related results see Bobkov and Götze [102].

3.2.2 FKG and Shearer's lemma. The simplest form of the FKG inequality states that two monotone events in Ω_n have a nonnegative correlation. There are many extensions, variations and applications; see [115], [106].

Let J_i be subsets of $[n]$ so that every element in $[n]$ is covered at least r times. Let $A \subset \Omega_n$ be an event, $H(A)$ its entropy and $H(A_{|J_i})$ the entropy function of A conditioned on the set of coordinates J_i. Shearer's lemma (for some applications, see [109], [117], [113], [127]) asserts:

$$\sum H(A) \leq \frac{1}{m} \sum_{i=1}^{m} H(X_{|J_i}) .$$

(When $J_i = [n] \backslash \{i\}$, Shearer's inequality is essentially the Loomis-Whitney inequality (Theorem 3.1, Part 1).)

3.3 Advanced theorems on influences. The following theorem describes in loose terms the main advanced general theorems about influences. Recall that $II(f)$ is the sum of the squares of the influences.

Theorem 3.7 (Vague formulations). 1. [Talagrand], [136] *For two monotone events A and B, a large inner product of their influence vectors implies a stronger FKG-inequality.*

2. [Talagrand], [137] *A small value of $II(f)$ implies a large Margulis-Talagrand boundary.*

3. [Bourgain and Kalai], [105] *High symmetry implies large total influence.*

4. [Friedgut, [111]] *For a constant p, if the total influence is small, then the function is determined (approximately) by few coordinates.*

5. [Friedgut, [112] and Bourgain, [104]] *When p tends to zero, bounded total influence implies that f is "local".*

6. [Benjamini, Kalai and Schramm], [97] *For monotone events, sensitivity to noise is equivalent to having a small value of $II(f)$.*

7. [Talagrand], [138] *Small total influence implies hyperconcentration in the mean for the Hamming metric.*

8. [Benjamini, Kalai and Schramm], [98] *A small value of $II(f)$ implies hyperconcentration in terms of the second moment for Talagrand's metric.*

Van Vu and Jeong Kim [143], [121] proved hyperconcentration results for events which can be expressed by low-degree polynomials with small coefficients.

3.4 Two basic problems.

PROBLEM 3.1. 1. Characterize Boolean functions for which $I(f)$ is small. Can such functions be always approximated by small-depth small-size Boolean circuits? (See section 3.5.10 below.)

2. Find general conditions for $I(f) \geq n^{\beta}$. Is this always the case when there is a notion of a scaling limit [90], [139], [164]?

3. Characterize Boolean functions whose Fourier coefficients have a small support. (For example, functions for which most of the 2-norm is concentrated on a polynomial number in n of Fourier coefficients.)

PROBLEM 3.2. 1. Let f be a real function on the discrete cube. Under which (combinatorial) conditions can we guarantee that the distribution of f is close to a normal distribution?

2. Under which conditions can you expect a distribution which is more concentrated than normal? What kind of other distributions can you encounter from "natural" functions?

3.5 Examples.

We consider now some examples of real and Boolean functions defined on the discrete cube. Given a real function f, consider the Boolean functions $S_T(f) = sign(f(x) - T)$. (Here, $sign(x) = +1$ if $x \geq 0$ and $sign(x) = -1$, otherwise.) Consider also the important special case $M(f) = S_T(f)$, where T is the median value of f.

3.5.1 Weighted majority. For weights $w_1, \ldots, w_n$ consider the functions $f(x) = \sum w_i x_i$. The functions $S_T(f)$ are called *weighted majority functions*. The usual majority function (all w_i's are equal and $T = 0$) is a special case. For this example the influence of each variable is $C/\sqrt{n}$. Another special case is the function $f(x_1, x_2, \ldots, x_n) = x_1$. Here $I_f(1) = 1$ and $I_f(k) = 0$ for $k \neq 1$. Weighted majority functions are uniformly noise-stable [97]. The Talagrand isoperimetric inequality is sharp for these functions and so is the Margulis-Talagrand inequality. More general examples are low degree polynomials and their signs. See also Bruck and Smolensky, [107].

3.5.2 Majority of majorities, tribes, runs. Of course, we can partition the set into parts and consider the majority of majorities with various parameters. An important example is the tribe example of Ben-Or and Linial [99]: Divide the variables into "tribes" of size $\log n - \log\log n$ and set $f = 1$ iff there is a tribe all of whose variables has the value 1.

A related function is: Let f be the size of the longest run of '1's' and consider $M(f)$. For these functions the influence of every variable is $\Theta(\log n/n)$

(which is optimal by Theorem 3.3).

3.5.3 Recursive majorities. Let n be a power of 3. Divide the set of variables into three parts, divide each part into 3 parts, and continue $\log_3 n$ steps. Let f be the majority of majorities ... of majorities of the 3-element sets and let $A \subset \Omega_n$ be the corresponding event [99]. If you start a SRW from a random vector of A you will come very close to a uniform distribution on Ω_n in $n^{\log 2/\log 3}$ steps. Mossel and Peres pointed out that replacing 3 by a larger (but constant) t the number of steps required is reduced to $n^{1/2+\delta}$ where δ tends to 0 as t tends to infinity.

3.5.4 Random subsets of Ω_n and error correcting codes. Consider A, which is not necessarily monotone. The parity function $f(x_1, x_2, \dots, x_n) = x_1 x_2 \cdots x_n$ is the most noise-sensitive with $\mathbf{P}(A) = 1/2$.

When $\log_2 |A| = sn$, $0 < s < 1$, we can ask: how quickly can a SRW from a random uniformly chosen point of A reach a distribution that is almost uniform on Ω_n? A reasonable guess is that the best choice would be to take A itself to be random. This is closely related to the conjecture that the Gilbert-Varshamov bounds for codes are optimal (see [142]).

3.5.5 Cliques in graphs. This time, let the variables correspond to the $n = \binom{m}{2}$ edges of the complete graph with m vertices. Every assignment of values to the variables corresponds to a graph: the graph whose edges correspond to the variables with value 1. Let f be the size of the largest clique in this graph.

Again consider $M(f)$. (The median value is $\Theta(\log n)$.) In this example the influence of each variable is $\Theta(\log^2 n/n)$. The threshold interval for having a clique of size $a \log n$ is $\Theta(1/\log^2 n)$.

3.5.6 General graph properties. More generally, every property of graphs on m vertices (connected, planar, have diameter ≥ 3, etc.) corresponds to a Boolean function on this set of $\binom{m}{2}$ variables. Bourgain and Kalai [105] used their study of influences under symmetry (Theorem 3.7, Part 3) to show that for every graph property the threshold interval is of size at most $c_\tau 1/\log^{2-\tau} n$, for every $\tau > 0$.

The most important cases for random graph properties are when the critical probability p_C itself depends on n. Already, Erdős and Renyi in their paper which introduced random graph theory showed that many graph properties (such as connectivity) have sharp threshold behavior, namely the threshold interval is $o(p_C)$.

Friedgut's theorem (Theorem 3.7, Part 5, [112]) for graph properties can be stated as follows:

"If a graph property does not have a sharp threshold then it can be approximated by the property of having a subgraph from a given finite list".

For example, the property "to have a complete subgraph with 4 vertices" has a coarse threshold. But the "connectivity" property has a sharp threshold since it cannot be approximated by having a subgraph from a finite list. Friedgut's theorem has many important applications for showing sharp threshold behavior.

3.5.7 Random formulas, the 3-SAT problem. Consider a Boolean formula with n variables of the form

$$(x_{11} \vee \bar{x}_{12} \vee x_{13}) \wedge (\bar{x}_{21} \vee \bar{x}_{22} \vee x_{23}) \wedge \cdots \wedge (x_{m1} \vee x_{m2} \vee \bar{x}_{m3})\,.$$

It is an **NP**-complete problem to determine if the formula is satisfiable. The problem of satisfiability of random formulas has drawn a lot of attention recently. Friedgut [112] used his general criteria for bounded influence to show that there is a sharp threshold between values of m for which the formula is almost surely satisfiable and those for which it is almost surely unsatisfiable.

3.5.8 Crossing events in percolation. Consider the graph of the k by m grid in the plane and let the variables correspond to the edges (so n is roughly $2km$). Assume that the ratio k/m is bounded and bounded away from zero. Consider the event of having a left-to-right crossing. The influence of a variable is known in percolation theory as "the probability for an edge to be pivotal". $I(f)$ is an important "critical exponent" of percolation. It is conjectured that $I(f) = \Theta(n^{3/8})$.

A principal problem in percolation theory is that the probability for having a crossing tends to a limit if the ratio of k and m is fixed and k tends to infinity. This is a special case of the (conjectured) existence of a "scaling limit". In [97] it is shown that the crossing event is (asymptotically) sensitive to noise.

3.5.9 First passage percolation. A simple (but representative) variant of first passage percolation (FPP) can be described as follows. Consider the plane grid on which the length of each edge is assigned the value 0 with probability 1/4 (any $p < 1/2$ will do) and 1 with probability 3/4. We would like to know: What is be the distribution of the distance D from the origin to the point (m, m)? Here the distance is the minimum over all paths from the origin to (m, m) of the sum of lengths of edges in that path.

D is a real random variable defined on n Boolean variables where n corresponds to all the edges of the grid in some large region containing $(0,0)$ and (m,m). There are many exciting geometric and probabilistic problems concerning FPP.

Kesten showed that the variance of D is $O(m)$, and another simple proof with sub-Gaussian tail estimates was given by Talagrand [133] using his isoperimetric inequality. Influences and noise sensitivity (parts 5 and 7 of Theorem 3.7) are used in [98] to show that for FPP "on the torus" (and on a large class of symmetric graphs) the variance of D is actually $o(m)$. It is suggested by physicists that the variance behaves like $m^{2/3}$ and it is even speculated that the distribution of D and the large deviation properties are related to the distribution of the largest eigenvalues of certain random matrices; see [116].

3.5.10 Boolean functions expressed by bounded depth Boolean circuits. Consider a function described by a Boolean circuit (whose gates are: and, or, and negation) of depth c and size M. Linial, Mansour and Nisan showed that the Fourier coefficients of such functions decay exponentially [123]. Boppana [103] showed that for such a function f

$$I(f) \leq \log^{c-1} M .$$

Note that this bound applies to the examples of tribes, runs and cliques in graphs mentioned above (see also [107], [97]). An important example based on certain random depth-3 circuits was given by Ajtai and Linial [91].

3.5.11 Determinants, eigenvalues. (Suggested by I. Benjamini.) Consider an m by m matrix with entries ± 1 and consider its determinant D, the sign of D, its largest eigenvalue, etc. All these can be regarded as (nonmonotone) real functions on Ω_n with $n = m^2$. The corresponding Boolean functions are also of interest. What can be said about influences, noise-sensitivity and the Fourier-Walsh coefficients? Is there a notion of scaling limit?

3.5.12 Signed combination of vectors. Given n vectors $v_1, ..., v_n$ in some Euclidean space, with $\|v_i\|_2 \leq 1$, write $f(\epsilon_1, \dots, \epsilon_n) = \|\sum \epsilon_i v_i\|_\infty$. The distribution of the values of f is of great interest and, in particular, a conjecture of Komlos asserts that f always attains a value below some absolute constant.

3.5.13 Linear objective functions. Consider a convex d-polytope which is combinatorially equivalent to the d-dimensional cube and let f be given by the values on the vertices of a linear functional on $\mathbb{R}^d$. Such functions extend weighted majority functions considered above and are of great importance in the theory of linear programming (see section 5).

3.6 Some links and references. The standard source for probabilistic combinatorics is [93], its second edition will treat various further topics discussed here. Various papers in [106] that we will not cite individually are good references for some probability topics discussed in this section and in the next one.

Influences and collective coin flipping [99], [119], [91], [92], [122]; Talagrand's method and applications [133], [135], [130]; extremal combinatorics [54]; Boolean circuit complexity [93], [107], [103], [123]; random graphs [100], [93]; combinatorial problems on the discrete cube [129]; noise sensitivity [97], [98], [140], [128], [139], [141] [Tsirelson][Schramm]; FKG variations and applications [115], [106]; entropy and Shearer's lemma [109], [113], [117]; percolation and first passage percolation [125], [126], [106]; random matrices and combinatorial connections [116], [118]; Komlos conjecture [94];

Examples

It is not unusual that a single example or a very few shape an entire mathematical discipline. Examples are the Petersen graph, cyclic polytopes, the Fano plane, the prisoner dilemma, the real n-dimensional projective space and the group of two by two nonsingular matrices. And it seems that overall, we are short of examples. The methods for coming up with useful examples in mathematics (or counterexamples for commonly believed conjectures) are even less clear than the methods for proving mathematical statements.

4 Enumerative Combinatorics and Probability: Counting Trees and Random Trees

4.1 Kirchhoff, Cayley, Kasteleyn and Tutte. Cayley proved that the number of trees on n labeled vertices is n^{n-2}. There are many beautiful proofs which demonstrate various principal techniques in enumeration theory and, amazingly, new proofs are still being found. See [162], [84] [Stanley]. The matrix tree theorem, asserting that the number of spanning trees for a graph G is (essentially) the determinant of the Laplacian of G, is even earlier and is attributed to Kirchhoff. Of the vast knowledge on tree enumeration, let me mention two additional results. Kasteleyn (see [17]) found fundamental relations between the number of perfect matchings of planar graphs and tree enumeration. Tutte [166] considered graphs drawn on the 2-dimensional sphere, which has the property that the antipodal map induces an order-reversing bijection between the faces (or dimensions 0, 1 and 2) of this graph. In particular, the graph is self-dual (but this is not sufficient). He proved that the number of spanning trees of such a graph is a perfect square and the square root is equal to the number of self-dual trees.

4.2 Random spanning trees and loop erased random walk. It is now understood that there is an intimate connection between exact or approximate enumeration of certain objects and between the problem of finding (exactly or approximately) a random element among them. What can be said about a random spanning tree of a graph G and how can you generate such an object?

Broder [148] and Aldous [145] proposed a very simple way to generate a random spanning tree in a finite graph: Start a random walk and add to the tree all edges in the walk which do not close a circle when first traversed. Wilson [168] found a remarkable algorithm with superior performances and important theoretical aspects. His algorithm is related to the discussion that follows.

Lawler [157], attempting to understand a self-avoiding random walk, considered the following (different) model. Given two vertices x and y, start from x a random walk until reaching y and erase all loops. Pemantle showed that the distribution on $x - y$ paths in Lawler's model of loop erased random walk is precisely the distribution of paths between x and y in a random spanning tree.

4.3 Random spanning trees II. Rick Kenyon [152], [153], [154], [155] was recently able to compute critical exponents, to prove conformal invariance and to establish the existence of scaling limits for models based on random spanning trees of planar grids. His ingenious and involved proofs use Kasteleyn's correspondence between matchings and spanning trees for planar grids in a crucial way. Here are two of Kenyon's results:

- The expected length of the path between the origin and the boundary of the n by n grid in a random spanning tree of the standard plane grid (equivalently, the expected length of the loop erased random walk) is $\Theta(n^{4/3})$.
- Consider a smooth planar figure F and three points x, y and z on its boundary, and the (unique) meeting point u of these three points in a random spanning tree of a fine planar grid. In the limit, u is a

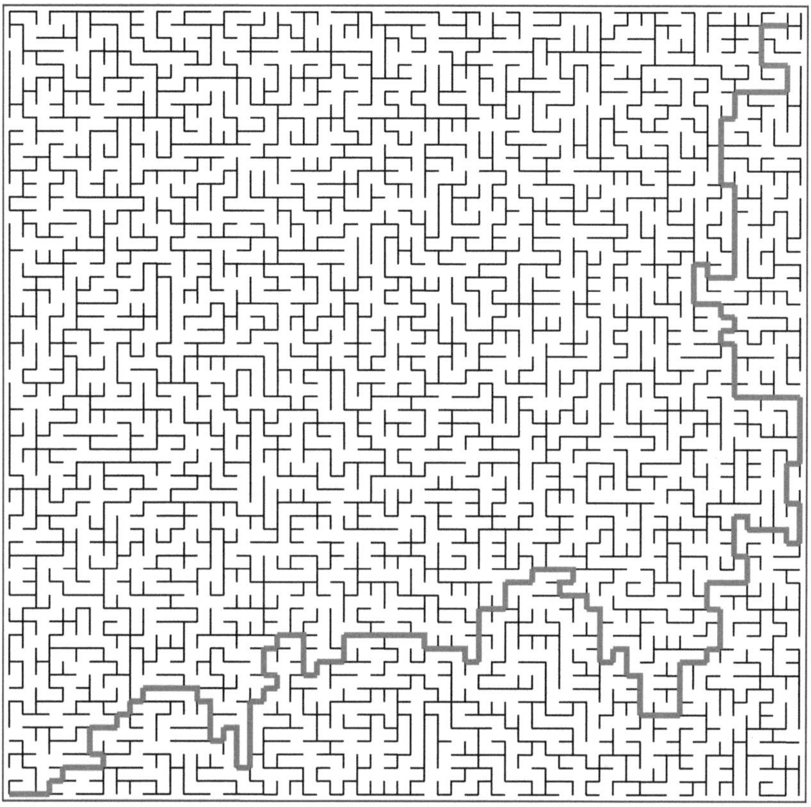

Figure 5: A random spanning tree and a loop erased random walk.

distribution on points inside F. Kenyon proved that this distribution is invariant under conformal maps of the plane.

4.4 Random spanning trees III. Oded Schramm [164] assumed a strong version of conformal invariance to show that various limiting objects for random spanning trees in planar grids are described by a certain stochastic process. Schramm constructed a remarkable class SLE_κ of stochastic processes depending on a parameter κ and showed that (assuming the existence of scaling limits and conformal invariance) these processes for $\kappa = 2$ describe the limiting paths of the loop erased random walk, and for $\kappa = 6$ the scaling limit of critical percolation cluster boundaries. For $\kappa = 8$ they are related to random Peano curves arising from random spanning trees, and for $\kappa = 4$ it is speculated that they describe the domino difference model (introduced and studied by Kenyon). This established surprising connections between objects which previously seemed different.

Here is a short description of Schramm's construction in his own words: "Consider a path γ in the closed unit disk $\mathbb{U}$ from the boundary to 0, which does not cross itself (and does not contain a nontrivial arc on the unit circle). Consider an initial arc β of γ. Let $q(\beta)$ be the endpoint of β which is not the initial point of γ. By Riemann's mapping theorem, there is a unique conformal map $g = g_\beta : \mathbb{U} - \beta \to \mathbb{U}$ normalized by $g(0) = 0$ and $g'(0) > 0$. Set $t(\beta) = -\log g'(0)$. The map g extends continuously to the closure of $\mathbb{U} - \beta$. In particular, $\delta = g(q(\beta))$ is a well defined point on the unit circle. We may think of β, and hence of δ as functions of t, $\delta : [0, \infty) \to \partial\mathbb{U}$. Loewner's slit mapping theorem shows that the collection of maps $g_\beta = g_t$ can be reconstructed from the function $\delta(t)$, by solving a differential equation with $\delta(t)$ as a parameter.

In the SLE_κ process, we take $\delta(t) = B(\kappa t)$, where κ is a constant and B is Brownian motion on the unit circle. By Loewner's theorem (and its extensions) this gives sufficient information to reconstruct g_t, hence γ."

Lawler, Schramm and Werner [159] related these objects to planar Brownian motion. Using conformal invariance which is known for Brownian motions and some earlier results of Lawler and Werner on the 'universality' of certain critical exponents for a large class of processes they managed to compute critical exponents anticipated by Duplantier-Kwon and Mandelbrot for Brownian motion in the plane.

These developments are a wonderful symphony of probabilistic, geometric, analytical and combinatorial reasoning.

4.5 Higher dimensions. Only a small fraction of the charm and importance of trees survives when we consider higher dimensional acyclic complexes. Yet in some contexts (e.g., buildings) such generalizations are useful. When it comes to tree enumeration it turns out that Cayley's formula does extend easily with a little twist: The weighted sum of $\mathbb{Q}$-acyclic k-dimensional acyclic complexes on n vertices with a complete $(k-1)$-dimensional skeleton is $n^{\binom{n-2}{k}}$, where the weight of a complex K is $|H_{k-1}(K)^2|$. Thus, for $n = 6, k = 2$ the formula gives 6^6 and there is a single type of complex, which is counted more than once (4 times): the 6-vertex triangulation of the real projective plane (Figure 3). The proof relies on extending the matrix-tree theorem and identifying the eigenvalues of certain Laplacians.

Very soon, as n grows, the weights in this formula become much larger than the number of summands. See Kalai [151] and Adin [144]. Analogs for Kasteleyn's and for Tutte's theorems mentioned above are expected but not known. Kenyon suggested that an appropriate extension of Kasteleyn's theorem to subcomplexes of the 3-dimensional grid may be useful in extending some of his explicit computations of critical exponents to 3 dimensions.

4.6 Haiman's diagonal harmonics. Consider the graded polynomial ring $H_n = \mathbb{C}[x_1, \dots, x_n, y_1, \dots y_n]/I$, where I is the ideal generated by all polynomials in the x_i's and y_i's which are invariant under the diagonal action of the symmetric group S_n on the variables.

Haiman [150], based on experimentation with Macaulay [158], conjectured that the dimension of H_n is $(n+1)^{n-1}$, the number of labeled trees on $n+1$ variables.

Further experimentation showed that finer statistics on the grading of H_n (the total degree or the degrees according to the variables x_i alone) turned out to be related to classical enumeration statistics of trees. Moreover, using the well known correspondence between trees and parking functions, it was possible to identify the representation of the symmetric group S_n on H_n. Haiman's conjecture turned out to be related to exciting issues in algebraic geometry and representation theory. Many tried to solve it, but very recently Haiman himself proved his conjecture [Haiman].

4.7 Some links and references. Tree enumeration [162], [84], [156]; trees and probability [161], [106]; random spanning trees and forests [146], [160], [163], [106]; enumerative combinatorics [83], [84], [147], [Zeilberger], [Stanley]; Schramm's processes, Brownian motion [164], [159], [167] Haiman's conjectures [150], [Haiman]; Macaulay [158], [Bayer]; approximate enumeration [183], [186]; eigenvalues of Laplacians of high dimensional complexes [Reiner].

Our community

Like musicians who can enjoy and understand complicated scores even in a world with no sound, for us mathematics is a source of delight, excitement and even controversy. This is hard to share with nonmathematicians.

In our small world we should seek new ways for communication and interaction and for the right balance between competition and solidarity, criticism and empathy, exclusion and inclusion.

5 Optimization: How Good is the Simplex Algorithm?

5.1 The simplex method. Linear programming is the problem of maximizing a linear objective function $\phi = b_1x_1 + b_2x_2 + \cdots + b_dx_d$ subject to n linear inequalities in the d variables $x_1, x_2, \ldots, x_d$. Linear programming and Danzig's simplex algorithm are among the most important applications of mathematics in the 20th century. The set Q of solutions for the inequalities is called the *feasible polyhedron.* The maximum of ϕ on Q (if ϕ is bounded on Q) is attained at a face of Q and, in particular, there is a vertex v for which the maximum is attained.

The *simplex algorithm* is a method to solve a linear programming problem by repeatedly moving from one vertex v to an adjacent vertex w of the feasible polyhedra so that in each step the value of the objective function is increased. The specific way to choose w, given v, is called the *pivot rule.* Klee and Minty [182], and later others, showed that various standard pivot rules may require exponentially many pivot steps in the worst case. On the other hand, Khachiyan [178], [190], [177] showed that linear programming is in **P** (namely, there is a polynomial-time algorithm for linear programming), and various authors (for several notions of a random linear programming problem) showed that the simplex algorithm requires a polynomial number of pivot steps on average [171].

It is an important open problem to decide if there is a variant of the simplex algorithm whose worst-case behavior takes polynomial time. Related problems are to show that there is a polynomial time algorithm for linear programming "over the reals" or to find a "strongly polynomial" algorithm (over the rationals) in the usual Turing model. These problems of fundamental importance in both complexity and optimization lie (as the problems of factorization of integers and graph isomorphism) in the grey area between **P** and **NP**, where surprising new results and insights are expected.

In the early 90s randomized subexponential simplex algorithms were found independently by Kalai [180] and by Matoušek, Sharir and Welzl [185]. These development as well as a related result on the diameter of graphs of polytopes [179] apply in a very general abstract combinatorial context. While it is possible that further improvements and even polynomial simplex algorithms can be found in this generality, the main point I would like to raise in this section is: can geometry help?

5.2 The combinatorics of linear programming. The following property is crucial:

- (local=global) ϕ takes its maximum on a vertex v of Q if and only if v is a *local maximum*, i.e., $\phi(v) \geq \phi(w)$ for every neighbor vertex w of v.

An ordering of the vertices of a polytope Q is an *abstract objective function* if the property (local=global) holds for Q and all its faces. (It is possible to consider also abstract linear programming problems in even greater generality; see [188], [179].)

For our purposes, there is no loss of generality in assuming that the feasible polyhedron Q is bounded, that the linear objective function is generic and that the problem is nondegenerate which, in other words, says that Q is a simple polytope: every vertex has d neighboring vertices, or equivalently every vertex belongs to exactly d facets.

The combinatorics of the problem involves the combinatorial structure of the polytope Q and the combinatorics of the total ordering on the vertices of Q induced by the linear objective function ϕ. The combinatorics of Q is relevant because the diameter of the graph of Q gives a lower bound on the number of pivot steps needed. However, I feel that the main difficulty lies in the combinatorics of objective functions and that understanding the case where Q is combinatorially isomorphic to the d-dimensional cube will go a long way towards solving the problem.

An interesting connection with section 2 is the following: Given a simple polytope Q (think about the cube!) and a linear objective function ϕ, we can consider for every vertex v its degree $deg(v)$, which is the number of neighbors u of v with $\phi(u) > \phi(v)$. It turns out that the distribution of the degrees of vertices does not depend on the objective function. The number of vertices of degree k is precisely the h-number h_k (considered in section 2). This applies to all abstract objective functions and, in fact, characterizes this class of orderings of the vertices of Q. This implies at once the Dehn–Sommerville relations $h_k = h_{d-k}$ (replace ϕ with $-\phi$), and relation (2.4).

The product measure $\mathbf{P}_p$ on the discrete cube (section 3) has a natural extension in this context for arbitrary simple polytopes Q by assigning to a vertex v the measure $(1-p)^{d-deg(v)}p^{deg(v)}$.

The combinatorics of the full arrangement of hyperplanes which correspond to points which satisfy one of the inequalities as equality, is also of great importance. In particular one can read from the hyperplane arrangement (and the ordering given by ϕ on its vertices) the combinatorics the *dual* linear programming problem.

5.3 Some classes of pivot rules. We will consider now various pivot strategies of a combinatorial nature for the simplex algorithm. Given a vertex v of the feasible polyhedron Q our aim is to reach the vertex of Q at the top.

1. (`"Random improving edge"`) Choose a random improving neighbor.
2. (`"Random facet"`) Choose a random facet F containing v and run the algorithm inside F until reaching the optimal vertex in that facet. Then repeat. The expected number of pivot steps is bounded above by $\exp(C\sqrt{n \log d})$ [185], [180].
3. (`"Universal instructions"`) For every vertex v of Q you are given an ordering of its neighbors. In each step of the algorithm you move to the first neighbor on the list which improves ϕ.

3(R). (`"Random universal instructions"`) The same as 3, except that you have a distribution on such an assignment of orderings and you choose a random one. (The best known randomized variant of the simplex algorithm in terms of worst case expected behavior is of this type [179].)

4. (Taking into account how well the neighbors improve) The same as 3 (or 3(R)) except the ordering in the vertex can depend on the ordering between the ϕ values of the neighbors of v. One of the earliest pivot rules using the most improving neighbor is, of course, a special case.
5. (Adaptive rule) The same as 3 (or 3(R) or 4 or 4(R)) except the ordering in the vertex can depend on the history of the algorithm up to this stage.
6. (`"Random walk"`) The basic operation is: Given two vertices v and u with $\phi(v) \geq \phi(u)$ start from a vertex v and perform a simple random walk on all vertices whose ϕ value is larger than $\phi(u)$. Then update v and u. (Note: here, we allow steps which decrease the value of the objective function.)

All these rules apply for abstract objective functions and for some of them (1-3(R)) all that is needed is the relation between the values of the objective function on neighboring vertices.

There are important pivot rules which depend in a stronger way on the geometry and cannot be described in terms of abstract objective functions. An important practical example is to always choose the steepest edge (towards the objective function) leaving the vertex. An important pivot rule from a theoretical point of view is the shadow-boundary rule, which is based on projecting the polyhedra on a two-dimensional space. For this (and only this) rule, it is known that the simplex algorithm is polynomial for an average problem [171].

5.4 Can geometry help? I: There are few objective functions. The first information about geometric objective functions is that there are not too many of them. It is not difficult to see that the number of abstract objective functions on the vertices of the d-cube is larger than 2^{2^d}. (For example, consider the orientation of edges of the discrete cube which corresponds to the linear objective function $x_1 + x_2 + \cdots + x_d$. Then consider a matching between the vertices of the two middle levels of the cubes. You may switch the orientation of any subset of edges of this matching and the property (local=global) will still hold (see, [89]).)

In sharp contrast,

Theorem 5.1. *The number of different possible geometric objective functions of the discrete d-dimensional cube is at most* $\exp(d^3 \log d)$. *Moreover, the number of combinatorial types of pairs (Q, ϕ), where Q is a d-polytope with n facets and ϕ a linear objective function on Q, is at most* $\exp(Kd^2 n \log n)$.

The proof of this theorem relies on a theorem of Warren from real algebraic geometry on the number of sign patterns determined by a set of polynomials. The basic argument is due to Goodman and Pollack, [176], [169]. (The result applies to the number of combinatorial types of arrangements of n hyperplanes in $\mathbb{R}^d$ and the orderings given by a linear objective function on the vertices of the arrangement.)

A hereditary class of abstract objective functions on discrete cubes is *small* if the number of distinct orderings on the vertices of the d-cube is only exponential in a polynomial of d.

PROBLEM 5.1. 1. Given a small hereditary class of abstract objective functions, is it possible to prove the existence of an oblivious process such as `"Random universal instructions"` which works well for all abstract objective functions in this class?

2. Is it possible, to improve algorithms like `"Random facet"` by "learning" information from the low level recursion, in a way which will dramatically reduce the running time for *all* small hereditary classes of abstract objective functions?

The second option we raised, that of learning, needs further explanation. The situation seems to resemble what is called a bounded VC-dimension [192], which is an extremely useful condition for various combinatorial and algorithmic applications. Specifically, it is possible that applying an algorithm (such as `"Random facet"`) where the choices of the algorithm higher up in the recursion are positively correlated with successful choices in lower levels will perform well on *every* small class of abstract objective functions.

A case worthy of study (even experimentally) is a remarkable class of abstract objective functions on the discrete d-cube described by Matoušek [184]. For an average abstract objective function in this class the expected number of pivot steps needed for `"Random facet"` is indeed $\exp(K\sqrt{d})$. This class is hereditary and small: the number of abstract objective functions of this class on the d-cube is exponential in d^2.

Gärtner [174] showed that for Matoušek's class the geometry does help as `"Random facet"` itself requires only a quadratic number of pivot steps for geometric objective functions in Matoušek's class. Gärtner used only the conditions for geometric objective functions on the 3-dimensional faces.

Abstract objective functions on 3-dimensional polytopes were recently characterized by Mihalisin and Klee [181]. The orientations of graphs of 3-dimensional polytopes induced by a geometric objective function (each edge is oriented from the smaller vertex to the larger) are precisely the acyclic orientations with a unique source and a unique sink which admits three disjoint independent monotone paths from the source to the sink.

5.5 Can geometry help? II: How to distinguish geometric objective functions. We should also try to find concrete ways to distinguish between abstract and geometric objective functions. Consider the set A_v of all vertices u in the feasible polyhedron with $\phi(u) \geq \phi(v)$. Which properties does A_v satisfy?

A recent important result by Morris and Sinclair [186] shows that for the standard cube and weighted majority functions, A_v has (mild) expansion properties which implies that a SRW on A_v reaches an approximate-uniform distribution in n^8 steps.

For an arbitrary linear programming problem, can the graph induced on A_v always be divided into a polynomial (in d and n) number of parts, each of which is (mildly) expanding? (Mildly expanding means that the expansion constant is $1/p(d,n)$ for some polynomial $p(d,n)$.) Since every ordering of the vertices of the cube given by a monotone real function is an abstract objective function and since monotone subsets of Ω_n may have dismal expansion properties, the geometry must be used.

Monotone functions on Ω_n are not a real challenge for the simplex algorithm as any (improving) pivot rule will reach the maximum vertex in at most n steps. But understanding the class of orderings and events of the type A_v which come from an objective function on a polytope combinatorially isomorphic to the n-cube is nevertheless of much interest also from the point of view of complexity theory. Going back to the examples in section 3 it seems that repeated weighted majorities of various types can be realized.

Perhaps the strongest known result towards a strongly polynomial algorithm for linear programming is by Eva Tardos [191]. Fixing the feasible polyhedron (in fact, only the matrix of coefficients), she described a strongly polynomial algorithm independent of the objective function. Proving this result with a simplex type algorithm (even randomized) will already be a major achievement (see [173] for a special case).

Added in proof. Spielman and Teng have recently made substantial progress towards a polynomial (not yet strongly polynomial) version of the simplex algorithm. They showed that the shadow-boundary pivot rule needs a polynomial number of steps for a small random gaussian perturbation of a linear programming problem.

5.6 Some links and references. Linear programming [190], average case behavior of the simplex method [171], randomized pivot rules [172], [173], [189], [175], [180], [188], [185], computational geometry [187] algorithmic applications of random walks [183], [186], the diameter problem for polyhedra [179].

Applications and Expectations

Judging from the conference on 'Vision in Mathematics', mathematicians have a strong desire to interact and influence other sciences, as well as technology, industry, and even economic life. The trends towards isolationism have been reversed, and there is a greater understanding of the subtleties of applying mathematics to and interacting with other fields.

The general public knows very vaguely what mathematicians do. At the same time people have quite clear expectations from mathematics. More than other sciences, and certainly much more than law, religion, politics and the media, mathematics is expected to be rigorous and precise in telling its uninteresting, irrelevant, and uncomforting truths. The value of mathematics for society goes far beyond its applications through technology; it is indeed a pillar of human culture. After a century of amazing technological development along with rising influence of pseudosciences and the occult, this value is important.

Acknowledgement. I would like to thank many friends who made useful comments and corrections to earlier versions of this paper and Nati Linial, Yuval Peres, Edna Ullman–Margalit and Günter Ziegler for helpful discussions.

Rodica Simion, to whose memory this paper is dedicated, was Professor of Mathematics at George Washington University until her untimely death on January 7, 2000.

Personal web sites

Alon,	http://www.math.tau.ac.il/~noga/
Barvinok,	http://www.math.lsa.umich.edu/~barvinok/
Bayer,	http://www.math.columbia.edu/~bayer/vita.html
Benjamini,	http://www.wisdom.weizmann.ac.il/~itai/
Friedgut,	http://www.ma.huji.ac.il/~ehudf/
Haiman,	http://math.ucsd.edu/~mhaiman/
Herzog,	http://www.uni-essen.de/~mat300/
Kalai,	http://www.ma.huji.ac.il/~kalai/
Kenyon,	http://topo.math.u-psud.fr/~kenyon/
Lovasz,	http://www.cs.yale.edu/~lovasz/
Lyons,	http://php.indiana.edu/~rdlyons/
Matoušek,	http://www.ms.mff.cuni.cz/~matousek
Peres,	http://www.ma.huji.ac.il/~peres
Reiner,	http://www.math.umn.edu/~reiner/
Schramm,	http://www.wisdom.weizmann.ac.il/~schramm/
Stanley,	http://www-math.mit.edu/~rstan/
Talagrand,	http://felix.proba.jussieu.fr/users/talagran/
Thomas,	http://www.math.gatech.edu/~thomas/
Tsirelson,	http://www.math.tau.ac.il/~tsirel/
Ziegler,	http://www.math.TU-Berlin.DE/~ziegler/
Zeilberger,	http://www.math.temple.edu/~zeilberg/

References

[An Invitation to Tverberg's Theorem]

[1] J.E. Goodman, J. O'Rourke (eds.), Handbook of Discrete and Computational Geometry, CRC Press, Boca Raton, New York, 1997.

[2] P. Gruber, J. Wills (eds.), Handbook of Convex Geometry, North-Holland/Elsevier Science Publishers, Amsterdam, 1993.

[3] R. Graham, M. Grötschel, L. Lovasz (eds.), Handbook in Combinatorics, North-Holland/Elsevier, Amsterdam, 1995.

[4] M. Aigner, G.M. Ziegler, Proofs from THE BOOK, Springer-Verlag, Heidelberg, 1998.

[5] N. Alon, M. Tarsi, Colorings and orientations of graphs, Combinatorica 12 (1992), 125–134.

[6] A. Andrzejak, E. Welzl, Halving point sets, Proceedings of the International Congress of Mathematicians, Vol. III (Berlin 1998), Doc. Math. Extra Vol. III (1998), 471–478.

[7] I. Bárány, Z. Füredi, L. Lovász, On the number of halving planes, Combinatorica 10 (1990), 175–183.

[8] I. Bárány, S. Shlosman, A. Szücs, On a topological generalization of a theorem of Tverberg, J. London Math. Soc. 23 (1981), 158–164.

[9] A. Björner, Topological methods, in: [3], 1819–1872.

[10] A. Björner, M. Las Vergnas, B. Sturmfels, N. White, G.M. Ziegler, Oriented Matroids, Encyclopedia of Mathematics 46, Cambridge University Press, Cambridge, 1993.

[11] B. Bollobás, Modern Graph Theory, Springer, New York, 1998.

[12] L. Danzer, B. Grünbaum, V. Klee, Helly's theorem and its relatives, in: "Convexity" (V. Klee, ed.), Proc. Symposia in Pure Mathematics, Vol. VII, Amer. Math. Soc., Providence, RI (1963), 101–180.

[13] T.K. Dey, Improved bounds on planar k-sets and related problems, Discrete Comput. Geom. 19 (1998), 373–382.

[14] J. Eckhoff, Helly, Radon, and Carathèodory type theorems, in : [2], 389–448.

[15] J.E. Goodman, R. Pollack, Allowable sequences and order types in discrete and computational geometry, in "New Trends in Discrete and Computational Geometry" (J. Pach, ed.), Springer, Berlin (1993), 103–134.

[16] T.R. Jensen, B. Toft, Graph Coloring Problems, Wiley, New York, 1995.

[17] L. Lovász, M.D. Plummer, Matching Theory, North-Holland, Amsterdam, 1986.

[18] L. Lovász, A. Schrijver, A Borsuk theorem for antipodal links and a spectral characterization of linklessly embeddable graphs, Proceedings of the American Mathematical Society 126 (1998), 1275–1285.

[19] D. Larman, On sets projectively equivalent to the vertices of a convex polytope, Bull. London Math. Soc. 4 (1972), 6–12.

[20] J. Matoušek, Piercing and selection theorems in convexity, Lecture notes available at [Matoušek].

[21] J. Matoušek, On combinatorial applications of topology, Lecture notes available at [Matoušek].

[22] J. Pach, P.K. Agarwal, Combinatorial Geometry, Wiley-Interscience, New York, 1995.

[23] J.P. Roudneff, Partitions of points into simplices with k-dimensional intersection, Part I: Tverberg's conic theorem, preprint.

[24] J.P. Roudneff, Partitions of points into simplices with k-dimensional intersection, Part II: Proof of Reay's conjecture in dimensions 4 and 5, preprint.

[25] K. Sarkaria, Tverberg's theorem via number fields, Israel J. Math. 79 (1992), 317–320.

[26] M. Sharir, S. Smorodinsky, G. Tardos, An improved bound for k-sets in three dimensions, manuscript, 2000.

[27] H. Tverberg, A generalization of Radon's Theorem, J. London Math. Soc. 41 (1966), 123–128.

[28] R. Thomas, An update on the four-color theorem, Notices Amer. Math. Soc. 45 (1998), 848–859.

[29] G. Toth, On sets with many k-sets, manuscript, 1999.

[30] A.Yu. Volovikov, On a topological generalization of the Tverberg theorem, Math. Notes 59:3 (1996), 324–326; Translation of Mat. Zametki 59:3 (1996), 454–456.

[31] A. Vučić, R. Živaljevic, Note on a conjecture by Seirksma, Disc. Comp. Geometry 9 (1993), 339–349.

[32] R. Živaljević, Topological methods, in [1].

[33] R. Živaljević, S. Vrećica, The colored Tverberg's problem and complexes of injective functions, J. Combin. Theory, Ser. A 61 (1992), 309–318.

[34] R. Živaljević, S. Vrećica, New cases of the colored Tverberg Theorem, Contemp Math. 178 (1994), 325–334.

[How general is the upper bound theorem?]

[35] R. Adin, A new cubical h-vector, Discrete Math. 157 (1996), 3–14.

[36] N. Alon, G. Kalai, A simple proof of the upper bound theorem, European J. Combin. 6 (1985), 211–214.

[37] A. Aramova, J. Herzog, Almost regular sequences and Betti numbers, Amer. J. Math. 122 (2000), 689–719.

[38] E. Babson, L.J. Billera, C. Chan, Neighborly cubical spheres and a cubical lower bound conjecture, Israel J. Math. 102 (1997), 297–315.

[39] V.V. Batyrev, L.A. Borisov, Mirror duality and string theoretic Hodge numbers, Invent. Math. 126 (1996), 183–203.

[40] M.M. Bayer, Equidecomposable and weakly neighborly polytopes, Israel J. Math. 81 (1993), 301–320.

[41] M.M. Bayer, An upper bound theorem for rational polytopes, J. Com-

binat. Theory, Ser. A 83 (1998), 141–145.

[42] M.M. Bayer, C. Lee, Convex polytopes, in: [2], Vol. A, 485–534.

[43] L.J. Billera, A. Björner, Face numbers of polytopes and complexes, in: [1], 291–310.

[44] L.J. Billera, A. Björner, C. Greene, R.E. Simion, R.P. Stanley (eds.), New perspectives in algebraic combinatorics, Mathematical Sciences Research Institute Publications, 38. Cambridge University Press, Cambridge, 1999

[45] L.J. Billera, C.W. Lee, A proof of the sufficiency of McMullen's conditions for f-vectors of simplicial convex polytopes, J. Combinat. Theory, Ser. A 31 (1981), 237–255.

[46] T. Bisztriczky, et al. (eds.), Polytopes: Abstract, Convex and Computational Kluwer, Dordrecht, 1994, 155–172.

[47] A. Björner, Partial unimodality for f-vectors of simplicial polytopes and spheres, Contemporary Math. 178 (1994), 45–54.

[48] A. Björner, G. Kalai, On f-vectors and homology, in "Combinatorial Mathematics" (G.S. Bloom, R.L. Graham, J. Malkevitch, eds.), Proc. NY Academy of Science 555 (1989), 63–80.

[49] B. Bollobás, Extremal Graph Theory, Academic Press, London, 1978.

[50] A. Borel (ed.), Intersection Homology, Birkhauser, 1984.

[51] T. Braden, R.D. MacPherson, Intersection homology of toric varieties and a conjecture of Kalai, Commentarii Math. Helv. 74 (1999), 442–455.

[52] R. Charney, M. Davis, The Euler characteristic of a nonpositively curved, piecwise Euclidean manifolds, Pacific J. Math. 171 (1995), 117–137.

[53] D. Eisenbud, Commutative Algebra with a View toward Algebraic Geometry, Springer-Verlag, Berlin, 1995.

[54] K. Engel, Sperner theory. Encyclopedia of Mathematics and its Applications, 65. Cambridge University Press, Cambridge, 1997.

[55] Z. Furedi, Turan type problems, in "Surveys in Combinatorics, 1991" (A.D. Keedwell, ed.), Cambridge University Press, Cambridge, 1991.

[56] M. Goresky, R. MacPherson, Intersection homology theory, Topology 19 (1980), 136–162.

[57] M. Green, Generic initial ideals, in "Six Lectures on Commutative Algebra" (J. Elias, J.M. Giral, R.M. Mir'o-Roig, S. Zarzuela, eds.), Birkhäuser, 1998, 119–185.

[58] B. Grünbaum, Convex Polytopes, Wiley Interscience, London, 1967.

[59] B. Grünbaum, Polytopes, graphs and complexes, Bulletin Amer. Math. Soc. 97 (1970), 1131–1201.

[60] M. Joswig, G.M. Ziegler, Neighborly cubical polytopes, in "Grünbaum Festschrift" (G. Kalai, V. Klee, eds.), Discrete Comput. Geometry, to ap-

pear.

[61] G. Kalai, Rigidity and the lower bound theorem, Invent. Math. 88 (1987), 125–151.

[62] G. Kalai, The diameter of graphs of convex polytopes and f-vector theory, in "Applied Geometry and Discrete Mathematics, The Klee Festschrift," DIMACS Series in Discrete Mathematics and Computer Science 4 (1991), 387–411.

[63] G. Kalai, Some aspects in the combinatorial theory of convex polytopes, in [46], 205–230.

[64] B. Kind, P. Kleinschmidt, Cohen–Macauley-Komplexe und ihre Parametrisierung, Math. Z. 167 (1979), 173–179.

[65] V. Klee, A combinatorial analogue of Poincaré's duality theorem, Canad. J. Math. 16 (1964), 517–531.

[66] V. Klee, P. Kleinschmidt, Convex polytopes and related complexes, in [3], 875–917.

[67] U.H. Kortenkamp, Every simplicial polytope with at most $d+4$ vertices is a quotient of a neighborly polytope, Discrete & Comput. Geometry 18 (1997), 455–462.

[68] W. Kühnel, Tight Polyhedral Submanifolds and Tight Triangulations, Springer LNM 1612, Berlin, 1995.

[69] W. Kühnel, T. Banchoff, The 9-vertex complex projective plane, Math. Intelligencer 5 (1983), 11–22.

[70] W. Kühnel, G. Lassman, The unique 3-neighborly 4 manifold with 9 vertices, J. Combin. Th. A 35 (1983), 173–184.

[71] N.C. Leung, V. Reiner, The signature of a toric variety, preprint, 1999.

[72] L. Lovász, M. Simonovits, On the number of complete subgraphs of a graph. II, Studies in Pure Mathematics, Birkhäuser, Basel-Boston, Mass. (1983), 459–495.

[73] P. McMullen, The maximum numbers of faces of a convex polytopes, Matematika 17 (1970), 179–184.

[74] I. Novik, Upper Bound Theorems for homology manifolds, Israel J. Math. 108 (1998), 45–82.

[75] G. Ringel, Map Color Theorem, Springer, Berlin, 1974.

[76] I. Shemer, Neighborly polytopes, Isr. J. Math. 43 (1982), 291–314.

[77] P. Siegel, Witt spaces: A geometric cycle theory for KO homology at odd primes, Amer. J. Math. 105 (1983), 1067–1105.

[78] R. Simion, Convex polytopes and enumeration, Adv. in Appl. Math. 18:2 (1997), 149–180.

[79] R.P. Stanley, The upper bound conjecture and Cohen–Macaulay rings, Studies in Applied Math. 54 (1975), 135–142.

[80] R.P. Stanley, The number of faces of simplicial convex polytopes, Ad-

vances Math. 35 (1980), 236–238.

[81] R.P. STANLEY, Subdivisions and local h-vectors, J. Amer. Math. Soc. 5 (1992), 805–851.

[82] R.P. STANLEY, Combinatorics and Commutative Algebra, Second Edition, Birkhäuser, Boston, 1994.

[83] R.P. STANLEY, Enumerative Combinatorics, Vol. 1, Corrected reprint of the 1986 original, Cambridge Studies in Advanced Mathematics 49, Cambridge University Press, Cambridge, 1997.

[84] R.P. STANLEY, Enumerative Combinatorics, Vol. 2, Cambridge Studies in Advanced Mathematics 62, Cambridge University Press, Cambridge, 1999.

[85] R.P. STANLEY, Positivity problems and conjectures in algebraic combinatorics, in "Mathematics: Frontiers and Perspectives" (V. Arnold, et al., eds.), American Mathematical Society, Providence, RI, 2000, 295–319.

[86] V.A. VASSILIEV, On r-neighbourly submanifolds in $\mathbb{R}^M$, Topol. Methods Nonlinear Anal. 11 (1998), 273–281.

[87] U. WAGNER, E. WELZL, Origin-Embracing distributions or a continuous analogue of the upper bound theorem, preprint, 1999.

[88] E. WELZL, Entering and Leaving j-Facets, Discrete & Computational Geometry, to appear, available at http://www.inf.ethz.ch/~emo/ps-files/ELjFacets.ps.

[89] G.M. ZIEGLER, Lectures on Polytopes, Graduate Texts in Math. 152, Springer-Verlag, New York, 1995.

[Influence of variables on Boolean functions]

[90] M. AIZENMAN, A. BURCHARD, C. NEWMAN, D. WILSON, Scaling limits for minimal and random spanning trees in two dimensions, Random Structures Algorithms 15 (1999), 319–367.

[91] M. AJTAI, N. LINIAL, The influence of large coalitions, Combinatorica 13 (1993), 129–145.

[92] N. ALON, M. NAOR, Coin-flipping games immune against linear-sized coalitions, SIAM J. Comput. 22 (1993), 403–417.

[93] N. ALON, J. SPENCER, The Probabilistic Method, Wiley & Sons, Inc., New York, 1992.

[94] W. BANASZCZYK, Balancing vectors and Gaussian measures of n-dimensional convex bodies. Random Structures Algorithms 12 (1998), 351–360.

[95] W. BECKNER, Inequalities in Fourier analysis, Annals of Math. 102 (1975), 159–182.

[96] I. BENJAMINI, O. SCHRAMM, Percolation beyond $\mathbb{Z}^d$, many questions and a few answers, Electronic Commun. Probab. 1 (1996), 71–82.

[97] I. Benjamini, G. Kalai, O. Schramm, Noise sensitivity of Boolean functions and applications to percolation, to appear, Publ. IHES.

[98] I. Benjamini, G. Kalai, O. Schramm, Noise sensitivity, concentration of measure and first passage percolation, in preparation.

[99] M. Ben-Or, N. Linial, Collective coin flipping, in "Randomness and Computation" (S. Micali, ed.), Academic Press, New York (1990), 91–115.

[100] B. Bollobaás, Random Graphs, Academic Press, Inc., London–New York, 1985.

[101] A. Bonami, Etude des coefficients Fourier des fonctiones de $L^p(G)$, Ann. Inst. Fourier 20 (1970), 335–402.

[102] S. Bobkov, F. Götze, Discrete isoperimetric and Poincaré-type inequalities, Probab. Theory Related Fields 114 (1999), 245–277.

[103] R. Boppana, The average sensitivity of bounded depth circuits, Inform. Process. Lett. 63 (1997), 257–261.

[104] J. Bourgain, Appendix to [112].

[105] J. Bourgain, G. Kalai, Influences of variables and threshold intervals under group symmetries, Geom. Funct. Anal. 7 (1997), 438–461.

[106] M. Bramson, R. Durrett, eds., Perplexing problems in probability, Progr. Probab. 44, Birkhäuser Boston, Boston, MA, 1999.

[107] J. Bruck, R. Smolensky, Polynomial threshold functions, AC^0 functions, and spectral norms, SIAM J. Comput. 21 (1992), 33–42.

[108] J.T. Chayes, L. Chayes, D.S. Fisher, T. Spencer, Finite-size scaling and correlation length for disordered systems, Phys. Rev. Lett. 57 (1986), 2999–3002.

[109] F.R.K. Chung, P. Frankl, R. Graham, J.B. Shearer, Some intersection theorems for ordered sets and graphs, J. Combinatorial Th. Ser. A. 48 (1986), 23–37.

[110] U. Feige, Noncryptographic Selection Protocols, FOCS 1999: 142–153.

[111] E. Friedgut, Boolean functions with low average sensitivity, Combinatorica 18 (1998), 27–35.

[112] E. Friedgut, Necessary and sufficient conditions for sharp thresholds of graphs properties and the k-sat problem, J. Amer. Math. Soc. 12 (1999), 1017–1054.

[113] E. Friedgut, J. Kahn, On the number of copies of one hypergraph in another, Israel J. Math. 105 (1998), 251–256.

[114] E. Friedgut, G. Kalai, Every monotone graph property has a sharp threshold, Proc. Amer. Math. Soc. 124 (1996), 2993–3002.

[115] G. Grimmett, Percolation, Springer-Verlag, Berlin, 1989.

[116] K. Johansson, Transversal fluctuations for increasing subsequences on the plane, math. PR/9910146 http://front.math.ucdavis.edu.

[117] J. Kahn, An entropy approach to the hard-core model on bipartite graphs,

preprint, 1999.

[118] J. KAHN, J. KOMLÓS, E. SZEMERÉDI, On the probability that a random ± 1-matrix is singular, J. Amer. Math. Soc. (1995), 223–240.

[119] J. KAHN, G. KALAI, N. LINIAL, The influence of variables on Boolean functions, Proc. 29-th Ann. Symp. on Foundations of Comp. Sci. (1988), 68–80.

[120] H. KESTEN, On the speed of convergence in first passage percolation, Ann. Applied Prob. (1993), 296–338.

[121] J.H. KIM, V.H. VU, Concentration of polynomials andits applications, preprint.

[122] N. LINIAL, Game-theoretic aspects of computing, in "Handbook in Game Theory" (R.J. Aumann, S. Hart, eds.), Elsevier Science, 1994.

[123] N. LINIAL, Y. MANSOUR, N. NISAN, Constant depth circuits, Fourier transform, and learnability, J. Assoc. Comput. Mach. 40 (1993), 607–620.

[124] R.J. MCELIECE, The Theory of Information and Coding, Addison-Wesley, London, 1977.

[125] C.M. NEWMAN, M. PIZA, Divergence of shape fluctuations in two dimensions, Ann. Probab. 23 (1995), 977–1005.

[126] R. PEMANTLE, Y. PERES, Planar first-passage percolation times are not tight, in "Probability and Phase Transition", Cambridge (1993), 261–264.

[127] J. RADHAKRISHNAN, An entropy proof of Bregman's theorem, J. Combinatorial Th. Ser. A 77 (1997), 161–164.

[128] O. SCHRAMM, B. TSIRELSON, Trees, not cubes: hypercontractivity, cosiness, and noise stability, Electronic Communication in Probability 4 (1999), 39–49.

[129] M. SAKS, Slicing the hypercube, Surveys in Combinatorics, 1993 (Keele), London Math. Soc. Lecture Note Ser. 187, Cabridge Univ. Press, Cambridge (1993), 211–255.

[130] M. STEELE, Probability theory and combinatorial optimization, CBMS-NSF Regional Conference Series in Applied Mathematics, 69, Society for Industrial and Applied Mathematics (SIAM), Philadelphia, PA, 1997.

[131] M. TALAGRAND, On Russo's approximate zero-one law, Ann. of Prob. 22 (1994), 1576–1587.

[132] M. TALAGRAND, Isoperimetry, logarithmic Sobolev inequalities on the discrete cube, and Margulis' graph connectivity theorem. Geom. and Funct. Anal. 3 (1993), 295–314.

[133] M. TALAGRAND, Concentration of measure and isoperimetric inequalities in product spaces, Publ. IHES 81 (1995), 73–205.

[134] M. TALAGRAND, A new look at independence, Ann. Probab. 24 (1996), 1–34.

[135] M. TALAGRAND, New concentration inequalities in product spaces, Invent.

Math. 126 (1996), 505–563.
[136] M. TALAGRAND, How much are increasing sets positively correlated? Combinatorica 16 (1996), 243–258.
[137] M. TALAGRAND, On boundaries and influences, Combinatorica 17 (1997), 275–285.
[138] M. TALAGRAND, On influence and concentration, Israel J. Math. 111 (1999), 275–284.
[139] B. TSIRELSON, Scaling limit of Fourier-Walsh coefficients (a framework), math.PR/9903121.
[140] B. TSIRELSON, Noise sensitivity on continuous products: an answer to an old question of J. Feldman, math.PR/9907011.
[141] B. TSIRELSON, Toward stochastic analysis beyond the white noise, Draft, 1999.
[142] J.H. VAN LINT, Introduction to Coding Theory, Third edition, Graduate Texts in Mathematics, 86. Springer-Verlag, Berlin, 1999.
[143] V.H. VU, On the concentration of multi-variable polynomials with small expectations, preprint.

[Counting trees and random trees]

[144] R. ADIN, Counting colorful multi-dimensional trees, Combinatorica 12 (1992), 247–260.
[145] D. ALDOUS, The random walk construction of uniform random spanning trees and uniform labelled trees, SIAM J. Disc. Math. 3 (1990), 450–465.
[146] I. BENJAMINI, R. LYONS, Y. PERES, O. SCHRAMM, Uniform spanning forests, Ann. Probab., to appear.
[147] D. BRESSOUD, Proofs and confirmations, The story of the alternating sign matrix conjecture, MAA Spectrum, Mathematical Association of America, Washington, DC; Cambridge University Press, Cambridge, 1999.
[148] A. BRODER, Generating random spanning trees, in "30th Annual Symp. Foundations Computer Sci.", IEEE, New York (1989), 442–447.
[149] P.G. DOYLE, J.L. SNELL, Random Walks and Electric Networks, Mathematical Assoc. of America, Washington, DC, 1984.
[150] M. HAIMAN, Conjectures on the quotient ring by diagonal invariants, J. Algebraic Combin. 3 (1994), 17–76.
[151] G. KALAI, Enumeration of $\mathbb{Q}$-Acyclic simplicial complexes, Israel J. Math. 45 (1983), 337–350.
[152] R.W. KENYON, Conformal invariance of domino tiling, Ann. Prob., to appear.
[153] R.W. KENYON, Long-range properties of spanning trees in $\mathbb{Z}^2$, J. Math. Phys. Probabilistic Techniques in Equilibrium and Nonequilibrium Statis-

tical Physics. 41 (2000), 1338–1363

[154] R.W. KENYON, Dominos and the Gaussian free field, preprint.

[155] R.W. KENYON, The asymptotic determinant of the discrete Laplacian, Acta Math., to appear.

[156] D. KNUTH, The Art of Computer Programming, Vol 1: Fundamental algorithms. Addison-Wesley Publishing Co., Reading, 1975.

[157] G. LAWLER, Loop-erased self-avoiding random walk in two and three dimensions, J. Stat. Phys. 50 (1988), 91–108.

[158] D. BAYER, M. STILLMAN, MACAULAY, [Bayer].

[159] G. LAWLER, O. SCHRAMM, W. WERNER, Values of Brownian intersection exponents I: Half-plane exponents, Acta Math., to appear.

[160] R. LYONS, A bird's-eye view of uniform spanning trees and forests, in "Microsurveys in Discrete Probability (Princeton, NJ, 1997), (D. Aldous, J. Propp, eds.), Amer. Math. Soc., Providence, RI (1998), 135–162.

[161] R. LYONS, Y. PERES, Probability on Trees and Networks, Cambridge University Press, in preparation.

[162] J.W. MOON, Counting labelled trees, Canadian Mathematical Congress, Montreal, Que. 1970.

[163] R. PEMANTLE, Uniform random spanning trees, in "Topics in Contemporary Probability and its Applications" (J.L. Snell, ed.), CRC Press, Boca Raton (1994), 1–54.

[164] O. SCHRAMM, Scaling limits of loop-erased random walks and uniform spanning trees, Israel J. Math. 118 (2000), 221–282.

[165] R. STANLEY, Hyperplane arrangements, parking functions, and tree inversions, in "Mathematical Essays in Honor of Gian-Carlo Rota" (B. Sagan, R. Stanley, eds.), Birkhauser, Boston/Basel/Berlin (1998), 359–375.

[166] W.T. TUTTE, On the spanning trees of self-dual maps, Annals of the NY Academy of Sciences 319 (1979), 540–548.

[167] W. WERNER, Critical exponents, conformal invariance and planar Browninan motion, Proc. 3rd ECM Barcelona 2000, Birkhauser, to appear.

[168] D. WILSON, Generating random spanning trees more quickly than the cover time, 1996 ACM Sympos. Theory of Computing, 296–302.

[How good is the simplex algorithm?]

[169] N. ALON, Tools from higher algebra in [3], 1749–1783.

[170] N. AMENTA, G.M. ZIEGLER, Deformed products and maximal shadows, in "Advances in Discrete and Computational Geometry" (B. Chazelle et al., eds.) Contemporary Mathematics 223, Amer. Math. Soc., Providence, RI (1998), 57–90.

[171] K.H. BORGWARDT, The Simplex Method, a Probabilistic Analysis, Algo-

rithms and Combinatorics 1, Springer-Verlag, Berlin, 1987.

[172] K.L. Clarkson, A Las Vegas Algorithm for linear programming when the dimension is small, J. ACM 42:2 (1995), 488–499.

[173] M.E. Dyer, A. Frieze, Random walks, totally unimodular matrices and a randomized dual simplex algorithm, Mathematical Programming 64 (1994), 1–16.

[174] B. Gärtner, Combinatorial linear programming: Geometry can help, Proc. 2nd Workshop "Randomization and Approximation Techniques in Computer Science" (RANDOM), Springer Lecture Notes in Computer Science 1518 (1998), 82–96.

[175] B. Gärtner, M. Henk and G.M. Ziegler, Randomized simplex algorithms on Klee–Minty cubes, Combinatorica 18 (1998), 349–372.

[176] J. Goodman, R. Pollack, There are asymptotically far fewer polytopes than we thought, Bull. Amer. Math. Soc. 14 (1986), 127–129.

[177] M. Grötschel, L. Lovàsz, A. Schrijver, Geometric Algorithms and Combinatorial Optimization, Springer-Verlag, Berlin–New York, 1988.

[178] L.G. Khachiyan, A polynomial algorithm in linear programming, Soviet Math. Doklady 20 (1979), 191–194.

[179] G. Kalai, Linear programming, the simplex algorithm and simple polytopes. Lectures on Mathematical Programming (ismp97) (Lausanne, 1997), Math. Programming, Ser. B 79 (1997), 217–233.

[180] G. Kalai, A subexponential randomized simplex algorithm, Proceedings of the 24-th Ann. ACM Symp. on the Theory of Computing, ACM Press, Victoria (1992), 475–482.

[181] J. Mihalisin, V. Klee, Convex and linear orientations of polytopal graphs, Discrete Comp. Geom., to appear.

[182] V. Klee, G.J. Minty, How good is the simplex algorithm, in "Inequalities III" (O. Shisha, ed.), Academic Press, New York (1972), 159–175.

[183] L. Lovász, Randomized algorithms in combinatorial optimization, in "Combinatorial Optimization" (New Brunswick, NJ, 1992–1993), DIMACS Ser. Discrete Math. Theoret. Comput. Sci. 20, Amer. Math. Soc., Providence, RI (1995), 153–179.

[184] J. Matoušek, Lower bounds for a subexponential optimization algorithm, Random Structures and Algorithms 5 (1994), 591–607.

[185] J. Matoušek, M. Sharir, E. Welzl, A subexponential bound for linear programming, Proc. 8-th Annual Symp. on Computational Geometry (1992), 1–8.

[186] B. Morris, A. Sinclair, Random walks on truncated cubes and sampling 0-1 napsack solutions, preprint.

[187] K. Mulmuley, Computational Geometry, An Introduction Through Randomized Algorithms, Prentice-Hall, Englewoods Cliffs, 1994.

[188] M. Sharir, E. Welzl, A Combinatorial bound for linear programming and related problems, Proc. 9-th Symp. Theor. Aspects of Computer Science, Lecture Notes in Comp. Sci. 577 (1992), 569–579.

[189] R. Seidel, Small-dimensional linear programming and convex hulls made easy, Discrete Comput. Geom. 6 (1991), 423–434.

[190] A. Schrijver, Theory of Linear and Integer Programming, Wiley-Interscience, New York, 1986.

[191] E. Tardos, A strongly polynomial algorithm to solve combinatorial linear programs, Oper. Res. 34 (1986), 250–256.

[192] V.N. Vapnik, Statistical Learning Theory, Adaptive and Learning Systems for Signal Processing, Communications, and Control, John Wiley & Sons, Inc., New York, 1998.

Gil Kalai, The Hebrew University, Givat Ram, Jerusalem 91904, Israel
kalai@math.huji.ac.il http://www.ma.huji.ac.il/~kalai/

GAFA, Geom. funct. anal.
Special Volume – GAFA2000, 792 – 815
1016-443X/00/S20792-24 $ 1.50+0.20/0

GAFA Geometric And Functional Analysis

TOPICS IN ASYMPTOTIC GEOMETRIC ANALYSIS

V. Milman

1 About the Subject

The term "Geometric Analysis" is a recent one but it has quickly become fashionable and is used too often and for very different mathematics. So, we added the adjective "asymptotic" to be more specific, and we will first explain the subject of this talk.

Some people use these words ("Geometric Analysis") to describe a mechanical use of Analysis in Geometry and vice versa. This is not what I mean. I mean the influence on the conceptual level: how the point of view of Analysis, or better to say, Functional Analysis, changes the questions we ask and the problems we consider in Geometry. This happens through the isomorphic study of geometric objects, instead of the isometric (or ε-isometric) as is usual in Geometry.

One of the main tools in Analysis is to reduce information (and complexity) through the use of inequalities, the main art being not to lose information during this process. We will show that an approach in this spirit may be applied also to some geometric problems.

Before I describe the subject let me mention that it was born inside Functional Analysis. In fact, during the 20th century, Functional Analysis was, perhaps, the leading force in the developing of Analysis and many different and rich directions in Mathematics germinated from Functional Analysis. This topic (Asymptotic Geometric Analysis) is just the last of them. Despite many achievements, and this article is about these achievements, I see it just as the start. It will "meld" with Complexity Theory, Statistical Physics and, even more than we see it already, with Geometry, and will become a significant part of the Mathematical Landscape.

In an article on such an occasion, it is difficult to avoid making a few brief historical remarks about Functional Analysis. If one were to ask what

Supported in part by a BSF grant, the Israel Science Foundation, and the Minkowski Minerva Center for Geometry at TAU.

Functional Analysis is, the short, and essentially correct answer should be: the study of infinite dimensional spaces ("dim $= \infty$"). But I am not sure it was so for some of the founders of Functional Analysis. The study of finite, but very high dimensional spaces and their asymptotic properties when dimension increases was, perhaps, one of the starting points for some of them. We see this in Minkowski, who for the purposes of Analytic Number Theory considered n-dimensional space from a geometric point of view (and before him, as well as long after him, geometry had to be 2 or 3-dimensional – see, e.g., the works of Blaschke); P. Lèvy [Lè]; von Neumann. I was very surprised when I discovered[1] the following paragraph, sounding so modern, in an old (1942) paper by von Neumann ([vN]; section 4 of the Introduction):

(Below H_n is an n-dimensional euclidean space and M_n is the space of $n \times n$ matrices.)

> "Our interest will be concentrated in this note on the conditions in H_n and M_n – mainly M_n – when n is *finite*, but *very great*. This is an approach to the study of the infinite dimensional, which differs essentially from the usual one. The usual approach consists in studying an actually infinite dimensional unitary space, i.e. the Hilbert space H, as done loc. cit. We wish to investigate instead the *asymptotic* behavior of H_n and M_n for finite n, when $n \to \infty$.
>
> We think that the latter approach has been unjustifiably neglected, as compared with the former one. It is certainly not contained in it,.....
>
> Since Hilbert space H was conceived as a limiting case of the H_n for $n \to \infty$, we feel that such a study is necessary in order to clarify to what extend H is or is not the only possible limiting case. Indeed we think that it is not, and that investigations on operator rings by F.J. Murray and the author show that other limiting cases exist, which under many aspects are more natural ones.
>
> Our present investigations originated in fact mainly from the desire to solve certain questions..... We hope, however, that the reader will find that they also have an interest of their own, mainly in the sense indicated above: as a study of the asymptotic behavior of H_n and M_n for finite n, when $n \to \infty$."

We know now how far the asymptotic approach, the theory of finite, but infinitely increasing dimension, can go, and how deep it is. However, we also know that Asymptotic Theory is very different from Infinite Dimensional Theory and a lot of the phenomena of infinite dimensional spaces cannot be

---

[1]In fact, I. Halperin gave me this reference after my colloquium talk in Toronto at the beginning of the '90s

understood through finite dimensional asymptotics (see the recent surveys [O] and [M]).

But, at the beginning of the century, the study of infinite dimensional spaces seemed to be a good "approximation" for asymptotic finite dimensional problems. The achievements and success of the theory were indeed great and the asymptotic approach was ignored and forgotten. Only in the 60s, and more intensively in the 70s, in a search for new approaches to the accumulation of unsolved problems of purely infinite dimensional nature, did we return to the study of finite dimensional spaces, with clear emphasis on *asymptotic* behavior when dimension increases to infinity.

But, before this was understood, we had two classical theories. One was the study of finite dimensional spaces, and I will add, of small dimension. And "small" is any but fixed dimension (it could be 3 or 10 or n, but fixed). This is the subject of geometry, if problems are geometric (or Linear Algebra, say). And another is infinite dimensional theory which is Functional Analysis.

However, as I already hinted, there is a third possibility, a theory "in between".

My first goal is to show that such a theory exists, not just by presenting this or that result, but by demonstrating a new and very different intuition it represents and the new structures it helps to discover. These are, I believe, the most valuable tests of new directions and theories. In the main body of the paper we will avoid terminology and notion born inside the theory to avoid unnecessary terminological "pollution". It will, hopefully, aid the reader in building his/her own intuition on the subject.

I will also describe and discuss one extremely powerful tool, a concept, the so called Concentration Phenomenon, which is used throughout the theory. My choice of this subject is twofold.

First, I believe it is the right tool, not only for Geometric Analysis, but also in all other problems where large parametric families are involved and under study (be this Complexity Theory or Statistical Physics). In fact, it is already intensively used in Asymptotic problems of discrete mathematics including Complexity (see, e.g. [AS], [MoR]).

But secondly, we will see that it is a typical example in the spirit of the theory and a consequence of an isomorphic view of classical geometric problems (more precisely – isoperimetric type problems). So, a change of point of view, a change of concept, made possible the discovery of the new phenomenon.

2 Essay on Asymptotic Theory

2.1 Entropy and volume behavior in high dimension. What is the expected diversity in high dimensional spaces and what happens when dimension increases to infinity? Do we expect some form of order with increasing dimension? Our intuition says "No", we do not expect ordered behavior and we expect exponential increase in diversity. What is the source of this intuition, which, as we will see, is not exactly correct? Most probably, exponential expansion (by dimension) of volume which leads to exponential expansion of entropy (coverings) is the source of this intuition. However, there is a compensation factor, a concentration of measure around "thin" sets, which always accompanies high parametric spaces, and this phenomenon is not taken into account by our intuition. At least it was not taken into account till very recently. We will discuss this phenomenon later, but now let us go through an outcome of the compensation effect.

Some notation. For a set $K \subset \mathbb{R}^n$ we write $|K|$ for its volume and let $N(K,T) = \min\{N \text{ such that } \exists\{x_i\}_1^N \subset \mathbb{R}^n \text{ and } K \subset \bigcup_{i=1}^N (x_i + T)\}$ be the covering number of K by T. Also, $K + T = \{x + y \mid x \in K,\ y \in T\}$ is the Minkowski sum of sets K and T and $f \sim \varphi$ means universal equivalence: there is an absolute number C, independent of dimension or anything else, such that $\varphi/C \le f \le C \cdot \varphi$. We use $|\cdot|$ for the standard euclidean norm in $\mathbb{R}^n$ and $D = \{x \mid |x| \le 1\}$ is the standard euclidean ball in $\mathbb{R}^n$.

To compare the geometry around the same volume level, we introduce the *volume radius*, v.r. $K = (|K|/|D|)^{1/n}$. Also $d(K,T) = \inf\{a{\cdot}b \mid K \subset aT$ and $T \subset bK\}$ is a multiplicative geometric distance.

Let us now measure differences in the geometry of two convex bodies K and T (in $\mathbb{R}^n$), of around the same v.r., by measuring $N(K,T)$ and $N(T,K)$. Of course, both numbers may be extremely, unboundedly large, not because of different geometry but for "incidental" reasons, because of a wrong *position* of comparison; consider, for example, D and a very "long" almost degenerated ellipsoid $\mathcal{E} = uD$, $u \in \mathrm{SL}_n$. So, we will measure

$$M(K,T) = \inf\left\{N(uK,T) \cdot N(T,uK) \mid u \in \mathrm{SL}_n\right\}.$$

Naturally, this number must behave exponentially by n for some bodies K and T and even for geometrically very similar bodies. Say, $M(2D,D)$ is C^n for some $C > 1$, and it is easy to build K of volume radius 1, $d(K,D) \le 2$, and covers $M(K,D)$ of the order C^n for $C > 1$. However, $\inf\{d(uK,T) \mid u \in \mathrm{SL}_n\}$ may be as large as $\sim n$ (as E. Gluskin showed in the 80's) for two centrally symmetric bodies (and, perhaps, significantly larger for not symmetric; this question is still under study now). So, the expected number $M(K,T)$ for some K and T of the same v.r. is of the order $\exp(cn\log n)$. However the actual result is the best imaginary:

There is a universal constant C, s.t. $M(K,T) \le e^{C\cdot n}$ for any two convex bodies of the same volume radius (i.e. $|K| = |T|$) in $\mathbb{R}^n$ (see [Mi3], [Pi2], [MiP2], for references, more details and extensions).

It follows from this fact that a family of ellipsoids $\{uD\}$, $u \in \mathrm{GL}_n$ is rich enough to represent (on a rough scale) ANY convex body in volume computations.

Precisely:

Let 0 be the barycenter of K. There is an ellipsoid $\mathcal{E}_K$ such that for any other convex body T with 0 being its barycenter

$$\text{v.r.}(K+T) \sim \text{v.r.}(\mathcal{E}_K+T)\,, \quad \text{v.r.}\, K\cap T \sim \text{v.r.}\, \mathcal{E}_K \cap T\,,$$
$$N(K,T)^{1/n} \sim N(\mathcal{E}_K,T)^{1/n} \quad \text{and} \quad N(T,K)^{1/n} \sim N(T,\mathcal{E}_K)^{1/n}\,.$$

(See the same references as above.)

2.1.1 Technical remarks. Convexity is not the main reason and source of the entropy and volume behavior we mentioned above. One may consider sets which are very far from being convex (but we will concentrate on centrally symmetric and star-shape bodies; the reason will be explained in 2.2.1). The natural notion here is quasi-convexity or p-convexity and the meaning is that our set is far from a "hedgehog-type" shape. See [BBP] for such an extension. We return to these extensions in the next subsection.

2.2 "Isomorphic" geometry. The previous discussion (and the results accompanying it) belongs to a broad direction and a large body of results which are called Geometric Inequalities. Brunn-Minkowski type inequalities, Alexandrov-Fenchel inequalities, and the many inequalities accompanying them, involving volumes, volume radius and so on, are today the classical examples of Geometric Inequalities which mostly "group" around Convexity Theory (see [BuZ], [S] on this subject). However, the difference which one immediately observes between the subject as described in these monographs and the direction outlined above is that the inequalities (equivalences) we described involved some universal constants. These are, of course, geometric type inequalities, but in "isomorphic" form.

Looking back, I see three periods in the development of what one may call "classical quantitative" convexity theory and what some experts, Rolf Schneider for one, rightly call "The Brunn-Minkowski Theory". The first was in the 1840s through Cauchy's and Steiner's works; then about fifty years later Brunn-Minkowski's inequality was discovered, and Minkowski shaped our view of convexity theory for most of the twentieth century. However, again about fifty years later, in 1936, inequalities proved by Alexandrov and also announced by Fenchel marked a whole new period of very deep finite dimensional convexity theory through the study of (exact) geometric inequalities of many different types, description of extremal cases, approximation theory and so on. I recommend R. Schneider's monograph [S] for a very detailed account on these three periods of the theory.

It took another fifty years after Alexandrov-Fenchel's works, in the middle of the 1980s, for the next step in the development of quantitative convexity theory to crystallize.

We now consider geometric problems from a functional analytic point of view. Consequently, typical for geometry "isometric" problems and views are substituted by "isomorphic" ones. This became possible with the *asymptotic* approach (with respect to dimension increasing to infinity) to the study of high dimensional convex bodies.

In particular, the meaning of geometric inequalities was extended. *Isomorphic* geometric inequalities which involved exact dependence of constants on dimension naturally accompany the asymptotic theory of convex bodies.

However, Isomorphic Geometry is not limited to what we called Isomorphic Geometric Inequalities, i.e. to some numerical inequalities.

Also the shape of any convex body is quickly regularized to an ellipsoid, but again, in an isomorphic sense. Let us first see some fact in its exact form:

Theorem. *There is a universal number C such that, for every convex compact body $K \subset \mathbb{R}^n$ with 0 being its barycenter and $|K| = |D|$, one may choose a* position *$\hat{K} = uK$, $u \in SL_n$, such that*

$$\exists u, v \in O(n) \text{ and } P := \hat{K} \cap u\hat{K},$$
$$Q := P + vP \underset{C}{\sim} D$$

(i.e. $\frac{1}{C} D \subset Q \subset CD$).

ε-version: $\exists v_i$, $v_i \in O(n)$, $i = 1, ..., [C/\varepsilon^2] = t$

$$\frac{1}{t} \sum_{1}^{t} v_i P \underset{1+\varepsilon}{\sim} rD$$

(dimension free estimate).

Actually, this theorem represents a reformulation in geometric language of a fact first formulated in the language of Functional Analysis and which is proved using methods of Functional Analysis (see the surveys [Mi3] or [LiM] for further references, and [MiP2] for a recent extension to the not centrally symmetric case). It is very natural from the point of view of Functional Analysis to consider a family of spaces uniformly isomorphic to euclidean spaces. There is a corresponding geometric notion of an *Isomorphic ellipsoid* which sounds less natural, but we regularly meet it in Asymptotic Theory. A family of convex bodies $\{K_\alpha\}$ of infinitely increasing dimension represents an "isomorphic ellipsoid" if there is a constant C and a family of ellipsoids $\{\mathcal{E}_\alpha\}$ such that $\mathcal{E}_\alpha \subset K_\alpha \subset C\mathcal{E}_\alpha$ for every K_α in the family, i.e. $d(K_\alpha, \mathcal{E}_\alpha) \leq C$.

So, we naturally derive *isomorphic* geometric objects as a family of objects in different spaces of increasing dimension. By *isomorphic* geometric properties, we mean a common property for such a family, and by *isomorphic* study we mean asymptotic behavior determined by some parameter (tending to infinite dimension, or another parameter having the meaning of dimension).

2.2.1 Remarks. 1. In many of the results we have described till now, convexity is not a crucial assumption (but *high* dimension is). One may substitute convexity by another, very weak condition. Define a convolution of two bodies $K \,\square\, T = \bigcup_{\substack{x\in K\\ y\in T}} [x,y]$, were $[x,y]$ is the interval joining points x and y.

Let K be a *centrally symmetric quasi-convex* star body, i.e.

(i) $tK \subset K$, $0 \leq t \leq 1$ (star-body condition); and $K = -K$;
(ii) $K \,\square\, K \subset C \cdot K$.

(We say that K is C-quasi-convex. Example: $K_{\ell_p^n}$, the unit ball of ℓ_p^n space for $0 < p < 1$, is a $C(p)$-quasi-convex for some constant $C(p)$ independent of dimension n.) Clearly, $C = 1$ iff K is convex.

Then the results of 2.1 and the theorem above are true for quasi-convex bodies with constant C in these facts depending on the quasi-convexity constant *only* (see [Mi3]).

2. The meaning of the extension of the theory to the quasi-convex setting may be explained by the following example: Let $f \in C(S^{n-1})$, $f > 0$ and even. Consider its homogeneous extension to $\mathbb{R}^n$ by $\hat{f}(x) = |x| f(x/|x|)$, for $x \neq 0$. Let $K = \{x \in \mathbb{R}^n \mid \hat{f}(x) \leq 1\}$ be a C-quasi-convex body. This means that the infimum convolution $(\hat{f} \,\square\, \hat{f})(x) = \inf\{\hat{f}(x_1) + \hat{f}(x_2) \mid x_1 + x_2 = x\}$ is $1/C$-equivalent to $\hat{f}$, i.e. $\hat{f} \,\square\, \hat{f} \geq \frac{1}{C}\hat{f}$. We say

that such an f belongs to the class $C(\alpha)$, for $\alpha = 1/C$. This condition should be seen as a different type (the global one) of "smoothness" type conditions. An application of the theorem in §2.2 states that, for $1 \geq \alpha > 0$, there is a constant K_α depending on $\alpha > 0$ only, such that for any n and $f(x)$ defined on S^{n-1} and from the class $C(\alpha)$, there are three operators $u_1, u_2, u_3 \subset SL_n$ such that the convolution $\psi(x) = \varphi(x) \square \varphi(u_3 x)$ for the function $\varphi(x) = \hat{f}(u_1 x) + \hat{f}(u_2 x)$ is K_α-equivalant to the euclidean norm x in $\mathbb{R}^n : r|x| \leq \psi(x) \leq K_\alpha r|x|$ (for some $r > 0$ and any $x \in \mathbb{R}^n$).

2.3 More asymptotic ideology (examples of *isomorphic* study). Next, we will see examples of asymptotic behavior which, in some cases, resemble very much phase transition type behavior in Statistical Physics while in some other cases look like thresholds in asymptotic combinatorial problems. We will also see that phase transition type behavior may be viewed in some problems as refining threshold type behavior. And I think this is a very promising observation. Needless to repeat that in our theory, we will have an isomorphic form of phase transitions. However, the thresholds are usually already described in Isomorphic form and, although we have different examples than, say, graph theory considers, there is no novelty in our point of view.

2.3.1 Example of phase transition; Local form. Now $A \subset \mathbb{R}^n$ is *any* bounded set, $d(A) =$ diameter A, and

$$w(A) = \int_{S^{n-1}} w(x) d\sigma(x)$$

is the mean-width; here $w(x) = sup_{y \in A}(x,y) - inf_{y \in A}(x,y)$. In Functional Analysis we use half the mean-width, $M^* := M^*(A) = \frac{1}{2} w(A)$. Let

$$d_m(A) = \mathbb{E}\{d(P_E A) \mid P_E\text{-orthoproj. onto subspace } E \text{ and } \dim E = m\}.$$

Fact. *For $m \geq k^* = n\left(\frac{w(A)}{d(A)}\right)^2$*

$$d_m(A) \sim \sqrt{\tfrac{m}{n}}\, d(A)$$

and for $m \leq k^*$

$$d_m(A) \sim w(A) \quad (\text{so } d_m(A) \text{ is stabilized}).$$

We see a phase transition type behavior at $m = k^*$.

Moreover, for any $\theta > 0$ there is (small) $c(\theta) > 0$ such that for $m \leq c(\theta) k^*$ the set $P_E A$ is a (θM^*)-net for $M^* \cdot D(E)$ (and $P_E A \subset (1+\theta) M^* \cdot D(E)$).

REFERENCES/REMARKS. The above result is an obvious extension of the corresponding fact for convex centrally symmetric bodies, which is in turn an interpretation of an old estimate [Mi5] in Dvoretsky-type theorem and its exactness shown in [MiS2]. These theorems are dealing with euclidean sections (or, by duality, projections) of convex bodies.

This is the traditional interest of Geometric Functional analysis. The influence of Classical Convexity was to turn these achievements (we call "Local") to the study of the properties of the original convex sets. This study is now called the "*Global Theory*". Many results of "Local Theory" have their precise analogue in the Global form. I think every local result should have its global analogue, and the dictionary between two theories is clear. However, proofs should be often invented from the start.

2.3.2 Let us demonstrate the *global form* of the phase transition we saw before.

Denote $A_t = \frac{1}{t}\sum_1^t u_i A$, $u_i \in O(n)$ (this is the Minkowski sum); of course,A_t depends on $\{u_i\}$ but we do not specify this in the notation,

$$\overline{d}_t(A) := \min \left\{ d\left(\frac{1}{t}\sum_1^t u_i A\right) \,\middle|\, u_i \in O(n) \right\}.$$

Fact. *For $t \le t_A = (d(A)/w(A))^2$*

$$\overline{d}_t(A) \sim \mathbb{E}\left\{ d\left(\frac{1}{t}\sum_1^t u_i A\right)\right\} \sim \frac{1}{\sqrt{t}}\, d(A)$$

and stabilizes for $t \ge t_A$:

$$\overline{d}_t(A) \sim w(A)\,.$$

(Again, we observe a phase transition type behavior at t_A.)

Moreover, for any $\theta > 0$ there is $C(\theta)$ (large) so that for $t \sim C(\theta) t_A$ and "random" $u_i \in O(n)$, A_t is (θM^*)-net in M^*D (and $A_t \subseteq (1+\theta)M^*D$).

Note also that $t_A \le n$ which means that very few rotations of any set A are enough for A_t to become almost a euclidean ball (see [BoLM1]).

REFERENCES. As in the Local case before, this is an easy extension of convex centrally symmetric case and the main fact we used is proved in [MiS2].

Two comments follow:

2.3.3 The first one is about *threshold view on phase transition* and we will demonstrate it in the following example:

Let the set A in the previous fact be an interval $I = [-1,1]$; then $M^*(I) \sim 1/\sqrt{n}$ and therefore

$$\min_{u_i \in O(n)} d\Big(\frac{1}{t}\sum_1^t u_i I\Big) \sim \frac{1}{\sqrt{t}} \quad \text{for} \quad t \le n\,.$$

Defining $A_t = \frac{1}{t}\sum_1^t u_i I$, we see that

$$d(A_t, D) = \begin{cases} \infty & t < n \\ \sqrt{n} & t = n \\ C(\lambda) & t = \lambda n\,,\ 1 < \lambda < 2\,. \end{cases}$$

(One may also show that $C(\lambda) \sim \min\big\{\sqrt{n}, \sqrt{\log(1/(\lambda-1))/(\lambda-1)}\big\}$ (E. Gluskin).)

So, the function $\operatorname{dist}(A_t, D)$ has a sharp threshold, around $t \sim n$. At the same time, if one measures changes in the set A_t by studying a different parameter, the diameter $d(A_t)$, then this function behaves more regularly, but not completely so: at the interval of the threshold for the distance-function, we have a phase transition type behavior for the diameter. (See another example of such behavior in [Kl].)

2.3.4 To come to our second comment, we first compare the result in 2.3.1 with one of a "baby-model" in Spinglasses theory. (I am indebted to M. Talagrand whose Bourbaki talk in March '99 triggered the comparison between the two theories, and to L. Pastur who explained this and other points of the connections to me.) For this consider a special set $A = [-1,1]^n \subset \mathbb{R}^n$, the unit cube in $\mathbb{R}^n$ (or, which is the same for our purpose, $A = \{\pm 1\}^n$ – the set of vertices of the cube $[-1,1]^n$). The standard question of Statistical Physics is the following: Let $E \in G_{n,k}$ be a k-dimensional subspace of $\mathbb{R}^n$ and $k = \lambda n$. Let P_E be the orthogonal projection on E. We introduce the following function:

$$\varphi(\lambda; n) = \mathbb{E}\frac{d(P_E A)}{\sqrt{n}} \tag{$*$}$$

(where expectation $\mathbb{E}$ is over the Grassmannian $G_{n,k}$). The problem is to decide if $\lim_{n\to\infty} \varphi(\lambda; n) := \varphi(\lambda)$ exists (supposedly "Yes") and to show the existence of phase transition(s) for the function $\varphi(\lambda)$. This means that we expect function $\varphi(\lambda)$ to be continuous and even with continuous derivative beside one (or two) value of non-continuity for the derivative $\varphi'(\lambda)$ where we have a "phase transition".

Comparing with what we stated in 2.3.1, we see that although the problems which are investigated in both theories are exactly the same Isomorphic Geometry provides the answers up to universal factors and this is not enough for studying limits. (However, it provides the answers for any sets.) I would like to emphasize that it is not a sign that the methods and results of Asymptotic Geometric Analysis are not strong enough to prove existence of limits. Just isomorphic type results were the right ones in a theory which interested in understanding the asymptotic behavior of any set in $\mathbb{R}^n$ and was not in another theory which concentrates its attention on the study of special sets with many distinguished symmetries.

I believe that the "isomorphic" point of view may be adapted also in Statistical Physics, but here I will show what kind of general statements on existence of limits follow from Asymptotic Geometric Analysis.

2.3.5 "Outside" of the isomorphic phase transition. The existence of a limit, as (*), may be stated for an arbitrary set A (as in 2.3.1) but unfortunately only outside an interval around the "isomorphic" phase transition point $k^* = n(w(A)/d(A))^2$. In the language we adapted in 2.3.4, outside $\lambda_0 = k^*/n = (w(A)/d(A))^2$.

Let us put the corresponding result precisely. Let $d = d(A)$ be the diameter of the set A, $w := w(A)$ be the meanwidth of A, $k = \operatorname{rk} P_E$ be the rank of the orthoprojection onto a subspace E (i.e. $k = \dim E$). Then *there are universal constant $C > 0$ and $c > 0$ such that for any $\epsilon > 0$ and for $k > Ck^*/\epsilon^2$ (k^*, $d(P_E A)$ were defined in* 2.3.1)

$$\frac{1}{1+\epsilon}\sqrt{\frac{k}{n}} \le \frac{1}{d}\mathbb{E}_{G_{n,k}}(d(P_E A)) \le (1+\epsilon)\sqrt{\frac{k}{n}}$$

(*where $\mathbb{E}_{G_{n,k}}$ is the averaging over grassmanian $G_{n,k}$) and for $k \le c\epsilon^2 k^*$*

$$\frac{1}{1+\epsilon} w \le \mathbb{E}(d(P_E A)) \le (1+\epsilon) w\,.$$

So, outside an interval of length $\sim k^*/\epsilon^2$ around k^* we have "almost", up to ϵ, limit *uniformly* on *all* sets A.

Again, one may ironize that Statistical Physics is interested exactly in the interval where we don't have much information. But the information we have is exactly what Geometric Analysis was interested to know.

2.4 Approximation; what we expected from our old intuition and reality of the new one. The example in section 2.3.3 needs more attention. How it happens that just λn intervals ($\lambda > 1$) enough for the approximation of the Euclidean ball by Minkowski sum up to a constant $C(\lambda)$

independent of dimensionof the (Isomorphic representation of a ball). And why the sum of just around $\sim n/\epsilon^2$ intervals can approximate a Euclidean ball up to $1+\epsilon$ (almost isometrically)?

To see why it looks impossible (even though this is a proven fact), one should realize that any set of Cn points $\{x_i\}$ on the Euclidean unit sphere belongs to some narrow strip between two parallel hyperplanes at a distance $\sim \sqrt{\log n/n}$. So, we expect a terrible dissymmetry in the sum of such intervals. From this point of view we would expect that we need an exponential (by dimension n) number of intervals to be able to approximate a Euclidean ball. However, the true answer is a logarithm of what our intuition expects. The same happened with many other problems of approximation. We saw in 2.3.2 that, starting any convex set K (or even any set A, but the understanding of the result should be adjusted), not more than Cn random rotations of K will average into $\sim$ a Euclidean ball. In Bourgain-Lindenstrauss-Milman [BoLM1,2], many examples of this nature were considered (the number of steps of Minkowski or Steiner symmetrizations needed to approximate Euclidean ball is of special interest). And in all examples we studied the results are of the order of logarithm (!) of previously expected estimates. As we already discussed in 2.1, the reason behind such unexpected behavior is Concentration Phenomenon which we will discuss next.

2.5 General references. I have just briefly outlined only a very few directions of a very rich theory, with many subjects of study and many unexpected turns. I recommend the following books and survey articles for different directions and points of view on this theory: [MiS1], [To], [Pi1], [Pi2], [LT], [LiM], [GiM1].

3 Isomorphic Form of Isoperimetric problems; Concentration Phenomenon

3.1 The standard form. Let (X, ρ, μ) be a metric (defined by ρ) probability (with respect to a measure μ) space. Let A be a measurable subset of X and $\mu(A \subset X) \geq 1/2$. Define $A_\epsilon = \{x \in X, \rho(x, A) \leq \epsilon\}$. We introduce the function $\alpha(X;\epsilon) = 1 - \inf_A \mu(A_\epsilon)$ which we call the *concentration function* of X.

Important observation: in many examples $\alpha(X,\epsilon)$ is small. But what does this mean? "Small" relative to which quantity or parameter? To continue our discussion let us consider a few typical examples:

The euclidean unit sphere $S^n \subset \mathbb{R}^{n+1}$, or the Stieffel manifold $W_{n,2}$, or the special orthogonal group $SO(n)$, all equipped with the natural Riemannian $SO(n)$-invariant metrics and the rotation invariant (Haar) probability measures;
the n-dimensional space $F_2^n = \{\pm 1\}^n$ over $\mathbb{Z}_2$, or symmetric (=permutation) group Π_n, equipped with the Hamming metric normalized so that the diameters of these sets are equal to 1 and with the counting probability measures;
$SL_2(\mathbb{Z}_p)$ with the counting probability measure and the word distance (i.e. Cayley-graph distance) defined through a special set of generators (see [AM] for precise description and discussion, or the book [AS]).

So we, in fact, consider "families" $\{X_n\}$ of spaces and not individual spaces. Taking this into account, we introduce the notion of a *Levi family* defined by the condition:

$$\alpha(X_n;\epsilon) \xrightarrow[n\to\infty]{} 0 \qquad \text{(for any but fixed } \epsilon > 0\text{)}\,. \tag{3.1}$$

Therefore, we add another parameter, n, a "natural" numeration of a "natural" family $\mathcal{X} = \{X_n\}$ (but what does "natural" mean?) Moreover, in the *natural* examples we often have the following estimates:

$$\alpha(X_n;\epsilon) \le c_1 \exp(-c_2\epsilon^2 n) \tag{3.2}$$

or

$$\le c_1 \exp(-c_2\epsilon\sqrt{n})\,. \tag{3.3}$$

Don't expect from me to explain what I mean by "natural". The theory of what we call Lèvy families, normal Lèvy families (satisfying estimate (3.2) above) is still, for me, the theory of examples. By the way, two types of estimates we observe correspond to families which one may like to identify as "elliptic type" families ($S^n; W_{n,2}; SO(n); F_2^n, \Pi_n$) and which are *normal Lèvy families* (i.e. satisfy the estimate (3.2)) and another one satisfying estimate (3.3), which clearly should be identified as "hyperbolic type" families ($SL_2(\mathbb{Z}_p)$, but also $SL_k(\mathbb{Z}_p)$ for a fixed $k > 2$ and "p" playing the role of "n"). I would like to see a justification for such (obvious) terminology and to find a reason behind the feeling.

In fact, the concept of a Lévy family (and especially a normal Lévy family) generalizes the concept behind the law of large numbers in two directions: a) the measures are not necessarily the product of measures (that is, we have no condition of "independence") and b) Lipschitz functions on the space are considered instead of linear functionals only.

Returning to our understanding of the concentration function $\alpha(X_n, \epsilon)$

being small, we see that it is connected with a family of spaces. For example, it means that for (any) $A_n \subset S^{n+1}$, $\mu(A_n) = 1/2$, and $\epsilon > 0$ but fixed, $\mu(A_n)_\epsilon$ tends to 1 when $n \to \infty$ (and exponentially quickly). This is not a geometric point of view. Classical geometry fixes a space (meaning "n") and considers dependence on $\epsilon > 0$ and then it does not observe the phenomenon! However, we don't consider one set "A" but a family of sets $\{A_n \subset S^{n+1}, \mu(A_n) \geq 1/2\}$ and the statement, which is an *isomorphic inequality*, gives a uniform bound for ϵ-extensions of sets of this family.

At the same time in many (most) interesting cases, we know how to estimate $\alpha(X;\epsilon)$ without having any idea of the structure of the extremal sets. The exact structure of extremal sets or exact value of $\alpha(X;\epsilon)$ are known only for two or three cases and are not known even for such "simple" cases as $X = \mathbb{T}^n = \prod_i^n S_i^1$ or the cube.

However, we don't need the exact values of $\alpha(X;\epsilon)$. An isomorphic form of inequalities is enough for all applications [and also the only one typically available]. So the isomorphic (=asymptotic) view on isoperimetric problems has freed us from the necessity of solving (exactly) the isoperimetric problems.

The above estimate (3.2) and (3.3) happen to be typical and imply the so-called "*concentration phenomenon*".

To explain the reason for such terminology and also outline why a bound of the form (3.1) is so crucial, let us consider a 1-Lip function $f(x)$ defined on (X,ρ,μ), i.e.

$$\left|f(x) - f(y)\right| \leq \rho(x,y)\,.$$

Denote by L_f the median of $f(x)$, which is defined by

$$\mu\{x \in X \mid f(x) \geq L_f\} \leq \tfrac{1}{2} \quad \text{and} \quad \mu\{x \in Z \mid f(x) \leq L_f\} \geq \tfrac{1}{2}\,.$$

Then

$$\mu\{x \in X \mid |f(x) - L_f| < \epsilon\} \geq 1 - 2\alpha(X,\epsilon)\,. \tag{3.4}$$

So, if the value of $\alpha(X,\epsilon)$ is very small, then the values of Lipschitz function "concentrate" in the measure around one value, meaning it is almost constant with high probability. This is the case when $X = S^n$ and dimension n is large, as well as for large n for $\mathbb{T}^n$ or SO_n or the other examples we mentioned above. It is, in fact, a general property of high dimensional metric probability spaces which is called "*concentration phenomenon*".

Such a "concentration" of measure (this type of estimates) balances the exponentially high entropy of n-dimensional spaces (or other n-parametric families) and leads to a "regularity" in high dimension, keeping "diversity" under control. The absolute constants involved in the examples we saw are needed to balance rates of exponential decay (coming from Concentration) and exponential expansion (coming from covering/entropy). Surprisingly, both exponents have "roughly" the same order of decay via expansion by dimension and only a factor is needed to compensate them.

On a slightly more technical level, let, for example, X_n be S^{n+1} (the euclidean unit sphere in $\mathbb{R}^{n+2}$). Then $\alpha(X_n,\varepsilon) \le c_1 \exp(-\varepsilon^2 n/2)$ and (3.4) implies that for any set $\mathcal{N} = \{x_i\}_1^N \subset S^{n+1}$, with $N < c_1 \exp(\varepsilon^2 n/2)$ one may find a rotation $u \in O(n)$ such that

$$f(ux_i) \underset{\varepsilon}{\sim} L_f \,. \tag{3.5}$$

So, our function $f(x)$ is almost (up to ε) constant on the configuration $u\mathcal{N}$. Choosing various functions and configurations we obtain various geometric consequences. (Naturally, similar applications for other Lèvy families are possible and used.)

[This is, of course, only a scheme. The whole theory was developed to use in a delicate and correct way the concept of "concentration", see [MiS1], [Pi2], [GiM1]. To study the concentration phenomenon itself, consult [Mi2], [T1], [T2], [Gr1]; one may also consult Gromov's article in this collection [Gr2].]

Today, the concentration phenomenon has many applications outside this Asymptotic Theory (Geometric Analysis) where it was actually born. It is used broadly in Probability Theory, Complexity Theory (construction of fast algorithms and so on), Discrete Mathematics (especially in its branch I would call Asymptotic Combinatorics) and even in PDEs and others. Because of its widespread applicatons let me comment briefly on two ways of extending this concept.

(i) (X,ρ) – metric spaces. We often want to apply the same technique and ideas as we learned from the use of the concept of concentration in situations where no natural measure may be introduced (say, for infinite dimensional spaces). Also, we often have a measure and use it, receiving answers "with very high probability" when in fact, we are just searching for "existence" results. So, the measure is not natural in these problems.

(ii) (X,μ) – probability spaces. In many examples of clearly "concentration" type results the metric we use is not natural and not connected with concentration. Also, in some situations, it cannot even exist (say, in the framework of non-commutative geometry).

So, let us check how one may define such "concentration properties".

For this let us turn to two main schemes of applicatons.

3.2 Metric G-spaces (X, ρ). Let $X = (X, \rho, \mu)$ be (still) a metric probability space and let X also be a G-space (meaning that a group G of metric and measure preserving maps acts on X). Note, that if $A \subset X$ has not too small measure, say $\mu(A) \geq 1/10$, but $\alpha(X, \epsilon)$ is very small then it is easy to see that $\mu(A_\epsilon)$ will be close to 1. Let, say, $\alpha(X, \epsilon)$ be so small that $\mu(A_\epsilon) > 1 - \frac{1}{100}$. Then, for any $\{g_i\}_1^{100} \subset G$ the intersection $\bigcap_1^{100} g_i(A_\epsilon) \neq \emptyset$. Therefore, for *any* partition of $X = \bigcup_1^{10} A_i$ one may find A_{i_0} such that for any 100 "rotation" g_i's from G intersection $\bigcap_1^{100} g_i((A_i)_\epsilon) \neq \emptyset$.

This is one of the important schemes of application of the concentration property. But it essentially deals only with a metric property of X and the action on the group G on it.

Now let (X, ρ) be a metric G-space (so without any measure structure on it). To introduce a notion of concentration on X with respect to G, I will deal with one example (and future generalizations are obvious).

Let $X = S^\infty = S(H)$ be the unit sphere of a Hilbert space H. Let G be a subgroup of unitary (orthogonal, if our field is $\mathbb{R}$) operators U.

We call a set $A \subset S^\infty$ *essential* with respect to G iff $\forall \epsilon > 0$ $\forall n$ and $\forall g_1, \ldots, g_n \subset G$

$$\bigcap_1^n g_i(A_\epsilon) \neq \emptyset .$$

We say that (S^∞, G) has a "*concentration property*" iff $\forall$ finite partition $S^\infty = \bigcup_1^N A_i$, $A_i \subset S^\infty$, there is A_{i_0} which is essential.

EXAMPLES. 1) Fix an orthonormal basis $\{e_i\}_1^\infty$ and let $U(n)$ be a subgroup of U which is the unitary group on span $\{e_i\}_1^n$ and the identity action on $\{e_i\}_{i=n+1}^\infty$. Let $G = \bigcup_1^\infty U(n)$. Then (S^∞, G) has the concentration property.

2) Let u be any unitary operator on H and $G = \{u^n\}_{-\infty}^\infty$. Then (S^∞, G) has the concentration property.

3) Let $\mathfrak{A}$ be a family of pairwise commuting unitary operators in H. Then $(S^\infty, \mathfrak{A})$ has the concentration property.

References. These notions appeared in our discussions with M. Gromov at the end of the 70s and beginning of the 80s. See [Mi2].

Later V. Pestov ([P1]) connected this property with the *amenability* of G-action on X (i.e. with the existence of invariant mean; i.e. with existence of some substitution for measure !). I will cite only one of Pestov's results in this direction (and recommend the interested reader to consult the original papers for deeper connections):

A locally compact group G is amenable iff the uniform action $(S_{L_2(G)}, G)$ *has the concentration property.*

3.3 Probability spaces (X, μ). Again, let (X, ρ, μ) be a metric probability space with small concentration function $\alpha(X, \epsilon)$. Then (3.4) implies that $\mathbb{E}|f|$ is close to $L_{|f|}$. Similarly $(\mathbb{E}|f|^p)^{1/p} :=$ L_p-norm of a Lip function f is around the median $(L_{|f|^p})^{1/p} = L_{|f|}$. So, different L_p-norms are, in fact, equivalent (up to a factor depending on $p > 0$ and Lip – constant of f) which means that Hölder inequality may be reversed:

$$\|f\|_{L_p(X,\mu)} \leq c(p)\|f\|_{L_1(X,\mu)} \tag{3.6}$$

and the order of the constant $c(p)$ as $p \to \infty$ reflects the degree of concentration.

Such inverse Hölder inequalities appear often in the context of probability spaces. For example, linear functionals on a convex body K with volume 1 satisfy the inequality $\|f\|_{L_p(K;dx)} \leq cp\|f\|_{L_1(K;dx)}$ where $c > 0$ is an absolute constant. More generally, Bourgain has shown that if $f : K \to \mathbb{R}$ is a polynomial of degree m, then $\|f\|_p \leq c(p,m)\|f\|_2$ for every $p > 2$, where $c(p,m)$ depends only on p and on the degree m of f. Talagrand showed that an analogous statement holds true for the class of convex functions on E_2^n. In a joint work with Giannopoulos [GiM2] we started a regular study of the level of concentration with respect to a given *class of functions*. Our goal was to understand concentrations of "random" sets in $\mathbb{R}^n$ with respect to linear functions on $\mathbb{R}^n$. Of course, there are many different meanings of randomness in $\mathbb{R}^n$ and also, depending on the cardinality of our subsets, we may achieve different levels of concentration. In a typical example, we have a log-concave probability measure on $\mathbb{R}^n$, a "random", with respect to this measure, set S of points and then we have very exact connection between the level of concentration of linear functions on S (measured by constants $c(p)$ in (3.6)) and the cardinality of S.

The subject is very fresh, possible applications and connections with other parts of the theory are very promising. But it is too early to discuss it here and I am referring to the original paper [GiM2].

3.4 Functional point of view. Let us return to (3.5). There is a way of vastly extending this consequence of concentration which we, again, will demonstrate on one example of Lèvy family. Consider a family of probability measures $\{\mu_i\}_1^N$ on S^{n+1} and let $f(x)$ be a 1-Lip. function on S^{n+1}.

Fact. *If* $N \sim e^{\varepsilon^2 n/8}$ *then there is a rotation* $u \in O(n)$ *such that* $\left| \int_{S^{n+1}} f(ux) d\mu_i(x) - L_f \right| < \varepsilon$ *for every* $i = 1, \ldots, N$.

Of course, when μ_i *are* δ*-measures, we return to* (3.5). *However, the opposite case of* μ_i *having good densities may be more interesting in many cases.*

This also opens the way to understanding how the purely geometric idea of concentration may be extended to a non-commutative setting.

4 Concluding Remarks

4.1 I considered in the paper behavior of sets (often convex sets) in $\mathbb{R}^n$ with the emphasis on asymptotics when dimension $n \to \infty$. But, in fact, I see them just as examples, better understood but typical, of features of the world of very high degree of freedom. One should realize that this world looks very different from what was expected when we entered it:

- The exponential estimates we expected in approximation procedures happened to be essentially linear (logarithmical of what "old intuition" prepared us to expect);
- In many problems, the expected diversity is essentially reduced to the orbit of GL_n. However, we have to take a new angle and adapt the isomorphic point of view in Geometry;
- Quite non-trivial functions depending on huge number of variables are, with high probability, almost constant ("concentration phenomenon").
- Also, in a geometric sense, a similar phenomenon may be described. I mean that complicated functions look almost constant on huge substructures (the concept of "spectrum" in my old terminology – see [Mi2], or "Ramsey type" theorem in Combinatorics – see [P2]).

We know of many more results which do not correspond to the intuition we inherit, and I believe even more surprising results will be discovered in the near future.

If one would ask what is the reason behind these phenomena (but not on the technical level, we discussed in the section on Concentration Phenomenon) I would refer to a perceived random nature of high dimension as being at the root of the reasons and I would add that the patterns such "high dimensional randomness" produces create the unusual phenomena we observe (see [Mi4] where this point of view persists).

4.2 Speculations. At what stage of its development is Asymptotic Geometric Analysis, as described in this article? What is the perspective of its development? To attempt to answer this almost philosophical question I will speculate even more than is required.

There are a few stages in the development of our knowledge at large which are repeated in all small (and even smaller) components, in the directions we discover, in problems and questions we ask and try to solve.

First, we are just *curious*. At this stage we ask natural questions ("bad" questions as Gromov described them in one of our discussions at the Conference). However, these "natural" questions force us to start to think on the subject, they slowly prepare us for the next stage.

Then we already *think* about the subject and our questions are already much less "natural", they are already deep and unexpected, and the answers we accumulate begin to piece together into a puzzle, but the picture is not yet complete.

At the next stage, comes *understanding*, and later on the stage of "*knowing*". Only at the end, do we finally *possess* our knowledge: we may use it, we may apply it, but, in fact, more that this, we *possess* it.

The subject I discussed in the paper has long ago passed the first stage of curiosity; we are deeply inside the "thinking" stage and even, sometimes, we have *hutzpa* to think we *understand*.

5 Some Open Problems of Asymptotic Geometric Analysis

(I advise also to consult [Mi1] for some list of open problems; only a few of them will be repeated below).

I General problem-directions.

1a. Randomness in the structure of high dimensional spaces. (See [Mi4] for a vague discussion of this direction.)

1b. What is the level of complexity (deterministic or "randomized") of different properties of n-dimensional normed spaces (or n-dimensional convex bodies)? (Of course, "complexity" is not a well defined notion, and, in fact, it is also interesting to develop different forms of understanding of the complexity involved.) An example of a very surprising fact (discovered and developed by Dyer, Frieze, Kannan, Lovasz and Simonovits, see, e.g. [Bol]) is that the volume of an n-dimensional body is computable (by a randomized algorithm) in polynomial time (say, $\sim n^5$ in dimension n). Although it cannot be computed in less than exponential time deterministically. We now know of more examples which show that a slight modification of the question, or the right point of view, changes problems with "exponential" complexity to similar ones with only "polynomial" complexity.

2. Different levels of "randomness". We often observe that "the best choice" (whatever that means in concrete problems) is equivalent to "random" choice. (But not in some other cases.) To give conditions when we should expect such equivalence.
3. Let K be a convex body of volume 1 in $\mathbb{R}^n$ and let $K = -K$. What may be said about the distribution of volume for a random projection of K onto a k-dim. space? One may expect it to approach a Gaussian distribution when $n \to \infty$ and k is small enough. I would think that k of order $O(k^*)$, k^* as defined in 2.3.1, should work. However, we don't know much even for $k = 1$. I would like to specially emphasise some interesting partial results obtained recently by K. Ball and his students (see, e.g. [AnBP]).
4. Fix $0 < \lambda < 1$. A generic $k = [\lambda n]$-dim. subspace of an arbitrary n-dim. space X has already some very special properties which do not hold for arbitrary k-dim. spaces. (This phenomenon was first observed in [BoM].) We know very little about these properties. Also, we know that "improvements" which accompany generic k-dim. subspaces and generic k-dim. quotients are different. But we don't know much about either of them.

II Some concrete central problems of the theory

5. Entropy duality problems. (Going back to Carl and Pietch) (see [BoPST]; or the more recent [MiSz] for discussion). On the geometric finite dimensional language the problem is equivalent to the following question:
   Do universal constants a and b exist such that the following is correct: If K and T are convex centrally symmetric compact bodies in $\mathbb{R}^n$; and K^0, T^0 are their polar bodies; then
   $$N(K, aT) \le N(T^\circ, K^\circ)^b\,?$$
6. Isotropic constant – problem (going back to J. Bourgain).
   Let K be convex centrally symmetric body in $\mathbb{R}^n$, v.r. $K = 1$ (i.e. $|K| = |D|$) and K be in the isotropic position, meaning that for $\forall y, z \in \mathbb{R}^n$
   $$\int_K (x, y)(z, x)\frac{dx}{|K|} = L_K(z, y)\,.$$

Does a universal constant C exist such that $L_K \le C$ (for any n and $K \subset \mathbb{R}^n$)? (The problem has many connections with different directions in classical convexity and Local Theory, see [MiP1]. Some experts would consider it to be the most attractive open question of Asymptotic Convexity Theory. The best known estimate by Bourgain is $L_k \lesssim n^{1/4}\sqrt{\log n}$). See [D] for a nice presentation of Bourgain's result and other references.

7. Does for every $\epsilon > 0$ a constant $C(\epsilon)$ exist such that for any n and n-dimensional normed space $X = (\mathbb{R}^n, \|\cdot\|)$ there is a quotient space qX of X such that

$$\dim qX \ge n/2 \qquad \text{and} \qquad C_{2+\epsilon}(qX) \le C(\epsilon)\ ?$$

($C_q(Y)$ is a cotype q constant of the space Y).
See [Mi6] for many connections of this problem with other open problems of Local Theory.

III Problems in the direction of Dvoretzky type theorems

8. It is an old and well known fact (originated in Kashin [K]) that for any $\lambda < 1$ and $k = [\lambda n]$ the space ℓ_1^n contains an isomorphic copy (up to a constant $C(\lambda)$ depending on λ only) of space ℓ_2^k. Moreover, it is shown in [FLM] that, for $k \sim \epsilon^2 n$ such ℓ_2^k-copy may be found almost, up to $1+\epsilon$, isometrically ℓ_2^k and with a very high probability k-dim subspace will be such copy. However, we cannot construct ("present") such ℓ_2^k subspace even for k just larger than $n^{1/2}$.
So, how to construct "explicitly" ("present", whatever our intuition says it means) ℓ_2^k copies in ℓ_1^n for k being proportional to n?
9. What is the best order of estimate in ϵ-version of Dvoretzky theorem? Precisely: Let $N_3(\epsilon)$ be the smallest integer such that any $N_3(\epsilon)$-dim. normed space contains an $(1+\epsilon)$-isometric copy of ℓ_2^3 (3-dim. subspace which is, up to $1+\epsilon$, Euclidean). What is order of growth of this function? Is it polynomial by $1/\epsilon$? Naturally, the same question for any $k \ge 3$ (see [Mi7], for discussion on this problem; the best known estimate is worse then exponential by $1/\epsilon$).
10. In fact, I would think that the worst, or almost worst, embedding of ℓ_2^k copy in a normed space $X = (\mathbb{R}^n, \|\cdot\|)$ should be for X being ℓ_∞^n space. I think, this may be true for any n and $k < n$.
11. Algebraic form of Dvoretzky type theorems (see the problem in [Mi7, Section 1].

References

[Books]

[AS] N. ALON, J.H. SPENCER, The Probabilistic Method, Wiley Interscience, 1992.

[BuZ] YU.D. BURAGO, V.A. ZALGALLER, Geometric Inequalities, Grundlehren der mathematischen Wissenschaften 285, Springer Verlag, 1980.

[Gr1] M. GROMOV, Metric Structures for Riemannian and Non-Riemannian Spaces, based on "Structures métriques des variétés Riemanniennes" (L. LaFontaine, P. Pansu, eds.), English translation by Sean M. Bates, Birkhäuser, Boston-Basel-Berlin, 1999 (with Appendices by M. Katz, P. Pansu and S. Semmes).

[LT] M. LEDOUX, M. TALAGRAND, Probability in Banach Spaces, Ergeb. Math. Grenzgeb. 3 Folge, vol. 23, Springer, Berlin, 1991.

[Lè] P. LÈVY, Problèmes concrets d'analyse fonctionelle, Gauthier-Villars, Paris, 1951.

[MiS1] V. MILMAN, G. SCHECHTMAN, Asymptotic Theory of Finite-Dimensional Normed Spaces, Springer Lecture Notes in Math. 1200, 1986.

[MoR] R. MOTWAIN, P. RAGHAVAN, Randomized Algorithms, Cambridge Univ. Press, 1995.

[Pi1] G. PISIER, Factorization of Linear Operators and the Geometry of Banach Spaces, CBMS, vol. 60, American Math. Soc., Providence, RI, 1986.

[Pi2] G. PISIER, The Volume of Convex Bodies and Banach Space Geometry, Cambridge Tracts in Math. 94, 1989.

[S] R. SCHNEIDER, Convex Bodies: The Brunn-Minkowski Theory, Encyclopedia of Mathematics and its Applications 44, Cambridge University Press, 1993.

[To] N. TOMCZAK-JAEGERMANN, Banach-Mazur Distance and Finite-Dimensional Operator Ideal, Pitman Monographs 38, Pitman, London, 1989.

[Surveys]

[Bol] B. BOLLOBAS, Volume estimates and rapid mixing, in "Flavors in Geometry" (S. Levy, ed.), MSRI Pub. 31, Cambridge University Press (1997), 151–182.

[GiM1] A.A. GIANNOPOULOS, V. MILMAN, Euclidean structures in finite dimensional spaces, Handbook on the Geometry of Banach Spaces, to appear.

[Gr2] M. GROMOV, Spaces and question, Proceedings of Visions in Mathematics – Towards 2000, Israel 1999, GAFA, GAFA2000, Special Volume, issue 1 (2000), 118–161..

[LiM] J. LINDENSTRAUSS, V. MILMAN, The local theory of normed spaces and its applications to Convexity, in "Handbook of Convex Geometry" (P.M. Gruber, J.M. Wills, eds.) 1149-1220 (1993).

[M] B. MAUREY, Quelques progrés dans la compréhension de la dimension infinie, in "Espaces de Banach classiques et quantiques", Societé Matmatiques de France, Journee Annuelle (1994), 1-29.

[Mi1] V.D. MILMAN, The concentration phenomenon and linear structure of finite dimensional normed spaces, Proceedings I.C.M, Berkeley (1986).

[Mi2] V. MILMAN, The heritage of P. Lèvy in geometric functional analysis, Asterisque 157/8, 73-141 (1988).

[Mi3] V. MILMAN, Surprising geometric phenomena in high-dimensional convexity theory, Proc. ECM2, vol. II, Birkhäuser Progress in Math. 196, 73-91 (1996).

[Mi4] V. MILMAN, Randomness and pattern in convex geometric analysis, Proceedings of ICM-98, Berlin, v. 2 (1998), 665-677.

[MiP1] V. MILMAN, A. PAJOR, Isotropic position and inertia ellipsoids and zonoids of the unit ball of a normed n dimensional space, GAFA-Seminar 87-88, Springer Lecture Notes in Math. 1376 (1989), 64–104.

[O] T. ODELL, in "Proceedings of Analysis and Logic Meeting, Mons, Belgium, August 1997" (C. Finet, C. Michaux, eds), London Math. Soc. Lecture Notes, Cambridge Press, to appear.

[T1] M. TALAGRAND, Concentration of measure and isoperimetric inequalities in product spaces, IHES Publ. Math. 81 (1995), 73-205.

[T2] M. TALAGRAND, A new look at independence, Ann. Probab. 24:1 (1996), 1-34.

[**Articles**]

[AM] N. ALON, V.D. MILMAN, λ_1, isoperimetric inequalities for graphs and superconcentrators, J. Combinatorial Theory, Ser. B 38:1 (1985), 73–88.

[AnBP] M. ANTTILA, K. BALL, I. PERISSINAKI, The central limit problem for convex bodies, Trans. Amer. Math. Soc., to appear,

[BBP] J. BASTERO, J. BERNUÉS, A. PEÑA, An extension os Milman's reverse Brunn-Minkowski inequality, GAFA 5:3 (1995), 572–581.

[BoLM1] J. BOURGAIN, J. LINDENSTRAUSS, V. MILMAN, Mikowski sums and symmetrization, GAFA-Seminar notes 86-87, Springer Lecture Notes in Math. 1317 (1988), 283–289.

[BoLM2] J. BOURGAIN, J. LINDENSTRAUSS, V. MILMAN, Estimates related to Steiner symmetrizations, Springer Lecture Notes in Math. 1367 (1989), 264–273.

[BoM] J. BOURGAIN, V.D. MILMAN, Distances between normed spaces, their subspaces and quotient spaces, Integral Equations and Operator Theory 9 (1986), 31–46.

[BoPST] J. BOURGAIN, A. PAJOR, S.J. SZAREK, N. TOMCZAK-JAEGERMANN,

On the duality problem for entropy numbers of operators, in "Geometric Aspects of Functional Analysis (1987–88) (J. Lindenstrauss, V.D. Milman, eds.), Springer Lecture Notes in Math 1376 (1989), 50–63.

[D] S. DAR, Remarks on Bourgain's problem on slicing of convex bodies, in "Seminar notes of Israel Seminar on Geometric Aspects of Functions Analysis", Birkhauser, Operator Theory: Advances and Applications 77 (1995), 61–66.

[FLM] T. FIGIEL, J. LINDENSTRAUSS, V.D. MILMAN, The dimension of almost spherical sections of convex bodies, Acta Math. 139:1-2 (1977), 59–94.

[GiM2] A.A. GIANNOPOULOS, V. MILMAN, Concentration property on probability spaces, Advances in Math. 156 (2000).

[K] B.S. KASHIN, Sections of some finite-dimensional sets and classes of smooth functions (in Russian), Izv. Akad. SSR Ser. Mat. 41 (1977), 334–351.

[Kl] B. KLARTAG, Remarks on Minkowski symmetrizations, GAFA Seminar Notes 1996-2000, Springer Lecture Notes in Mathematics 1745 (2000, 109-117.

[Mi5] V.D. MILMAN, A new proof of the theorem of Dvoretzky on sections of convex bodies, Functional Analysis and its Applications 5:4 (1971), 28–37.

[Mi6] V. MILMAN, Proportional quotients of finite dimensional normed spaces, Springer Lecture Notes in Math. 1573 (1994), 3–5.

[Mi7] V.D. MILMAN, A few observations on the connections between Local Theory and some other fields, GAFA-Seminar Notes 86-87, Springer Lecture Notes in Math. 1317 (1988), 283–289.

[MiP2] V. MILMAN, A. PAJOR, Entropy and asymptotic geometry of non-symmetric convex bodies. Advances in Math., 152 (2000), 314–335.

[MiS2] V. MILMAN, G. SCHECHTMAN, Global vs. local asymptotic theories of finite dimensional normed spaces, Duke Math. J. 90: (1997), 73–93.

[MiSz] V. MILMAN, G. SZAREK, A geometric lemma and duality of entropy numbers, GAFA Seminar Notes 1996-2000, Springer Lecture Notes in Math. 1745 (2000), 191–222.

[vN] J. VON NEUMANN, Approximative properties of matrices of high order rank, Portugal. Math. 3 (1942), 1–62.

[P1] V.G. PESTOV, Amenable representations and dynamics of the unit sphere in an infinite- dimensional Hilbert space, GAFA, Geom. funct. anal. 10 (2000), 1171–1201.

[P2] V.G. PESTOV, Ramsey-Milman phenomenon, Urysohn metric spaces and extremely amenable groups, preprint.

VITALI MILMAN, School of Mathematical Sciences, Sackler Faculty of Exact Sciences, Tel Aviv University, Tel Aviv 69978, Israel milman@post.tau.ac.il

GAFA, Geom. funct. anal.
Special Volume – GAFA2000, 816 – 838
1016-443X/00/S20816-23 $ 1.50+0.20/0

GAFA Geometric And Functional Analysis

QUANTUM INFORMATION THEORY: RESULTS AND OPEN PROBLEMS

PETER SHOR

1 Introduction

The discipline of information theory was founded by Claude Shannon in a truly remarkable paper [Sh] which laid down the foundations of the subject. We begin with a quote from this paper which is an excellent summary of the main concern of information theory:

> The fundamental problem of communication is that of reproducing at one point either exactly or approximately a message selected at another point.

Quantum information theory is motivated largely by the same problem, the difference being that either the method of reproduction or the message itself involves fundamentally quantum effects. For many years, information theorists either ignored quantum effects or approximated them to make them susceptible to classical analysis; it was only in the last decade or so that the systematic study of quantum information theory began. We next give a quote from John R. Pierce which shows roughly the state of quantum information theory a quarter century ago. In a 1973 retrospective [P], celebrating the 25th anniversary of Shannon's paper, Pierce says

> I think that I have never met a physicist who understood information theory. I wish that physicists would stop talking about reformulating information theory and would give us a general expression for the capacity of a channel with quantum effects taken into account rather than a number of special cases.

In retrospect, this quote seems both optimistic and pessimistic. It was certainly pessimistic in that there are now many physicists who understand

A large part of this paper is included in the paper "Quantum Shannon Theory," which appeared in the IEEE Information Theory Society Newsletter 50:3 (September 2000), 3–5 and 28–33.

information theory, and I believe that even when Pierce wrote this, there were several who did. Ironically, one of the first fundamental theorems of quantum information theory was proved in the same year [Ho1]. On the other hand, Pierce was quite optimistic in that he seems to have believed that finding the capacity of a quantum channel would be fairly straightforward for a physicist with the right background. This has not proven to be the case; even now, we do not have a general formula for the capacity of a quantum channel. However, there have been several recent fundamental advances made in this direction, and I describe these in this paper.

2 Shannon Theory

Shannon's 1948 paper [Sh] contained two theorems for which we will be giving quantum analogs. The first of these is the *source coding* theorem, which gives a formula for how much a source emitting random signals can be compressed, while still permitting the original signals to be recovered with high probability. Shannon's source coding theorem states that n outputs of a source X can be compressed to length $nH(X) + o(n)$ bits, and restored to the original with high probability, where H is the entropy function. For a probability distribution with probabilities $p_1, p_2, \ldots, p_n$, the entropy H is

$$H(\{p_i\}) = \sum_{i=1}^{n} -p_i \log p_i\,, \tag{1}$$

where information theorists generally take the logarithm base 2 (thus obtaining bits as the unit of information).

The second of these theorems is the *channel coding* theorem, which states that with high probability, n uses of a noisy channel N can communicate $Cn - o(n)$ bits reliably, where C is the channel capacity given by

$$C = \max_{p(X)} I\big(X; N(X)\big) \tag{2}$$

Here the maximum is taken over all probability distributions on inputs X to the channel, and $N(X)$ is the output of the channel given input X. The *mutual information* I is defined as:

$$\begin{aligned} I(X;Y) &= H(Y) - H(Y|X) & (3)\\ &= H(X) + H(Y) - H(X,Y)\,, & (4) \end{aligned}$$

where $H(X,Y)$ is the entropy of the joint distribution of X and Y, and $H(Y|X)$ is the conditional entropy of Y, given X. That is, if the possible values of X are $\{X_i\}$, then the conditional entropy is

$$H(Y|X) = \sum_i \Pr(X = X_i) H(Y|X = X_i)\,. \tag{5}$$

In this paper, I outline the progress that has been made in extending these formulae to quantum channels, while also taking a few side detours that address related problems and results in quantum information theory. I will keep this paper at a fairly low technical level, so I only sketch the proofs for some of the results I mention.

When the formula for mutual information is extended to the quantum case, two generalizations have been found that both give capacities of a quantum channel, although these capacities differ in both the resources that the sender and receiver have available and the operations they are permitted to carry out. One of these formulae generalizes the expression (3) and the other the expression (4); these expressions are equal in the classical case.

3 Quantum Mechanics

Before we can start talking about quantum information theory, I need to give a brief description of some of the fundamental principles of quantum mechanics. The first of these principles that we present is the *superposition principle*. In its most basic form, this principle says that if a quantum system can be in one of two distinguishable states $|\,x\rangle$ and $|\,y\rangle$, it can be in any state of the form $\alpha\,|\,x\rangle + \beta\,|\,y\rangle$, where α and β are complex numbers with $|\alpha|^2 + |\beta|^2 = 1$. Here $|\,\cdot\rangle$ is the notation that physicists use for a quantum state; we will occasionally be using it in the rest of this paper. Recall we assumed that $|\,x\rangle$ and $|\,y\rangle$ were distinguishable, so there must conceptually be some physical experiment which distinguishes them (this experiment need not be performable in practice). The principle says further that if we perform this experiment, we will observe $|\,x\rangle$ with probability $|\alpha|^2$ and $|\,y\rangle$ with probability $|\beta|^2$. Furthermore, after this experiment is performed, if state $|\,x\rangle$ (or $|\,y\rangle$) is observed the system will thereafter behave in the same way as it would have had it originally been in state $|\,x\rangle$ (or $|\,y\rangle$).

Mathematically, the superposition principle says that the states of a quantum system are the unit vectors of a complex vector space, and that two orthogonal vectors are distinguishable. In accordance with physics usage, we will denote quantum states by column vectors. The Dirac *bra-ket* notation denotes a column vector by $|\,v\rangle$ (a *ket*) and its Hermitian transpose (i.e., complex conjugate transpose) by $\langle v\,|$ (a *bra*). The inner product between two vectors, v and w, is denoted $\langle w|v\rangle = w^\dagger v$, where $w^\dagger$ is the conjugate transpose of w. Multiplying a quantum state vector by a complex phase factor (a unit complex number) does not change any properties of the system, so mathematically the state of a quantum system is a point in projective complex space. Unless otherwise stated, however, we will denote quantum states by unit vectors in a complex vector space $\mathbb{C}^d$.

We will be dealing solely with finite dimensional vector spaces. Quantum information theory is already complicated enough in finite dimensions without introducing the additional complexity of infinite-dimensional vector spaces. Many of the theorems we will be discussing do indeed generalize naturally to infinite-dimensional spaces.

A *qubit* is a two-dimensional quantum system. Probably the most widely known qubit is the polarization of a photon, and we will thus be using this example in the remainder of the paper. For the polarization of a photon, there can only be two distinguishable states. If one sends a photon through a birefringent crystal, it will take one of two paths, depending on its polarization. By re-orienting this crystal, these two distinguishable polarization states can be chosen to be horizontal and vertical, or they can be chosen to be right diagonal and left diagonal. In accordance with the superposition principle, each of these states can be expressed as a complex combination of basis states in the other basis. For example,

$$\begin{aligned}
|\nearrow\rangle &= \tfrac{1}{\sqrt{2}}\,|\leftrightarrow\rangle + \tfrac{1}{\sqrt{2}}\,|\updownarrow\rangle \\
|\searrow\rangle &= \tfrac{1}{\sqrt{2}}\,|\leftrightarrow\rangle - \tfrac{1}{\sqrt{2}}\,|\updownarrow\rangle \\
|\circlearrowright\rangle &= \tfrac{1}{\sqrt{2}}\,|\leftrightarrow\rangle + \tfrac{i}{\sqrt{2}}\,|\updownarrow\rangle \\
|\circlearrowleft\rangle &= \tfrac{1}{\sqrt{2}}\,|\leftrightarrow\rangle - \tfrac{i}{\sqrt{2}}\,|\updownarrow\rangle\,.
\end{aligned}$$

Here, $|\circlearrowright\rangle$ and $|\circlearrowleft\rangle$ stand for right and left circularly polarized light, respectively; these are another pair of basis states for the polarization of photons. For example, when diagonally polarized photons are put through a birefringent crystal oriented in the $\updownarrow, \leftrightarrow$ direction, half of them will behave like vertically polarized photons, and half like horizontally polarized photons.

If you have two quantum systems, their joint state space is the tensor product of their individual state spaces. For example, the state space of two qubits is $\mathbb{C}^4$ and of three qubits is $\mathbb{C}^8$. The high dimensionality of the space for n qubits, $\mathbb{C}^{2^n}$, is one of the places where quantum computation attains its power.

The polarization state space of two photons has as a basis the four states

$$|\updownarrow\updownarrow\rangle, \quad |\updownarrow\leftrightarrow\rangle, \quad |\leftrightarrow\updownarrow\rangle, \quad |\leftrightarrow\leftrightarrow\rangle.$$

This state space includes states such as an EPR (Einstein, Podolsky, Rosen) pair of photons

$$\tfrac{1}{\sqrt{2}}(|\updownarrow\leftrightarrow\rangle - |\leftrightarrow\updownarrow\rangle) = \tfrac{1}{\sqrt{2}}(|\nearrow\searrow\rangle - |\searrow\nearrow\rangle), \tag{6}$$

where neither qubit alone has a definite state, but which has a definite state when considered as a joint system of two qubits. In this state, the two photons have orthogonal polarizations in whichever basis they are measured in. Bell [Be] showed that the outcomes of measurements on the photons of this state cannot be reproduced by joint probability distributions which give probabilities for the outcomes of all possible measurements, and in which each of the single photons has a definite probability distribution for the outcome of measurements on it, independent of the measurements which are made on the other photon. In other words, there cannot be any set of hidden variables associated with each photon that determines the probability distribution obtained when this photon is measured in any particular basis.

I will present here another demonstration of this impossibility of local hidden variables; namely, the proof involving the *GHZ state* (named for Greenburger, Horne and Zeilinger) [GrHSZ]. Many fewer people have seen this than have seen Bell's inequalities, probability because it is much more recent; however, the demonstration for the GHZ state is in some ways simpler because it is deterministic. From now on, instead of using $|\updownarrow\rangle$ and $|\leftrightarrow\rangle$ for qubits, we will use $|0\rangle$ and $|1\rangle$, as these are equivalent and probably more comfortable for our audience. The GHZ state is

$$\tfrac{1}{\sqrt{2}}(|000\rangle + |111\rangle). \tag{7}$$

The thought experiment demonstrating the impossibility of hidden variables involves measuring each of the qubits in either the C basis $\frac{1}{\sqrt{2}}(|0\rangle \pm i|1\rangle)$ or in the D basis $\frac{1}{\sqrt{2}}(|0\rangle \pm |1\rangle)$. For photon polarization, the C basis corresponds to circularly polarized light and the D basis to diagonally polarized light. We will first suppose that each of the qubits is measured in the D basis. This projects the joint state of our three qubits onto one of the eight mutually orthogonal vectors

$$\tfrac{1}{\sqrt{8}}(|0\rangle \pm |1\rangle)(|0\rangle \pm |1\rangle)(|0\rangle \pm |1\rangle). \tag{8}$$

Let us consider the state formed by taking all plus signs in the superpositions above. This is equivalently

$$\tfrac{1}{\sqrt{8}}\big(\,|\,000\rangle+|\,001\rangle+|\,010\rangle+|\,011\rangle+|\,100\rangle+|\,101\rangle+|\,110\rangle+|\,111\rangle\,\big). \tag{9}$$

The inner product of this state with the GHZ state (7) is 1/2, so the probability of observing the state (9) when measuring all three qubits in the D basis is $(1/2)^2 = 1/4$. It is easy to check that similarly, the probability of observing any of the states of (8) with an even number of $-$'s is 1/4 and that the probability of observing any state of (8) with an odd number of $-$'s is 0.

We now consider measuring two of the qubits in the C basis and one (say the third) in the D basis. This measurement projects onto the eight basis states

$$\tfrac{1}{\sqrt{8}}\big(\,|\,0\rangle \pm i\,|\,1\rangle\,\big)\big(\,|\,0\rangle \pm i\,|\,1\rangle\,\big)\big(\,|\,0\rangle \pm |\,1\rangle\,\big)\,. \tag{10}$$

Here, it is easy to check that if we measure the GHZ state (7) in this basis, we will always observe an odd number of $-$'s.

We can now show that it is impossible to assign measurement outcomes to each of the qubits independent of the basis that the other qubits are measured in, and remain consistent with the predictions of quantum mechanics. Consider the following table

qubit 1	qubit 2	qubit 3	parity
D	D	D	even
D	C	C	odd
C	D	C	odd
C	C	D	odd

(11)

The last entry in each row gives the parity of the number of $-$'s if the three qubits are measured in the bases given by the first three entries of the row. Suppose there is a definite outcome assigned to each qubit for each of the two possible measurement bases. Since each basis appears for each qubit exactly twice in the table, the total number of $-$'s in the table would thus have to be even. However, the results predicted by quantum mechanics (the fourth column) are that the total number of $-$'s in the table is odd. This implies that the outcome of at least one measurement on one qubit must depend on the measurements which are made on the other qubits, and that this must hold even if the qubits are spatially separated. It can be shown, however, that this correlation cannot be used to transmit any information between people holding these various qubits; for example, the probability that a qubit is found to be $+$ ($-$) is one-half independent of the measurements on the other qubits, so which measurements are chosen for the other qubits do not affect this probability (although the outcomes of these measurements may).

The next fundamental principle of quantum mechanics we discuss is the *linearity principle.* This principle states that an isolated quantum system undergoes linear evolution. Because the quantum systems we are considering are finite dimensional vector spaces, a linear evolution of these can be described by multiplication by a matrix. It is fairly easy to check that in order to make the probabilities sum to one, we must restrict these matrices to be unitary (a matrix U is unitary if $U^\dagger = U^{-1}$; unitary matrices are the complex matrices which take unit vectors to unit vectors).

Although many explanations of quantum mechanics restrict themselves to pure states (unit vectors), for quantum information theory we need to treat probability distributions over quantum states. These naturally give rise to objects called density matrices. For an n-dimensional quantum state space, a *density matrix* is an $n \times n$ Hermitian trace-one positive semidefinite matrix.

A rank one density matrix ρ corresponds to the pure state $|\,v\rangle$ where $\rho = |\,v\rangle\,\langle v\,|$. Recall $\langle v\,|$ was the complex conjugate transpose of $|\,v\rangle$, and for most of this paper we denote $\langle v\,|$ by $v^\dagger$. Density matrices arise naturally from quantum states in two ways.

The first way in which density matrices arise is from probability distributions over quantum states. Suppose that we have a system which is in state v_i with probability p_i. The corresponding density matrix is

$$\rho = \sum_i p_i v_i v_i^\dagger\,. \tag{12}$$

An important fact about density matrices is that the density matrix for a system gives as much information as possible about experiments performed on the system. That is, any two systems with the same density matrix ρ cannot be distinguished by experiments, provided that no extra side information is given about these systems.

The other way in which density matrices arise is through disregarding part of an entangled quantum state. Recall that two systems in an entangled (pure) state have a definite quantum state when considered jointly, but each of the two systems individually cannot be said to have a definite state. Suppose that we have a pure state ρ_{AB} on a tensor product system $\mathcal{H}_A \otimes \mathcal{H}_B$. If we can only see the first part of the system, this part behaves as though it is in the state $\rho_A = \mathrm{Tr}_B \rho_{AB}$. Here, Tr_B is the partial trace operator. Consider a joint system in the state

$$\rho_{AB} = \begin{pmatrix} B_{11} & B_{12} & B_{13} \\ B_{21} & B_{22} & B_{23} \\ B_{31} & B_{32} & B_{33} \end{pmatrix}. \quad (13)$$

In this example, the dimension of $\mathcal{H}_A$ is 3 and the dimension of $\mathcal{H}_B$ is the size of the matrices B_{ij}. The partial trace of ρ_{AB}, tracing over $\mathcal{H}_A$, is

$$\mathrm{Tr}_A\ \rho_{AB} = B_{11} + B_{22} + B_{33}\,. \quad (14)$$

Although the above formula also determines the partial trace when we trace over $\mathcal{H}_B$, through a change of coordinates, it is instructive to give this explicitly:

$$\mathrm{Tr}_B\ \rho_{AB} = \begin{pmatrix} \mathrm{Tr}\, B_{11} & \mathrm{Tr}\, B_{12} & \mathrm{Tr}\, B_{13} \\ \mathrm{Tr}\, B_{21} & \mathrm{Tr}\, B_{22} & \mathrm{Tr}\, B_{23} \\ \mathrm{Tr}\, B_{31} & \mathrm{Tr}\, B_{32} & \mathrm{Tr}\, B_{33} \end{pmatrix}. \quad (15)$$

The final ingredient we need before we can start explaining quantum information theory is a *von Neumann measurement.* We have seen examples of this process before, while explaining the superposition principle and the GHZ non-locality proof; however, we have not yet given the general mathematical formulation of a von Neumann measurement. Suppose that we have an n-dimensional quantum system $\mathcal{H}$. A von Neumann measurement corresponds to a complete set of orthogonal subspaces $S_1, S_2, \ldots, S_k$ of $\mathcal{H}$. Here, complete means that the subspaces S_i span the space $\mathcal{H}$, so that $\sum_i \dim S_i = n$. Let Π_i be the projection matrix onto the subspace S_i. If we start with a density matrix ρ, the von Neumann measurement corresponding to the set $\{S_i\}$ projects ρ into one of the subspaces S_i. Specifically, it projects ρ onto the i'th subspace with probability Tr $\Pi_i\rho$, the state after the projection being $\Pi_i\rho\Pi_i$, renormalized to be a unit vector. A special case that is often encountered is when the S_i are all one-dimensional, so that $S_i = w_i w_i^\dagger$, and the vectors w_i form an orthogonal basis of $\mathcal{H}$. Then, a vector v is taken to w_i with probability $|w_i^\dagger v|^2$, and a density matrix ρ is taken to w_i with probability $w_i^\dagger \rho w_i$.

4 Von Neumann Entropy

We are now ready to consider quantum information theory. We will start by defining the entropy of a quantum system. To give some intuition for this definition, we first consider some special cases. Consider n photons, each being in the state $|\updownarrow\rangle$ or $|\leftrightarrow\rangle$ with probability 1/2. Any two of these states are completely distinguishable. There are thus 2^n equally probable states of the system, and the entropy is n bits. This is essentially a classical system.

Consider now n photons, each being in the state $|\updownarrow\rangle$ or $|\nearrow\rangle$ with probability 1/2. These states are not completely distinguishable, so there are effectively considerably less than 2^n states, and the entropy should intuitively be less than n bits.

By thermodynamic arguments involving the increase in entropy associated with the work extracted from a system, von Neumann deduced that the entropy of a quantum system with density matrix ρ (now called *von Neumann entropy*) should be

$$H_{\mathrm{vN}}(\rho) = -\mathrm{Tr}\rho \log \rho \,. \tag{16}$$

Recall that ρ is positive semidefinite, so that $-\mathrm{Tr}\rho\log\rho$ is well defined. If ρ is expressed in coordinates in which it is diagonal with eigenvalues λ_i, then in these coordinates $-\rho\log\rho$ is diagonal with eigenvalues $-\lambda_i\log\lambda_i$. We thus see that

$$H_{\mathrm{vN}}(\rho) = H_{\mathrm{Shan}}(\lambda_i) \,, \tag{17}$$

so that the von Neumann entropy of a density matrix is the Shannon entropy of the eigenvalues. (Recall $\mathrm{Tr}\rho = 1$, so that $\sum_i \lambda_i = 1$.) This definition is easily seen to agree with the Shannon entropy in the classical case, where all the states are distinguishable.

5 Source Coding

Von Neumann developed the above definition of entropy for thermodynamics. One can ask whether this is also the correct definition of entropy for information theory. We will first give the example of quantum source coding [JS], [S], also called *Schumacher compression*, for which we will see that it is indeed the right definition. We consider a memoryless quantum source that at each time step emits the pure state v_i with probability p_i. We would like to encode this signal in as few qubits as possible, and send them to a receiver who will then be able to reconstruct the original state. Naturally, we will not be able to transmit the original state flawlessly. In fact, the receiver cannot even reconstruct the original state perfectly most of the time, which is the situation that is possible in classical communication theory. Unlike classical signals, however, quantum states are not completely distinguishable theoretically, so reconstructing the original state most of the time is too stringent a requirement. What we will require is that the receiver be able to reconstruct a state which is almost completely indistinguishable from the original state nearly all the time. For this we need a measure of indistinguishability; we will use a measure called *fidelity.* Suppose that the original signal is a vector

$$u = v_1 \otimes v_2 \otimes \ldots \otimes v_n \,.$$

Then the fidelity between the signal u and the output ρ (which is in general a mixed state, i.e., a density matrix, on n qubits) is $F = u^\dagger \rho u$ and the average fidelity is this fidelity F averaged over u. If the output is a pure state v, the fidelity $F = u^\dagger v v^\dagger u = |u^\dagger v|^2$. The fidelity measures the probability of success of a test which determines whether the output is the same as the input.

Before I can continue to sketch the proof of the quantum source coding theorem, I need to review the proof of the classical source coding theorem. Suppose we have a memoryless source, i.e., a source X that at each time step emits the i'th signal type, S_i, with probability p_i, and where the probability distribution for each signal is independent of the previously emitted signals. The idea behind classical source coding is to show that with high probability, the source emits a *typical sequence,* where a sequence of length n is typical if it contains approximately np_i copies of the signal S_i for every i. The number of typical sequences is only $2^{nH(X)+o(n)}$. These can thus be coded in $nH(X) + o(n)$ bits.

The tool that we use to perform Schumacher compression is that of *typical subspaces.* Suppose that we have a density matrix $\rho \in \mathcal{H}$, where $\mathcal{H} = \mathbb{C}^k$, and we take the tensor product of n copies of ρ in the space $\mathcal{H}^n$, i.e., we take $\rho^{\otimes n} \in \mathbb{C}^{nk}$. There is a typical subspace associated with $\rho^{\otimes n}$. Let $\hat{v}_1, \hat{v}_2, \ldots, \hat{v}_k$ be the eigenvectors of ρ with associated eigenvalues $\lambda_1, \lambda_2, \ldots, \lambda_k$. Since $\mathrm{Tr}\rho = 1$, these λ_i form a probability distribution. Consider typical sequences of the eigenvectors $\hat{v}_i$, where λ_i is the probability of choosing $\hat{v}_i$. A typical sequences can be turned into a quantum state in $\mathcal{H}^{\otimes n}$ by taking the tensor products of its elements. That is, if a typical sequence is $\hat{v}_{i_1}, \hat{v}_{i_2}, \ldots, \hat{v}_{i_n}$, the corresponding quantum state is $w = \hat{v}_{i_1} \otimes \hat{v}_{i_2} \otimes \cdots \otimes \hat{v}_{i_n}$. The typical subspace $\mathcal{T}$ is the subspace spanned by typical sequences of the eigenvectors. The subspace $\mathcal{T}$ has dimension equal to the number of typical sequences, or $2^{H_{\mathrm{vN}}(\rho)n+o(n)}$.

We can now explain how to do Schumacher compression. Suppose we wish to compress a source emitting v_i with probability p_i. Let the typical subspace corresponding to $\rho^{\otimes n}$ be $\mathcal{T}$, where $\rho = \sum_i p_i v_i v_i^\dagger$ is the density matrix for the source, and where we are using a block length n for our compression scheme. We take the vector $u = v_{i_1} \otimes v_{i_2} \otimes \cdots \otimes v_{i_n}$ and make the von Neumann measurement that projects it into either $\mathcal{T}$ or $\mathcal{T}^\perp$. If u is projected onto $\mathcal{T}$, we send the results of this projection; this can be done with $\log \dim \mathcal{T} = nH_{\mathrm{vN}}(\rho) + o(n)$ qubits. If u is projected onto $\mathcal{T}^\perp$, our compression algorithm has failed and we can send anything; this does not degrade the fidelity of our transmission greatly, because this is a low probability event.

Why did this work? The main element of the proof is to show that the probability that we project u onto $\mathcal{T}$ approaches 1 as n goes to ∞. This probability is $u^\dagger \Pi_{\mathcal{T}} u$. If this probability were exactly 1, then u would necessarily be in $\mathcal{T}$, and we would have noiseless compression. If the probability that the state u is projected onto $\mathcal{T}$ is $1-\epsilon$, then $u^\dagger \, \Pi_{\mathcal{T}} \, u = 1 - \epsilon$, and when u is projected onto $\mathcal{T}$, the fidelity between the original state u and the final state $\Pi_{\mathcal{T}} \, u$ is thus $|\langle u | \Pi_{\mathcal{T}} \, u \rangle|^2 = (1-\epsilon)^2$.

Now, recall that if two density matrices are equal, the outcomes of any experiments performed on them have the same probabilities. Thus, the probability that the source v_i with probabilities p_i projects onto the typical subspace is the same as for the source $\hat{v}_i$ with probabilities λ_i, where $\hat{v}_i$ and λ_i are the eigenvalues and eigenvectors of $\rho = \sum_i p_i v_i v_i^\dagger$. We know from the classical theory of typical sequences that $w = \hat{v}_{i_1} \otimes \hat{v}_{i_2} \otimes \cdots \otimes \hat{v}_{i_k}$ is in the typical subspace at least $1-\epsilon$ of the time; because the $\hat{v}_i$ are distinguishable, this is essentially the classical case, and w is in the typical subspace exactly when the sequence of $\hat{v}_i$ is a typical sequence.

6 Accessible Information

The next concept is that of *accessible information*. Here, we again have a source emitting state ρ_i with probability p_i. Note that now, the states ρ emitted may be density matrices rather than pure states. We will ask a different question this time. We now want to obtain as much information as possible about the sequence of signals emitted by the source. That is, we wish to maximize the mutual information $I(X;Y)$ where X is the variable telling which signal ρ_i was emitted, and Y is the variable giving the outcome of a measurement on X. This gives the capacity of a channel where at each time step the sender must choose one of the states ρ_i to send, and must furthermore choose ρ_i a fraction p_i of the time; and where the receiver makes a separate measurement on each signal sent.

To find the accessible information, we need to maximize over all measurements. For this, we need to be able to characterize all possible quantum measurements. It turns out that von Neumann measurements are not the most general class of quantum measurements; the most general measurements are the *positive operator valued measurements,* or *POVM's.* One way to describe these is as von Neumann measurements on a quantum space larger than the original space; that is, by supplementing the quantum state space by an *ancilla* space and taking a von Neumann measurement on the joint state space.

For a POVM, we are given a set of positive semidefinite matrices E_i satisfying $\sum_i E_i = I$. The probability of the i'th outcome is then

$$p_i = \mathrm{Tr}(E_i \rho)\,. \tag{18}$$

For a von Neumann measurement, we take $E_i = \Pi_{S_i}$, the projection matrix onto the i'th orthogonal subspace S_i. The condition $\sum_i \Pi_{S_i} = I$ is equivalent to the requirement that the S_i are orthogonal and span the whole state space. To obtain the maximum information from a POVM, we can assume that the E_i's are pure states; if there is an E_i that is not rank one, then we can always achieve at least as much accessible information by refining that E_i into a sum $E_i = \sum_j E_{ij}$ where the E_{ij} are rank one.

We now give some examples of the measurements maximizing accessible information. The first is one of the simplest examples. Suppose that we have just two pure states in our ensemble, with probability 1/2 each. For example, we could take the states $|\updownarrow\rangle$ and $|\nearrow\rangle$. Let us take $v_1 = (1,0)$ and $v_2 = (\cos\theta, \sin\theta)$. We will not prove it here, but the optimal measurement for these is the von Neumann measurement with two orthogonal vectors symmetric around v_1 and v_2. That is, the measurement with projectors

$$w_1 = \left(\cos\left(\tfrac{\pi}{2} + \tfrac{\theta}{2}\right), \sin\left(\tfrac{\pi}{2} + \tfrac{\theta}{2}\right)\right) \tag{19}$$

$$w_2 = \left(\cos\left(-\tfrac{\pi}{2} + \tfrac{\theta}{2}\right), \sin\left(-\tfrac{\pi}{2} + \tfrac{\theta}{2}\right)\right). \tag{20}$$

This measurement is symmetric with respect to interchanging v_0 and v_1, and it leads to a binary symmetric channel with error probability

$$\cos^2\left(\tfrac{\pi}{2} + \tfrac{\theta}{2}\right) = \tfrac{1}{2} - \tfrac{\sin\theta}{2}\,. \tag{21}$$

The accessible information is thus $1 - H\left(\frac{1}{2} - \frac{\sin\theta}{2}\right)$.

For the ensemble containing v_1 and v_2 with probability 1/2 each, the density matrix is

$$\rho = \frac{1}{2}\begin{pmatrix} 1+\cos^2\theta & \sin\theta\cos\theta \\ \sin\theta\cos\theta & 1-\cos^2\theta \end{pmatrix}, \tag{22}$$

which has eigenvalues $\frac{1}{2} \pm \cos\theta$, so the von Neumann entropy of the density matrix is $H\left(\frac{1}{2} - \frac{\cos\theta}{2}\right)$. The values of I_{acc} and H_{vN} are plotted in Figure 1.

One can see that the von Neumann entropy is larger than the accessible information.

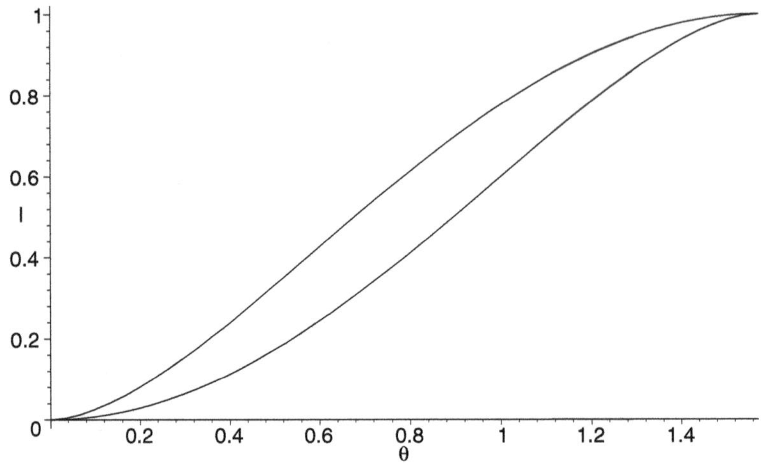

Figure 1: A plot of the von Neumann entropy of the density matrix and the accessible information for the ensemble of two pure quantum states with equal probabilities and that differ by an angle of θ, for $0 \leq \theta \leq \pi/2$. The top curve is the von Neumann entropy and the bottom the accessible information.

Note that in our first example, the optimum measurement was a von Neumann measurement. If there are only two states in an ensemble, it has been conjectured that the measurement optimizing accessible information is always a von Neumann measurement, mainly because extensive computer experiments have not found a counterexample [FP]. This conjecture has been proven for quantum states in two dimensions [L2]. Our next example shows that this conjecture does not hold for ensembles composed of three or more states.

Our second example is three photons with polarizations that differ by $60°$ each. These are represented by the vectors

$$\begin{aligned} v_0 &= (1, 0) \\ v_1 &= \left(-\tfrac{1}{2}, \tfrac{\sqrt{3}}{2}\right) \\ v_2 &= \left(-\tfrac{1}{2}, -\tfrac{\sqrt{3}}{2}\right) \end{aligned}$$

The optimal measurement for these states is the POVM corresponding to the vectors w_i where $w_i \perp v_i$. We take $E_i = \frac{2}{3} w_i w_i^\dagger$, in order for $\sum_i E_i = I$. If we start with vector v_i, it is easy to see that we never obtain w_i, but do obtain the other two possible outcomes with probability 1/2 each. This gives $I_{\text{acc}} = \log 3 - 1$. For these three signal states, it is also easy to check that the density matrix $\rho = \frac{1}{2}I$, so $H_{\text{vN}} = 1$. Again, we have $I_{\text{acc}} < H_{\text{vN}}$.

This leads to a conjecture: that $I_{\text{acc}} \leq H_{\text{vN}}$. The correct theorem is somewhat stronger, and we will shortly state it. The first published proof of this theorem was given by Holevo [Ho1]. It was earlier conjectured by Gordon [G] and stated by Levitin with no proof [L1].

Theorem (Holevo). *Suppose that we have a source emitting a (possibly mixed) state ρ_i with probability p_i. Let*

$$\chi = H_{\text{vN}}\Big(\sum_i p_i \rho_i\Big) - \sum_i p_i H_{\text{vN}}(\rho_i)\,. \tag{23}$$

Then

$$I_{\text{acc}} \leq \chi\,. \tag{24}$$

The conditions for equality in this result are known. If all the ρ_i commute, then they are simultaneously diagonalizable, and the situation is essentially classical. In this case, $I_{\text{acc}} = \chi$; otherwise $I_{\text{acc}} < \chi$.

7 The Classical Capacity of a Quantum Channel

One can ask the question: is this quantity I_{acc} the most information that one can send using the three states of our second example? The answer is, surprisingly, "no". Suppose that we use the three length-two codewords $v_0 \otimes v_0$, $v_1 \otimes v_1$, and $v_2 \otimes v_2$. These are three pure states in the four-dimensional quantum space of two qubits. However, since there are only three vectors, they lie in a three-dimensional subspace. The inner product between any two of these states is 1/4. One can show that the optimal measurement is attained by the von Neumann measurement having three basis vectors obtained by "pulling" the three vectors $v_i \otimes v_i$ apart until they are all orthogonal. This measurement gives $I_{\text{acc}} = 1.369$ bits, which is larger than $2(\log 3 - 1)$ bits $= 1.170$ bits. In fact, 1.369 bits is larger than twice the maximum accessible information attainable by varying both the probability distribution and the measurement on the three states v_0, v_1 and v_2. This maximum is attained using just two of these states, and is $1 - H\big(\frac{1}{2} - \frac{\sin(\pi/3)}{2}\big) = .6454$. We thus find that block coding lets us achieve a better information transmission rate than I_{acc}.

Having found that length two codewords work better than length one codewords, the natural question becomes: as the lengths of our codewords go to infinity, how well can we do. The answer is:

Theorem (Holevo [Ho2], Schumacher–Westmoreland [SW]). *The classical capacity obtainable using codewords composed of signal states ρ_i, where the probability of using ρ_i is p_i, is*

$$\chi = H_{\mathrm{vN}}\Big(\sum_i p_i\rho_i\Big) - \sum_i p_i H_{\mathrm{vN}}(\rho_i)\,. \tag{25}$$

We will later give a sketch of the proof of this formula in the special case where the ρ_i are pure states. We will first ask: Does this formula give the capacity of a quantum channel $\mathcal{N}$?

Before we address this question (we will not be able to answer it) we should give the general formulation of a quantum channel. If $\mathcal{N}$ is a memoryless quantum communication channel, then it must take density matrices to density matrices. This means $\mathcal{N}$ must be a trace preserving positive map. Here, trace preserving is required since it must preserve trace 1 matrices, and positive means it takes positive semidefinite matrices to positive semidefinite matrices. For $\mathcal{N}$ to be a valid quantum map, it must have one more property: namely, it must be completely positive. This means that $\mathcal{N}$ is positive even when it is tensored with the identity map. There is a theorem [HeK] that any such map can be expressed as

$$\mathcal{N}(\rho) = \sum_i A_i\rho A_i^\dagger \tag{26}$$

where A_i are matrices such that $\sum_i A_i^\dagger A_i = I$.

A natural guess at the capacity of a quantum channel $\mathcal{N}$ would be the maximum of χ over all possible distributions of channel outputs, that is,

$$\chi_{\max}(\mathcal{N}) = \max_{\{p_i\},\{\rho_i\}} \chi\big(\{(\mathcal{N}(\rho_i), p_i)\}\big)\,, \tag{27}$$

since the sender can effectively communicate to the receiver any of the states $\mathcal{N}(\rho_i)$. We do not know whether this is the capacity of a quantum channel; if the use of entanglement between separate inputs to the channel helps to increase channel capacity, it might be possible to exceed this $\chi_{\max}$. This can be addressed by answering a question that is simple to state: Is $\chi_{\max}$ additive [AHW]? That is, if we have two quantum channels $\mathcal{N}_1$ and $\mathcal{N}_2$, is

$$\chi_{\max}(\mathcal{N}_1 \otimes \mathcal{N}_2) = \chi_{\max}(\mathcal{N}_1) + \chi_{\max}(\mathcal{N}_2)\,. \tag{28}$$

Proving subadditivity of this quantity is easy. The question is whether strictly more capacity can be attained by using the tensor product of two channels jointly than by using them separately.

We now return to the discussion of the proof of the Holevo–Schumacher–Westmoreland theorem in the special case where the ρ_i are pure states. The proof of this case in fact appeared before the general theorem was proved [HJSWW]. The proof uses three ingredients. These are

1. random codes,
2. typical subspaces,
3. the square root measurement.

The square root measurement is also called the "pretty good" measurement, and we have already seen an example of it. Recall our second example for accessible information, where we took the three vectors $v_i \otimes v_i$, where $v_i = (\cos 2\pi i/3, \sin 2\pi i/3)$ for $i = 0, 1, 2$. The optimal measurement for I_{acc} on these vectors was the von Neumann measurement obtained by "pulling" them farther apart until they were orthogonal. This is, in fact, an example of the square root measurement.

Suppose that we are trying to distinguish between vectors $u_1, u_2, \ldots, u_n$, which appear with equal probability (the square root measurement can also be defined for vectors having unequal probabilities, but we do not need this case). Let $\phi = \sum_i v_i v_i^\dagger$. The square root measurement has POVM elements $E_i = \phi^{-1/2} v_i v_i^\dagger \phi^{-1/2}$. We have

$$\sum_i E_i = \phi^{-1/2} \Big(\sum_i v_i v_i^\dagger \Big) \phi^{-1/2} = I \,, \tag{29}$$

so these E_i do indeed form a POVM.

We can now give the coding algorithm for the capacity theorem for pure states. We choose N codewords $u_j = v_{i_1} \otimes v_{i_2} \otimes \cdots \otimes v_{i_n}$, where the v_i are chosen at random with probability p_i. We then use the codewords u_j to send information; we need to show that each codeword can be identified with high probability.

To decode, we perform the following steps:

1. Project into the typical subspace $\mathcal{T}$. Most of the time, this projection works, and we obtain $\tilde{u}_j = \Pi_{\mathcal{T}} u_j$, where $\Pi_{\mathcal{T}}$ is the projection matrix onto the subspace $\mathcal{T}$.
2. Use the square root measurement on the $\tilde{u}_j$.

The probability of error is

$$1 - \frac{1}{N} \sum_{j=1}^{N} |\tilde{u}_j \phi^{-1/2} \tilde{u}_j|^2 \,. \tag{30}$$

The intuition for why this procedure works (this intuition is not even close to being rigorous; the proof works along substantially different lines) is that for this probability of error to be small, we need that $\phi^{-1/2}\tilde{u}_j$ be close to

$\tilde{u}_j$ for most j. However, the $\tilde{u}_j$ are distributed more or less randomly in the typical subspace $\mathcal{T}$, so $\phi = \sum_j \tilde{u}_j \tilde{u}_j^\dagger$ is moderately close to the identity matrix on its support, and thus $\phi^{-1/2}\tilde{u}_j$ is close to $\tilde{u}_j$. Note that we need that the number N of u_j be less than $\dim \mathcal{T}$, or otherwise it would be impossible to distinguish the $\tilde{u}_j$; by Holevo's bound (24) a d-dimensional quantum state space can carry at most d bits of information.

8 Quantum Teleportation and Superdense Coding

In this section, we will first describe *quantum teleportation,* a surprising phenomenon which is an unusual means of transmitting a quantum state. It is impossible to send a quantum state over a classical channel. Quantum teleportation lets a sender and a receiver who share an EPR pair of qubits send two classical bits and use this EPR pair in order to communicate one qubit [BenBCJPW]. (See Figure 2.)

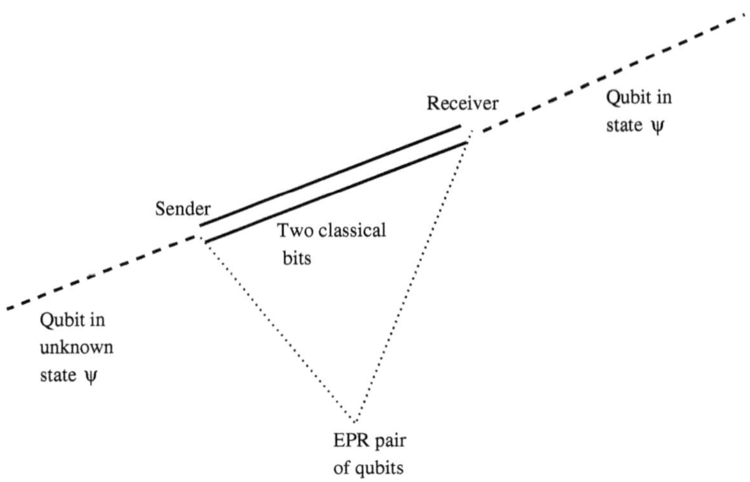

Figure 2: A schematic drawing of quantum teleportation. The sender has a qubit in an unknown state ψ that he wishes to send to the receiver. He also has half of an EPR state which he shares with the receiver. The sender makes a joint measurement on the unknown qubit and half of his EPR state, and communicates the results (2 classical bits) to the receiver. The receiver then makes one of four unitary transformations (depending on the two classical bits he received) on his half of the EPR state to obtain the state ψ.

To perform teleportation, the sender starts with a qubit in an unknown state, which we take to be $\alpha\,|\,0\rangle + \beta\,|\,1\rangle$, and a shared EPR pair, which we assume is in the state $\frac{1}{\sqrt{2}}(|\,00\rangle + |\,11\rangle)$, with the sender holding the first qubit of the EPR pair and the receiver holding the second qubit. The joint system is thus in the tensor product of these two states, which is

$$\tfrac{1}{\sqrt{2}}\big(\alpha\,|\,0\rangle + \beta\,|\,1\rangle\big)\big(\,|\,00\rangle + |\,11\rangle\,\big)\,. \tag{31}$$

Note that the sender has posession of the first two qubits, and the receiver posession of the third one. Using the distributive law, we can rewrite the above state (31) as

$$\begin{aligned}\tfrac{1}{\sqrt{8}}\big[\ &(|\,00\rangle + |\,11\rangle) \quad (\alpha\,|\,0\rangle + \beta\,|\,1\rangle)\\ +&(|\,00\rangle - |\,11\rangle) \quad (\alpha\,|\,0\rangle - \beta\,|\,1\rangle)\\ +&(|\,10\rangle + |\,01\rangle) \quad (\beta\,|\,0\rangle + \alpha\,|\,1\rangle)\\ +&(|\,10\rangle - |\,01\rangle) \quad (\beta\,|\,0\rangle - \alpha\,|\,1\rangle)\ \big]\,.\end{aligned} \tag{32}$$

The sender can now perform the von Neumann measurement that projects the state onto one of the four lines of Eq. (32), as the four states

$$\tfrac{1}{\sqrt{2}}\big(\,|\,00\rangle \pm |\,11\rangle\,\big)\,,\quad \tfrac{1}{\sqrt{2}}\big(\,|\,10\rangle \pm |\,01\rangle\,\big)$$

are all orthogonal. This leaves the receiver with one of the four states

$$\alpha\,|\,0\rangle \pm \beta\,|\,1\rangle\,,\quad \beta\,|\,0\rangle \pm \alpha\,|\,1\rangle\,,$$

all of which can be transformed into $\alpha\,|\,0\rangle + \beta\,|\,1\rangle$ by the appropriate unitary transform. The sender needs to communicate to the receiver which of the four measurement outcomes was obtained (using two bits), and the receiver can then perform the appropriate unitary transform to obtain the original quantum state.

Quantum teleportation is a counterintuitive process, which at first sight seems to violate certain laws of physics; however, upon closer inspection one discovers that no actual paradoxes arise from teleportation. Teleportation cannot be used for superluminal communication, because the classical bits must travel at or slower than the speed of light. While a continuous quantum state appears to have been transported using two discrete bits, by Holevo's bound (24) one qubit can be used to transport at most one classical bit of information, so it is not possible to increase the capacity of a classical channel by encoding information in the teleported qubit. Finally, there is a theorem of quantum mechanics that an unknown quantum state cannot be duplicated [WZ]. However, the original state is necessarily destroyed by the measurement, teleportation cannot be used to clone a quantum state.

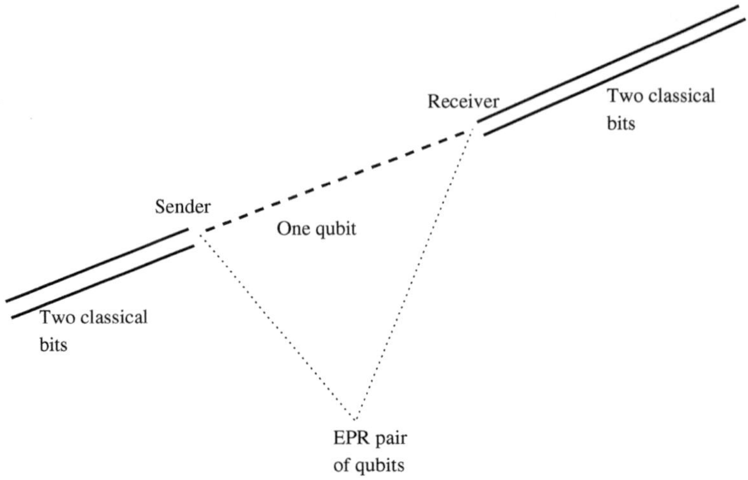

Figure 3: A schematic drawing of superdense coding. The sender can communicate two classical bits to the receiver using one qubit and a shared EPR pair. Here, the sender makes the same unitary transformation that the receiver would make in quantum teleportation, and the receiver makes the joint measurement that the sender would make in quantum teleportation.

There is a converse process to teleportation, *superdense coding,* which uses a shared EPR pair and a single qubit to encode two classical bits [BenW]. In this protocol, the sender and receiver use the same operations as teleportation, but reverse their roles; the sender performs the unitary transformation and the receiver performs the measurement. (See Figure 3.)

9 Other Results from Quantum Information Theory

In this final section, I briefly survey some other results of quantum information theory which were unjustly neglected by the previous sections of this paper.

Using teleportation, the sender can send the receiver qubits over a classical channel if they possess shared EPR pairs. Thus, *shared EPR pairs* (an instance of quantum entanglement) can be seen as a resource that lets these two parties send quantum information over a classical channel, a task that would otherwise be impossible. This leads to the question: how do you quantify entanglement? If two parties have n copies of an entangled state ρ, how many EPR pairs does this let them share? We will let the two parties use classical communication and perform local quantum operations on their own states, but no quantum communication and no quantum operations on the joint state space will be allowed.

If ρ is a pure state, then the answer is known and quite nice [BenBPS]. Let the two parties' quantum state spaces be A and B. Then if $\rho \in A \otimes B$ is a pure state, n copies of ρ can be made into

$$nH_{\mathrm{vN}}(\mathrm{Tr}_A \rho) + o(n) = nH_{\mathrm{vN}}(\mathrm{Tr}_B \rho) + o(n) \tag{33}$$

nearly perfect EPR pairs, and vice versa, where the fidelity of the actual state with the desired state goes to 1 as the block length n goes to infinity.

If ρ is not a pure state, the situation becomes much more complicated. In this case, we can define entanglement of formation (E_F), which is asympotitically the number of EPR pairs that we need to form ρ; and distillable entanglement (E_D), which is asymptotically the number of EPR pairs which can be created from ρ. If ρ is pure, then these two quantities are equal, but this does not appear to be true if ρ is mixed.

Much like the classical capacity of a quantum channel, there is a nice expression which would be equal to the entanglement of formation if it could be proved to be additive. We call it the one-shot entanglement of formation, and it is the minimum average entanglement over ensembles of pure states whose density matrix is ρ. That is,

$$E_{F,1}(\rho) = \min_{\sum_i p_i \rho_i = \rho} \sum_i p_i H_{\mathrm{vN}}(\mathrm{Tr}_A \rho_i)\,. \tag{34}$$

We now give another capacity for quantum channels, one which has a capacity formula which can actually be proven. Suppose that we have a quantum channel $\mathcal{N}$. Recall that if $\mathcal{N}$ is a noiseless quantum channel, and if the sender and receiver possess shared EPR pairs, they can use superdense coding to double the classical information capacity of $\mathcal{N}$. If $\mathcal{N}$ is a noisy quantum channel, using shared EPR pairs can also increase the classical capacity of $\mathcal{N}$. We define the entanglement assisted capacity, C_E, as the quantity of classical information that can asymptotically be sent per channel use if the sender and receiver have access to a sufficient quantity of shared entanglement.

Theorem (Bennett, Shor, Smolin Thapliyal [BenSST1,2]). *The entanglement assisted capacity is*

$$\begin{aligned} C_E(\mathcal{N}) = \max_{\rho \in \mathcal{H} \otimes \mathcal{S}} & H_{vN}\big(\mathrm{Tr}_{\mathcal{R}}(\mathcal{N} \otimes \mathcal{I})\rho\big) \\ & + H_{vN}\big(\mathrm{Tr}_{\mathcal{S}}(\mathcal{N} \otimes \mathcal{I})\rho\big) - H_{vN}\big((\mathcal{N} \otimes \mathcal{I})\rho\big) \end{aligned} \tag{35}$$

where $\mathcal{R}$ and $\mathcal{S}$ stand for receiver and sender, respectively. Here ρ is maximized over pure states on the tensor product of the input state space $\mathcal{H} = \mathbb{C}^d$ of the channel and a quantum space $\mathcal{S}$ (which may be assumed also to be of dimension d) that the sender keeps.

The quantity being minimized in the above formula (35) is called quantum mutual information, and it is a generalization of the expression for mutual information in the form of Eq. (4). The proof of this result uses typical subspaces, superdense coding, the Holevo–Schumacher–Westmoreland theorem on the classical capacity of a quantum channel, and the strong subadditivity property of von Neumann entropy.

Finally, we briefly mention the problem of sending quantum information (i.e., a quantum state) over a noisy quantum channel. In this scenario, several of the theorems that make classical channel capacity behave so nicely are not true. Here, a back channel from the receiver to the sender increases the quantum channel capacity, leading to two quantum capacities, Q_2 where the receiver has a classical back channel from himself to the sender, and $Q \le Q_2$, where all communication is from the sender to the receiver over the noisy quantum channel $\mathcal{N}$. There is a conjectured capacity formula for Q. It is essentially the last two terms of the expression (35) for entanglement-assisted capacity

$$Q(\mathcal{N}) = \lim_{n\to\infty} \max_{\rho\in(\mathcal{H}\otimes\mathcal{S})^n} H_{vN}\big(\mathrm{Tr}_{\mathcal{S}}(\mathcal{N}\otimes\mathcal{I})\rho\big) - H_{vN}\big((\mathcal{N}\otimes\mathcal{I})\rho\big) \quad (36)$$

where ρ, $\mathcal{H}$ and $\mathcal{S}$ are defined as in (35). The quantity being maximized is called the *coherent information.* We now need to take the maximum over the tensor product of n uses of the channel, and let n go to infinity, because unlike the classical (or the quantum) mutual information, the coherent information is not additive [DSS]. The quantity (36) is an upper bound for the quantum capacity of a noisy quantum channel $\mathcal{N}$ [BST], and is conjectured to be equal to this capacity [HorHH].

There are many more results in quantum information theory, including several large areas that I have not discussed at all. I have not mentioned *quantum error-correcting codes,* which are the tools one needs to send quantum information over a noisy channel [Go]. I have also not mentioned quantum cryptography, in connection with which there exist several recent security proofs [BiBBMR], [LoC], [M], [ShoP], and associated results on tradeoffs between disturbing a quantum state and extracting information from it. Finally, I have not mentioned a large literature on entangled quantum states shared among more than two parties. I hope that this paper stimulates some readers to learn more about quantum information theory.

References

[AHW] G.G. Amosov, A.S. Holevo, R.F. Werner, On some additivity problems in quantum information theory, LANL e-print math-ph/0003002, available at `http://xxx.lanl.gov`.

[BST] H. Barnum, J.A. Smolin, B.M. Terhal, Quantum capacity is properly defined without encodings, Phys. Rev. A 58 (1998), 3496–3501.

[Be] J.S. Bell, On the Einstein–Podolsky–Rosen paradox, Physics 1 (1964), 195–200.

[BenBPS] C.H. Bennett, H.J. Bernstein, S. Popescu, B. Schumacher, Concentrating partial entanglement by local operations, Phys. Rev. A 53 (1996), 2046–2052.

[BenBCJPW] C.H. Bennett, G. Brassard, C. Crépeau, R. Jozsa, A. Peres, W.K. Wootters, Teleporting an unknown quantum state via dual classical and Einstein–Podolsky–Rosen channels, Phys. Rev. Lett. 70 (1993), 1895–1899.

[BenSST1] C.H. Bennett, P.W. Shor, J.A. Smolin, A.V. Thapliyal, Entanglement-assisted classical capacity of noisy quantum channels, Phys. Rev. Lett. 83 (1999), 3081–3084.

[BenSST2] C.H. Bennett, P.W. Shor, J.A. Smolin, A.V. Thapliyal, manuscript in preparation.

[BenW] C.H. Bennett, S.J. Wiesner, Communication via one- and two-particle operators on Einstein–Podolsky–Rosen states, Phys. Rev. Lett. 69 (1992), 2881–2884.

[BiBBMR] E. Biham, M. Boyer, P.O. Boykin, T. Mor, V. Roychowdhury, A proof of the security of quantum key distribution, in "Proceedings of the Thirty-Second Annual ACM Symposium on Theory of Computing," ACM Press, New York, (2000), 715–724; longer version LANL e-print quant-ph/9912053, available at `http://xxx.lanl.gov`.

[DSS] D. DiVincenzo, J.A. Smolin, P.W. Shor, Quantum-channel capacity of very noisy channels, Phys. Rev. A 57 (1998), 830–839.

[FP] C.A. Fuchs, A. Peres, personal communication.

[G] J.P. Gordon, Noise at optical frequencies; information theory, in "Quantum Electronics and Coherent Light; Proceedings of the International School of Physics Enrico Fermi, Course XXXI (P.A. Miles, ed.), Academic Press New York (1964), 156–181.

[Go] D. Gottesman, An introduction to quantum error correction, LANL e-print quant-ph/0004072, available at `http://xxx.lanl.gov`.

[GrHSZ] D.M. Greenburger, M.A. Horne, A. Shimony, A. Zeilinger, Bell's theorem without inequalities, Am. J. Phys. 58 (1990), 1131–1143.

[HJSWW] P. Hausladen, R. Jozsa, B. Schumacher, M. Westmoreland, W.K. Wootters, Classical information capacity of a quantum channel, Phys. Rev. A 54 (1996), 1869–1876.

[HeK] K. Hellwig, K. Krauss, Operations and measurements II, Communications in Mathematical Physics 16 (1970), 142–147.

[Ho1] A.S. Holevo, Information theoretical aspects of quantum measurements, Probl. Info. Transm. (USSR) 9:2 (1973), 31–42 (in Russian); [translation: A.S. Kholevo, Probl. Info. Transm. 9 (1973), 177–183].

[Ho2] A.S. Holevo, The capacity of the quantum channel with general signal states, IEEE Trans. Info. Theory 44 (1998), 269–273.

[HorHH] sc M. Horodecki, P. Horodecki, R. Horodecki, Unified approach to quantum capacities: Towards a quantum noisy coding theorem, LANL e-print quant-ph/0003040, available at `http://xxx.lanl.gov`.

[JS] R. Jozsa, B. Schumacher, A new proof of the quantum noiseless coding theorem, J. Modern Optics 41 (1994), 2343–2349.

[L1] L.B. Levitin, On the quantum measure of information, in "Proceedings of the Fourth All-Union Conference on Information and Coding Theory, Sec. II", Tashkent, 1969.

[L2] L.B. Levitin, Optimal quantum measurements for pure and mixed states, in "Quantum Communications and Measurement," (V.P. Belavkin, O. Hirota, R.L. Hudson, eds.), Plenum Press, New York and London (1995), 439–448.

[LoC] H.-K. Lo, H.F. Chau, Unconditional security of quantum key distribution over arbitrarily long distances, Science 283 (1999), 2050–2056.

[M] D. Mayers, Unconditional security in quantum cryptography, J. ACM, to appear; also LANL e-print quant-ph/9802025, available at `http://xxx.lanl.gov`.

[P] J.R. Pierce, The early days of information theory, IEEE Trans. Info. Theory 19 (1973), 3–8.

[S] B. Schumacher, Quantum coding, Phys. Rev. A 51 (1995), 2738–2747.

[SW] B. Schumacher, M. Westmoreland, Sending classical information via a noisy quantum channel, Phys. Rev. A 56 (1997), 131–138.

[Sh] C.E. Shannon, A mathematical theory of communication, The Bell System Tech. J. 27 (1948), 379–423, 623–656.

[ShoP] P.W. Shor, J.A. Preskill, Simple proof of security of the BB84 quantum key distribution protocol, Phys. Rev. Lett. 85 (2000), 441-444.

[WZ] W.K. Wootters, W.H. Zurek, A single quantum cannot be cloned, Nature 299 (1982), 802–803.

Peter Shor, AT&T Labs–Research, Florham Park, NJ 07932, USA

GAFA, Geom. funct. anal.
Special Volume – GAFA2000, 839 – 858
1016-443X/00/S20839-20 $ 1.50+0.20/0

GAFA Geometric And Functional Analysis

UNIVERSALITY, PHASE TRANSITIONS AND STATISTICAL MECHANICS

Thomas Spencer

1 Introduction

This survey will describe some mathematical results and conjectures related to phase transitions and statistical mechanics. In particular, we will focus on the principle of universality which asserts that singularities associated with second order phase transitions are universal. This means that they do not depend on the details of the model. Hence, although the temperature at which a phase transition occurs is typically model dependent, the macroscopic or long distance behavior at the transition is believed to depend on only a few general features such as the dimension of space and the symmetry of the model. Universality explains why relatively simple mathematical models can give quantitatively accurate information about transitions for a wide class of physical and mathematical systems. After a brief digression about random walks and self-avoiding walks we shall focus on the Ising model for interacting spins. The phase transition of the Ising model is believed to describe liquid-gas transitions, coupled chains of quantum anharmonic oscillators, certain quantum field theories and models of probabilistic cellular automata. The last section is devoted to a review of some problems in random Schrödinger operators and GOE and GUE matrix ensembles. The universality of eigenvalue correlations is explored.

General notions of phase transitions and critical exponents. Phase transitions are typically transitions from a disordered state to an ordered state as some parameter, such as temperature, is varied. The classic example is that of an interacting spin system on a d-dimensional lattice $\mathbb{Z}^d$ with spins s_j indexed by sites j of the lattice. In the Ising approximation, these spins simply take values ± 1. We shall describe this class of models in more detail later. For the moment we will attempt to give an intuitive description of the Ising model. We assume that the model is ferromagnetic, which means that neighboring spins tend to point in the same direction and that it is Z_2 symmetric under spin flip, $s_j \to -s_j$. At high temperature T, the spins will be disordered and nearly independent because the effects of thermal noise. The correlation of the spins at $x, y \in \mathbb{Z}^d$ is probability that they are parallel, minus the probability that they are anti-parallel and is given by

$$\langle s_x s_y \rangle(T) \cong \exp -|x-y|/\ell(T) \tag{1.1}$$

where $\ell(T)$ is the correlation length. This is the length beyond which spins are essentially independent. As the temperature decreases this length increases and in two or more dimensions there is a temperature, called the critical temperature T_c, at which this length diverges

$$\ell(T) = \text{Const}|T - T_c|^{-\nu}. \tag{1.2}$$

The critical exponent ν is expected to be universal but not the prefactor. Real experiments as well as numerical simulations support this conjecture. At T_c, the spins lose their exponential independence and collective behavior appears at all scales. One expects that for some critical exponent η

$$\langle s_x s_y \rangle(T_c) = \text{Const}|x-y|^{-(d-2+\eta)}. \tag{1.3}$$

The scaling limit (or continuum limit) of an Ising model, is defined to extract the asymptotic long distance behavior of correlations. In this limit, the lattice $\mathbb{Z}^d$ is replaced by $\mathbb{R}^d$ and new symmetries, such as Euclidean and scale invariance, should emerge. For example, the continuum limit of random walk is Brownian motion. In two dimensions, the scaling limit of statistical mechanics models at T_c is expected to be a conformal field theory. However, the rigorous control of the scaling limit remains one of the major analytical problems in statistical mechanics. For the 2D nearest neighbor Ising model, the massive scaling limit is controlled [PT], [ScO] but there are still problems with the scaling limit at T_c.

For temperatures below T_c, the spins become aligned and there is a net magnetization M given by the limit as $|x-y|$ gets large:

$$\lim \langle s_x s_y \rangle(T) = M(T)^2 > 0. \tag{1.4}$$

A second order phase transition may characterized by the divergence of the length $\ell(T)$ as T approaches T_c from above and the continuous vanishing of $M(T)$ as T approaches T_c from below. Critical exponents, universality and scaling limits are only meaningful notions for models with second order transitions.

Our mathematical understanding of the phase transitions and universality is still quite primitive. There has been substantial progress when the dimension of the lattice is 4 or more and some partial results in dimension 2. In three dimensions, phase transitions are known to occur but critical phenomena and critical exponents are not well understood. In dimension 4 or more, the long distance behavior at T_c is described by a Gaussian process called the free field [A], [Fr], [GaK1], [FMRS1]. The two-dimensional Ising model with nearest neighbor coupling was solved by Onsager [O] in 1944. However, it was only recently [PiS] that universality for small perturbations of the nearest neighbor were proved. These topics are discussed in section 3. In section 4 we explain the relation of Ising models and spectral properties anharmonic oscillators in many variables. In section 5 we conclude with some conjectures and results for random matrices and random Schrödinger operators.

2 Random Walk and Self-avoiding Walk

One of the simplest examples of critical phenomena and universality is given by a discrete time random walk $W(n)$ on $\mathbb{Z}^d$, $n \in \mathbb{Z}$. We shall assume that the steps, $W(n+1) - W(n)$, are independent and bounded by $R \geq |W(n+1) - W(n)|$. Note that $W(n)$ just denotes the position in $\mathbb{Z}^d$ of the walk at time n and we set $W(0) = 0$. Although there is no temperature or phase transition for random walk, we introduce a parameter T analogous to temperature in the following generating function,

$$G(0, x, T) = \sum_{n=0}^{\infty} T^{-n} \operatorname{prob}\big(W(n) = x\big) \tag{2.1}$$

where $x \in \mathbb{Z}^d$. $G(0, x, T)$ is the analogue of the spin correlation (1.1) and is the Greens function for a finite difference Laplacian.

Series (2.1) obviously converges for $T > 1$ and we define $T_c = 1$. For $|x|$ larger than $\ell(T)$ and $T > T_c = 1$ we have

$$G(0, x, T) = \text{Const} \frac{e^{-|x|/\ell(T)}}{|x|^{(d-1)/2}} \tag{2.2}$$

where $\ell(T)$ satisfies (1.2) with $\nu = 1/2$. When $T = T_c$ and $d \geq 3$, then

$$G(x, T_c) = \frac{\text{Const}}{|x|^{d-2}}. \tag{2.3}$$

The critical exponents $\eta = 0$ in (1.3) and $\nu = 1/2$ are universal in the sense that they do not depend on the probability distribution of the steps as long as the distribution is symmetric. The exponent $\nu = 1/2$ implies that after N steps, the average distance traveled by the walk is proportional to $N^{1/2}$. Universality is basically a consequence of the central limit theorem. The same results hold if we do not assume that the steps are independent but depend on some finite number of preceding steps. The continuum limit of random walk is Brownian motion: $W(t) \in \mathbb{R}^d$ with t real. This is both scale invariant and Euclidean invariant.

Self-avoiding walk. A more interesting generating function may be defined for self-avoiding walk on $\mathbb{Z}^d$. Consider paths $W(n) \in \mathbb{Z}^d$ which self-avoid, $W(n) \neq W(n')$ for $n \neq n'$ and whose steps have lengths at most R, $|W(n) - W(n+1)| \leq R$. Let $N(n, x)$ be the number of such paths with $W(0) = 0$ and $W(n) = x$. Now define the generating function by

$$G(x,T) = \sum_{n=0}^{\infty} T^{-n} N(n,x). \tag{2.4}$$

There is a T_c, which depends on R and d, such that for $T > T_c$ (2.7) decays exponentially fast and (2.2) holds for large $|x|$. Moreover, as $T \to T_c$; $\ell(T)$ diverges. We expect that $\nu = \frac{3}{4}$, $\eta = \frac{5}{24}$ for $d = 2$ and for $d = 3$, $\nu \cong .59$; $\eta \cong .03$. See the book of Madras and Slade [MS] for a review. However, only the most trivial bounds are known in two or three dimensions. There are rigorous results for $d \geq 5$ where $\nu = \frac{1}{2}$, $\eta = 0$ [BrS], [HS1], [MS] and for $d = 4$ (R large) $\eta = 0$ [IM]. Thus at long distances a self-avoiding walk in high dimensions looks like a random walk. For $d = 5$ there are similar results for spin systems which will be described later.

Intersection exponents for Brownian paths in two dimensions. Recently, Lawler, Schramm and Werner [LSW] have proved some remarkable results about intersections of Brownian paths in the plane using a stochastic Löwner equation. To give a sample of their results, consider the image of a Brownian path, $B_0(T)$ starting at the origin and running for time T.

A) With probability one, the Hausdorff dimension of outer boundary of $B_0(T)$ is 4/3. More precisely, the boundary of the unbounded connected component of $\mathbb{R}^2 \backslash B_0(T)$ has Hausdorff dimension 4/3.
B) Let $B_1(T)$ be an independent path starting at $x_1 \neq 0$. Then the probability that $B_0(T)$ and $B_1(T)$ do not intersect is proportional to $T^{-5/8}$.
C) The probability that a union of 3 Brownian paths does not intersect an independent union of 3 Brownian paths up to time T is $(73 - 2\sqrt{73})/12$.

Many of the results of [LSW] should apply to percolation and self-avoiding walk in two dimensions once conformal symmetry of the scaling limit is established. The advantage of working with Brownian paths is that for this case, the scaling limit is well defined and has simple transformation laws under conformal mappings.

3 Ising Model

The Ising model describes the collective behavior of interacting spins $s_j = \pm 1$ on a regular lattice. It is also used to model interacting particle systems, where $n_j = (1 + s_j)/2$ indicates the presence or absence of a particle at site j. The energy of a spin configuration $\{s_j\}$ in a box $\Lambda \subset \mathbb{Z}^d$ is given by

$$H_\Lambda(s) = -\sum_{j_1 j_2 \epsilon \Lambda} J_{j_1,j_2} s_{j_1} s_{j_2} \tag{3.1}$$

where J_{j_1,j_2} is translation invariant and decays exponentially as $|j_1 - j_2| \to \infty$. If $J_{j_1,j_2} = 1$ when $|j_1 - j_2| = 1$ and equals 0 otherwise, we say that the model has *nearest neighbor* interactions. When $J \geq 0$ i.e. it is clear that the minimum energy spin configuration completely aligned and $M^2 = 1$, see (1.4). This is the 0 temperature state. At positive temperature $T > 0$ fluctuations from the minimum are allowed. We set $\beta = 1/T$. The Gibbs weight is

$$W_\Lambda^\beta(s) = e^{-\beta H_\Lambda(s)}/Z_\Lambda(\beta) \tag{3.2}$$

where

$$Z_\Lambda(\beta) = \sum_{s_j = \pm 1} e^{-\beta H_\Lambda(s)}.$$

The free energy per unit volume is defined to be

$$F(\beta) = -\lim_{\Lambda \uparrow \mathbb{Z}^2} \frac{\log Z_\Lambda(\beta)}{|\Lambda|} \tag{3.3}$$

and correlations are given by summing over all spin configurations with weight W_Λ^β

$$\langle s_A \rangle_\Lambda(\beta) \equiv \sum W_\Lambda^\beta(s) s_A \tag{3.4}$$

where $s_A = \underset{j \epsilon A}{\pi} s_j$ and A is a finite subset of $\mathbb{Z}^d$. When $\Lambda \uparrow \mathbb{Z}^d$, we drop the subscript Λ.

General inequalities. We now briefly review some general inequalities for correlations and critical exponents under the assumption that J in (3.1) is non-negative (ferromagnetic).

A) Griffiths inequalities.

$$\langle s_A\rangle_\Lambda \geq 0 \qquad (3.5)$$
$$\langle s_A s_B\rangle_\Lambda - \langle s_A\rangle_\Lambda\langle s_B\rangle_\Lambda \geq 0.$$

This last inequality implies that the correlations of the form (3.4) increase as β or the box Λ increases. Hence the limit $\Lambda \uparrow \mathbb{Z}^d$ is well defined.

B) The Lebowitz inequality

$$\begin{aligned} U_4(x_1,x_2,x_3,x_4) \equiv{}& \langle s_{x_1}s_{x_2}s_{x_3}s_{x_4}\rangle_\Lambda \\ &-\big[\langle s_{x_1}s_{x_2}\rangle_\Lambda\langle s_{x_3}s_{x_4}\rangle_\Lambda + \langle s_{x_1}s_{x_3}\rangle_\Lambda\langle s_{x_2}s_{x_4}\rangle_\Lambda \\ &+\langle s_{x_1}s_{x_4}\rangle_\Lambda\langle s_{x_3}s_{x_4}\rangle_\Lambda\big] \leq 0. \end{aligned} \qquad (3.6)$$

C) There is a critical temperature T_c such that for $T > T_c$, the spin-spin correlation decays exponentially fast and using (B) it can be shown that the magnetic susceptibility χ satisfies

$$\chi \equiv \sum_x \langle s_0 s_x\rangle(T) \geq \text{Const}(T-T_c)^{-1} \qquad (3.7)$$

and the correlation length, (1.1), satisfies

$$\ell(T) \geq \text{Const}(T-T_c)^{-1/2},$$

thus $\nu \geq 1/2$. In dimension $d \geq 4$, Aizenman [A] and Fröhlich [Fr] proved that

$$\begin{aligned} \chi(T) &= \text{Const}(T-T_c)^{-1} \quad d>4 \\ &\leq \text{Const}(T-T_c)^{-1}\big|\log(T-T_c)\big| \quad d=4. \end{aligned} \qquad (3.8)$$

Moreover by the nonperturbative estimates of [A], [Fr] we know that when $d > 4$ any continuum limit of Ising Models is Gaussian. In particular U_4 vanishes in the continuum limit and all higher lattice correlations asymptotically factor into products of pair correlations at long distances. These results rely on correlation inequalities which estimate U_4 from below and infrared bounds.

D) Infrared bounds [FrSS]:

$$0 \leq \sum_x \langle s_0 s_x\rangle e^{ixp} \leq \frac{\text{Const}}{p^2} \qquad (3.9)$$

for nonzero p and for $T > T_c$ we have

$$0 \leq \langle s_0 s_x\rangle \leq \text{Const}|x|^{-(d-2)},$$

thus $\eta \geq 0$. These bounds hold for nearest neighbor systems or more generally those satisfying reflection positivity.

For certain weakly coupled lattice field models (defined in the next section) or Ising systems with with long range $J_{i,j}$ there are more detailed results due to Gawedzki and Kupiainen [GaK1] and Feldman et al. [FMRS1] which hold for $d \geq 4$. In particular they prove that $\eta = 0$ for $d \geq 4$ and $\nu = 1/2$ for $d \geq 5$. For $d = 4$ there is a logarithmic correction [HT] to $\nu = 1/2$

$$\ell(T) \cong (T - T_c)^{-1/2} \left|\log(T - T_c)\right|^{1/6}. \tag{3.10}$$

For $d = 4$ the scaling limit is a Gaussian free field. These results are established by rigorous renormalization group methods. They are believed to hold without the assumption of weak coupling. There are related results for percolation due to Hara and Slade [HS2, HS3] except one must require that $d > 6$.

Universality for the 2-dimensional Ising model. The nearest neighbor Ising model in two dimensions was solved by Lars Onsager in 1944 [O]. It was the first example of a model with a phase transition which could be studied in detail and which yielded critical exponents different from those of mean field theory. Since Onsager's seminal work, the model has been analyzed and re-expressed in many different ways. However, all solutions rely heavily on the nearest neighbor interaction together with the absence of a magnetic field. See the book of McCoy and Wu, [McW], for a review. Next to nearest neighbor perturbations or 4 spin plaquette perturbations destroy the integrability of the model. Nevertheless, it is widely believed that such perturbations do not change the critical exponents. Thus the long distance behavior of correlations near T_c should be proportional to those calculated by Onsager. This is a special case of the principle of universality which asserts that critical exponents and scaling limits depend only on a few general features of the model such as symmetry and dimension. However, one must be careful about such assertions since it is known that two independent Ising models in 2D coupled with a four spin interactions is equivalent to Baxter's eight vertex model [Ba] which has non-universal behavior even for small coupling.

Recent results with Haru Pinson [PiS] establish a form of universality for perturbations of the 2D nearest neighbor model. These perturbations are required to be even and local. Under these conditions, we show that the specific heat and energy-energy correlations near T_c behave as they do in the nearest neighbor case. Moreover, the partition function on N by N a periodic box is universal at T_c, after the free energy is extracted. We cannot yet treat spin-spin correlations.

To state our results more precisely, let $\Lambda \subset \mathbb{Z}^2$ be a large box with periodic boundary conditions. Let H_I denote the perturbation of the nearest neighbor model and define

$$H^\epsilon_\Lambda(s) \equiv - \sum_{\substack{|j-j'|=1 \\ j,j' \in \Lambda}} s_j s_{j'} + \epsilon H_I . \tag{3.11}$$

For illustration we consider perturbation H_I of the form

$$H_I = a \sum_{i,j\in\Lambda} J_{i,j} s_i s_j + b \sum_p \prod_{j\in p} s_j \tag{3.12}$$

where p is a plaquette and $j \in p$ are the 4 vertices of the plaquette. The letters a and b denote constants. The couplings $J_{i,j}$ are assumed be translation invariant and decay exponentially fast in $|i-j|$.

Let $F^\epsilon(\beta)$ denote the free energy of H^ϵ, see (3.3). For $\epsilon = 0$, Onsager proved that F is analytic in β except for one singularity β_c at which there is a logarithmic divergence in the second derivative. In [PiS] we prove that for small ϵ, there is a $\beta_c(\epsilon)$ depending smoothly on ϵ such that $F^\epsilon(\beta)$ is smooth except at $\beta_c(\epsilon)$

$$\frac{\partial^2}{\partial\beta^2} F^\epsilon(\beta) \approx C \log\big(|\beta - \beta_c(\epsilon)|^{-1}\big). \tag{3.13}$$

Moreover, energy-energy correlations at $\beta_c = \beta_c(\epsilon)$ behave as in the case $\epsilon = 0$ solved by Onsager:

$$\big|\langle s_x s_{x'} s_y s_{y'}\rangle(\beta_c) - \langle s_x s_{x'}\rangle(\beta_c)\langle s_y s_{y'}\rangle(\beta_c)\big| \cong \frac{C_1}{(|x-y|+1)^2} \tag{3.14}$$

for some positive constant C_1. Here, x and x' denote nearest neighbor sites and $< \cdot >$ is the expectation with respect to (3.11).

For $\beta \neq \beta_c$ the energy-energy correlations decay exponentially $\exp -|x-y|/\ell(\beta)$ with a correlation length exponent $\nu = 1$, (see (1.2)). Our result on ν is proved only in the even sector for systems with a selfadjoint transfer matrix. The spin-spin correlation length is expected to have the same exponent. The exponent $\nu = 1$ differs from that of random walk or Ising models in dimension greater than four where in both cases $\nu = 1/2$.

Let $Z^\epsilon_\Lambda(\beta)$ denote the partition function corresponding to (3.12) in an N by N periodic box Λ and let $F^\epsilon(\beta)$ be the free energy per unit area (3.3). Then, assuming that the interaction is invariant under certain lattice rotations and reflections, we prove [PiS] that the partition function at β_c is universal after its bulk contribution is extracted. Thus

$$\log Z^\epsilon_\Lambda\big(\beta_c(\epsilon)\big) + F^\epsilon\big(\beta_c(\epsilon)\big) N^2 \tag{3.15}$$

is of order 1 and independent of ϵ as N goes to infinity.

Problems in PCA – probabilistic cellular automata. Probabilistic cellular automata are models for the time evolution of spin systems on a lattice. One is typically interested in the long time or steady state behavior of such systems and their dependence on initial data. We shall consider simple deterministic updating rules in which a spin at time time $t+1$, $s_j(t+1)$, $j \in \mathbb{Z}^d$, is determined by the spins $s_{j'}(t)$ for j' near j. Then we add a small noise to the system, independently in space and time, so that with probability p the spin is randomly assigned a value ± 1 . We shall suppose that the spins are updated simultaneously, and that the rule is time independent.

One of the simplest examples of such a system is that of majority rule. The spin $s_j(t+1)$ is given by the majority of itself and its nearest neighbors at time t. However, with probability p the rule is ignored and it takes values ± 1 with equal probability. Clearly, if $p = 0$, an initial state of all $+$ is preserved. However, in one dimension, for any $p > 0$, noise dominates the majority rule and the initial condition is forgotten in the long time limit. Thus if we started with all $+$ or with all $-$, after a long time the spin configurations would be statistically identical and we would see an equal number of $+$ and $-$ over long distances. This result was established for sequential updating by Gray [Gr]. On the other hand, P. Gacs [G] has developed much more complicated rules in one dimension for which there are uncountably many invariant measures. His rule has finite range and uses ideas of error correcting codes.

In two dimensions it is conjectured that the majority rule model has a magnetization for small p. This means that if the initial condition is all $+$, our long time state will be primarily $+$ with rare pockets of $-$. Unfortunately, the standard tools of statistical mechanics do not seem adequate to establish this conjecture in general. There are some special PCA which are known to have phase transitions. See [BiCLS].

Finally, we turn to a conjectured universality for the majority rule. In two dimensions as we vary p a phase transition should occur. If p is near 1 then the system is disordered and the initial condition is forgotten. The parameter p is like temperature in equilibrium statistical mechanics. As p is decreased there should be a p_c at which order begins to emerge. We expect that the invariant measure for this case is again governed by the Onsager's Ising model, [GriJH].

Remarks on the effects of noise on universality. Let us return to the Ising model but now consider the case where the nearest neighbor interaction $J_{i,j}$ is modified by some small noise which is independent for each nearest neighbor pair. For example one might consider sparse random defects in the lattice where $J_{i,j} = 0$. There is still a phase transition with probability one. However, another universality class is expected to govern this system. It is conjectured that the specific heat singularity should go from a log to a log log singularity. See, [DoD]. If the randomness is strongly correlated along vertical lattice lines, see [McW]. For the nearest neighbor case, the analysis reduces to the study of the Greens function of a Euclidean Dirac operator with a random mass fluctuating about 0. In three dimensions, the effect of weak noise is expected to be even more pronounced but this case is much more difficult to analyze.

The role of symmetry. The Ising model described above has a Z_2 symmetry. It may be generalized by considering spins s_j with values on the sphere S^{n-1}. The sum over $s_j = \pm 1$ is replaced by integration over the sphere. The interaction has the form $H = -\sum_{i,j} s_i \cdot s_j$ where the sum is taken, for example, over nearest neighbors. This defines the O(n) symmetric Heisenberg model. There is a different universality class associated with each n. In three dimensions, models of super-fluids have an $O(2)$ symmetry and should be described by the $O(2)$ Heisenberg model. In two dimensions, one of the major open conjectures due to Polyakov asserts that for $n \geq 3$ the $O(n)$ Heisenberg models have no phase transition and at all positive temperatures, spin correlations decay exponentially fast at long distances. This conjecture is widely accepted in the theoretical physics community, although no proof has been found. For the 2D $O(2)$ model there is a phase transition characterized by an interval of temperatures for which spin correlations have power law decay. This is called the Kosterlitz-Thouless [KT] transition which was rigorously established in [FrS]. The correlation length exponent ν, see (1.3), is expected to be infinite: $\ell(T) \cong e^{C(T-T_c)^{-1/2}}$ but details of the transition have not yet been rigorously established.

Ideas of the proof. The nearest neighbor Ising model in two dimensions can be expressed in terms of "Gaussian" Grassmann integrals. See [S], [ItD], [PiS]. This means that the action is quadratic in anti-commuting Grassmann fields indexed by $j \in \mathbb{Z}^2$ which satisfy

$$\psi_{j_1}\psi_{j_2} + \psi_{j_2}\psi_{j_1} = 0, \qquad \psi_j^2 = 0. \tag{3.16}$$

There are well defined rules for calculating Grassmann integrals which provides a very convenient algebraic formalism for dealing with determinants and Pfaffians. The covariance of the "Gaussian" Grassmann integral is given by the Greens function of a finite difference Euclidean Dirac operator.

Perturbations of the nearest neighbor case can also be expressed in terms of Grassmann integrals but the perturbation introduces terms of degree 4 and higher into the action. Thus the model is no longer solvable. Nevertheless, we prove that such perturbations are irrelevant provided that there are no more than 2 independent Grassmann fields per lattice site, [PiS].

REMARK. Baxter's [Ba] eight vertex model, can be expressed in terms of 4 independent Grassmann fields per site but in this case universality is violated since even small quartic perturbations are known to change critical exponents.

Our proof uses a rigorous renormalization group analysis and follows the formalism developed by Feldman, Knörrer and Trubowitz [FKT] which seems well suited to our problem. These techniques have their roots in the earlier work of [GaK2], [FMRS2] on the Gross-Neveu quantum field theory. See also [BeG]. Since we are working with Grassmann variables, there are no "large field" problems and the Hadamard inequality is used to estimate remainders and large products of fields.

4 Lattice Field Models and Anharmonic Oscillators

Many of the results described in the previous section hold for lattice field theories with fields or spins s_j taking real values. Let us consider the Hamiltonian

$$H_\Lambda = \tfrac{1}{2}\sum_{j\in\Lambda}(\nabla s)_j^2 + \sum_{j\in\Lambda} V(s_j) \tag{4.1}$$

with V a symmetric double well potential and ∇ a finite difference gradient. The potentials

$$V_1(s_j) = \lambda(s_j^2 - b^2)^2 \tag{4.2}$$

$$e^{-V_2(s_j)} = e^{-\lambda(s_j+b)^2} + e^{-\lambda(s_j-b)^2} \tag{4.3}$$

both give rise to local Gibbs distributions that are sharply concentrated near $s_j = \pm b$ when λ is large. In the case of V_2, the partition function

$$Z_\Lambda = \int e^{-H_\Lambda(s)} \prod_{j\in\Lambda} ds_j \tag{4.4}$$

can be evaluated since for each choice of $\pm b$ in (4.3) the integral is Gaussian. The result is an Ising sum proportional to

$$\sum_{s_j=\pm 1} \exp\left[\sum_{\Lambda} J_{i,j} s_i s_j\right] \tag{4.5}$$

where J_{ij} is the Greens function

$$J_{ij} = \frac{b^2\lambda^2}{-\Delta + 2\lambda}(i,j) \tag{4.6}$$

and Δ denotes the lattice Laplacian. Thus we have mapped the lattice field model corresponding to V_2 to an Ising model. Note that the parameter b plays the role of β. When λ is large, J_{ij} is a small perturbation of the nearest neighbor case since for $|i-j| > 1$, J_{ij} is proportional to $\lambda^{-(|i-j|-1)}$. Hence, the results of the preceding section can be applied. We believe that V_1 can be analyzed in a similar way except that there will also be multi-spin interactions generated.

If we let Λ be an N by T cylinder and take the continuum limit of (4.1) in the T direction with periodic boundary conditions, the partition function (4.4) becomes the Feynman-Kac representation for the semigroup generated by a chain of quantum anharmonic oscillators denoted H_q:

$$H_q^N = \tfrac{1}{2}\sum_i^N \left[-\frac{d^2}{dx_i^2} + (x_i - x_{i+1})^2 + V(x_i)\right] \tag{4.7}$$

and

$$Z_\Lambda = Ctre^{-TH_q^N}.$$

Let $E_0^N(b) < E_1^N(b) < E_2^N(b) \cdots$ denote the eigenvalues of H_q^N in the subspace of even functions. Note that these eigenvalues depend on b through (4.2). Then for large λ we expect that in the limit of large N, $E_0^N(b)/N$ has a logarithmic singularity in its second derivative at some b_c and

$$E_1^N(b) - E_0^N(b) = \text{Const}|b - b_c|. \tag{4.8}$$

Thus as N goes to infinity, the gap vanishes at b_c. This signals the phase transition and the linear dependence, $|b - b_c|$, implies $\nu = 1$. To establish such a result, one must map the lattice field system onto an Ising model, which is then expressed in Grassmann variables. The continuum limit in one direction produces anisotropy in the Grassmann action which can hopefully be handled by standard methods.

The Hamiltonian given by (4.7) can be generalized to higher dimensions by letting the index i range over a box in $\mathbb{Z}^d$. The quadratic coupling term becomes $(x_i - x_{i'})^2$ where i and i' are nearest neighbors. In dimensions four and above, the eigenvalue separation (4.8) is proved [GaK1], [FRMS1] to be proportional to $|b - b_c|^{1/2}$ for small $\lambda > 0$ and $b \leq b_c$. Hence, $\nu = 1/2$ and $\eta = 0$ is also established. These results rely on renormalization group methods. Thus in high dimension, lattice spin systems and lattice field models governed by a free Gaussian bosonic field whereas in two dimensions it is governed by the free Fermi field. In both cases universality should hold for all $\lambda > 0$.

5 Random Schrödinger, Random Matrices and Supersymmetry

In this section we shall describe problems and conjectures concerning universality of eigenvalue distributions and the effects of randomness on the time evolution of a quantum particle. Eigenvalue spacings for Gaussian matrix ensembles are believed have a wide universality class that even extends to 3-dimensional random Schrödinger operators.

Consider a Schrödinger operator $H = -\Delta + a\,V$ with a random potential $V(x)$. Let us assume that the potential is stationary, and that $V(x)$ is independent for distant values of x. For simplicity we shall often suppose that we are on the lattice $\mathbb{Z}^d$ with independent, identically distributed random variables for each site. The random potential is supposed to model impurities or defects in a crystal. If $|a|$ is small and $|V(x)| \leq 1$, one might imagine the randomness would have little effect on the spectrum of the operator and time evolution of a wave packet. This is not true. In one dimension it was proved that such an operator has pure point spectrum. The spectrum consists of a dense set of eigenvalues and the eigenstates decay exponentially fast with probability one. This is called localization. For a review see [CL]. In any dimension, there are intervals of pure point spectrum near the bottom of the spectrum. One of the major conjectures, which is analogous to Polyakov's conjecture for the 2D Heisenberg model, is that for $a \neq 0$, H has pure point spectrum in two dimensions.

In three dimensions the Hamiltonian H is expected to exhibit a phase transition in the following sense. For small $|a|$ there should exist an energy E_m called the mobility edge below which there is pure point spectrum and above which there is absolutely continuous spectrum. The pure point spectrum is quite well understood but the absolutely continuous spectrum seems much more challenging to analyze mathematically. Moreover, the behavior of the localized eigenstates of H for energies near the mobility edge is not even understood in high dimensions.

The time evolution of a 3D random Schrödinger operator is quite different from the case without randomness, even if we project onto the absolutely continuous spectrum. Let $\Psi_0(x)$ denote a function which decays exponentially fast away from the origin. Define

$$R^2(t) = Av \int \left|e^{itH}\Psi_0(x)\right|^2 x^2 dx \tag{5.1}$$

where Av denotes the average over randomness. $R^2(t)$ represents the mean square displacement of a particle after time t. If H has pure point spectrum (as in 1D), $R^2(t)$ is bounded. When there is no randomness, $R^2(t) \cong$ Const t^2 corresponding to ballistic motion - the particle travels a distance proportional to t.

DIFFUSION CONJECTURE. *For $|a|$ small and $d \geq 3$, $R^2(t) = Dt$ for large t. On a lattice, D is proportional to $|a|^{-2}$.*

REMARKS. Thus randomness induces a change in type in the effective behavior of the differential equation from dispersive or ballistic to diffusive. The above conjecture is supported by singular perturbation theory in which many terms must be resumed. It is also supported by a more systematic approach of supersymmetric field theory. The diffusive nature of the time evolution may be intuitively understood as follows. If a classical particle was scattered by random impurities in the R^3 it reasonable to suppose that its trajectory would look like that of random walk after a long time. This suggests that a semi-classical approach to this problem is natural. However, I strongly suspect that the classical problem is far more difficult than its quantum counterpart.

DENSITY OF STATES CONJECTURE. *For all $a \neq 0$. The density of states*

$$\rho(E) = \frac{1}{\pi} \lim_{\epsilon \downarrow 0} Im\, Av \frac{1}{H - E + i\epsilon}(0,0) \tag{5.2}$$

is smooth for all E. Moreover

$$\left| Av \frac{1}{H - E + i\epsilon}(0,x) \right| \leq C\, e^{-\gamma|x|} \tag{5.3}$$

where $\gamma \cong a^2$.

REMARKS. Smoothness of the density of states and exponential decay is known for the Cauchy distribution and for analytic distributions for energies in the localized region - e.g. in 1D. Perhaps one should be more cautious and conjecture a rapid power law decay rather than exponential decay.

In any case, rapid decay of the average Greens function does not reflect the localized or diffusive nature of the spectrum. For E in the absolutely continuous spectrum, the decay presumably reflects the rapid decorrelation of the phases of eigenstates at long distances.

Note that one expects for $d \geq 3$ and E in the "middle" of the spectrum

$$Av \int e^{ix\cdot p} \left| \frac{1}{H - E + i\epsilon}(0, x) \right|^2 dx \cong \frac{\rho(E)}{Dp^2 + \epsilon} \tag{5.4}$$

for small values of p. Thus the order which averages and absolute values are taken is important. The p^2 pole on the right reflects diffusive character of H and $D = D(E)$ is the diffusion constant.

Eigenvalue correlations. Eigenvalue correlations have been extensively studied in both the mathematics and physics literature. The Gaussian unitary ensemble, GUE, and the Gaussian orthogonal ensemble, GOE, are N by N matrices $H_{i,j}$ whose matrix elements are independent Gaussian random variables with mean 0 and variance $1/N$ subject to the constraints $H_{i,j} = \bar{H}_{j,i}$ for GUE and $H_{i,j} = H_{j,i}$ for GOE. If U is an arbitrary N by N unitary matrix then H and U^*HU have identical distributions for H in GUE. Similar results hold for GOE.

Let $\delta = \delta(E)$ denote the average spacing between eigenvalues near E. Inside the spectrum, δ is proportional to $1/N$ and we define $x = \pi\omega/\delta$. For large N, the correlation function for two eigenvalues near E separated by ω is given by

$$R_U(E, E + \omega) = 1 - \frac{\sin^2(x)}{x^2} \tag{5.5}$$

and

$$R_O(E, E + \omega) = 1 - \frac{\sin^2(x)}{x^2} - \frac{d}{dx}\left(\frac{\sin x}{x}\right) \int_1 \frac{\sin(sx)}{s} ds \tag{5.6}$$

for the unitary and orthogonal cases, respectively. Higher order correlations and level spacings can also be rigorously be calculated for GUE and GOE. See [Me] for precise definitions and references for GUE and GOE.

If the matrix elements are independent but non-Gaussian, numerical results and supersymmetric methods suggest the above correlations do not change. See the recent preprint of K. Johansson. Similar results are expected to hold for sparse random matrices in which there is an average of p nonzero matrix elements per row, [MirF]. There are rigorous results [BlI], [DKMVZ], [PaS] which prove universality of GUE correlations if the hermitian matrix elements are distributed by the weight $\exp -[NtrV(H)]$ for a wide class of analytic functions V. This is a achieved using properties of orthogonal polynomials in y with respect to the weight $e^{-V(y)}$. We emphasize that eigenvalue correlations (5.5-5.6) are only valid for energies inside the spectrum of H. There are different scalings associated with energies near the edge of the spectrum. In particular the statistics of the lowest eigenvalue of a GUE ensemble has been calculated by Tracy and Widom [TW]. These statistics appear in problems related to random permutations and directed polymers in two dimensions [BDJ], [J]. Universality for the lowest eigenvalue distribution was established by Soshnikov [So] for independent non-Gaussian matrix elements.

CONJECTURE. *If H is a random Schrödinger operator in a box $\Lambda \subset Z^d$ of side N then in three or more dimensions, eigenvalue correlations inside the spectrum are given by GOE provided $|\omega| \leq N^{-2}$. See* [E].

For energies away from spectral edges, the level spacing δ is proportional to N^{-d} and the normalized eigenstates are believed to have amplitudes proportional to $N^{-d/2}$ when $|a|$ is small. If we are in one dimension or E is in an interval of localized spectrum, then the distribution is not GOE but rather Poisson due to the localized nature of the eigenstates [Mi]. Finite volume corrections to GOE statistics have been formally calculated and involve the diffusion modes. These corrections allow for larger values of ω but depend on the effective diffusion constant D in (5.4). If one adds a magnetic field or magnetic randomness then GUE statistics should emerge.

Remarks on the supersymmetric formalism. A natural approach to the problems described above is through statistical mechanics. A lattice field model consisting of continuous spins as well as Grassmann fields ψ_j, enables one to formally integrate over the randomness. The price that we pay is that after averaging out the randomness, we obtain an interacting lattice field theory that resembles (4.1) but with a non-compact symmetry group. In the case of GUE and GOE one can make a change of coordinates so that the field theory reduces to a finite dimensional integral which can be calculated explicitly. This field theoretic approach has its roots in the work of physicists F. Wegner and K. Efetov. We briefly sketch some of the ideas below and refer to the book of Efetov [E] for details.

Let $\phi_j = (s_j, t_j, \psi_j, \bar{\psi}_j)$ where the ψ fields are Grassmann variables satisfying (3.16) and the s_j and t_j are real fields. The Greens function is a spin-spin correlation

$$\frac{1}{H - E + i\epsilon}(0, x) = \langle s_o s_x \rangle \tag{5.7}$$

where the the expectation on the right is with respect to the quadratic weight

$$\exp i \left[\sum_{j,k\epsilon\Lambda} \phi_k \cdot (H - E + i\epsilon)(k,j)\phi_j \right]. \tag{5.8}$$

Note that this is a Gaussian integral, so the spin-spin correlation yields the covariance which is the Greens function or the inverse of the above quadratic form. Also notice that the integral is oscillatory but is well defined for positive ϵ. The role of the Grassmann fields is to cancel a determinant which also produced by the Gaussian integral in t and s variables. Thus the cancellation of these determinants by the ψ integrals means that the integral of (5.8) over ϕ is one. The integral over the disorder in H can be computed in terms of the Fourier transform of the probability distribution. If we have an random potential on the lattice with an independent Gaussian distribution at each site, then the average over disorder produces a quartic interaction of the form $\exp - \sum_j (\phi_j \cdot \phi_j)^2$. In addition there is a quadratic action coming from the Laplacian which couples the fields at different sites.

In order to study eigenvalue correlations we must consider Greens functions and their complex conjugates at energies E and $E + \omega$. This requires introducing an additional field ϕ'_j. The interaction now has the form

$$\exp - \sum_{j\in\Lambda} (\phi_j \cdot \phi_j - \phi'_j \cdot \phi'_j)^2. \tag{5.9}$$

The minus sign above arises from the complex conjugate. The indefinite nature if this action is reflects for a noncompact, hyperbolic symmetry. It is the precise nature of this symmetry which ultimately governs universality classes for eigenvalue correlations. It is argued in [E] that when calculating $R_O(\omega)$ via the field theoretic formalism, the integral over the field can be approximated by the 0 mode contribution in suitable variables. Unfortunately, this is not an easy approximation to control mathematically. The resulting integral is over a finite dimensional saddle manifold and can be evaluated. As mentioned earlier, ω must be suitably small so that the diffusion modes do not contribute.

References

[A] M. Aizenmann, Geometric analysis of φ^4 fields and Ising models. I, II, Comm. Math. Phys. 86 (1982), 1–48.

[BDJ] J. Baik, P. Deift, K. Johansson, On the distribution of the length of the longest increasing subsequence of random permutations, J. Amer. Math. Soc. 12 (1999), 1119–1178.

[Ba] R. Baxter, Exactly Solved Models in Statistical Mechanics, Academic Press, 1982.

[BeG] G. Benfatto, G. Gallavotti, Renormalization Group, Princeton University Press, 1995.

[BiCLS] S. Bigelis, E.N.M. Cirillo, J.L. Lebowitz, E.R. Speer, Critical droplets in metastable states of probabilistic cellular automata, Phys. Rev. E (3) 59:4 (1999), 3935–3941.

[BlI] P. Bleher, A. Its, Semiclassical asymptotics of orthogonal polynomials, Riemann-Hilbert problem, and universality in the matrix model, Ann. of Math. 150 (1999), 185–266.

[BrS] D. Brydges, T. Spencer, Self-avoiding walk in 5 or more dimensions, Comm. Math. Phys. 97 (1985), 125–148.

[CL] R. Carmona, J. Lacroix, Spectral Theory of Random Schrödinger Operators", Birkhäuser, Boston, 1990.

[DKMVZ] P. Deift, T. Kriecherbauer, K. McLaughlin, S. Venakides, X. Zhou, Uniform asymptotics for polynomials orthogonal with respect to varying exponential weights and applications to universality questions in random matrix theory, Comm. Pure Appl. Math. 52 (1999), 1335–1425.

[DoD] V. Dotsenko, V. Dotsenko, Critical behaviour of the phase transition in the 2D Ising model with impurities, Adv. in Phys. 32 (1983), 129–172.

[E] K. Efetov, Supersymmetry in Disorder and Chaos, Cambridge University Press, Cambridge, 1997.

[FKT] J. Feldman, H. Knörrer, E. Trubowitz, A representation for fermionic correlation functions, Comm. Math. Phys. 195 (1998), 465–493.

[FMRS1] J. Feldman, J. Magnen, V. Rivasseau, R. Sénéor, Construction and Borel summability of infrared ϕ_4^4, Comm. Math. Phys. 109 (1987), 437–480.

[FMRS2] J. Feldman, J. Magnen, V. Rivasseau, R. Sénéor, A renormalizable field theory: the massive Gross-Neveu model in two dimensions, Comm. Math. Phys. 103 (1986), 67–103.

[Fr] J. Fröhlich, On the triviality of $\lambda\varphi_d^4$ theories and the approach to the critical point in $d > 4$ dimensions, Nuc. Phys. B200 (1982), 281–296.

[FrSS] J. Fröhlich, B. Simon, T. Spencer, Infrared bounds, phase transitions and continuous symmetry breaking, Comm. Math. Phys. 50 (1976), 79–95.

[FrS] J. Fröhlich, T. Spencer, The Kosterlitz–Thouless phase transition in two-dimensional abelian spin systems and the Coulomb gas, Commun. Math. Phys. 81 (1981), 527–602

[G] P. Gacs, Reliable computation with cellular automata, J. Comput. System Sci. 32(1) (1986), 15–78 and preprint submitted to J. Stat. Phys.

[GaK1] K. GAWEDZKI, A. KUPIAINEN, Massless lattice ϕ_4^4 theory: Rigorous control of a renormalizable asymptotically free model, Comm. Math. Phys. 99 (1985), 197–252.

[GaK2] K. GAWEDZKI, A. KUPIAINEN, Gross-Neveu model through convergent perturbation expansions, Comm. Math. Phys. 102 (1985), 1–30.

[Gr] L. GRAY, The positive rates problem for attractive nearest neighbor spin systems on Z.Z. Wahrsch, Verw. Gebiete 61(3) (1982), 389–404.

[GriJH] G. GRINSTEIN, C. JAYAPRAKASH, Y. HE, Statistical mechanics of probabilistic cellular automata, Phys. Rev. Lett. 55 (1985), 2527–2530.

[HS1] T. HARA, G. SLADE, Self-avoiding walk in five or more dimensions. I. The critical behaviour, Comm. Math. Phys. 147 (1992), 101–136.

[HS2] T. HARA, G. SLADE, Mean-field critical behaviour for percolation in high dimensions, Comm. Math. Phys. 128(2) (1990), 333–391.

[HS3] T. HARA, G. SLADE, The scaling limit of the incipient infinite cluster in high-dimensional percolation. I. Critical exponents, math-ph/9903042.

[HT] T. HARA, H. TASAKI, A rigorous control of logarithmic corrections in four-dimen–sional φ^4 spin systems. II. Critical behavior of susceptibility and correlation length, J. Stat. Phys. 47 (1987), 99–121.

[IM] D. IAGOLNITZER, J. MAGNEN, Polymers in a weak random potential in dimension 4, Commun. Math. Phys. 162 (1994), 85–121.

[ItD] C. ITZYKSON, J. DROUFFE, Statistical Field Theory: 1, Cambridge Univ. Press, 1989.

[J] K. JOHANSSON, Shape fluctuations and Random matrices, Commun. Math. Phys. 209 (2000), 437–476

[KT] J. KOSTERLITZ, D. THOULESS, Order, metastability and phase transitions in two dimensional systems, J. Phys. C 6 (1973), 1181–203.

[LSW] G. LAWLER, O. SCHRAMM. W.WERNER, Values of Brownian intersection exponents I, II, III, preprint.

[MS] N. MADRAS, G. SLADE, The Self-avoiding Walk, Birkhäuser, 1996.

[McW] B. MCCOY, T. WU, The Two-dimensional Ising Model, Harvard Univ. Press, 1973.

[Me] M. MEHTA, Random Matrices, Academic Press, Boston, 1991.

[Mi] N. MINAMI, Local fluctuation of the spectrum of a multidimensional Anderson tight binding model, Comm. Math. Phys. 177 (1996), 709–725.

[MirF] A. MIRLIN, Y. FYODOROV, Universality of level correlation function of sparse random matrices, J. Math. Phys. A24 (1991), 2273–2286.

[O] L. ONSAGER, Phys. Rev. 65 (1944), 117.

[PT] J. PALMER, C. TRACY, Two-dimensional Ising correlations: Convergence of the scaling limit, Adv. App. Math. 2 (1981), 329–388.

[PaS] L. PASTUR, M. SHCHERBINA, Universality of the local eigenvalue statistics for a class of unitary invariant random matrix ensembles, J. Stat. Phys. 86(1-2) (1997), 109–147.

[PiS] H. Pinson, T. Spencer, Universality and the two dimensional Ising model, preprint, to appear in CMP.

[S] S. Samuel, The use of anticommuting variable integrals in statistical mechanics, J. Math. Phys 21 (1980), 2806.

[ScO] R. Schor, M. O'Carroll, The scaling limit and Osterwalder- Schrader axioms for the two-dimensional Ising model, Comm. Math. Phys. 84 (1982), 153–170.

[So] A. Soshnikov, Universality at the edge of the spectrum in Wigner random matrices, Comm. Math. Phys. 207 (1999), 697–733.

[TW] C. Tracy, H. Widom, Level-spacing distributions and the Airy kernel, Commun. Math. Phys. 159 (1994), 151–174.

Thomas Spencer, School of Mathematics, Institute for Advanced Study, Princeton, NJ 08540, USA spencer@math.ias.edu

GAFA, Geom. funct. anal.
Special Volume – GAFA2000, 859 – 879
1016-443X/00/S20859-21 $ 1.50+0.20/0

GAFA Geometric And Functional Analysis

HOW CLASSICAL PHYSICS HELPS MATHEMATICS

Vladimir Zakharov

1 Introduction

The history of the relations between Physics and Mathematics is a long and romantic story. It began at the time of Archimedes, and up to the seventeenth and eighteenth centuries the relations were quite cordial. Mathematics supplied the tools for the solution of physical problems, and in its turn, the necessity to develop proper tools was a very strong factor in stimulating progress in mathematics itself. The problem of the brachistochrone, which was a starting point in the creation of variational calculus, is a classic example. In those times most outstanding mathematicians were also physicists.

In the nineteenth century the relations were still close, but some tendency to alienation and separation had become visible. Riemann was both a mathematician and a physicist, while Weierstrass was a pure mathematician and Faraday was a pure physicist. Until the middle of the last century physics was not divided into theoretical and experimental branches. In the second half of the century the efforts of giants like Maxwell and Boltzmann gained for theoretical physics the status of an independent power. What they created was classical theoretical physics.

The profession of the theoretical physicist was new for that time. Like mathematicians, theoretical physicists use only paper and pen. However, they did not identify themselves with mathematicians. They were sure that what they study is not a world of abstract mathematical concepts, but real nature. On the other hand, pure mathematicians rightly considered results obtained by theoretical physicists as not rigorously justified. Only a few outstanding mathematicians, like Poincaré, were at the same time qualified theoretical physicists.

The rift between physics and mathematics widened after the First World War, which took the lives of so much young talent educated by the old masters. The creation of quantum mechanics boosted this process tremendously. The logic and intuition of quantum physics was so dramatically different from the "classical" intuition that those who studied the quantum world usually lost interest in other parts of science and preferred to stay in that field forever. An explosive progress of experimental atomic and nuclear physics stimulated enormously the development of theoretical physics. The subject was so hot that physicists just had no time to follow the progress in contemporary mathematics.

At the same time mathematicians after the First World War were too busy to think about physics. At hand were the problems of the "axiomatic revolution", the time of rethinking and reassessment of the very foundations of mathematics. During this time mathematics took its contemporary shape. Set theory, mathematical logic, abstract algebra, topology, and functional analysis were created at that time. Thus mathematicians had a good excuse for paying very limited attention to physicists and their theoretical activity.

I think that the rift reached its widest in the middle of the fifties. At that time physicists had in their possession a substantial mathematical apparatus, including special functions, complex analysis, representation of finite groups and Lie groups. They knew how to perform sophisticated asymptotic expansion. If they felt a lack of mathematical tools, they invented new ones and used them boldly, not caring too much about their rigor. The δ-function, offered by Dirac, is the most impressive invention of this sort. However, the use of sophisticated machinery was not necessary in many cases. As Laurent Schwarts said bitterly in the fifties, the development of the perturbation technique diminished the volume of mathematics used by physicists to elementary algebra and the knowledge of the Greek and Latin alphabets.

I began my scientific career at the beginning of the sixties and was one of the few students who were equally keen on physics and mathematics. All my life I have been unable to make a real choice between these two sister branches of science. Since my youth I have had close friends in both scientific communities, and I testify that at the beginning of the sixties these two communities were almost completely divided. In fact, by that time, the maximum of separation was already over.

Some important steps, if not towards reunification than to convergence of physics and mathematics, were taken by mathematicians. In the middle of the century the great axiomatic revolution was over, and mathematicians changed the focus of their interest to more "pragmatic" objects, such as Partial Differential Equations (what are they good for?), or infinite representations of Lie groups (can they be found useful by anybody?). Finally, some mathematicians started to express an interest in the relentless activity of physicists, who every day performed unjustified, risky but unquestionably efficient operations, making possible to obtain quite reasonable results.

Dirac's δ-function is an especially intriguing object. Its mathematical nature was understood by Laurent Schwarts in 1950 ([S]). This was an event of tremendous importance. It led to explosive development of the theory of generalized functions, the theory of linear topological spaces, to a real breakthrough in the theory of PDE. The accurate use of the theory of generalized functions was important for theoretical physics as well. It made possible to develop a consistent theory of renormalization in quantum electrodynamics.

Discovery of generalized functions was the first move to the renewal of the romance between physics and mathematics. I think that the δ-function, born inside the world of quantum physics, was the most valuable gift presented by physicists to mathematicians in twentieth century. Since that time quantum physics presented to mathematicians several such gifts. Quantum groups and topological quantum field theory are the recent ones.

However, in this article I would like to discuss a quite different subject. Quantum physics dominated in the physical world until the middle of sixties, but then classical physics came out of the shadow and started to grow steadily and persistently. I dare to say that today it has a status equal with quantum physics.

The rise of classical physics in the last four decades was a direct consequence of the general technical progress in those years. The invention and fast development of lasers led to the creation of nonlinear optics. Massive use of satellites for monitoring oceans and the atmosphere stimulated the development of physical oceanography and geophysical hydrodynamics. Extensive efforts towards the realization of controlled thermonuclear fusion together with progress in observational astronomy caused an explosive growth of plasma physics and magnetohydrodynamics. All these disciplines became parts of a renewed, mostly nonlinear, classical physics, which also covers an essential part of the theory of superfluidity and magnetism. The enormous progress of computers made possible the numerical solution of certain vital problems in classical physics, giving another boost to its progress.

One can say that the "old" classical physics gave to mathematics the linear partial differential equations. All their three basic classes – elliptic (Laplace equation), parabolic (heat transport equation), and hyperbolic (wave equation) – were born in the classical physics of the eighteenth century. Needless to say, how important a role they played in the progress of even the most pure of mathematics. The "new" classical physics opened for mathematicians the magic world of nonlinear PDE. In a sense, history repeats itself. Of the whole variety of linear PDE, only a few basic ones (Laplace, heat transport and wave equation) play a really fundamental role. If linear PDE form a sea, nonlinear PDE form an immense ocean. But again, only a few selected systems, such as Korteveg–de Vries, Nonlinear Schrödinger, and Sine–Gordon equations are really interesting both from a physical and mathematical point of view. It would be very difficult to pull these equations out of the ocean without understanding their fundamental role in classical physics.

One can say that classical physics made mathematics. Two of the most valuable presents – solitons and fractal sets, appearing in the theory of turbulence. In fact, these objects appeared before in pure mathematics, but their fundamental role was not properly estimated. It is natural to add to this list the discovery of nonlinear integrable Hamiltonian systems with an infinite number of degrees of freedom, but this subject is closely connected with the mathematical theory of solitons. Any relatively complete review of these subjects will take at least two full-scale monographs.

In this article we discuss an application of ideas of classical physics to several important problems of pure mathematics. One idea runs through our examples: classical physics can help mathematicians to handle Nonlinear Partial Differential Equations.

2 n-wave Equations and n-orthogonal Coordinate Systems

Classical physics is a rich source of "good" nonlinear PDE, but mathematics has its own source – in Differential Geometry. What is remarkable, is that the "best" equations generated by these two quite different sources sometimes become closely related or even identical. Of course, any comment on this phenomenon belongs to metaphysics, and is beyond the scope of this purely scientific article.

A following situation is typical for different physical applications: three wave trains, possibly of a different physical nature, propagate in a weakly nonlinear conservative medium. Their leading wave vectors $\vec{k_1}$, $\vec{k_2}$, $\vec{k_3}$ and corresponding frequencies ω_1, ω_2, ω_3 satisfy the resonant conditions

$$\begin{aligned}\omega_1 &= \omega_2 + \omega_3\,,\\ \vec{k_1} &= \vec{k_2} + \vec{k_3}\,.\end{aligned} \tag{2.1}$$

The wave trains are described by complex-valued functions $\psi_i(\vec{x},\, t)$, $\vec{x} = (x_1,\, x_2,\, x_3)$, obeying Hamiltonian equations

$$\frac{\partial \psi_k}{\partial t} = i\frac{\delta H}{\delta \psi_k^*}\,, \tag{2.2}$$

$$H = \sum_{k=1}^{3} \int Im\,\psi_k(\vec{v_k}\nabla)\psi_k^*\,d\vec{x} + \lambda \int (\psi_1^*\psi_2\psi_3 + \psi_1\psi_2^*\psi_3^*)d\vec{x}\,. \tag{2.3}$$

Here $\vec{v_k}$ are group velocities and λ is an interaction coefficient. One can put $\lambda = 1$, then the equations read

$$\begin{aligned}\frac{\partial \psi_1}{\partial t} + (v_1\,\nabla)\psi_1 &= i\,\psi_2\,\psi_3\,,\\ \frac{\partial \psi_2}{\partial t} + (v_2\,\nabla)\psi_2 &= i\,\psi_1\,\psi_3^*\,,\\ \frac{\partial \psi_3}{\partial t} + (v_3\,\nabla)\psi_3 &= i\,\psi_1\,\psi_2^*\,.\end{aligned} \tag{2.4}$$

From the physical view-point system (2.4) is fundamental. It describes an important phenomena – stimulated Raman scattering as well as a three-wave resonant interaction of wave packets. System (2.4) is usually called the "three wave system". To make this system looking more "mathematical" one should introduce new variables

$$\frac{\partial}{\partial u_i} = \frac{\partial}{\partial t} + (v_i\,\Delta)\,, \tag{2.5}$$

and put

$$\begin{aligned}\psi_1 &= -i\,Q_{23} = i\,Q_{32}^*\,,\\ \psi_2 &= -i\,Q_{13} = i\,Q_{31}^*\,,\\ \psi_3 &= -i\,Q_{12} = i\,Q_{21}^*\,.\end{aligned} \tag{2.6}$$

Then system (2.4) takes a form

$$\frac{\partial Q_{ij}}{\partial x_k} = Q_{ij}\,Q_{kj}\,, \quad i \neq j \neq k\,,\ i = 1,2,3\,. \tag{2.7}$$

One can generalize system (2.7) to an n-dimensional case just by putting in (2.7) $i, j, k = 1, \ldots, n$. For $n > 3$ system (2.7) is overdetermined. A further generalization can be done as follows.

Let A be an associate algebra (for instance, algebra of $N \times N$ matrix $N > n$) and I_k ($k = 1, \ldots, n$) is a set of commuting projectors

$$I_i\,I_k = I_k\,\delta_{ik}\,. \tag{2.8}$$

A generalization of (2.7) reads

$$I_i \frac{\partial Q}{\partial x_k} I_j = I_i Q I_k Q I_j . \tag{2.9}$$

System (2.9) can be called a "general n-wave system".

In the physical case (2.4) matrix Q is Hermitian, $Q^+ = Q$. This is an example of "reductions" – additional restrictions imposed on Q and compatible with system (2.9). This is an example of a more general reduction

$$Q^+ = J Q J , \quad [J, I_k] = 0 , \quad J^2 = 1 . \tag{2.10}$$

For $n = 3$ a choice $J = diag(-1, 1, 1)$ leads to a so-called "explosive three-wave system".

It is remarkable that systems (2.7), (2.9) can be applied for solution of some important problems in Differential Geometry. The most famous one is the problem of n-orthogonal coordinate systems.

Suppose S is a domain in R^n. How to find all orthogonal curvilinear coordinate systems in S? Let $x = (x_1, \dots, x_n)$ be such coordinates. In this coordinate system the matrix tensor is diagonal

$$ds^2 = \sum H_i^2 \, dx_i^2 . \tag{2.11}$$

Hamiltonian $H_i = H_i(x)$ and the Lamé coefficients are subjects for determination. They satisfy a heavily overdetermined system of nonlinear PDE, the Gauss–Lamé equations. These equations read:

$$\frac{\partial Q_{ij}}{\partial x_k} = Q_{ik} Q_{kj} , \quad i \neq j \neq k , \tag{2.12}$$

$$E_{ij} = \frac{\partial Q_{ij}}{\partial x^j} + \frac{\partial Q_{jk}}{\partial x^i} + \sum_{n \neq i,j} Q_{ik} Q_{jk} = 0, \ i \neq j . \tag{2.13}$$

Here

$$Q_{ij} = \frac{1}{H_j} \frac{\partial H_i}{\partial x^j} . \tag{2.14}$$

One can see that the first group in the Gauss–Lamé equations (2.12) exactly coincides with the n-wave equations (2.7), (2.9). The second group of equations (2.13) can be treated as a reduction. As far as Q_{ik} is real, one more reduction is imposed

$$Q_{ij} = \bar{Q}_{ij} . \tag{2.15}$$

To understand the origin of the Gauss–Lamé equations, one can consider that a domain is not in R^n but in some Riemann space, admitting introduction of "diagonal" coordinate (2.11), and calculate the Riemann curvature tensor R_{ijkl}. One finds that by the virtue of (2.11)

$$R_{ij,kl} = 0\,, \quad i \neq j \neq k \neq l\,,$$
$$R_{ik,jk} = -H_i\,H_j\left(\frac{\partial Q_{ij}}{\partial x^k} - Q_{ik}\,Q_{kj}\right),$$
$$R_{ij,ij} = -H_i\,H_j\,E_{ij}\,. \tag{2.16}$$

The curvature tensor is a symmetric matrix in the space of bivectors in TS. If only (2.12) is satisfied, this matrix is *diagonal.* A corresponding Riemann space S can be called a space of "diagonal curvature". Riemann spaces of diagonal curvature are a very interesting class of objects. They include, for instance, homogeneous spaces as well as conformal flat spaces. If the second system (2.13) is satisfied, $E_{ij} = 0$, the space is *flat* and S is a domain in R^n.

We see that the Gauss–Lamé equations differ from the n-wave system only by a choice of reduction. All these systems are completely integrable and can be efficiently solved by the use of the method of Inverse Scattering Transform, elaborated in the theory of solitons. We will present here the most advanced version of this method known as a "Dressing Method". It makes possible to construct solitonic, multisolitonic, and more general solutions of integrable equations locally in X-space. We will do this in a maximally general form assuming that all unknown functions belong to some associative algebra A over the complex field.

We introduce again a set of projectors I_k satisfying the condition (2.8) and construct an element $\Phi \in A$:

$$\Phi = \sum_{i=1}^{n} x_i\,I_i\,. \tag{2.17}$$

Let λ is a point on the complex plane, $\chi = \chi(\lambda, \bar{\lambda})$ is an A-valued quasianalytic function on C. Suppose that $\chi(\lambda, \bar{\lambda}, x)$ is a solution of the following non-local $\bar{\partial}$-problem:

$$\frac{\partial \chi}{\partial \bar{\lambda}} = \chi \times R = \int \chi(\nu,\, \bar{\nu}, x)\, R(\nu,\, \bar{\nu},\, \lambda,\, \bar{\lambda},\, x)\, d\lambda\, d\bar{\lambda}\,, \tag{2.18}$$

normalized by the condition

$$\chi \to 1 \quad \text{as} \quad \lambda \to \infty\,. \tag{2.19}$$

In (2.18)

$$R(\nu, \bar{\nu}, \lambda, \bar{\lambda}) = e^{\Phi\,\nu}\,T\,e^{-\Phi\,\lambda}\,, \tag{2.20}$$

where $T = T(\nu,\, \bar{\nu},\, \lambda,\, \bar{\lambda})$ does not depend on x. Function T is a "free parameter" of the theory. It should be chosen by such a way that (2.18) has a unique solution for all $x \in S$. At $\lambda \to \infty$ this solution has an asymptotic expansion

$$\chi = 1 + \tfrac{Q}{\lambda} + \tfrac{P}{\lambda^2} + \cdots . \tag{2.21}$$

According to Freidholm's alternative, any solution $\tilde{\chi}$ of $\bar{\partial}$-problem vanishing at infinity is identically zero

$$\tilde{\chi} \equiv 0 \quad \text{if} \quad \tilde{\chi} \to 0 \quad \text{as} \quad \lambda \to \infty . \tag{2.22}$$

The solution of $\bar{\partial}$-problem (2.21) is called "dressing", while a free kernel T is called a "dressing function".

The following statement is a cornerstone of the theory:

Theorem 1. *For $x \in S$ the term Q in (2.21) is a solution of n-wave system (2.9).*

Proof. Let us construct a set of differential operators L_{ij} acting on χ as follows

$$L_{ij}\,\chi = I_i\left(\tfrac{\partial \chi}{\partial \chi^j} + \lambda\,\chi\,I_j\right) - I_i\,Q\,I_j\,\chi . \tag{2.23}$$

A straightforward calculation shows that $L_{ij}\,\chi$ satisfies the same $\bar{\delta}$-problem

$$\tfrac{\partial}{\partial \bar{\lambda}}\,L_{ij}\,\chi = L_{ij}\,\chi \times R . \tag{2.24}$$

Substitution of asymptotic (2.21) into $L_{ij}\,\chi$ shows that $L_{ij}\,\chi \to v(1/\lambda)$ as $\lambda \to \infty$. Hence

$$L_{ij}\,\chi = 0 , \tag{2.25}$$

and

$$L_{ij}\,\chi\,I_k = 0 , \quad i \neq j \neq k . \tag{2.26}$$

Substituting asymptotic expansion (2.21) into (2.24) and taking into account only the leading nonvanishing terms of order $1/\lambda$, one can see that Q satisfies the equation (2.9).

The solution of (2.9) can be found in closed algebraic form if the kernel T is degenerated:

$$T(\nu,\,\bar{\nu},\,\lambda,\,\bar{\lambda}) = \sum_{k=1}^{N} A_k\,(\nu,\bar{\nu})\,B_k(\lambda,\,\bar{\lambda}) . \tag{2.27}$$

In the most simple case

$$T = A(\nu,\,\bar{\nu})\,B(\lambda,\,\bar{\lambda})$$

the solution has a compact form

$$Q = \langle A\rangle\big(1 - (B|A)\big)^{-1}\langle B\rangle . \tag{2.28}$$

Here

$$\langle A\rangle = \int e^{\lambda\Phi}\, A(\lambda,\, \bar\lambda)\, d\lambda\, d\bar\lambda\,,$$

$$\langle B\rangle = \int B(\lambda,\, \bar\lambda)\, e^{-\lambda\,\Phi}\, d\lambda\, d\bar\lambda\,,$$

$$\langle B|A\rangle = \frac{1}{\pi}\int \frac{B(\nu,\, \bar\nu)\, e^{(\lambda-\mu)\,\Phi} A(\lambda,\, \bar\lambda)}{\nu-\lambda}\, d\nu\, d\bar\nu\, d\lambda\, d\bar\lambda\,. \tag{2.29}$$

The equation (2.27) is a general solitonic solution of the n-wave system. It is a quite nontrivial solution describing a set of interesting physical phenomena. In a general case the equation (2.27) leads to N-solitonic solution. The role of solitonic solutions in Differential Geometry has not yet been studied to the proper degree.

So far we did not impose on the solution of the system (2.13) any additionally restrictions (reductions). They can be imposed by imposing some additional constrains on the "dressing function" $T(\nu,\, \bar\nu,\, \lambda,\, \bar\lambda)$. Imposing of condition

$$\bar T(\bar\nu,\, \nu,\, \bar\lambda,\, \lambda) = T(\nu,\, \bar\nu,\, \lambda,\, \bar\lambda) \tag{2.30}$$

makes Q real:

$$\bar Q = Q\,.$$

Condition

$$T^{+}(\bar\nu,\, \nu,\, \bar\lambda,\, \lambda) = J\, T(\nu,\, \bar\nu,\, \lambda,\, \bar\lambda) J\,, \quad J^2 = 1\,, \quad [J,\, \Phi] = 0 \tag{2.31}$$

leads to the reduction

$$Q^{+} = J\, Q\, J\,, \tag{2.32}$$

the most important from a physical view-point. Finally, condition

$$T^{tr}(-\mu, -\bar\mu, -\lambda, -\bar\lambda) = \tfrac{\mu}{\lambda} T(\lambda, \bar\lambda, \mu, \bar\mu) \tag{2.33}$$

provides satisfaction of the last set of equations (2.13).

Formula (2.14) shows that the n^2 elements of matrix Q are expressed through n Lamé coefficients $H^i(x)$. This might give the impression that we are looking for some special solution of the equation (2.12). This is not actually true. Any solution of this system can be presented in a form (2.14) by many different ways.

Indeed, one can introduce a new function,

$$\psi = \chi\, e^{\lambda\Phi}\,, \tag{2.34}$$

which satisfies the equation

$$I_i\, \tfrac{\partial\psi}{\partial x^j} - I_i\, Q\, I_j\, \psi = 0\,. \tag{2.35}$$

Let $A_l(\lambda,\bar{\lambda})$, $l=1,\dots,n$ is an arbitrary set of functions of variables λ, $\bar{\lambda}$, and

$$H_i = \int \sum_{l=1}^{n} \psi_{il}(\lambda,\bar{\lambda},x)\, A_l(\lambda,\bar{\lambda}) d\lambda\, d\bar{\lambda}\,, \tag{2.36}$$

one can see that

$$\frac{\partial H_i}{\partial x^j} = Q_{ij}\, H_j\,. \tag{2.37}$$

A different choice of $A_l(\lambda,\ \bar{\lambda})$ leads to a different set of H_i. All these sets are called *Combescure* equivalent. One can see that a classical problem of classification of all Combescure equivalent arrays of n-orthogonal coordinate systems is solved efficiently in this "solitonic" formalism.

3 Theory of Surfaces as a Chapter of Theory of Solitons

In the previous section we saw how easily the method of "mathematical theory of solitons", elaborated in the classical theory of integrable systems, makes it possible to solve a classical problem of differential geometry. In this chapter we will develop this success and find a way to solve another important problem in differential geometry – the classification of surfaces in R^3.

Let Γ be a surface in R^3. One can introduce coordinates x_1, x_2 on Γ such that both the first and the second quadratic forms are diagonal:

$$\begin{aligned} \Omega_1 &= p^2\, dx_1^2 + q^2\, dx_2^2\,, \\ \Omega_2 &= pA\, dx_1^2 + qB\, dx_2^2\,. \end{aligned} \tag{3.1}$$

Coordinates x_1, x_2 are defined up to trivial transformations $x_1 = x_1(u_1)$, $x_2 = x_2(u_2)$. Coefficients of the two quadratic forms Ω_1, Ω_2 cannot be chosen independently. Four functions p, q, A, B are connected by three nonlinear PDE known as Gauss–Codazzi equations (GCE). We will now show that these equations are simply a very degenerate case of a classical three-wave system. They can be integrated by a minor modification of the dressing method used for the integration of n-orthogonal coordinate system.

Let us imbed the surface Γ in a special three-orthogonal coordinate system in R^3 in vicinity of F

$$ds^2 = H_1^2\, dx_1^2 + H_2^2\, dx_2^2 + dx_3^2\,, \tag{3.2}$$

where

$$H_1 = p + A\, x_3\,, \quad H_2 = q + B\, x_3\,, \quad H_3 = 1\,. \tag{3.3}$$

Obviously,

$$Q_{31} = Q_{32} = 0\,,$$

and other coefficients of matrix Q_{ij} do not depend on x_3. Indeed,

$$Q_{13} = \tfrac{1}{H_3}\tfrac{\partial}{\partial x_3} H_1 = A\,, \quad Q_{23} = \tfrac{1}{H_3}\tfrac{\partial}{\partial x_3} H_2 = B\,, \tag{3.4}$$

so

$$\begin{aligned}\tfrac{\partial}{\partial x_2}(p + A\,x_3) &= Q_{12}(q + B\,x_3)\,,\\ \tfrac{\partial}{\partial x_1}(q + B\,x_3) &= Q_{21}(p + A\,x_3)\,.\end{aligned}$$

Hence

$$\begin{aligned}Q_{12} &= \tfrac{1}{B}\tfrac{\partial A}{\partial x_2} = \tfrac{1}{q}\tfrac{\partial p}{\partial x_2}\,,\\ Q_{21} &= \tfrac{1}{A}\tfrac{\partial B}{\partial x_1} = \tfrac{1}{p}\tfrac{\partial q}{\partial x_1}\,.\end{aligned}$$

In this case only two equations survive in the system (2.12):

$$\tfrac{\partial Q_{13}}{\partial x_2} = Q_{12}\,Q_{23}\,, \quad \tfrac{\partial Q_{23}}{\partial x_1} = Q_{21}\,Q_{13}\,. \tag{3.5}$$

The reduction condition (2.13) leads to one more equation,

$$\tfrac{\partial Q_{12}}{\partial x_2} + \tfrac{\partial Q_{21}}{\partial x_1} + Q_{13}\,Q_{23} = 0\,. \tag{3.6}$$

Let us denote $Q_{12} = \alpha$, $Q_{21} = \beta$. The Gauss–Codazzi equations read

$$\begin{aligned}&\tfrac{\partial \alpha}{\partial x_2} + \tfrac{\partial \beta}{\partial x_1} + A\,B = 0\,,\\ &\tfrac{\partial A}{\partial x_2} = \alpha\,B\,, \quad \tfrac{\partial B}{\partial x_1} = \beta\,A\,.\end{aligned} \tag{3.7}$$

The system (3.7) should be accompanied by equations for elements of first quadratic form, p and q:

$$\tfrac{\partial p}{\partial x_2} = \alpha\,q\,, \quad \tfrac{\partial q}{\partial x_1} = \beta\,p\,. \tag{3.8}$$

Comparing (3.7), (3.8) one can realize that A, B and p, q are Combescure equivalent pairs of Lamé coefficients. Physicists know the equation (3.8) as the two-dimensional Dirac system.

System (3.7) consists of three equations imposed on four unknown functions α, β, A, B. Hence, its general solution should be parametrized by some functional parameters. To perform the solution one should remember that system (3.5), (3.6) is a special case of Gauss–Lamé equations. Thus, one can use the standard scheme described in the previous chapter. In another words, one can solve the $\bar{\partial}$-problem

$$\begin{aligned}&\tfrac{\partial \chi}{\partial \bar{\lambda}} = \chi \times R\,,\\ &R_{ij}(\lambda, \bar{\lambda}, \mu, \bar{\mu}) = e^{\lambda \chi_i - \mu \chi_j}\,T_{ij}(\lambda, \bar{\lambda}, \mu, \bar{\mu})\,.\end{aligned} \tag{3.9}$$

The dressing function T should satisfy the condition (2.33) and condition of reality,

$$\bar{T}(\bar{\lambda},\lambda,\bar{\mu},\mu)=T(\lambda,\bar{\lambda},\mu,\bar{\mu})\,, \tag{3.10}$$

and must satisfy one more condition,

$$\tfrac{\partial R_{ij}}{\partial x_3}\equiv 0\,. \tag{3.11}$$

One can assume also that

$$R_{11}=R_{22}=R_{33}=0\,,\quad T_{11}=T_{22}=T_{33}=0\,.$$

Under these assumptions the matrix function is defined uniquely.

Let

$$\begin{aligned} f_1(\lambda,\bar{\lambda}) &= \bar{f}_1(\bar{\lambda},\,\lambda)\,,\\ f_2(\lambda,\bar{\lambda}) &= \bar{f}_2(\bar{\lambda},\,\lambda)\,,\\ R(\mu,\,\bar{\mu},\lambda,\bar{\lambda}) &= \bar{R}(\bar{\mu},\,\mu,\,\bar{\lambda},\,\lambda)\,, \end{aligned} \tag{3.12}$$

be arbitrary functions. Then all non-zero elements of the dressing matrix $T_{ij}(\lambda,\bar{\lambda},\mu,\bar{\mu})$ are the following:

$$\begin{aligned} T_{12} &= \mu\, R(\mu,\bar{\mu},\lambda,\bar{\lambda})\,,\\ T_{21} &= -\mu\, R(-\lambda,-\bar{\lambda},-\mu,-\bar{\mu})\,,\\ T_{13} &= -\mu\, f_1(-\mu,-\bar{\mu})\delta(\lambda)\,\delta(\bar{\lambda})\,,\\ T_{23} &= -\mu\, f_2(-\mu,-\bar{\mu})\,\delta(\lambda)\,\delta(\bar{\lambda})\,,\\ T_{31} &= \mu\,\delta(\mu)\,\delta(\bar{\mu})\,f_1(\lambda,\bar{\lambda})\,,\\ T_{32} &= \mu\,\delta(\mu)\,\delta(\bar{\mu})\,f_2(\lambda,\bar{\lambda})\,. \end{aligned} \tag{3.13}$$

The $\bar{\partial}$-problem is equivalent to the integral equation

$$\chi_{ij}(\lambda,\bar{\lambda},x)=\delta_{ij}+\frac{1}{\pi}\int\frac{\chi_{ik}(\mu,\bar{\mu},x)\,R_{kj}(\mu,\bar{\mu},\xi,\bar{\xi},x)}{\lambda-\xi}d\mu\,d\bar{\mu}\,d\xi\,d\bar{\xi}\,. \tag{3.14}$$

According to (3.13) the kernel R is partly degenerated. One can obtain from (3.14) the following relations:

$$\begin{aligned} \chi_{31}=\chi_{32}=0\,,\quad \chi_{33}=1\,,\\ \chi_{13}=\tfrac{1}{\lambda}A\,,\quad \chi_{23}=\tfrac{1}{\lambda}B\,. \end{aligned} \tag{3.15}$$

$$\begin{aligned} A &= -\frac{1}{\pi}\int\mu\sum_{k=1}^{\infty}\chi_1 k(\mu,\bar{\mu},x)\,f_k(-\mu,-\bar{\mu})e^{-\mu x_k}d\mu\,d\bar{\mu}\,,\\ B &= -\frac{1}{\pi}\int\mu\sum_{k=1}^{\infty}\chi_2 k(\mu,\bar{\mu},x)\,f_k(-\mu,-\bar{\mu})e^{-\mu x_k}d\mu\,d\bar{\mu}\,. \end{aligned} \tag{3.16}$$

From (3.15), (3.16) one can observe that it is possible to construct a closed system of integral equations for the left upper block of the matrix χ_{ij}. Omitting intermediate calculations we present only the result:

Theorem 2. *The solution of the Gauss–Codazzi equation is given by the solution of the integral equation imposed on a 2×2 complex matrix $Q_{ij}(\lambda, \bar{\lambda}, x)$:*

$$Q_{ij}(\lambda, \bar{\lambda}, x) = \delta_{ij} + \frac{1}{\pi} \int \frac{\mu\, Q_{ik}(\mu, \bar{\mu}, x)\, S_{kj}(\mu, \bar{\mu}, \xi, \bar{\xi})\, e^{\mu x_k - \xi x_l}}{\lambda - \xi}\, d\mu\, d\bar{\mu}\, d\xi\, d\bar{\xi}\,. \tag{3.17}$$

Here

$$S_{ij} = S^{(1)}_{ij} + S^{(2)}_{ij}\,,$$

$$S^{(1)}_{ij}(\mu, \bar{\mu}, \lambda, \bar{\lambda}) = \begin{bmatrix} 0 & R(\lambda, \bar{\lambda}, \mu, \bar{\mu}) \\ -R(-\mu, -\bar{\mu}, -\lambda, -\bar{\lambda}) & 0 \end{bmatrix},$$

$$S^{(2)}_{ij}(\mu, \bar{\mu}, \lambda, \bar{\lambda}) = -\tfrac{1}{\pi}\, f_i(-\mu, -\bar{\mu})\, f_j(\lambda, \bar{\lambda})\,. \tag{3.18}$$

If this equation uniquely resolves, then

$$Q_{12} \to \tfrac{\alpha}{\lambda}\,, \quad Q_{21} \to \tfrac{\beta}{\lambda} \quad \text{at} \quad \lambda \to \infty\,. \tag{3.19}$$

Formulae (3.16), (3.19) *generate a solution of the Gauss–Codazzi system.*

One should also mention that a solution of the Gauss–Codazzi equation does not define the surface uniquely. It defines a whole class of surfaces with different elements of the first quadratic form p^2, q^2. One can find them using the formulae

$$p = \int \big[Q_{11}(\lambda, \bar{\lambda}, x) e^{-\lambda x_1} u(\lambda, \bar{\lambda}) + Q_{12}(\lambda, \bar{\lambda}, x) e^{-\lambda x_2} v(\lambda, \bar{\lambda})\big]\, d\lambda\, d\bar{\lambda}\,,$$

$$q = \int \big[Q_{21}(\lambda, \bar{\lambda}, x) e^{-\lambda x_1} u(\lambda, \bar{\lambda}) + Q_{22}(\lambda, \bar{\lambda}, x) e^{-\lambda x_2} v(\lambda, \bar{\lambda})\big]\, d\lambda\, d\bar{\lambda}\,. \tag{3.20}$$

Here

$$u(\lambda,\, \bar{\lambda}) = \bar{u}(\bar{\lambda},\, \lambda)\,, \quad v(\lambda,\, \bar{\lambda}) = \bar{v}(\bar{\lambda},\, \lambda)\,, \tag{3.21}$$

are arbitrary functions. In particular, one can choose the case

$$u(\lambda,\, \bar{\lambda}) = -\tfrac{1}{\pi}\, \lambda\, f_1(-\lambda,\, -\bar{\lambda})\,,$$

$$v(\lambda,\, \bar{\lambda}) = -\tfrac{1}{\pi}\, \lambda\, f_2(-\lambda,\, -\bar{\lambda})\,, \tag{3.22}$$

where $p = A$, $q = B$, and the surface has a constant curvature.

The theory of surfaces in R^3 is a classical chapter in differential geometry, still actively developing. We can offer a "solitonic" program to the systematic study and classification of surfaces. It includes the following three steps.

1. Classification of solutions of the Gauss–Codazzi system. Each solution define the whole class of Combescure-equivalent surfaces.
2. Study of surfaces in the framework of a given class.
3. Embedding the surface into R^3. As far as we know the first and second quadratic terms, according to the Bonnet theorem this embedding is unique up to Eucledean motion. The presented method for solution of the Gauss–Codazzi equation makes it possible to perform embedding efficiently. In this article we just announce the result, the details will be published separately.

The solution of the integral equation (3.17) can be done explicitly in a closed algebraic form if the kernel R in $\bar{\partial}$-problem (3.9) is degenerated

$$R(\mu,\bar{\mu},\lambda,\bar{\lambda}) = \sum_{k=1}^{n} A_k(\mu,\bar{\mu})\, B_k(\lambda,\bar{\lambda})\,. \tag{3.23}$$

The class of surfaces arising from the solution of the $\bar{\partial}$-problem can be called $n+1$ solitonic surfaces. In the simplest case $R = 0$, and the surface is one-solitonic. In this case the integral equation (3.17) can be easily solved. The results are:

$$\begin{aligned}
\alpha &= -\frac{2h_1'(x_1)\,h_2(x_2)}{1+h_1^2(x_1)+h_2^2(x_2)}\,,\\
\beta &= -\frac{2h_2'(x_2)\,h_1(x_1)}{1+h_1^2(x_1)+h_2^2(x_2)}\,,\\
A &= -\frac{2h_1'(x_1)}{1+h_1^2(x_1)+h_2^2(x_2)}\,,\\
B &= -\frac{2h_2'(x_2)}{1+h_1^2(x_1)+h_2^2(x_2)}\,,
\end{aligned} \tag{3.24}$$

where

$$\begin{aligned}
h_1(x_1) &= \sqrt{2\pi}\int f_1(\mu,\bar{\mu})\, e^{-\mu x_1}\, d\mu\, d\bar{\mu}\,,\\
h_2(x_2) &= \sqrt{2\pi}\int f_2(\mu,\bar{\mu})\, e^{-\mu x_2}\, d\mu\, d\bar{\mu}\,.
\end{aligned}$$

One can see that the choice of f_1, f_2 is just a redefinition of variables x_1, x_2.

Even this simplest class of "solitonic" surfaces has not been studied properly. The simplest representative of this class arises if one puts $p = A$, $q = B$. This is a sphere in stereographic projection.

4 Long-time Asymptotics in the Hamiltonian PDE Equation

In the previous section we formulated more or less rigorous mathematical statements. Now we will speak the language of theoretical physics. This language has its own logic, convincing for physicists, who I hope, will agree with the final conclusions of our considerations.

Mathematicians quite rightfully will consider these statements as not proved rigorously enough. We would be quite satisfied, if they treat them as plausible conjectures, stimulating their curiosity to examine and finally either to prove or to refute them. One has just to remember that the path from the "physical" and the "mathematical" versions of scientific truth might be long and difficult.

In this chapter we will discuss long-time asymptotics of certain nonlinear PDE equations. In as much as nonlinear PDE are the most common tool for the study of very different mathematical phenomena, from black holes to the dynamics of population, nobody can believe that a kind of general theory can be anticipated here. Thus we restrict our considerations only by an evolutionary equation of Hamiltonian type. These equations preserve energy and, possibly, other constants of motion. We will ask ourselves the following question. Suppose we impose initial data, with a certain level of smoothness, on our equation or system of equations. Will this smoothness improve or deteriorate in time?

In dissipative systems, like heat transport equations, rough initial data have the tendency to become as smooth as is compatible with the boundary condition. In Hamiltonian systems the situation is quite the opposite. One would expect that their solutions will lose their smoothness and become, in process of evolution, more and more rough. Only in exceptional cases, for linear and for integrable systems (also for systems asymptotically linear or integrable), the smoothness will tend in time to a certain finite limit.

This very general statement is just a consequence of the second law of thermodynamics. Conservative PDE describe Hamiltonian systems with infinite number of degrees of freedom. As all evolving Hamiltonian systems in the world, they tend to thermodynamic equilibrium.

However, thermodynamic equilibrium for classical continuous systems means equipartitioning of energy between all the degrees of freedom and, consequently, excitation of all possible spatial harmonics. This is exactly the lack of smoothness. Any initial data tending to thermodynamic equilibrium become more and more rough. Such phenomena as formation of the "islands of stability" or KAM tori can slow this process but cannot completely stop it.

In other words, the right question is about the rate – how soon a system loses its smoothness and relaxes into thermodynamic equilibrium. Theoretical physicists offer several ways to answer these important and interesting questions.

The first is using a statistical description of the initial system. One can conjecture that the system being still far from thermodynamic equilibrium, nevertheless displays chaotic behaviour which should be described in terms of correlating functions. One can try to find closed equations for these functions and find their solution asymptotically in time. This might be much easier than to follow directly the lack of smoothness in the initial dynamic system.

We illustrate this idea on one basic example. We will study the "defocusing" nonlinear Scrödinger equation in infinite three-dimensional space

$$i\Psi_t + \Delta\,\Psi - |\Psi|^2\,\Psi = 0\,, \quad x \in R^3\,. \tag{4.1}$$

We will assume that no boundary condition is imposed and that the initial data

$$\Psi|_{t=0} = \Psi_0(2)$$

is an infinitely smooth function. More exactly we will assume for any finite domain $\Omega \in R^3$

$$W_2^l(\Omega) < A_2^l\, V_\Omega\,. \tag{4.2}$$

Here V_Ω is the volume of Ω, A_2^l is a set of positive constants. We now ask the question: how fast does A_2^l grow in time?

Equation (4.1) is a Hamiltonian system and can be written in a form

$$i\Psi_t = \frac{\delta H}{\delta \Psi^*} \tag{4.3}$$

$$H = \int \{|\nabla\,\Psi|^2 + \tfrac{1}{2}|\Psi|^4\}d\vec{r}. \tag{4.4}$$

Hamiltonian H is a constant of motion, another constant of motion is "number of particles" $N = \int |\Psi|^2\, d\vec{r}$.

One can assume that the initial data $\Psi_0(r)$ is homogeneous in space stochastic process described by a pair of correlation functions in Fourier space

$$\langle \Psi_{0k}\, \Psi_{0k'}^* \rangle = n_0(k)\, \delta_{k-k'}\,. \tag{4.5}$$

Now estimate (4.2) reads

$$\int k^{2l}\, n_0(k)\, d < A_2^l\,, \tag{4.6}$$

and $n_0(k)$ decays at $k \to \infty$ faster than any powerlike function.

In Fourier space Hamiltonian is

$$H = H_0 + H_{int}\,,$$
$$H_0 = \int \omega_k |\Psi_k|^2\, dk\,,$$
$$H_{int} = \frac{1}{2}\int \Psi_{k_1}^*\, \Psi_{k_2}^*\, \Psi_{k_3}\, \Psi_{k_4}\, \delta_{k_1+k_2-k_3-k_4}\, dk_1\, dk_2\, dk_3\, dk_4\,, \tag{4.7}$$

where $\omega_k = k^2$ is a symbol of linear part of equation (4.1), known in physics as the "dispersion law". Equation (4.4) is a nonlinear wave equation describing interaction of spatial harmonics with different wave vectors $\vec{k}$.

A thermodynamic equilibrium in a system of harmonics can be reached on the Rayligh–Jeans spectrum in presence of condensate

$$n_k^T = A_0\, \delta(k) + \tfrac{T}{\omega_k}\,, \tag{4.8}$$

where T is temperature and A intensity of condensate.

Conservation of particle number N means that L_2-norm (A_2^0) is a constant of motion. Meanwhile for spectrum (3.8) $A_2^0 = \infty$ at any finite T. This is a completely general fact.

Thermodynamic equilibria in a classical wave system can be reached only if constants of motion (wave number, energy) are infinite. Hence any initial data with finite integrals can evolve only into the state with $T = 0$. The second law of thermodynamics leads to the following plausible

Statement 1. Let initial data for equation (4.1) be a stationary homogeneous field described by spectral density $n_0(k)$ and

$$\int n_0(k)\, dk = A_2^0\,. \tag{4.9}$$

Then in the Hilbert space L_2

$$n_0(k) \to A_2^0\, \delta(k)\,, \tag{4.10}$$

in other words, if you wait long enough, "almost all" particles from the initial distribution will be concentrated in the condensate.

From the mathematical point of view "almost all particles" mean that convergence (4.10) takes place only in L_2, not in higher Sobolev spaces. To find plausible estimates for the behavior of higher norms A_2^l one should use a kinetic equation for n_k. It reads

$$\frac{\partial n_k}{\partial t} = 4\pi \int |T_{kk_1k_2k_3}|^2\, \delta(\vec{k}+\vec{k_1}-\vec{k_2}-\vec{k_3})\delta(\omega_k+\omega_{k_1}-\omega_{k_2}-\omega_{k_3})$$

$$\times\,(n_{k_1}n_{k_2}n_{k_3}+n_kn_{k_2}n_{k_3}-n_kn_{k_1}n_{k_2}-n_kn_{k_1}n_{k_3})\,dk_1\,dk_2\,dk_3\,, \quad (4.11)$$

and is written for a more general Hamiltonian system, when

$$H_{int} = \frac{1}{2}\int T_{kk_1k_2k_3}\,\Psi_{k_1}^*\,\Psi_{k_2}^*\,\Psi_{k_2}\,\Psi_{k_3}\,\delta_{k+k_1-k_2-k_3}\,dk\,dk_1\,dk_2\,dk_3\,. \quad (4.12)$$

In our case $T\equiv 1$, and in the presence of condensate equation (4.11) is valid in the limit of high wave numbers

$$k^2 \gg A_0\,. \quad (4.13)$$

Equation (4.11) has the following constants of motion

$$N = \int n_k\,d\vec{k}\,, \quad (4.14)$$

$$\vec{p} = \int \vec{k}\,n_k\,dk\,, \quad (4.15)$$

$$E = \int \omega_k\,n_k\,dk\,, \quad (4.16)$$

which can be interpreted as densities of number of particles, momentum and energy. It has, for $T=1, \omega_k = k^2$, a family of self-similar solutions

$$n = \frac{1}{t^{\frac{4a+1}{2}}}\,f\left(\frac{\vec{k}}{t^a}\right), \quad (4.17)$$

where a is an as yet unknown constant. To find a, one should use constants of motion (4.14)–(4.16). It is reasonable to assume that $n(\vec{k})$ is spherically symmetric, hence $\vec{p}=0$. Almost all the particles concentrate in the condensate, hence the conservation law (4.16) should be disregarded. Conservation of energy reads

$$E = \int k^2\,n_k\,d\vec{k} = const\,, \quad (4.18)$$

from which we obtain $a=1/6$, and solution (4.17) takes a form

$$n = \frac{1}{t^{5/6}}\,f\left(\frac{\vec{k}}{t^{1/6}}\right)\,. \quad (4.19)$$

For higher Sobolev norms one can obtain

$$A_2^l = \frac{1}{t^{\frac{4a+1}{2}}}\int k^{2l}\,f\left(\frac{k}{t^a}\right)\,dk \simeq t^{(2l+1)a-1/2}\,, \quad (4.20)$$

assuming $a = 1/6$ one obtains

$$A_2^l \simeq t^{1/3(l-1)} . \tag{4.21}$$

Formula (4.20) is the central result of these considerations. According to our assumption the first Sobolev's norm A_2^1 is constant and all higher norms grow. L_2-norm, which is a number of particles outside the condensate decreases in time, $L_2 \simeq t^{-1/3}$. The whole picture in nontrivial and even tricky. Decreasing the number of particles outside the condensate provides increasing roughness of the solution.

One should note that infinity of the domain and statistical homogeneity of $\Psi_0(x)$ are essential. If equation (4.1) would be put in a finite domain, for instance in the box $0 < x_i < 2\pi$ with zero or periodic boundary conditions, the situation would be completely different. The equation (3.1) in this case can be treated as a discrete system of infinite number of oscillators, and kinetic equation (3.11) cannot be applicable. One can expect that in a finite domain the rate of relaxation to thermodynamic equilibrium and growing of roughness will be in the discrete case much slower than in the continuous.

The branch of theoretical physics studying statistical properties of nonlinear waves is known as "weak turbulence". It is an actively developing field having important applications in physical oceanology, meteorology and astrophysics. On our opinion, methods of weak turbulence could be very helpful for the solution of pure mathematical questions on nonlinear partial differential equations.

5 Briefly on Collapses

Another subject equally interesting both from a mathematical and a physical point of view is the formation of finite-time singularities in nonlinear PDE. A classical example of the system having a collapsing solution is the focusing nonlinear Schrödinger equation in R^3

$$i\,\Psi_t + \Delta\,\Psi + |\Psi|^2\,\Psi = 0\,, \quad \Psi|_{t=0} = \Psi_0(r)\,. \tag{5.1}$$

The Cauchy problem for (5.1) explodes in a finite time of

$$H = \int \{|\Delta\,\Psi_0|^2 - \tfrac{1}{2}|\Psi_0|^4\}\,d\vec{r} < 0\,, \tag{5.2}$$

and this collapse leads to formation of integrable singularities $|\Psi|^2 \to c/r^2$. All Sobolev's norms in the moment of singularity become infinite.

Existence of singularities in (5.1) is a rigorous mathematical fact. Existence of finite-time singularities in another fundamental system, the Navier–Stokes equation,

$$\tfrac{\partial v}{\partial t} + v\,\nabla\, v + \nabla p = v\Delta v\,, \quad div\, v = 0\,, \tag{5.3}$$

is an open question. Moreover, as much as one million dollars will presented to a lucky mathematician, who will manage to prove the existence or absence of finite-time singularity in (5.2). Note that the question about singularities is open only for Euler equation arising from (5.2) if $v = 0$. Today both Euler and Navier–Stokes systems are very hot business.

Here we will show that in the limit of "almost two-dimensional" hydrodynamics, the Euler equations have solutions collapsing in finite time. Let us study a system of almost parallel vorticies, crossing the perpendicular plane in point

$$\omega = x + iy\,, \quad \omega = \omega(\vec{s})\,,$$

where $\vec{s}$ is two-dimensional marker of the vortex line. Let z is a coordinate along the vortex, $\Gamma(\vec{s})$ is the distribution of vorticity which can be treated as a measure of the s-plane. If the bending of a vortex line is small, one can describe the system of vortices by the following Nonlinear Schrödinger equation [Z2],

$$-i\,\frac{\partial w}{\partial t} + \Gamma(s)\,\frac{\partial^2 w}{\partial z^2} + \int \frac{\Gamma(\vec{s'})\, d\vec{s'}}{\bar{w}(s) - \bar{w}(s')} = 0\,. \tag{5.4}$$

Equation (5.4) has the following self-similar solution

$$w = (t_0 - t)^{1/2+i\epsilon}\, F\left(\frac{z, \vec{s}}{(t_0 - t)^{1/2}}\right)\,, \tag{5.5}$$

where $F(\xi, \vec{s})$ satisfies the equation

$$-\left(i\frac{1}{2} + \epsilon\right)\, F + \frac{i}{2}\xi\, F_\xi + \Gamma(s)\, F_{\xi\xi} + \int \frac{\Gamma(s')}{\bar{F}(s) - \bar{F}(s')}\, ds' = 0 \tag{5.6}$$

Here ϵ is an arbitrary real number and $\Gamma(s)$ is an arbitrary, not necessarily positive measure. Let us take

$$\epsilon = 0\,, \quad \Gamma(s) = \Gamma\,\delta(s+1) - \Gamma\delta(s-1)\,, \quad F(1) = A + iB\,, \quad F(-1) = A - iB\,,$$

with A, B satisfying the system of equations

$$\begin{aligned} \tfrac{1}{2}(-A + \xi\, A') &= -B'' + \tfrac{1}{2B} \\ \tfrac{1}{2}(-B + \xi\, B') &= A''\,. \end{aligned} \tag{5.7}$$

Asymptotically,

$$A \to \alpha z\,, \quad B \to +\beta z \text{ as } z \to \infty$$
$$A \to -\alpha z\,, \quad B \to \beta z \text{ as } x \to -\infty\,. \tag{5.8}$$

This solution describes a collapse of two antiparallel vortex tubes. There is some hope that a similar collapsing solution still exists in presence of a very small viscosity.

In conclusion the author expresses deep gratitude to Tel Aviv University and to the organizers of the conference "Visions in Mathematics – Towards 2000" for this extremely interesting and intellectually brilliant event.

References

[B] J. Bourgain, On the growth in time of higher Sobolev norms of smooth solutions of Hamiltonian PDE, IMRN 6 (1996), 277–304.

[S] L. Schwartz, Theorie des Distributions, Hermann & Co, Editeurs, Paris, 1950.

[Z1] V. Zakharov, Description of the n-orthogonal curvilinear coordinate systems and Hamiltonian integrable systems of hydrodynamic type, Duke Mathematical J. 94:1 (1998), 103–139.

[Z2] V. Zakharov, Quasi-two-dimensional hydrodynamics and interaction of vortex tubes, in "Nonlinear MHD Waves and Turbulence: Proceedings of the Workshop, Nice, France, 1998," Springer Lecture Notes in Physics 536 (1999), 369–385.

[ZFL] V. Zakharov, G. Falkovich, V. L'vov, Kolmogorov Spectra of Turbulence. I. Wave Turbulence, Series in Nonlinear Dynamics, Springer Verlag, 1992.

[ZM] V. Zakharov, S. Manakov, Reductions in systems integrable by the method of inverse scattering problem, Doklady Mathematics 57:3 (1998),471–474.

Vladimir Zakharov, Landau Institute of Theoretical Physics, Moscow
and
Department of Mathematics, University of Arizona, Tucson, AZ 86721, USA

GAFA, Geom. funct. anal.
Special Volume – GAFA2000, 881 – 977
1016-443X/00/S20881-97 $ 1.50+0.20/0

GAFA Geometric And Functional Analysis

ADDENDUM: DISCUSSIONS AT THE DEAD SEA

27-28 August, 1999

Introduction

In this Addendum we present extracts from four of the discussions we had at the Dead Sea Resort. Discussion is not a traditional form of exchange for mathematical ideas and thoughts. However, the organizers decided to try such a format on this occasion, and we think it developed into a very interesting exchange of opinions especially towards the end of the conference. Only the first four discussions are presented below because they were directed at a very general audience. We discussed what are "good" and "bad" mathematical problems, the role of mathematics in the Real World, public relations of the Mathematical Community with the World around us, the future attractiveness of Mathematics to the younger generation, new opportunities (and dangers) connected with the explosion of Computer Science, our traditional contacts with Mathematical Physics and many other subjects.

We would like to emphasize that participants were not prepared for the subjects of the discussions and their thoughts do not necessarily reflect a deep analysis but rather their "gut feeling". As a result, the discussions reflect, perhaps, the concerns of the mathematical community, but not what we would write in our articles, especially in the very "dry" framework in which mathematicians usually present their thoughts. The organizers tried to "relax" the audience, and often provoked the discussion by controversial statements and suggestions. (As one participant said privately, "with such provocative statements I may be convinced to defend two opposing opinions". We believe that this may be true, but that the reasoning could be of value anyway.) We suggest the reader takes this information into account and adjusts his/her expectations before reading this addendum. We think that such an open and spontaneous flow of subjects, thoughts and opinions is a useful experience (at least once a century!). Although one may not agree with some opinions one may find the list of subjects that arose in these discussions interesting.

As a part of this Addendum, we also include an article by D. Kazhdan. He could not be present at two of the discussions and expressed his opinion on, "How is mathematics possible?", in a short essay.

Finally, we would like to thank I. Scherbak for her tremendous help in the scientific editing of these discussions, Miss Asya Scherbak for her careful notes extracted from the taped discussions, and Mrs. Diana Yellin for her essential contribution to the editing of the final write-up.

N. Alon and V.D. Milman, Tel Aviv, November 2000.

DISCUSSION on MATHEMATICAL PHYSICS

with introduction by A. Connes

A. CONNES:

Let me first show you a well-known cartoon[1] due to Robbert Dijkgraaf which gives a rather sobering view of the interaction between mathematics and physics in the 30 years from 1968 to 1998. The two pairs of mathematicians and physicists are unchanged, same guys, just older and balding but now the blackboards have been swapped and they are greatly puzzled by what the other team had discovered thirty years ago!

To make a really provocative statement (meaningful but politically incorrect) I would say that (true) physicists really treat mathematics like a harlot. When they need it they just use it, they drop it when they are done and they don't care if it complains. The amazing fact is that it works! Feynman said "Mathematics is masturbation, physics is real sex". Of course we

[1]Reproduced with the consent of the creator, Robbert Dijkgraaf.

should not complain about it, given for instance the great pay-off that the ideas of E. Witten brought from string theory to mathematics.

One difference which is a bit worrisome is in the sociology of the two camps. One way to stress this difference is the saying "Physicists are Bosons, Mathematicians are Fermions". By the Pauli exclusion principle, no two fermions can occupy the same state and this type of repulsion is common in mathematics, where joint papers rarely involve more than two authors. In physics there are papers in which the list of authors is longer than the text itself. But what I find worrisome is that this "bosonisation" of crowds of theoretical physicists does not make them tolerant to new ideas that don't fit with the common hopes of the "believers".

Of course this is only a small worry as opposed to the actual decline in the number of students who choose mathematics or physics at University.

Y. NE'EMAN:

I have to make a pessimistic remark - but this could be salutary if it arouses us to do something to forestall the recurrence of such dangerous developments. I reached these conclusions when I tried to understand the decay and end of Greek Science, after its having reached such peaks as the Mathematics and Physics of Archimedes, the Planetary and Earth Science of Erathostenes, Heron's invention of the Steam engine, etc. Partly, this was played out as Greek tragedy, with two most dramatic episodes: (1) the bishop-instigated 415 AD assassination of Hypatia, the woman mathematician who headed the School of Alexandria (see Maria Dzielska's study "Hypatia of Alexandria" published by Harvard University Press) and the burning of the School - and (2) Justinian's edict (550 AD) closing the School of Athens - an act instigated by the bishops via the Empress Theodora. The latter blow was mitigated by the saving of much of the accumulated scientific texts and oral heritage with the transfer of nine teachers headed by the Rector Damascinus to a Mesopotamian site under Sassanid Persia.

So much for drama, but the decay started with a gradual contraction of the system from the end of the Second Century AD on. The intellectual elites of the Hellenistic circum-Mediterranean culture, from which the student body derived, gradually lost interest, being attracted by the new social ideas and the apocalyptic promises surrounding the spread of Christianity and Judeo-Christian ethics. Less and less new students also meant after a while less and less researchers and teachers. You can now understand why I am so worried by the global phenomenon of a falling student intake in

Physics, for instance. This is presumably due to the attraction of the business and economical disciplines (the only redeeming feature is the interest in Computer Science, and we should exploit this remaining opening and enrich the relevant teaching programs and make it provide a solid mathematical education. One further comment: beyond the attraction of the economical disciplines, there is also the negative impact of Post-Modernism and its caricature of Science - and the calls for the deletion of the teaching of Mathematics "now that computers can do the calculations". We had such a call in Israel some four years ago - see *Ha'aretz Supplement (Mussaf)* of 5 January 1996, "Why do we still teach Mathematics" - and my answer in a letter published in the 19 January 1996 issue. However, the same call was recently made in France by no less a personality than the Minister of Science and Education, Mr. Pierre Allegre, himself a geologist (see *Le Monde*, 20 March 2000). This is one facet of the "naive" "End of Science" movement, a new doctrine, according to which we should now forego scientific activities altogether, because we have discovered everything we have to know (or according to some, even everything there is to know) and any further expenditure would be a waste. For a sample of this intellectual message, see John Horgan, *The End of Science*, Addison-Wesley Helix Books Series, Reading (Mass.) 1996, 309 pp.

[A QUESTION ABOUT THE SITUATION IN BIOLOGY]

Y. NE'EMAN:

I don't know what the situation there is, I apologize, I don't know. But in the exact sciences, in physics, we have a very strong decay and that's all over the rest of the world, everywhere. In Israel we had for a while a Renaissance when we had a million emigrants from the Soviet Union arriving, and they were not yet conscious of or corrupted by Western notions. And now it's getting that way again.

A. CONNES:

During the talk by Arthur [Jaffe], there was one question which was sort of implicitly asked to Sergui [Klainerman]. So maybe the person who asked it should ask it again.

[A QUESTION ABOUT CLASSICAL FIELD THEORY]

S. KLAINERMAN:

What is known about classical field theory? In classical field theory (CFT) just like in quantum field theory (QFT) there is a classification based on the scaling properties of the equations. But instead of being done relative to the action, as in QFT, it's done relative to the energy. Thus the Yang-Mills equations, which are critical from the point of view of QFT in three space dimensions, are in fact critical in four space dimensions from the CFT point of view. In $3+1$ dimensions the Yang-Mills equations are subcritical; one can in fact prove global regularity for any finite energy data. So the classical problem becomes hard in $4+1$ and higher dimensions.

[A QUESTION ABOUT THE NUMBER 137 - AS A CONTINUATION OF NE'EMAN'S TALK]

Y. NE'EMAN:

There exists an extensive literature about 137. Its role in physics is due, as I mentioned, to its emergence as the inverse of the numerical value of the "charged-squared" coefficient in the Coulomb force between two electrons - and in a plethora of results involving the electron in Quantum Mechanics, since the probability is obtained by squaring the quantum amplitude. It is a dimensionless quantity $e^2/4\pi hc = 1/137$ ($-e$ is the electric charge on the electron, h is Planck's constant and c is the velocity of light - which together produce a dimensionless number). Sir Arthur Eddington is the astronomer who provided the first verification of Einstein's General Theory of Relativity, by organizing an expedition to the island of Principe in the Southern Atlantic and measuring the deflection of light during the 1919 solar eclipse. He was also responsible more than anybody else for the propagation of Einstein's image and prestige in the lay public, in the presentation and explanations which he gave at the press conference he held upon the expedition's return. He was indeed one of the first to grasp the importance and implications of the new theory and tried to extend it. Eddington was a Pythagorean and when 137 entered physics, he tried to derive its value from "first principles" - counting degrees of freedom - and failed. In December 1930, three Cambridge graduate students from Germany, George Beck, Hans Bethe and W. Riezler, sent the following letter to the prestigious journal *Die Naturwissenschaften* under the title "Comments on the Quantum Theory of Absolute Zero".

"Imagine a hexagonal crystal lattice. Zero temperature is thereby characterized by the freezing out of all degrees of freedom of the system, i.e. the cessation of all inner motion. Let each electron be endowed - following Eddington - with $1/\alpha$ degrees of freedom, where α denotes Sommerfeld's *fine-structure constant.* Aside from the electrons, the crystal contains protons - for which the counting of the degrees of freedom is the same as for the electrons, since according to Dirac, the proton represents a hole in the electron-gas. To achieve our purpose of attaining absolute zero, we have to ensure that our crystal be electrically neutral. We thus extract $(2/\alpha) - 1$ degrees of freedom, since we also have to preserve one degree of freedom for the overall collective motion. This yields for the Absolute Zero of temperature the equation $T_0 = -(2/\alpha - 1)$ degrees. Inserting the experimental value $T_0 = -273^0$ we get $1/\alpha = 137$, which fits within the observational error bars with the value obtained from completely different and independent data. Note also that our result does not involve the specific structure of the crystal."

Of course, the authors played a trick in which they exploited a confusion between "degrees of freedom" and "degrees (centigrade) of temperature". The editor did not detect the hoax and published the letter (*Die Naturwissenschaft* **8** (1931) 39). Eddington was extremely unhappy and broke with the journal.

Several of the stories about 137 involve Pauli. A true one, which I heard from Fierz, Pauli's colleague at the Zurich ETH, relates to Pauli's final illness. They had arrived at the local hospital and Pauli - accompanied by Fierz - was shown to his room. Lo and behold - it was room 137! Pauli looked at Fierz and declared "I shall not come out alive from this room", which turned out to fit the tragic facts.

This is the place where I can add an apocalyptic sequel. When Pauli died, he was well-received in Heaven and considering his standing in Science, he was granted an interview with the Lord. "Why 137?" asked Pauli. The Lord went to the blackboard and started proving his point and became very involved in calculation. All the while, Pauli was preserving his scepticism and nodding his head in the negative - until God became impatient and challenged Pauli to show him a better way of calculating this constant. Pauli then went to the blackboard and showed the Lord that a mistake had indeed occurred and that 137 was just the result of an error.

My next story I heard from Gershon Scholem, the great Cabala scholar and at the time also President of the Israeli National Academy of Sciences and Humanities. Scholem was visiting the USA and had been invited to lecture on the Cabala in Cambridge (Mass.) during a visit to the American Academy of Arts and Sciences. Its President was then our physicist colleague Victor Weisskopf, who introduced Scholem. After the lecture, Weisskopf asked: "Any role for the number 137 in the Cabala?" - "Of course", answered Scholem, "the Cabala is 137!". Indeed, applying the standard Cabala procedure of adding up the numerical values assigned to the letters of the Hebrew alphabet, you find - remembering that only consonants rate as letters, in Hebrew (the vowels are represented - if necessary - by diacritical signs), for the word Cabala itself the sum is

$$\{Q = 100\} + \{b = 2\} + \{l = 30\} + \{h = 5\} = 137.$$

Here there is no mistake, whatever the implications (the Cabala assumes that the biblical text is heavily loaded with messages, aside from the literal one). That's some of the folklore on 137. Dan Amir seems to have another one.

V. ZAKHAROV:

I can add one just very simple thing I heard from Zeldovich. Zeldovich told me that he had a bad memory and always forgot the number where he put his coat in the lobby room. So to be sure that he would never forget it, he always used number 137.

D. AMIR:

I happen to have a cassette-tape of a lecture by a famous Israeli rabbi who does missionary work trying to attract people to Jewish orthodoxy and Cabala - a piano student of my son is a student of that rabbi. The cassette is named "The holy number or the important number 137, found in the Bible". There is in it a lot of numerology leading to 137. For instant, 137 is the most frequent age mentioned in the Bible - three Biblical figures lived up to 137: Ishmael (Abraham's son), Levi (Jacob's son), and Amram (Moses' father). Another one: 137 is the average of the age of Sara (127) and Jacob (147) and so on...

QUESTION:

But how do they prove that there are no other numbers with such properties?

D. AMIR:

Of course you can do similar tricks for almost any number. Anyhow, at the end of the cassette he says: "The scientists have only one explanation about only one role of 137, but we have got so many of them"!

Y. NE'EMAN:

Following my previous remarks about the legendary 137 (which, in more recent physics has turned out to be an energy-dependent variable, rather than a constant), I now suggest we look at other examples, those in which the Pythagorean approach has been vindicated. This holds for the various quantities involved in $SU(3)$, $U(3)$ and the quark-anti-quark counting $[U(3) \otimes U(3)]_\beta$ and also in the algebraic completion of their tensor-multiplication with the $SU(2)$ of spin, yielding $SU(6)$, $U(6)$. As a matter of fact, any *quark-model* constructs will generally do. As against Eddington's failure with 137, we find such successful, very Pythagorean results in Hadron Physics, results based on the quark-contents in each particular process. I shall cite here three such beautiful examples of Pythagoreanisms.

My first example consists in the evaluation of the ratio between total cross-sections for electron-positron annihilation into hadrons, as compared with the production of muon pairs under the same conditions. This ratio, denoted as R, has a staircase-like shape, in its energy dependence, with a new stair occurring each time a new threshold is reached for a new channel (i.e. a new flavor of quark). QED theory plus the quark-model tell that the process involves an electron-positron pair annihilating into a virtual photon, which then re-interacts and creates a physical particle pair. The first vertex is thus common to all particles created in this process, whereas it is the second vertex which varies every time a new threshold is reached. This fixes the amplitude and for a probability we have to square the result. The prediction is thus a number R resulting from *counting and adding up the squared electrical charges at the quark level, for those quarks for whose on-mass-shell creation there is enough energy.* R thus exhibits a sudden growth each time a new such threshold opens up. The standard result is given as

$$R = \frac{\sigma(e^+ + e^- \to \text{hadron pairs})}{\sigma(e^+ + e^- \to \mu^+ + \mu^-)},$$

and this is proportional to the sum of the squares of the electrical charges of the quarks allowed at that threshold. Thus, between 0.5 and 2.5 GeV, we have enough energy for $\overline{u}u$ or $\overline{d}d$ and $\overline{s}s$, with a hadron-to-muon ratio

$$R(0.5 - 2.5\ GeV) = \frac{3 \times (2/3)^2 + 3 \times (-1/3)^2 + 3 \times (-1/3)^2}{(1)^2} = 2\ ,$$

but at 2.8 GeV we reach the threshold for the creation of a *charmed* pair and we thus reevaluate the ratio, finding that it should now take the value 10/3. At 10 GeV, we can make a $\overline{b}b$ and the predicted ratio again changes to 11/3. These values indeed show up experimentally.

My next example relates to the high-energy asymptotic region. The Pomeranchuk theorem ensures that particle and anti-particle beams when scattered over the same target have equal cross-sections. Since a nucleon is made of 3 quarks, and a meson (such as the pion) represents a quark-anti-quark combination, and assuming that the asymptotic cross-section is dominated by the Pomeranchuk trajectory (i.e. with the quantum numbers of the vacuum), we get

$$\frac{\sigma(\pi N)}{\sigma(NN)} \to 2/3$$

with the experimental (measured) values of 26mb (millibarns) and 39 mb respectively. Without the quark model, this ratio could have taken on any value.

My last example is an $SU(6)$ result - and I refer the audience or future reader to the original paper by Gursey, Pais and Radicati or to my monograph *Algebraic Theory of Particle Physics* in which I gave a heuristic argument as proof. I am pointing to the prediction with respect to the ratio between the magnetic moments of the proton and neutron, predicted to be

$$\frac{\mu(p)}{\mu(n)} = -3/2 \ .$$

The value of the two magnetic moments was known since the late '40s, but nobody had noticed that they indeed had this simple ratio between their values.

S. KANIEL:

This conference is about the end of the 20th century and the beginning of the 21st. I would like to compare it to the 19th century. In the 19th century mathematicians knew physics and physicists knew mathematics. No, not all. But the great physicists knew mathematics very well and most of the great mathematicians knew physics also very well, from the very beginning, from Gauss to Poincaré throughout. Moreover they looked for principles. The action principle, the constancy of the speed of light - this is actually a principle...

S. KLAINERMAN:

Did they look or did they discover?

S. KANIEL:

No, no, this is a principle. I mean, the greatness of Einstein was that he said that this is a principle. It's not just a coincidence.

S. KLAINERMAN:

I would say he is from our century.

S. KANIEL:

No, I don't agree. He belongs to the 19th century. Einstein's postulate that gravity is really the metric tensor - this is also a principle.

[DISCUSSION ON WHAT CENTURY EINSTEIN BELONGS TO]

S. KANIEL:

It's not so important. OK, I can say - before 1920 and after 1920. OK? Fine. Because when quantum mechanics started it was equations and mathematical models, no more principles. Just recall Jaffe's lecture today - only equations, lots of equations that we actually didn't see. But you said there are equations and we sort of... One after the other, equations and models - $SU(3)$, $SU(6)$ and so on. All of them are models.

[HUBBUB]

I have a definite question that I asked Jaffe privately and he suggested that I ask it publicly: is there any physical theory that explains Coulomb's law? First of all, Coulomb's law is on a quantum scale... It is a combination of electromagnetism and mass. I mean, the gravity is small but the inertional mass is not small at all. And as far as I know (maybe I don't know) there is no good explanation to this law. I think there are two different levels - electromagnetism and inertia...

[HUBBUB]

A. JAFFE:

The best known field theory, the one that I quoted in the predication theory of quantum electrodynamics, the Dirac equation with Maxwell field theory, predicts the Coulomb force. But one very mysterious feature of this set of equations - by the way, we don't yet know whether there are mathematical solutions of these equations - is that physicists now believe (and I attempt to think it's great) that these equations have no solutions, that they have the same defect as the wave (equations in four dimensions) that I mentioned. And yet the perturbation theory of these equations gives this most accurate number that we know in nature and in calculating the perturbation theory. So you can ask whether that's a philosophical paradox. The answer in physics is that unless you make the equations more complicated and introduce them out of abelian gauge theory, the equations remain inconsistent. But is that a real way out? It could be that this theory that we know best has no mathematical meaning. That cannot have a meaning without being impaired into a larger theory.

Y. FRÖHLICH:

One of the stories about 137 is that just because 1 divided by 137 is such a small number, the cut-off at large energy that makes quantum electrodynamics well defined, is at 10^{137} electron volt. So if you are willing to introduce a cut-off at these very high energies, this theory is perfectly well defined and of course reproduces the perturbation theory. So in this sense there is no problem.

Since I'm already here I shall ask you. I thought that it was very important that you mentioned first the openness of the information and second the informational mess. Of course physics has lived with an informational mess ever since the beginning of quantum theory, I would say. I think it was a little bad in the last century because not much was happening. I mean after Maxwell, in the last part of the 19th century, not much was happening, so we had this informational mess. Now, I think as long as there is a lot of progress and lots of new discoveries we can live with it. Now apparently we are reaching another period and so it becomes a little painful, and I think it's one of the reasons young people turn away from physics... I think we tend to be a little greedy in not being willing to give credit to young people who just fill in a little piece in a puzzle. And I think we should do this because we cannot expect wonderful discoveries at the pace we saw them during the last 50 years. So this would be a sort of plea to be more generous with each other.

Maybe another observation. It seems to me this discussion shows that mathematicians like to be a little anecdotal. I think there are real issues to be discussed. Maybe we should really discuss them?

A. CONNES:

There is very little time to do that, but the following question was repeatedly raised: "Why do we need to quantize gravity?" Somehow if I may interpret what he says, Sergei [Novikov] doesn't believe it's really needed. To answer this question one must first understand why do we need quantized fields, such as the quantized electromagnetic field. The point, which is very clearly made in a small paper that Feynman wrote for the Dirac memorial volume, in 1984, is that you can't have coexistence of special relativity with quantum mechanics without quantum fields. The two basic requirements of causality (the corner stone of special relativity) and positivity of the energy are in contradiction if we stay within quantum mechanics. Mathematically this boils down to the nonexistence of (nonzero) functions f of one real variable whose support is contained in $[0, \infty]$ and whose Fourier transform has the same property, namely its support is contained in $[0, \infty]$. In physics it means that if you try to enclose a single particle in a very small box, you automatically create pairs of particles so that you stop being able to treat the question within quantum mechanics. The number of particles present is not conserved and you are outside quantum mechanics. The only known way out is quantum field theory. The effect of quantum field theory is precisely, just by writing the quantum field, to reconcile causality, which becomes commutation of observables at space like points, with positivity of the energy (which is easily formulated in terms of representations of the Poincaré group). On the other hand, we have gravity, I mean there is no way out, we have gravity, and when you look at small perturbations of the Minkowski flat space, what you find out is that the way to parameterize these small fluctuations is by waves. These waves are called gravitational waves; they have not been observed directly but thanks to the incredible richness of the data from binary pulsar observations we know they are there, as the basic way such systems spend their enormous gravitational energy. I mean there is an implicit observation of them through the theory of binary pulsars. Now what happens is that once you have these waves, you have this gravitational field, and there is absolutely no way out from the point of view of physics and they have to be quantized. The issue is that if you go ahead and try to quantize this gravitational field as one was quantizing the electromagnetic field, you meet a new wall. This time you can't reconcile renormalizability with unitarity, i.e. the facts that probabilities are numbers between zero and one. Renormalizability is a basic requirement in order to be able to make predictions (otherwise you would have infinitely many parameters to fix before knowing the theory which would hence loose most of its predictive power). The outcome is that we have two great theories, quantum field theory on the one hand and gravity on the other, both perfectly in agreement with experiment so far, but as far as we know they are not compatible. If I am correct this fact that the two theories are contradictory has been called, I think by Brian Green, the biggest cover-up of the century. He then goes on in his book on string theory to explain the type of solution proposed by string theorists. One of my witty friends described the content of one of the chapters of his book by saying "It shows that the M-theory of M-theory is M-theory". Major issues such as the enormous cosmological constant created by the breaking of supersymmetry are still right there, and it would really be unfair to let laymen believe that the final theory has been found. It has not yet been and we are faced with a major challenge. But we can't say that the problem does not exist, it is a basic intellectual challenge of coherence.

S. KLAINERMAN:

Just one more remark. It's clear that Alain [Connes] is right and this is the problem that absolutely has to be solved, but one may wonder whether the time is now. It has been observed many times that in the 20th century Physics underwent not just one but two revolutions. One was, of course, quantum mechanics and the other general relativity. Each one of them has deep mathematical implications which may require many years of intense activity on our part to uncover. While it true that the mathematical implications of non-relativistic quantum mechanics are relatively well understood this is definitely not the case with Quantum Field Theory and General Relativity. If history is a good guide we might infer that we have first to understand both at a far deeper level before making the next big revolution.

At a classical level General Relativity has been almost completely ignored by us mathematicians. I mean, with few exceptions, there have been very few...

[Some obstructions of A. Connes]

Alain, you are talking about an experimental level but I'm talking about a mathematical level...

This is perturbation theory and doing calculations, but I'm talking about the mathematical principles of general relativity, to develop it in the way the Classical Mechanics was developed in previous generations. While the mathematical community has shown, in this century, an enormous amount of interest in quantum mechanics, this did not quite happen with general relativity. You may wonder whether we don't need, at least in parallel, to develop general relativity before undertaking the far harder problem of quantizing it. I don't know, it may be that a full general quantum theory of gravity is not even a problem for the next century. Maybe it will take another millennium. Unlike physicists, we mathematicians are very patient; we have the experience of problems that took hundreds of years, even millennia, to be solved.

A. CONNES:

Thank you. Unfortunately it is time to stop.

Transcribed by A. Scherbak

DISCUSSION on GEOMETRY

with introduction by M. Gromov

M. GROMOV:

We provoke a discussion. We are waiting for definite statements.

V. ZAKHAROV:

I'd like to.

M. GROMOV:

So something "outrageous" about geometry.

V. ZAKHAROV:

Thanks! Misha encouraged me, and I really will say something outrageous about geometry. I probably will say something very, how do you say, bold. I will make two statements.

SOMEBODY:

Two outrageous statements.

V. ZAKHAROV:

An outrageous statement, and then some statement in support of this outrageous statement. People say there is a classical part of differential geometry, which is the theory of two-dimensional surfaces in three-dimensional space. And there is the theory of solitons in the non-linear wave theory including the solitons of Korteweg-de Vries equations or solitons of Kadomtsev-Petviashvili equations, in wave equations. My statement is that the whole theory of surfaces is a part of the theory of solitons. And now I add only two words. Probably people have heard about the classical problem of classification of three-orthogonal coordinate systems. This problem was formulated in 1813 by Dupin and Binet and the problem is how to describe all all three-orthogonal coordinate systems in three-dimensional space, or, in other words, how to find a set of three orthogonal functions, $u_1(x_1, x_2, x_3)$, $u_2(x_1, x_2, x_3)$, $u_3(x_1, x_2, x_3)$, such that $(\nabla u_i, \nabla u_j) = h_i^2 \delta_{ij}$. This problem was considered as one of the central problems in differential geometry. Ninety years later, in 1910, Darboux published a very thick 600 page book on this problem, and the central part of this problem is the so called Gauss-Lamke equations. My statement is that these equations are completely integrable by the inverse scattering method, and all their solutions can be found in a regular way, and this result is published by myself in Duke Math. J., Vol.94, No.1 (1998), p.103-139.

M. GROMOV:

That's really outrageous.

V. ZAKHAROV:

That's outrageous. Thank you. And the way you immerse the theory of the surfaces in the theory of solitons is, a very natural way to immerse any surface in a three-orthogonal system. You just use a normal, and Lie transformation, and you construct a natural three-orthogonal system, and by using this method you can solve the Gauss-Codazzi equation. The Gauss-Codazzi equation is a completely integrable system. This is my second statement. The same could be done for the surfaces in symmetric spaces.

M. GROMOV:

It was outrageous but not sufficiently provocative. So, somebody else wants to volunteer? Otherwise I will elect someone. Yasha [Eliashberg], come here. The question, which was asked about your [Eliashberg] talk: what do you think is the way we are going? We have this huge system of invariants in contact geometry; what development is expected?

Y. ELIASHBERG:

I am just afraid it is not sufficiently outrageous for this discussion. All these kinds of ideas were already discussed in these past two days from the point of view of topology. What we want from all these kinds of physical light constructions is to define some invariants. I was coming from this topological sight and I was thinking in these terms, but really - and I still think this is a kind of important question in this part of the world - this is some kind of universal structure which just exists by itself, and what I like about it is the kind of picture which keeps developing all the time. Everything fits together in growing, so it is a science which should be good for something maybe more serious than just counting of invariants, but I do not know what it is!

M. GROMOV:

You have to say that contact geometry is going to absorb all geometry, all physics, all mathematics and everything else. That's what we expect from you! Why don't you say this?

Y. ELIASHBERG:

What do *you* expect?

M. GROMOV:

That's what I was asking you! I don't tell what I think. Actually I partially have this feeling. There is a structure, which may overcome differential topology eventually. It has some edge, which mixes geometry and topology. In topology except for dimension four, three and two there is no non-trivial geometry. But symplectic and contact geometry have geometry *and* topology in all dimensions.

Y. ELIASHBERG:

Yes, if we just speak more generally, not as I was talking at this moment, but in general. You want outrageous statements, so then you should say: indeed, we would go in this direction.

Y. FRÖHLICH:

May I ask something as a thesis? There is this idea which I feel, at least very partly, should be responsible for non-commutative geometry in string theory being logical trident. One of the things that has a sort of impeded implementation of the idea is that we don't know what Lorentzian non-commutative geometry is. And I would like to ask if anybody has an idea what it would look like? [He gives the microphone to A. Connes.]

A. CONNES:

I think I will pass the microphone to my neighbor, Sergui [Klainerman], because I think he has even a stronger opinion about this. So why don't you repeat what you told me?

S. KLAINERMAN:

Well, actually instead of answering, maybe I'll ask Misha [Gromov] a related question. Why don't you consider Lorentzian geometry as being a part of Geometry?

M. GROMOV:

I do, but I just have not come to that yet. About Lorentzian geometry I may say something which will be irrelevant to the kind of things you [Klainerman] do, because we don't have simple geometric terminology or a language to speak about Lorentzian geometry. So apparently my view is you can only speak about special Lorentzian manifolds. There is no geometry on a general Lorentzian manifold. And I think I can give a good justification for that.

S. KLAINERMAN:

But many of the most interesting objects in mathematics are special. Why do you look for generality? Why don't you restrict yourself to some objects?

M. GROMOV:

But we are made this way. We are mathematicians. We look for general principles. I am not saying we should not look at that. I am just saying that we have to change this stand, which we had in Riemannian geometry, where we could, in principle, say something meaningful about the geometry of Riemannian manifolds. Apparently we cannot say anything meaningful about all Lorentzian manifolds. And there should be a theorem stating precisely why we could not do it, essentially because the Lorentz group is non-compact. That comes from it. But I do not know exactly how to formulate it. This is something similar to the invariants theory. The space of these metrics is a kind of non-separable one; maybe it can be treated immediately as a non-commutative space. That's a possibility. It is more non-commutative in this sense and less separable than the space of a Riemannian metric.

S. KLAINERMAN:

But what would take Lorentzian geometry to be considered part of Geometry? Most geometers I know do not consider Lorentzian geometry as part of real Geometry.

M. GROMOV:

What do you mean: "most"?

S. KLAINERMAN:

I'm not going to give names, but I have talked to many who say it's part of Physics, not Geometry.

M. GROMOV:

You haven't given names. Then we cannot discuss them name by name.

S. KLAINERMAN:

Even you have told me that.

M. GROMOV:

No, I can't say it. Actually I don't know what physics is, so I couldn't say it. The only thing I think it is, is very different from Riemannian geometry in spirit, and it cannot be studied using the same principles. We have to change our stand, and it is unclear how to change it, but it is clear that we cannot study all Lorentz manifolds in a meaningful way.

S. KLAINERMAN:

I want to say a few words in defense of the view that Lorentzian geometry is as much part of Mathematics as Riemannian Geometry. Certainly the questions which will appear in Lorentzian geometry are not the same as the ones we might expect in view of our experience with Riemannian geometry. But nevertheless there *is* this well-defined Lorentzian world. It just happens to play an important role in Physics, through the relativity theory, but at the same time it is based on very, very simple geometric mathematical principles. It's a basic object, just like the basic objects you [Gromov] talked about at the beginning of your lecture. It has been investigated very little in comparison to the other major structure in mathematics which is the Euclidean structure. As far as I am concerned there exist two fundamental geometric structures - Euclidean and Lorentzian. Though the Lorentzian one has been investigated far less, we should not expect that the questions, which will arise, will be the same as the ones of Riemannian geometry.

M. GROMOV:

Yes, exactly, I agree. There are different kinds of questions. You don't know the questions and this what I'm saying. If you ask naive questions extrapolating from the Riemannian geometry, these just don't work here. That's absolutely clear.

S. BLOCH:

Just a little anecdote. I'm not really an expert in this, but... Some years ago Atiyah gave lectures to mathematicians and physicists, and the first thing he said in his lectures was that he was going to have an analytical continuation from the Lorentzian to the Euclidean metric. And what was interesting this was the reaction of the audience. The audience there was roughly 50:50 mathematicians and physicists. The physicists all said: "Oh, yes, yes, we do this, yes, absolutely". All the mathematicians were completely outraged, they said: "You can't do that, that's not positive definite, that's going to be completely different". So, I don't know, maybe there is a cultural reason why we don't...

A. CONNES:

Yes, this is quite true. When you begin to look at quantum field theory, you take this problem very seriously. You start working in Minkowski space and so on. But after a while the physicists start to convince you that things are cleaner and simpler after doing the Wick rotation to Euclidean space-time. This has been exploited systematically in constructive quantum field theory and the relation between the Euclidean and Minkowskian points of view is well established by the Osterwalder-Shrader reconstruction for instance. Here is a related thing which is very rarely mentioned nowadays. There is a framework of quantum gravity by J. Wheeler and B. de Witt, in which the basic object is not space-time, but it is the 3-geometry of space-like slices. In this model "time" only appears as a semi-classical variable, as a semi-classical parameter. This approach didn't succeed in constructing quantum gravity, but it gave a very striking hint about the naivety of our standard idea of time. In fact when we talk about Lorentzian geometry or about Lorentzian space-time we are imposing on ourselves the idea that there is a complete universe whose story has been written. This model could be totally wrong. In fact it could well be that we have a totally wrong idea about time. For instance, in my talk I explained that time could very well appear just from non-commutativity and from quantum mechanics, I mean from the lack of equality between xy and yx. And we cannot a priori decide that time is on the same footing as the other space coordinates. I think this is a very delicate question, and before embarking on deliberate work on the Lorentzian side, one has to reflect on this basic problem, which is that we really don't understand the role of time somehow. Somehow physicists are extremely pragmatic people, so they decide that if it's easier to do computations by putting $i = \sqrt{-1}$ in front of t, they go ahead and do it then. They do it, they get beautiful formulas reminiscent of statistical mechanics and they can use all their familiarity with statistical mechanics. So then they are happy with this. And some people, like S. Hawking even take it as a fundamental thing. They think that the functional integral in quantum gravity should be done in the space of Euclidean space-time metrics. And for many reasons you do get a lot of mileage out of that. To be truthful this question of time is, I think, completely open.

S. KLAINERMAN:

But, Alain [Connes], just in case, we should also look at Lorentzian geometry on its own. Just in case your picture is not right. Just to make a point about this business of complexifying. If you look at hyperbolic equations, which is the subject I know reasonably well, if you restrict yourself to the issue of finding the fundamental solution of the wave equation in Minkowski space, analytic continuation makes perfect sense. You can take the Euclidean Green function and obtain from it, by a Wick rotation, the fundamental solution of the wave equation. However if you want to go deeper into the study of the properties of wave equations, analytic continuation makes no sense whatsoever. You can consider, for example, the case of the Strichartz estimates. In the case of the Laplacian operator, on the Euclidean space-time, there exist many estimates, known under the name of Hardy-Littlewood-Sobolev inequalities. Most of them have no counterpart for the wave equation in Minkowski space-time. There are however a few inequalities which survive. They go under the name of Strichartz inequalities. Can you derive them by a Wick rotation? The answer is: I don't know. I have no idea. I don't think anybody has an idea. And I think it's probably impossible. One thing is clear; you cannot derive Strichartz type inequalities directly from the form of the fundamental solution in physical space. They are much deeper than the H-L-S inequalities. There are many other examples of this type. I suspect that the usefulness of analytic continuation is quite limited.

A. CONNES:

It might be that we just don't know it now, but somehow for instance you see that causality translates into finite propagation speed. So there are some things which do translate nicely. Another one is the Feynman prescription for the Green's functions. It becomes beautiful in the Euclidean formulation.

S. KLAINERMAN:

Yes, but all these are related to the explicit form of the fundamental solution.

S. NOVIKOV:

I would like to make some remarks. First of all, purely Lorentzian geometry does not exist, because you always need locally Euclidean topology. Therefore, some invisible Riemannian metric presents also. Local topology is generated by some local Euclidean structure. Otherwise you cannot write differential equations. My second point is the following. I would just like to inform you - some people with physics education, who are present here, certainly know this. In the quantum electrodynamics (and in all other interactions known now) the possibility to make a "Wick Rotation" from Lorentzian to Euclidean geometry is more or less experimentally confirmed. There is the so-called CPT invariance principle. Let me remind you that the simplest fundamental topological difference between the groups $O(3,1)$

and $SO(4)$ is that the Lorentzian group has four connectivity components $1, P, T, PT$, and the Euclidean group has two components only: $PT \sim 1$. The CPT invariance principle confirms that everything is still trivial if you are in the same component from the viewpoint of the Euclidean geometry. The charge conjugation C is non-geometrical. This is a confirmed fact.

Y. FRÖHLICH:

It's true that if your space-time is Minkowski, then you can do this Wick rotation, as described in his [Novikov's] lecture. But, of course, if you want to do gravity or so, then Minkowski space is a little bit too limited. We have no analogs of these theorems for general space-time, as far as I know.

P. SARNAK:

I agree with Sergui in the following sense that it's more a result of style. If you'd like the theory of the symmetric spaces that was mentioned a few times already, that's G/K, where K is a maximal compact subgroup, and it has been studied, probably, mostly by Harish-Chandra. And I think his influence is such that it was studied for many, many years. However, recently, it has become absolutely clear that all affine symmetric spaces are just as important in the theory of automorphic forms when one studies G_H, where H is a fixed point set of an involution. And the reason that it was not studied formerly in America (in fact only Japanese and Danish people studied this for some strange reason) is purely style. So, I think I agree with Gromov that there are certain things in Riemannian geometry of the style that he does, where it's really difficult to imagine what thc analogue is in the Lorentzian case. But if you talk about harmonic maps, or partial differential equations, or wave maps they are analogues, and the reason it's not studied in the Lorentzian case is style. People do what they feel comfortable with. "He was born to do Riemannian geometry first." The answer to Sergui is: he must get up in his talk and try to influence people to move in this direction by making interesting problems and some kind of influence in style, it's nọt really intrinsic in the case of affine symmetric spaces.

M. GROMOV:

There is a difference for compact and non-compact groups...

P. SARNAK:

But for certain problems it is not so different. For affine symmetric spaces in general it's true that the Riemannian case is easier, but in the end all the same theorems are true: Plancherel formula and so on.

M. GROMOV:

Here is another important point: what do you call geometry?

P. SARNAK:

It fits into my definition of geometry.

V. MILMAN:

I would just like perhaps to turn our discussion from physical geometry to geometry and ask you [Gromov] what couple of directions you think is the most interesting in geometry now.

M. GROMOV:

I refuse to answer that in two words - it needs two lectures and a justification. The gradient lies in symplectic and contact geometry. That's obvious. Symplectic geometry and contact geometry, certainly, by far pass Riemannian geometry, Lorentzian geometry, or what you call geometry. And this will continue at least for two decades. That's an obvious prediction.

L. POLTEROVICH:

How do you see the future of these subjects?

M. GROMOV:

I don't know. Some time ago I wrote an article on symplectic geometry, that was called "Part One". And I still haven't written "Part Two", because I couldn't do it, so I don't know. Because I didn't know what do next or I couldn't do what I wanted to do, how can I speak about what I don't know?

M. KONTSEVICH:

Do you consider super-manifolds as a part of geometry?

M. GROMOV:

This is just terminology, some emotion. It depends on how you think about that, and who cares what I think personally. It goes on and that's it. But of course, again, geometry has several aspects to that. You can have something very algebraic and call it geometry, if you can apply some geometric intuition. You can make it in a wide sense and in a narrow sense. It becomes a question of terminology, because in the wide sense geometric intuition is used, and in the narrow sense it is not.

N. ALON:

You mentioned in your first lecture one problem that you think is not interesting. It was the densest packing problem. Is this not interesting, or is this not a problem in geometry, or is this not a problem in interesting geometry?

M. GROMOV:

Actually, to me it looks like a historical problem, which came from history, raised by Kepler. Just erase Kepler from the story and the question would disappear. Like erasing Fermat from Fermat's theorem and there wouldn't be interest in the theorem. It is a sociological phenomenon.

N. ALON:

It seems that the interest also came from the solution.

M. GROMOV:

The solutions were great but before the solution there was a long development, influenced by that. The focus on Fermat theorem was due to the history, not due to internal interest in the subject.

N. ALON:

But that's true for most of mathematics.

M. GROMOV:

No, it's not true at all. Problems come from history but the interest may still be focused on the internal, like the Riemannian hypothesis. Everybody believed that this was a central problem, not because Riemann raised it. It's completely different.

SOMEBODY:

But it was natural.

M. GROMOV:

"Natural" is a good word for food marketing, but natural problems are usually bad, raised without understanding them. A natural problem is a bad problem, because "natural" covers an uneducated mind. You know, people in the street ask natural questions; these are stupid questions. Deep questions, like the Riemannian hypothesis; these are interesting questions. "Natural" is a bad word in mathematics.

Y. NE'EMAN:

Alain [Connes] mentioned Hawking's book "A brief history of time", which is supposed to represent the status of cosmological theory. I believe I should point out that the situation and problematics have changed more then considerably since the publication of that book. There has been a "revolution" in cosmology, of which mathematicians are perhaps not aware

- and let me bring you up to date. Twenty years ago, after Hawking and Penrose had proved the "singularity theorem", stating that in *classical GR, every time-like line has to have a singularity either in its past or in its future*, the cosmologists worried about the multiplications for the Big Bang and the lack of a physical description for the "start" of time. The book is an exposition of a solution suggested by Jim Hartle and Steven Hawking, namely that the Universe was born with a Euclidean metric and only "later" (whatever then the meaning of this word) underwent a "change of signature" - like a transsexual operation. Moreover, at birth, there was no singular point in any other sense, because the Euclidean 4-manifold was really a sphere S^4 (you only add a point at infinity - and the Big Bang becomes the "South Pole" - and there is no true singularity at the South Pole, only a matter of coordinate patches). This was the status in 1982. All of this appears irrelevant now.

The new program is "Eternal Inflationary Cosmology", developed in 1982 – 1987 by A. Guth, A. Linde, A. Vilenkin, P. Steinhard and others. The singularity theorem becomes irrelevant because the treatment is fully quantum-mechanical. If you check in detail how this is achieved by QM, you find in the examples that *quantum tunneling* does it: a tiny (Planck-size) blob of Planck-density vacuum-energy (i.e. a cosmological constant provided by QM) triggers an exponentially expanding de-Sitter solution. To start with, it is a *black hole* whose singularity awaits it in the future, but in its (very slow in a spectator) collapse, it *tunnels* to a trajectory which is that of a *white hole* which has thus already had its singularity in its past! This is an example of how QM can circumnavigate the obstacles set by Classical Mechanics. This example appears in an article entitled "Creation of a Universe in the Laboratory" (with tongue in cheek?) by Farhi, Guth and Guven in *Nuclear Physics* **B339**, 417(1990). In this new picture, the universe is infinite in time and space. At some irregular intervals (of the order of 10^{1000} or more) a new Big Bang is triggered and generates a new expanding "sub-universe". In any case, the motivation for a Euclidean origin has vanished in the Quantum treatment.

V. MILMAN:

If it's possible, I will ask my second question. You answered only half of my question, about the one direction, which you think is interesting. And I won't ask about the second. I ask you [Gromov] now, please, if you may, to give a couple of directions, which are *not interesting* in geometry. You've already said something, answering Noga's [Alon] question, but perhaps you can add something. It's also very interesting.

M. GROMOV:

There was, I think, tremendous confusion, not only in geometry but in other branches of mathematics, when, for example, sociological facts have taken over, or when people were mixing words. There is a Riemannian manifold, there is topology, so we ask the question (which I mentioned in my lecture): which topology supports what kind of metric? This is a kind of very natural, very naive question. And that's exactly what you should avoid; we should avoid asking natural questions. So if you have a natural question, change direction...

S. KLAINERMAN:

What about quantum gravity? That's a natural question.

M. GROMOV:

Not at all. People in the street, do they ask this question about quantum gravity? Never!

S. KLAINERMAN:

But somebody who knows quantum mechanics, who knows that there is a contradiction between...

M. GROMOV:

No, it's a highly unnatural sophisticated question. It comes from deep thinking and conceptualizing, not from natural thinking. I don't want to make specific...

V. MILMAN:

Can you be a little bit more concrete?

M. GROMOV:

If I'm concrete... I only can say that I then have to invent stupid directions or what? Or I have to indicate directions, which are stupid; it will be abrasive, I shouldn't say it.

G. KALAI:

If you regard the natural problem, which you can not solve, as a bad direction, as something ridiculous...

M. GROMOV:

I realize that's certainly an outrageous opinion, but I think it's correct. Look at the natural world: all common opinions of people, on any subject are nonsensical, anything.... people think about astronomy, about physics, about chemistry, biology, life. Everything is sheer nonsense, so-called folk-wisdom. It's the extension of the folk-wisdom, it's exactly what you have to shy away from.

[HUBBUB-LAUGHTER]

N. ALON:
Why do you talk about a man in the street?

M. GROMOV:
I mean a mathematician in a street. A man is a mathematician. If two people are agreed that this is a good question, this means that it is a bad question!

N. ALON:
There is an excerpt from the introduction of Gary Larson's book (slightly modified), which says: what I am asking is interesting, what you are asking is not. We are trying to say something provocative. I'm saying that usually a lot of the good questions are good questions *because* they have a history. So Fermat's last theorem was a good question because it created the great area of algebraic number theory. The four-color theorem (one may think that that's another stupid problem) was one of the main reasons for the creation of graph theory and a great deal of combinatorics. Usually there are these questions that are natural, at least according to my definition - are natural, and are often indeed stupid questions, but still they often do lead to very serious mathematics.

G. KALAI:
The densest packing of spheres is more natural than any other question. This is a challenge; this clearly deals with the asymptotic question in high dimension. If you think it is stupid then solve it. It relates to very good mathematics.

M. GROMOV:
I don't think it's a stupid question. In my hierarchy of questions, this question lies low. There are infinitely many such questions.

[HUBBUB-LAUGHTER]

G. KALAI:
There are not so many questions of this level. It's not clear why it lies so low.

V. MILMAN:

I propose a compromise between these two opinions. The point is that, of course, Misha [Gromov] wanted to say that the things, which are already obviously on the surface (and everyone understands what should be around, although we cannot yet prove them), they are already, in a sense, questions of yesterday's fashion, which continue today to be still considered fashionable, even though they are already of secondary interest. However it was not so when we started to ask these questions; not when Fermat invented this question, but perhaps when the main development, which developed mathematics... Let me be more provocative... You see, the four-color problem was important...

[HUBBUB-LAUGHTER]

M. GROMOV:

I want to answer Noga [Alon], about this relevance of Fermat's theorem to number theory (certainly nobody argues) and also the four-color theorem to graph theory (nobody argues). There is a ready-made Russian tale, exactly describing the situation. It's about a soldier who comes to somebody's house and he is hungry. The people in the house don't want to feed him, and he says: "You know, I can make a beautiful soup out of an axe. Just give me a metal axe and I'll do the soup."

SOMEBODY:

It's called "stone soup".

M. GROMOV:

OK, you know the story? They are very much surprised and want to know how to do this, and the soldier says: "Put the axe into the water and just boil it". They are boiling and boiling, and then the soldier says: "Now you have to add some cabbage, just to make it better." And then they have to add some meat and so on, so in the end they have a beautiful soup. That's the same, exactly the same. Number theory was developing, and by the way, Kummer was little concerned with Fermat's theorem, contrary to popular books. I don't know about graph theory. These remain the issues.

P. SARNAK:

I think the one thing you haven't pointed out yet is, that the solution of Fermat's last theorem supposed that God has a very easy proof, which uses a lot of mathematics, where mathematics looks in a content way... What happens? Yes, you might complain, as you did in your lecture about the nature of the proof. You don't see any development, and in that sense this is what he [Milman] says. This is something more than you have complained about the problem itself.

M. GROMOV:

Concerning specifically the theorem of Fermat, there is a real theorem proved, there is a special case of Langlands conjecture; this was an achievement, this was a development. Why you speak about Fermat? It was a minor side-effect.

P. SARNAK:

I agree with you.

N. ALON:

So, you think that you care more about the solution than about the question.

M. GROMOV:

No, no, no. This fantastic question, Langlands conjecture, whose name the conjecture, Taniyama-Weil, used to be called. That was a great question and it stands a great question. It's not a question that anybody can ask; it's a very deep question based on understanding of the structure. But Fermat's theorem like irrationality of $2^{\sqrt{2}}$ (though it was asked by Hilbert) appears a stupid question. You have to admit it.

P. SARNAK:

I think a little more . I said that it smells of a bad problem. It doesn't seem to have a feature, which stimulates work.

M. GROMOV:

Yes, absolutely. That's quite possible, and that's what happened. I completely agree with that.

S. KLAINERMAN:

Don't you think that there exist in Mathematics these simple minded questions which may seem boring at first glance but whose solutions may be far reaching? Here, in this question, we obviously should know the answer. I mean, it's so simple, so obvious. And the fact that we don't know the answer says that we are missing something very important.

M. GROMOV:

OK, but this is not the case in Fermat's theorem.

G. KALAI:

But the densest three-packing belongs a little bit to this category.

[Hubbub-Laughter]

M. GROMOV:

OK. I may be wrong, and it may have some depth, and I am missing that...

S. KLAINERMAN:

Is it possible to define a good question as one which appears out of nothing at a certain moment due to the insight of a great mathematician, and which is immediately recognized as being very natural by everybody else. Is it natural...

M. GROMOV:

"Natural" is a bad word.

S. KLAINERMAN:

But once the question is asked, maybe one can allow it to be called...

M. GROMOV:

OK, let's begin with a simple example. Is it natural for you, when you rotate a stone on a rope and when you let it free, that it goes in a straight line? Everybody learned it at school, but most people still believe it goes in a circle. That's our natural belief. Right? We just say it goes in a straight line, but no matter what we say, in certain situations we behave as if it goes in a curve.

SOMEBODY:

It is not a natural question.

M. GROMOV:

I want to say that it is a natural perception, question or no question. Natural perception always deceives us. You can conduct experiments with yourself and you'll see, that you have rudimentary beliefs coming from our animal ancestors. And the same is in mathematics. And we should accept it, acknowledge it. Then we can make better questions. If we believe our feelings naturally have a deep sense, whatever... It's absurd. We have to assume that we are very stupid and our natural questions are stupid, and only by hard work, by conceptualizing, working hard, calculating, whatever, we can make good questions or good mathematics. And it's naive to think that we all have intuition or something. It's a stupid opinion. That's what I believe.

S. BLOCH:

It's a shame we spent so much time on what people feel are stupid questions. Can I ask a question slightly differently: where is geometry blocked? In other words, for example in number theory, which I know rather better, I can point to three or four of its directions that are like a wall, where the subject is blocked. So I'm interested in your perception or the experts' perception on where geometry is blocked?

M. GROMOV:

Geometry has a structure which is so very different from number theory. It just doesn't go a definite way; it is spread. There are particularly difficult questions, some of them are very good and unnatural. We cannot solve them, that's for sure. But there is no one point where it is blocked. It was never like that. Geometry never goes as far. Compared to other branches of mathematics it depends on a different part of your brain. It's not the consecutive part of your brain, exercising long sequences; it's spread like visual perceptions, so it cannot be blocked. When you see something you cannot be blocked; when you go you are blocked. In geometry you don't go far, ever.

D. ZAGIER:

Maybe number theory is blocked, but geometry is *gromoved*!

V. MILMAN:

I would like to finish this discussion and to ask another question. And I'll finish this with one compromise: a question hasn't the same value in time. When the four-color problem was invented, it helped mathematics a lot, it was good. When it was solved, it was bad, and it was solved badly. In any sense it's bad, and it may be in another discussion, on mathematics in the real world, say.

S. NOVIKOV:

I should answer in this case about the "bad" solution of the famous problem. I would like to explain my personal opinion. The great topologist Haken who started a very deep 3D topology (and anticipated the hyperbolic topology in fact) solved important problems before. In particular, he proved that the recognition problem of the trivial knot is algorithmically solvable. He was immediately attacked by people. They claimed that this algorithm cannot be realized practically. I cannot tell you about its type from the viewpoint of the abstract complexity theory (if it is P or NP), but you have to input so much data from the beginning, that it is simply impossible to start any computation. People have been looking for effective algorithms in the knot problem until now. It is a very important problem. Haken started to look at the practical algorithms for a different goal. As everybody knows, a few years later Haken used a huge numerical job for the solution of the 4-color problem. Unfortunately, no proper credit has been given to this result in the math community, for example, in comparison with the proof of the last Fermat's theorem. Many pure mathematicians are saying even now that this solution is ugly. In my opinion, it is in fact a much greater achievement than any normal type of beautiful proof, like for example the solution of the Fermat's last theorem. It is in fact the logic of the 21-st century: we even cannot control it without a computer. You may think it is ugly, but it is absolutely unique. It was a great achievement. So it was my defense of this great work. It was not badly solved. It was the logical future.

P. SARNAK:

In fact, I agree with you, I agree with you completely. I think it's actually Haken and Appel, is it? Firstly, since you like to give credit to the right guy you should get the name. I don't know about Appel, but Haken, I agree with you, was a great guy. I think one point that is coming up here, and it came up in your [Gromov's] discussion, is, and I think this is going to be very relevant in the future, is the role of computers in proofs and in mathematics, and it doesn't seem to have been discussed. It's inevitable that computers will appear in mathematics, I mean, computers in proofs, not in science, not in numerical analysis, the role of computers in pure math, in something like the proof of the four-color theorem or the proof of the sphere packing problem.

[HUBBUB-LAUGHTER]

M. GROMOV:

I have a more radical opinion. I believe, that may be in 50 years mathematicians will be less relevant than computers.

V. MILMAN:

You see, Peter [Sarnak], if the Riemannian hypothesis would be solved with the help of computers, I would consider this to be a great achievement, because it doesn't matter how it will be solved. When you talk about the four-color problem, the only interest, which exists in such a problem, is that if this question is so simple and we don't understand how to answer it, then I would expect that some very young guy will find a missing structure here, and this structure will be more important for mathematics than the four-color problem. And when you put in ten thousand hours of computer time just to tell me that I will be able to find the right colors, indeed I don't care. If some mathematics was developed by it, that is the only thing I care about. The problem is not solved well, because someone killed a good problem by ten thousands hours of computer work. And this is what I want to say. Do we still continue with this subject? I have another question for Gromov.

SOMEBODY:

I wanted to say that actually there was a matter going into it. I mean we just have to realize that there will be certain steps in proofs, which are just not worth doing manually.

V. MILMAN:

It means that the proof was not clear enough, that structure was not found, that something was missing.

[HUBBUB-LAUGHTER]

Y. ELIASHBERG:

I just want to say something slightly different. It seems to me that a kind of great breakthrough and a kind of great development happens, when a subject isn't very fashionable yet. It seems to me that especially in the States (I mean the West compared to how it was in Russia), some area becomes from time to time extremely fashionable and everybody rushes in there. For instance, I feel kind of uncomfortable with what Misha [Gromov] says about the next 10-20 years, when symplectic and contact geometry will be in focus, and I really think that I should already look for some place to hide.

V. MILMAN:

It becomes so natural that it becomes not so good.

V. ZAKHAROV:

I must say that the problem of the orthogonal coordinate system was very fashionable in the last century and is completely forgotten now.

V. MILMAN:

My question is: it's now 18:02. We may try to provoke something else but I don't know if anyone wants to stay here any more... Perhaps we should stop, I don't know. No? Then I have the third question. Perhaps some geometry is not so great, not the best direction, and perhaps some bad geometry you [Gromov] don't want to mention. But what geometry is it *necessary* to develop? Perhaps it is not the best, perhaps it is not that good, but what is it *necessary* to develop?

M. GROMOV:

You know, to answer this question I have to think because this is committing myself.

V. MILMAN:

OK, so we will give food for some other discussion.

M. GROMOV:

For example, if we write proceedings, I can, say. But this requires hours of thinking before answering this question.

V. MILMAN:

So, you can write this in proceedings.

M. GROMOV:

Written commitment, exactly. As you have said, Don [Zagier], the conjecture is the most responsible thing one can do and sometimes people make conjectures when they absolutely have no right to make conjectures. A conjecture really comes hard. I agree with you, Don, that one can make a serious conjecture once or twice in one's life, after deep thinking. You come to a deep understanding, and you cannot finish it, and you make a conjecture. You just cannot turn any question into a conjecture.

P. SARNAK:

What about publishing your conjecture?

M. GROMOV:

This is exactly the way of making commitments - to indicate directions and make conjectures. Do you have conjectures?

Transcribed by A. Scherbak

DISCUSSION on MATHEMATICS in the REAL WORLD

with introduction by R. Coifman

R. COIFMAN:

This could be a really fun discussion. We didn't have many opportunities to speculate, and since it's only a discussion we can say anything we want. It's actually essential that we do, I think it's a matter of survival. Not survival of mathematics, that's independent of us, but survival of certain traditions and certain ways of doing, of practicing our field.

The first issue here is the topic of discussion. You know we are supposed to discuss mathematics in the real world. It can take forever since mathematics is the real world on some level, at least for some of us, and the discussion is endless. But there are really several aspects that maybe it would be good to address. Some of them are intellectual and some of them are political. They are both important.

We have a tendency as a community to be quite reticent to have a big view of the world like the astronomers or others. And maybe it's a good idea that we try at least to go in that direction, but for that we have to put our house in order and understand what really are the challenges.

I think that this meeting has as one of its missions to basically define the areas in which we are weak rather than strong. And the weaknesses are much more interesting. I mean what we don't know how to do is infinitely more interesting and challenging. Unfortunately we like to talk about what we just did. Well, that's fine but not very productive.

I distributed a blurb around. There was no particular intention in it in terms of topics that we need to concentrate on or anything, it was just an indication of possible challenges.

This was a blurb, which was given to a mathematical-physical science advisory board, including chemists, physicists, astronomers and others. And my intent was to basically to describe to them on some very rudimentary level to what extent we don't have the mathematics they need. And it actually had some impact, I mean, I was surprised by how receptive they were. On the other hand most of my colleagues were not interested at all. But in that blurb the point was made that the state of science and technology around us is reaching a certain level of paralysis, in the sense, that complex phenomena are handled with very ad hoc methods, which are invented on the spot. Nobody is putting in the intellectual effort necessary to build the mathematical infrastructure for the language and the definitions of the geometries, and the definitions of the interactions between objects that are needed in order to actually go ahead with the science.

And this goes all the way from Newtonian mechanics to biology. Now we would think that Newtonian mechanics is well understood. I mean some of us think that having the axioms for the theory of all would solve a lot of problems. It's true, it will on some intellectual level. On the other hand simple problems that were understood by Newton are still not understood by us. The reason that they are not understood is that we just don't have the mathematical tools to describe them and to deal with them. So in the blurb there is a specific example given, having to do with computation of gravitational interactions. It's a simple example but it illustrates the point at some rough level, showing that although we understand the basic infinitesimal laws of physics we don't understand how they work or at least we can't describe how they work. So I think rather than be abstract I would, maybe, just try and give an impressionistic summary of this half-page, that you got, and use it as a scaffold on which maybe we can build a discussion.

By the way, again, my intention is just to have something concrete in mind, it's not that this is a particular topic we need to work on. I mean the issue is on what level mathematics interacts with the real world. Mathematics has always been the language for the description of physical law or biological law, any law, in which there are specific rules of operation, and mathematics was used both as a language to formulate problems or to define structures and as a language to describe relations and interactions between objects. And the real issue is, and it has always been like this until the second half of this century where we started to focus much more on our internal needs and somewhat disconnected from the real world.

We could all become applied mathematicians, but that's not the point. The point is that the wealth of mathematics actually was essentially nourished by outside interaction and I think that the real question is what's in it for us as pure mathematicians. What kind of insight are we going to get into our own internal core of mathematics by actually viewing external problems or problems that we think are external although they are just manifestations of things that we have seen all our life. Tomorrow, in my schedule lecture, I'll try to give some examples in complex function theory, which are very simple, which when viewed from the point of view of music suddenly make sense. But viewed as a formula in complex analysis it is just a complicated formula.

So let me, for a minute, be a little bit more precise, I just want to summarize briefly what was in the blurb. The issue is the most elementary problem of all. You have a tremendous number of gravitational masses as in the universe and you just want to try to compute where they are going to be maybe a year later or maybe a month later and so on. So those masses are distributed. We have Newton's law at our disposal, it's not a big deal. We know that if we have, say, n masses, there are n^2 interactions. And we can just compute it. When you do that this way, which is the way people have been doing it, nothing works because you have too many masses and the computation is intractable. And you really haven't gained any insight into the problem. There is no insight whatsoever, I mean all you have is all those little masses interacting. On the other hand there are a lot of structures out there, there are galaxies and there are other groups and clusters of stars and other masses around. And you want to understand how the various objects interact with each other and work together. Again naively there is nothing hard, right? Just put it in the computer and let the computer do it and everybody is satisfied and happy with this. In fact it's clear that this is a completely dumb and impossible way to compute.

By the way when I say impossible and I know there are some people who don't normally look at numbers, you know when you have a million points or ten million points and you square it, it becomes a pretty big number. And very quickly, if you have to do this every tenth of a second and you have to do it for a year, it's a very, very big number. And it's just intractable. On the other hand if you could do the computation in order, say, corresponding to the number of points that you have there, well, you have a chance.

Coming back, before I describe it slightly more, let me just say I view the issue of being able to compute effectively as a problem "a la Gromov". It's a good way to ask questions about the structures of things. If you can do it you've understood exactly how objects are organized and interact, you've understood geometry and so on. If you are incapable of doing it but just do it naively, well, that's fine, you've understood it on some Newtonian level, but that's it.

So returning to this gravitational masses issue, there is an observation that was made by Newton that if you want the impact of the moon on the earth or vice versa you don't add together and compute interactions of all atoms of one object with all the atoms of the other object, but you learn that the whole earth is one mass, the moon is another mass, and it gives you some approximation of the interaction.

When you have the full universe to do this kind of thing you can do it too: you break it up in clusters, the clusters have to be broken at different scales again, the mathematics is not an issue here, it's just a scaffold. By doing this organization you know that if you have a galaxy here, and you have a star there, the whole galaxy is a single object, and interaction of the galaxy with a particular mass or a cluster of masses here is given by basically clustering the whole thing. When you get within a galactic diameter from the galaxy you do something else, you look at the neighboring clusters, at different scales and so on. This by the way is very easy to formally organize mathematically so that you don't even have to write an algorithm with the geometric organization, just described. There is a language to do it. That language is capable of doing it automatically. Basically it corresponds to what people called multipole algorithms or you can find very special basis expansions to do that. That's fine. But what has happened here beyond Newtonian mechanics is that we suddenly see that the organization in clusters of this universe actually tells you how physical interactions occurred, and this organization is really a much more precise description of how the gravitational fields work than the Newtonian microscopic description.

Again, this is baby-stuff. If you were to deal like this with more complex interactions like the acoustic vibration in this room, again you are dead. Classical physics is incapable of doing it. Classical mathematics is incapable of doing it because the interactions occurring here are dependent on so many different parameters, and acoustic fields are extremely variable. How are you going to that kind of stuff? There is really no language to describe it that we are aware of. There is no traditional mathematical formulation, that's what I'm trying to say. There is really no mathematics because people haven't invested, I would say, the time to do it.

Let me finish with this brief introduction by just giving you a picture that you may have seen before. Could we switch off the light for a moment? Here you have a picture of the real world.

SEE PICTURE 1

It's a threatening animal in the picture of the real world. By the way what I'm telling you here and we may discuss it tomorrow, the analysis I'm going to describe to you or the transcription of this object is one that you need in fact in order to compute exactly the acoustical interference in this room. So, that's not just a picture, it's really a serious question. So, the question is how do I describe the geometry of this mandrill? That's really

Picture 1

Picture 2

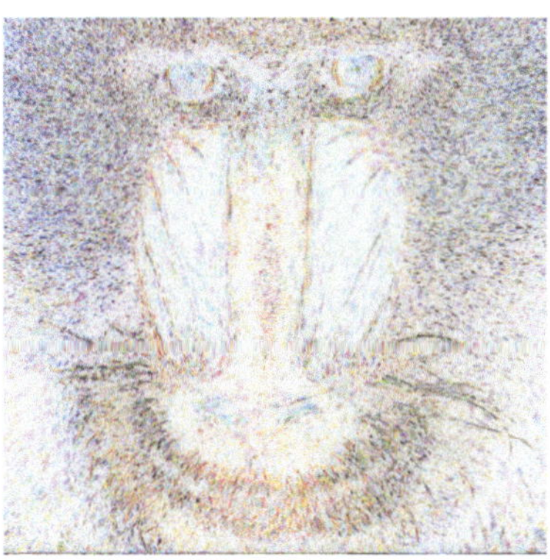

Picture 3

Picture 4

the issue. The geometry of the mandrill and the structures out of which this image is made. It's really quite complex; there are a lot of mechanisms affecting it. And you raise your hands and you say: "Surrender". So here is the transcription of the object: this is the same object, which is described with basically less than one third of one percent of the number of parameters.

SEE PICTURE 2

So if the original object has 250 thousand parameters to describe it, this second one has one third of one percent, and it's a transcription which is done by using essentially watercolor. So, the transcription here is done by mathematics, completely automatically, using wavelets. It's done by essentially writing it as the syntheses of various instruments. So, here is watercolor, here is a second layer, which is a Van Gogh type layer.

SEE PICTURE 3

This is done with a paintbrush. And the last residual is done with a pointillistic version, by using pencils so to speak.

SEE PICTURE 4

So, what we have is that the original image is the sum of the three. By the way, it is a sum of a synthesis of the image as a sum of three different instruments - one is a pencil, one is a paintbrush, and the first one is just a watercolor. And if our projector were stronger I would show that the original is the sum of three stills.

So, it's an orchestration. Again, my point is that this is an automatic transcription. How it's done and so on, maybe we'll describe tomorrow, that's not the issue. The issue is that somehow there has to be an efficient transcription. We start with the fact that we don't have the language to describe nature, and because we don't have the language to describe nature we don't have a way of thinking about it; we don't have a way of doing anything about it, because we are dealing with primitive tools like formulas and modes of thinking that we have been ingrained with. And those primitive tools just are insufficient. We are dumb because we don't have the words, so to speak. And so the words have to be invented, and we are not, at least I'm not, smart enough to invent them. They have to be discovered in some sense, discovered by interaction, by fighting with real problems. As Gromov said, for example, how do we describe the geometry that occurs in biology. Well, you can do it, but the problem is it's done by scientists, amateur mathematicians, very often, who don't really care. I mean I'm not sure if they really care to invest the intellectual effort necessary, the years of work or analysis necessary to understand the actual internal structures of the objects, they want to get an answer. So, you have computer scientists and physicists who invent neural nets, in order to be able to describe complex objects like this, or signal processors, or a machine who do various tricks in one form or another, or financial people doing computations, all of them doing it, by whatever tool they have at their disposal, sometimes in an ad hoc fast way because they want an answer.

They want an answer, that's great. But I think it's our profession, which should invest the time and effort and understanding of the structures and develop a language necessary to actually do the scientific modeling (if we are talking science). I think I talked too much already. The main question is what can we as a group do about our field. We are really at a completely pre-Newtonian stage, you know. Before Newton there were all kinds of insights on calculus issues, versions of calculus and so on. And Newton formulated the language needed to describe mechanics, very nicely, and then everything evolved.

At this point any time you have a phenomenon with ten parameters or more interfering with each other you find out you don't have any tool to describe the problem. You may think you have but it doesn't work, you can't compute it, and formulas are not rich enough.

When you are dealing with pictures which have 250 thousand parameters, there is just no way. So, somehow we need a Newton, which I could say is a new transcription of nature as a way to proceed forward. Biology is stuck on some level. Chemistry, material science it's the same problem. I mean we know the basic laws and that's it.

OK, enough said. I would like to actually open the floor really to discussion. I'm sure everyone of us has some relation to the real world so to speak, in terms of their professional life. And the real question is what can we do.

First of all not to have a canyon between mathematics as it is in the 20th century and as it is going to be in the 21st century.

Kids who are coming in are quite a bit more excited by microbiology or by various other fields which look like virgin territory, where there is a lot to do and you don't have to be a genius in order to make progress. It's much harder for us to convince kids that this is a profession that they want to pursue and make a lousy living at. So we have to address this issue, I mean we have to address it both on the level of telling them that this is a critical profession, it's the core of everything that happens nowadays. We are really at the golden age of mathematics in a variety of ways. But we managed beautifully to hide it. How to do the propaganda is a basic issue. Not just for the kids who are coming to study, also for what they will study. I mean to convince somebody that he needs to understand something in algebraic geometry, it's much more effective to show him that he can use it in a variety of situations, and that it's beautiful and relates to a variety of different fields, whether analysis or algebra or geometry, any one of the classical activities that we are involved in and whether we want to call it classical or not. So I think a good outcome of a meeting like this would be if we got a good set of suggestions on how for us as professionals to interact with a future world.

M. RABIN:

I have two or three comments about the introduction. The main theme that I hear from what you say is that you are advocating, let us say, a new mathematics for dealing with complexity or maybe asking the question how one can present mathematics whether you want to call it classical or modern, but present day mathematics, how that can be extended or is it sufficient to deal with complexity. Now, I think that a basic question here is whether there is a unifying generalization. You spoke about Newton. Newton mechanics came about in part to explain and to prove if you wish Kepler's laws, and he beautifully explained the solar system, which is a system not too complex if you write the equations and use perturbations and some reasonable approximations. You really get a very good computational description of the solar system, enabling you even to say where it is going to be, let us say, in a million years from now. Now, the problem that you illustrated about the universe is of course much bigger. By the way it has points of contact with other issues such as protein folding, where again you have these clusters of electrostatic attractors that cause proteins to fold, and this is a problem of enormous importance, maybe more important than, you know, predicting where the universe will be in two billion years from now, and people are, as you say, applying ad hoc methods. I think the basic question, one basic question is whether there is a unifying or there is a very small number of unifying principles. So you spoke in a somewhat denigrating form about those people who are doing this and that in order to solve the problems, the question is whether the world is not so complex and the phenomena are fundamentally so different from each other that you would need maybe even a large number of methods. I think this is a very basic question. And this afternoon, in the discussion of computer science, I'm going to describe another extremely complicated situation where classical mathematics is not sufficient. It has to do with the Internet and there you have exactly the same issue arising again.

The other comment, the final comment I want to make, is about the picture of the mandrill that you have shown. And that again illustrates the complexity involved. So the question is how do you parameterize and describe a picture like that, concisely, which is in a sense a question of compression, dealt with very extensively again in computer science and information theory. But then there are other questions relating to - this is computer graphics. So for example, what computer graphics people do is they describe an object, such as, let's say, an airplane or a person, but it's not just to render the front image. They want to have a coding of that image which will enable one also for example to rotate it and view it from various sides, and in a sense if you wish to also look inside and so on, and all that is being done. In your picture there is also the problem, and they solve it, actually getting the whiskers, say, of the cat or of the mandrill around the mouth. So that's again a different issue, and it is solved in another way from the one that you described. Again an example that the situation is more complex and that the kind of solution that we like, which is prevalent and catches everything in what we call an elegant way, I would think is probably not available.

R. COIFMAN:

I actually meant to really describe an intermediate situation between the multimedia and the mathematics situation. And that is that there is the transcription of the mandrill which was designed in order to do computation on it and not designed in order to look at it. So it's really a very structural situation there, it's not a perceptual situation. Should I summarize the comment or you [Rabin] want to summarize it?

V. MILMAN:

He [Sullivan] wants to summarize it.

D. SULLIVAN:

His summarizing of the comment would be complicated really to get... What are the underlying principles? What is the answer?

R. COIFMAN:

Yes, the answer is that there are hints of several underlying principles both in geometry and in analysis, and this is what we, Peter [Jones] and I, just intend to describe in some sense. There are universal principles, so to speak, underlying some complex structures that permit us to answer a lot of those questions. I completely agree with you [Rabin] about the way that we have to define matters. It seems that the universe has so many different structures that for each one we will have to invent its own mathematics, so to speak, or find its own internal structure. And that's probably true. The question is: we want to have underlying basic principles that permit us at least to deal with the physical world, say just the mechanical physical world. That's something that we understand much better than the biological world.

M. RABIN:

Could I ask one question? The question is roughly the following: do you think that when we try to describe this natural phenomenon and not only that, also the financial world and others, neural phenomena, there are going to be three or four underlying frameworks or are there are going to be twenty or thirty?

R. COIFMAN:

I think we are a little stuck with the paradigms of the past. I think that meta-frameworks may be very few. In other words, guides on how to proceed in order to attack a certain class of problem might exist, but then the framework will be very specific to a situation. I don't know, I wish I knew. Nobody knows.

D. KAZHDAN:

I don't understand the word that was used. When you say: inventing new types of mathematics for dealing with different phenomena, I don't really understand what do you mean by the word "mathematics".

R. COIFMAN:

I wish I knew that too.

M. GROMOV:

May I make some kind of remark. As mathematicians when we express our wishes we also consider the possibility that what you want to do is impossible. And so I want to make a statement and I wish you to disprove that. First, there is no general framework when describing the world and there are no arguments for this fact, and secondly and really most importantly, as a community, mathematicians are unable to participate meaningfully in that. So, let me start with this statement which is an obvious sociological issue, because as a community we developed under very particular conditions after the Second World War. There was a boom of development in mathematics, and people were drawn there. We were drawn to the particular kind of features, structural mathematics, far from the real world. We built our community like that, there is no chance that we can adjust to the new conditions. We'll be a different community of different people with different ideology. There is no way we can meaningfully participate in that.

This was the first statement. And the second is: why there is really little chance for a general theory. First we will try many times outside physics to do it. Look, attempts in biology - mathematicians all thought about biology, all attempts failed, absolutely. No chance to realize the structure. And also if you look how our system or how we have organized in this world. We cannot describe it but, though I will never see someone do it, we have some idea how he does it; it's absolutely perpendicular to how mathematics works. It's kind of broad and very shallow. And mathematics cannot be like that. So there are two good reasons why your problem absolutely has no chance. Can you defend it?

R. COIFMAN:

Sure.

[LAUGHTER]

P. JONES:

What are these two reasons?

M. GROMOV:

One is that as a community we are unable to do this, because we are built absolutely opposite. And secondly we have an instant in our system as a visual system that's also going in the opposite direction to what mathematics is doing.

P. JONES:

Let me respond, maybe in two points, which is that many of us have experienced exactly this phenomenon of great resistance when people try to turn to do new kinds of problems and express them in new ways, and your first reaction is to get rejection from your colleagues, because they've been trained in a different ideology and they don't recognize even the question. So, I'd like to say in response to one of the previous comments that there are certain hints that one sees of smaller numbers of structures than one might guess at. And three examples which I'm going to talk about tomorrow, very simple examples, are, first, theory of image processing and, second, not specifically Newtonian mechanics in the sense of gravitation, but electrostatics, wherein, for example, three-dimensions, we have very, very little understanding of how an electrostatic potential distributes itself on a surface. If you look at what we know it's in fact nothing, it's very elementary. And you have the same kinds of structure turning out to be present if you change your view of what an image is instead of perhaps viewing an image as a vector lying in some very high dimensional space. Anyway they look quite promising and fruitful. And another example is exactly financial data where in the halls of academia there is a great belief that we know a lot and there are also such fancy courses taught with Brownian motion for example, and if you look under the hood of this car you'll find no motor. On the other hand we now see a lot of structures that feed into these other kinds of frameworks very well, and there is a lot of hope for understanding them on a deeper level.

M. GROMOV:

I want to make one objection because it lacked historical perspective. Take another instance where there were some people looking for artificial intelligence and they were also very hopeful and they thought there was a structure. By the way their approach in the Minsky school in fact was preceded by philosophers at the turn of the century. They already tried all that and came to the conclusion that it's all impossible. How to carefully analyze the possibility of impossibility of this approach? Of course you are hopeful, you want to do something, you see hints. But you can see a wide historical negative experience showing that it's impossible. And we have to face the fact that there is no consistent mathematical approach to the problem you consider. And this is quite realistic.

S. KLAINERMAN:

To me this sounds like a very defeatist attitude. After all, before Kepler and Newton, people may have thought that the n-body problem, or rather the motion of planets around the sun, would be impossible to describe.

But also I want to make another point, which is that Mathematics is broad enough to allow different points of view. I cannot accept what Misha [Gromov] said about the social structure of mathematics, after the Second World War, being such that it cannot accommodate new points of view. Changes have in fact happened all the time. People may have been intensely interested in certain types of problems which were later, after fifty years maybe, completely abandoned as a new generation of mathematicians started to look at new things. Thus historically there is no reason to affirm that changes cannot happen. So, again, I don't think we should be so defeatist. Finally, I want to make one more comment which has to do with what is meant by mathematical physics. Many define it today as being whatever has to do with quantum gravity and finding out the equations of the ultimate unified theory. There is however an alternate point of view, which holds that the consequences of equations which are already established are at least as interesting, as important and as difficult as writing down the equations in the first place. Just as an example, take Euclidean geometry, whose axioms were written down two thousand years ago and yet we are continuing to do geometry.

Y. FRÖHLICH:

Since Misha [Gromov] started to discuss the sociological aspects of the problems we are discussing, I would like to make a remark on that. I don't believe it's inherent in our community that we cannot innovate or work on something completely new; however there is a problem with the way we are financed. We are under far too much pressure to produce so if we have to produce constantly and prove our quality and our performance then of course we tend to work on relatively risk-free problems and we do what we can. So I think we should all work in the direction that we are, under a little less pressure to publish twenty papers a year.

Maybe one further comment. I'm always a little unhappy if people think that the only challenge is to understand problems that are as complex as possible. I think in the old days when I was young they wanted to find simple phenomena that illustrate basic principles. Nowadays apparently people want to do extremely complicated stuff.

R. COIFMAN:

Nobody wants to.

Y. FRÖHLICH:

Are we forced to? I mean... things worked pretty much better then.

R. COIFMAN:

Let's be serious. If you go to your doctor and he has ten parameters he knows about you: your blood pressure, your age and various other things, and he needs to reach a diagnostic conclusion. He doesn't have any choice. That's exactly the problem.

[LAUGHTER]

V. VOEVODSKY:

I spent a lot of time during last two or three years thinking about exactly the things you were mentioning at the very beginning, and yes - it's a real problem. Things which are happening in mathematics are so far from things which are happening elsewhere in sciences and applications. Yes, it's really troubling. However last year I even took a course in physical chemistry, a lab course. I really did the spectroscopy with my own hands and learned quite a bit of quantum mechanics this way. The conclusion which I came to, however, is completely opposite from what you are suggesting. I think that we should let the kids do the complexity theory and wish them well. We can do nothing about this, that's my opinion... We have a completely different approach to things, and we should do what we know how to do. And if we cannot do this then we should just all retire. And the best thing we can do in order to make our life better and in order to somehow close this gap between mathematics and natural sciences is to make our mathematics as simple as possible and as accessible to people on a more applied level instead of going down to their level and trying to invent some kind of new mathematics. We have great ideology, which developed in the 50's, and in the 60's, we've made great progress and we should made it accessible to the people who can use it.

L. LOVÁSZ:

I support very much what you are saying. I think there is a lot of great mathematics, but if you look around, almost nothing can be described by classical means. We are all, of course, biased by the way we learn and what we think is deep mathematics, and we are just blind to other questions, which are also mathematical. We had a discussion here about geometry: there are at least two or three other branches of geometry, which haven't even come up. Even branches with very prominent representatives sitting here, having organized this conference, even these branches haven't really come up into discussion. And that's because we have some of our own views. My own view of the world is discrete, so I could go on now and talk a lot about that. Unfortunately often a discrete element is missing. But I'll talk more about this of course in my talk.

Anyway, it is not our choice, I feel, to say what we should do, because the world is full of terribly exciting questions, very basic questions like what is the genetic code, and how does it work, or how the galaxies arose, and a number of other things. We just don't know how to describe them in a mathematical way.

Now, I feel that indeed there are some basic principles, which may be sometimes above the horizon. One of these is organizing data (Raphy [Coifman] and I had interesting discussions about this). Similar ideas arise in analysis and in discrete mathematics about organizing large data sets, large graphs and so on; another such topic is introducing complexity through iteration. But we understand very little of these general questions and so I really support that we should try to extend the science in this direction (although this probably will be the job of the next generation).

P. SHOR:

Let's look at history. There is this argument that mathematics and the real world are too far apart and we can't learn anything about the real world and we can't teach anything to people about the real world. But if you look at history, look at, say, quantum field theory, I mean, 20 years ago people probably would throw up their hands and say that quantum field theory is just impossible and yet since then it has given a lot of new mathematics. And also look at dynamical systems 30 years ago, I think, about chaos and turbulence. I don't know the history of the dynamical systems enough, but I know that a lot of mathematics has come from real world issues, come from the real world study of dynamical systems and vice versa. And 40 years ago I don't think anybody could predict either of these things. We look at biology, we look at the many-body problem, maybe there is really very good mathematics coming from it and maybe one can tell scientists something about their real problems. But we are not able to learn this unless we try.

V. MILMAN:

In fact, I did not consider participating in the discussion, nevertheless I'll make some dramatic step in a different direction. But you may turn it back. First, it is perhaps surprising for me how much I agree with all the opinions even if they sometimes look very different, and it is, perhaps, because we are not one-dimensional and so there is room for very different opinions to be correct at the same time, just different people emphasizing different things. But none of the things that worry people here, worry me. This is the only thing I would like to say trying another direction for the discussion. The point is: I believe that whatever problem is discussed, it will be solved by this or another generation just because the problem exists... The problem which needs to be solved, will be solved just because it appears. And saying it once more does not add anything to it. However, when organizing this discussion, some different problem was also in mind. Kidding, we could call it "a problem of microphone", which, as I have mentioned before, we don't know how to use! Unfortunately, we are very far from being understood outside a very small community of mathematicians. Through the dramatic times of big wars the great political leaders called upon us and we returned to a good life. But then, again, there is a huge misunderstanding between us and the contemporary world on what the output of our work is. That's a concern. Of course, these days Computer Science and a bit of Applied Mathematics are in a better position. But when it comes to pure or, better to say, basic Mathematics we encounter a lack of understanding of its importance. It is not understood that the resolution of problems in Mathematics influences the resolution of problems in many other aspects of our life. As a result, we are less attractive to the most talented young people. And similar problems will be felt also in Computer Science.

Computer Science loses students too, because often students don't need much knowledge to produce the software that allows them to earn big bucks, sometimes - millions of dollars. But our salaries do not compete with the fast pace of development in this World and we are not well recognized by neither media nor politicians. And, yes of course, salaries are a secondary thing. The problem starts at another point. The primary issue is that the World just does not get it: that the critical point of the development of our Civilization is Mathematics, that any other significant technical or scientific achievement, development, breakthrough follows the corresponding breakthrough in Mathematics. As it used to happen with the big time-gaps (fifty years or more, - nowadays much less though), it used to be left unnoticed. The vast majority of talented mathematicians start getting significant results while being too young and socially immature, or simply don't have time to think over the place of their work within the global picture. Time runs fast and when you do not feel young any more - you often turn over sixty. And then you start writing memoirs. This is how it becomes our, mathematicians', fault that the extraordinary significance of Mathematics is so much underestimated by our World. Thus the issue to discuss is: how to propagate to the World the importance of mathematical achievements. Perhaps, I may suggest some way to do it, probably it is a very naive one, I don't care. If it could do the job or if a wrong way can do the job, - let it do it.

I myself started thinking about those things accidentally at around the age of 50 (not young already), when I was invited to talk on Mathematics to my University Board of Governors. I just had no choice. I had to explain the importance of Mathematics to people who had money, but whose brains were not overloaded with thoughts in this direction. The experience was dramatic. When one works hard one discovers a number of things. Let me talk about that, and then we might have a discussion on the issue of our connection to the Real World.

Human Civilization needs Mathematics not because the human brain is very sophisticated, but the opposite, because it is not sophisticated enough. The Lord does not need to practice mathematics because He knows all answers at once. In a sense, mathematics is a lever to increase the power of the human brain quite similarly to how a real lever increases the power of muscles. Mathematics improves the very knowledge of understanding, of thinking. If this fact would be known to society, then we could expect much support given to mathematics. Everyone knows the importance of the use of the brain and that it would be better to use it efficiently.

I could stop at this point. But because I am not sure that this simple thought is acknowledged, even in our mathematical community, let me elaborate on it just a little bit more. You see, our ability to use the brain, this "little piece of meat", as Misha [Gromov] put it once, is very limited. For instance, our capability of comparing allows us to process concurrently only seven different facts. Perhaps I am mistaken and it is six or ten, - it does not change a thing. (A psychologist, Prof. Uri Shafrir from Toronto, told me that the number is seven and that it is the same for all people. He also explained to me a few points that I will use.) As result, the consequence of the events based on more than seven observations cannot be comprehended by us. We just don't see the consequences. As an example, our experience tells us that sometimes we forget most important things that were planned ahead, which probably happens because the number of things to deal with is more than seven. Whatever in your mind exceeds this number of seven is not under your control. Your brain skips it. If the last theorem of Fermat would be a trivial consequence of, say, eight observations - it could take three hundred years to develop the theory to avoid this difficulty. It would not be simple.

Now, how to avoid this complication? Clearly, most of the things we do depend on a much larger number of parameters, and we cannot observe them simultaneously. That is where mathematics is needed. What is the approach it uses? Strangely perhaps, but to deal with the problem we need to develop the most basic mathematics; we develop theories, absolutely abstract theories.

And what is the abstractization? We connect different things and create one (abstract) notion which accounts for all of them. So, five (say) originally different things collapse into one. By the way, some people fail to make their colloquiums understandable for the same reason. They say easy things and easy consequences of a couple, even not too many definitions, which they put at the start and which are new for the audience. So, the listeners don't have time to get used to them and "to collapse" these new notions into a single one. Consequently, they accumulate more than seven new things and they don't see the desired consequence. It is triviality! Spend some time on the definitions, make sure they collapse into one notion and then you will be able to explain easily what is going on. It is also the reason why students need so much time to learn new things. On a study that might need just two weeks to present, they need sometimes three months, because they need time to get used to the new knowledge, "to collapse" it into a smaller number of points to see the picture. In this auditorium I need not expand this idea much.

The basic mathematics, the basic science is the critical place, a corner stone for the development of the Civilization. It builds a language that allows one to talk, a small number of pieces with a huge number of connections. It builds a tree of links. And this is an action of abstractization. Those who do Applied Mathematics simply follow down the line of the structure built by Pure Mathematics. One may now use different routes in the constructed tree to understand different connections to produce different computations. Also this may be very non-trivial despite the fact that the discovery of the right connections is already done.

But the understanding that the development of basic mathematics is so crucial, I believe, must be expanded, brought to the society, to the media, to other sciences, because if we cannot solve today a problem in biology, or in economics, or wherever large systems are involved, it is just because Pure Mathematics has not yet developed the appropriate language. The right level of abstractization has not yet been built. Not enough funds have been brought into basic mathematics. That is all I wanted to say, and I hope you have your opinions on what I said, perhaps totally contradicting one another's opinions.

M. GROMOV:

However what was said here, I imagine a similar meeting called by fish, several billion years ago who want to remain fish and conquer the land. You cannot have both. I mean a fish having a meeting one billion years ago under the water and saying: "We are fish, we have all this power, now it is time to conquer the land." But you can't conquer the land while remaining fish. You can't go to the real world remaining mathematicians; that's absurd. Either you study real problems in the real world - it's a remarkable intellectual challenge, or you remain a mathematician.

V. MILMAN:

You are answering me or Raphy [Coifman]?

M. GROMOV:

Both of you. We have all these wishes, all these beautiful projects, but the contradictory experience we have in history, in the history of humanity or in the history of the animal kingdom, we cannot have both.

V. MILMAN:

It's not understandable.

M. GROMOV:

You wish to remain mathematicians, to extend mathematics and to apply it to the real world. You wish it. It doesn't mean it's possible.

V. MILMAN:

No, Misha, if you're answering Raphy [Coifman], he will tell you. If you're answering me, it just sometimes isn't so. I don't want to go into the history of mathematics.

M. GROMOV:

There was a couple of fish...

SOMEBODY:

Explain your attitude...

S. KLAINERMAN:

Then Newton was a fish...

[HUBBUB]

V. MILMAN:

Look Misha [Gromov], it has no relation to what I am discussing. Your remark is going in the tenth dimension with respect to what I talked about, I agree with this, but it is...

S. KLAINERMAN:

But as far as I know all great mathematicians up to maybe the 1950's seemed to be looking at the real world for inspiration.

V. MILMAN:

I'm talking about a different "real world". We may do our Pure Mathematics... just a moment, Misha, I will answer. I will say only one small thing. In the time of Viette, the French King was clever enough to understand that to decode Spanish code he needed the help of mathematicians and indeed Viette did it, and mathematicians had a very good life. Also, I think, algebra was developed from this later (but this is another matter). In the time of the Second World War, the German code was decoded by Turing.

M. GROMOV:

All of these fellows were just fishes...

V. MILMAN:

Just a moment. And, as result, politicians supported mathematics during the next forty years, they just forgot this already. So this is a different thing.

You do exist, and do what you want to do, what you like to do, but at the same time, having meetings, having discussions, giving interviews (which you give sometimes unfortunately for you and we know you don't like this, but you are anyway forced), you should always remember that you are not ashamed of being a mathematician, because mathematics plays such an important key role. Not you [to Y.Eliashberg] but some people are slightly ashamed... I have in mind many examples of mathematicians, also great ones. For example, Gelfand and many others very much like to do biology... as a compensation for their understanding that mathematics looks less important for Humanity. Many people have this complex of doing pure mathematics.

S. NOVIKOV:

Very nice discussion indeed. I'd also like to participate in it. First of all I would like to make the following remark that fifty years ago the main tool of our sciences, I mean mathematics, theoretical physics, and so on, was to understand the fundamental laws of nature. And everybody believed that if you would understand them, anything else would immediately be correlated after that. Some people like Fermi appeared after that and then did engineering... But I think that approximately after 1990 the situation changed and this is... I don't know whether the fundamental law of nature exists or not but this part even of physics became abstract, I would say more or less like part of pure mathematics. Maybe the fundamental law of nature even does not exist, maybe not, but certainly there was a huge crisis in that since the early 80's. Our ideology is therefore now in a state of crisis, that part of ideology, and, besides, mathematics has become more engineering. I believe this engineering mathematics, which certainly had huge achievements (which our chairman told us about partly, or at least had in mind to, yes? [to R.Coifman]), certainly is today the number one achievement in humanity. I completely do not agree with the idea that we should do that and we should not do that. What I heard from my old friends, theoretical physicists of a great generation, - they said that they never thought about it. If something is needed we try to do that. We never said that we shouldn't do that. This kind of opinion's itself some kind of thermometer showing the state of decay, intellectual decay of society. And therefore such things are at least something, which I myself don't like, and it does not correspond to the mentality of great scientists of our areas of, say, fifty years ago. There was no problem for people like Hilbert to combine, say, mathematical logic and mathematical physics and geometry and so on, in analysis. Has this time passed on or not? In my opinion I was always aware that such beautiful great science as abstract mathematics, it IS beautiful and great, but for humanity at least now... living in humanity when we need to raise money and so on, we need to have some maybe smaller community of leading people, who are visibly cleverer than others. Otherwise how do we get money? In that case I completely agree with our chairman - we have to increase our efforts to make really applied modern area open to the best young mathematicians.

But there is a huge informational mess in society. For example, an algebraic geometer starts to do a colloquium talk normally. As he is absolutely aware that everybody in the room knows what motivic structure is, he does not give this definition. Vitali [Milman] said that they start with definitions. No, it is wrong. They think that everybody knows things exactly, as in some narrow seminars which we are living in. It is only an example; not only this area is like that. Algebraic geometers of more complex areas have completely stopped, have lost the ability to write any analytical formulas; we are capable only of speaking in some geometric language. There is also a question in analysis; almost 99% of people doing applied PDEs, I believe Sergui [Klainerman], who were trained in mathematics, completely cannot give a mathematically-exact definition, for example, of such a fundamental notion like energy momentum tensor, as I found out. You spoke about that. No, you can because you worked there for Yang-Mills, but I didn't say 100% looking at you, I said 99% only. This is the real situation and somehow we think that people have the right to do what they can do, but we have to do a huge job to open all informational gates for younger people, to make mathematics and applications transparent, visible for them. And this is what is concerning: what we should do and what we should not do. For example, everybody in abstract mathematics certainly thought at least twenty years ago that they should not learn and do quantum field theory. Quantum field theory came, started to get good results in abstract mathematics. Almost nobody can learn it except a few people - one, two, three, four - of the best people.

So this is how I understood the idea of our discussion today. I think its also the idea of our chairman, to make this, our science, transparent and unify its applications. In my opinion there is nothing contradictory in hierarchical structures like fundamental laws of nature. And people in statistical mechanics, in dynamical systems, sometimes introduce good understanding of which is associated with normalization groups and so on. This language and these techniques work only partially in real applications; you have to unify them informally and so on. But I don't see any contradiction between that and old Newtonian or quantum mechanics. It is a new development of it and it is part of our sciences. And mathematics was unique, and was unique with its applications, and it is unique. It will only be, say, a period of decay in society if we forget about that. That's what I have in mind. It is unique.

Y. ELIASHBERG:

I just have one or two comments to make. Our chairman told us that this kind of interaction of mathematicians with other fields goes in two ways. Some people I know in physics, in biology, in engineering, have problems with their amateurish tools trying to apply mathematics and this may sound bad because they really don't know great mathematics. They have to do some simple things which somehow nevertheless correspond to what they are doing. Or some clever mathematician comes into the area and imposes there what he knows without really much knowledge of what is going on there. I think actually it's not too bad when... I think it's better when people in some engineering or whatever... it's not only about mathematicians. For instance, like physicists when they go to biology without really understanding what is going on there, doing the same things about brain structure and this kind of stuff. I think it's maybe good that these people without really profound knowledge of some deep mathematics do this thing, and we should just try to understand and then do the mathematics after that, [we are] not necessarily leading the way.

And the second thing is just about this funding problem. Of course it's the structure of funding of mathematics that is really crucial. It's really a problem. One thing is that there are some established fields and established areas, which are well funded, and of course anything beyond this is... Maybe in mathematics it's not so but in some fields, maybe closer to computer science, where bigger money is involved and then there are some groups fighting for the source of money and there is maybe nothing really clever... Also another problem with funding is that sometimes too much money goes to a certain area and it completely destroys it, like in the States, we see it with math education. This education in the States is really being destroyed, in my point of view, just by the amount of money which was put into it.

D. KAZHDAN:

There are many topics floating around. I'm not reacting to the previous one... A century ago Hilbert decided to formalize mathematics. His assumptions were that it's possible to do. In a sense, one can solve any problem. But by formalizing the question it was possible to answer it, not in the way in which Hilbert thought, but we have an answer. The answer is that not every problem can be solved. We have a paradox - of course I don't think anybody understands, why so many problems are solvable. We almost can... Of course there are some things like continuum hypotheses and a number of others which we know are independent, but we almost always know now, when we see a problem - it could be wrong but it's clear feeling - when a problem can be independent and expected to be solved. But there was some analysis of the situation done and we came to some conclusions. When we are discussing now the relation between mathematics and, say, complex systems, - the real world is something different - somehow what I don't see and what I would be glad to see is some attempts to analyze the problem, because whether we say without analyzing that we are sure we can solve or we say without analyzing that we cannot be sure we can solve, it doesn't say much. And what I see as a problem is to create some way of thinking when we could analyze the problem.

Y. ELIASHBERG:

To analyze the problem whether we can solve it?

D. KAZHDAN:

Whether we can solve it, what kind of formalism we could produce, what could be done in principle. The same for Hilbert who developed logic to analyze how we approach problems. And then we can ask the question: what can or cannot be done inside this framework? I would definitely appreciate if I would see some approach to development.

D. SULLIVAN:

What ideas do we have to begin with?

R. COIFMAN:

Well, I wish I knew but there are some things. I mean there are hints. This is what Peter alluded to before. One thing that is rather clear is that there is a collection of questions, which are raised from the sciences, which are basically just the description, the botanical description if you wish, of the structures which arise in nature, various structures that arise in nature. When you start to look at that you realize that the basic mathematical principles behind this description, and the description should be from the point of view of a mathematician, not from the point of view of multimedia. You want to describe it as an object that you can compute with, manipulate with, do transformation on and so on. Then you realize that there are very few basic principles. These are more than effective theories; there are two things that happen. For example, the traditional thing that has happened in classical mechanics, that you have a moving frame. Now, objects that we often deal with in the real world are moving frames in infinite dimensions or in high dimensional space. So the question is how do you deal with moving frames. In the situation involving turbulence you may have better luck doing it in such a context. Turbulence is too complicated for us to describe. So it's not clear whether the problem of handling the turbulence issue stems from our inability to describe the object that we want to describe or stems from inherent complexity of the phenomenon. Maybe we are really stuck because the language is wrong. We just can't describe it because we don't know how to describe a rapidly varying rough function, for example. And we are going to get very rough functions. It's possible that in a moving frame this rough function doesn't look that rough, and the whole physics is going to be described by the frame motion, as well by organizational principles.

V. ZAKHAROV:

This moving frame is a local moving frame?

R. COIFMAN:

No, a moving frame in function space. Not in three-dimensional, in function space. Infinite dimension where infinite is ten thousand or a million. Beyond ten it is infinite, from a computational point of view.

So we see hints, at a variety of different places, that such attempts to describe this kind of complex phenomenon could lead to some success. Again, I'm not asserting it, I'm just saying that there are a lot of hints in a variety of fields that there are unifying principles. We'll hear some tomorrow. Peter is going to talk on some unifying algorithms. I'll give some examples.

In some sense the talk Alain [Connes] gave... it's sort of the beginning of unifying principles possible for a variety of descriptive modes. I mean, why did this whole theory develop? To describe, say, various complicated geometric foliations that you get from ergodic actions, the description, the language to do it is critical in order to pursue various kinds of physics. So what I'm saying and what you were saying in some sense are very close in spirit. This will emerge providing us more insights into mathematics. I think this is the beautiful thing about what we call applied. Actually the separation between applied and real, and non-applied, I think is completely artificial.

D. SULLIVAN:

In the example you gave, why should mathematicians try to look for an effective theory? In the example, you gave many methods... Why should they look for it? Why should they be optimistic that there is an effective theory they can deal with?

R. COIFMAN:

But the point is there is. There is one.

D. SULLIVAN:

Why are you sure that there is one?

R. COIFMAN:

It's an algorithm that describes how to do it, in order to understand the organization of computation. There is one. And there is one for such examples as acoustics. By the way, the acoustic case and the electromagnetic are much harder. I think he [Sullivan] is objecting very specifically.

Y. FRÖHLICH:

It seems to me that there are still huge problems in simulating Newtonian gravity so that you can get reliable results.

R. COIFMAN:

The problem stems from the inability of the astronomers very often to be in touch with the computational argument. Yes, the particular problem is the baby problem. The example is just to show this baby problem. It was just pointed out that the astronomers are not even using it. They are trying to re-invent the wheel. It was actually done in mathematics fifty years ago. The translation was not done. The acoustical and electromagnetic issues are much more modern. And again 99% of the community decide to ignore it because it doesn't follow the beaten track, so to speak. So this happens all the time.

M. KATZ:

I want to return to the final remarks you [Coifman] made in your introduction. When you spoke about the need to attract young kids to mathematics you came very close to my area of interest. I might be the only person in this room who in the last few years has been dealing less with mathematics and more with mathematical education. I'm currently the chairman of the Education Department of Haifa University in the north of Israel. And I think that if we speak about the real world, if we speak about sociology, if we speak about propaganda, if we speak about making mathematics more transparent, we have to start with education. And the problem with mathematical education nowadays is that it's hard to believe how little it has changed in the last decades. It's true, for university mathematics as well, but I'm talking about, let's say, high school mathematics. We teach now exactly the same things that we taught forty or fifty years go.

SOMEBODY:

Hundred years ago.

M. KATZ:

OK, a hundred years ago. The methods may have changed a little - we now have computerized edit programs to handle various topics in mathematics and so on. But when we think about the content, very little has changed. At the turn of the century we still don't teach ideas from the beginning of the century, the ideas that came into the forefront with mathematical physics. I mean, a person in Israel can graduate from high school, and I presume it's the same in other countries, one can graduate from high school doing what we call four or five units in mathematics without ever hearing the term "Hilbert space", and there are many other examples like this. So, what I want to say is that I think Vitali [Milman] is right in saying that our brains are weak. But we have one thing going for us, and that is - we have imagination. And especially kids have imagination. And we have to bring imagination into mathematics in high school. We have to show kids in grades 10, 11 and 12, if not before, how abstract notions in mathematics can become visualized and can be used in various areas of life. Kids would love it and would know how to built on it. So maybe it's the task of mathematicians to think about the curriculum of mathematics in high schools. They should think not only about what are the new ideas to be studied in mathematics, what are the new areas to be pursued in mathematics, but also about how mathematics should be handled in high school, because it's high time to take new avenues.

E. SIMEONOV:

I would like to add to this statement and to what Vitali Milman said: It's not only education which has to be taken into account by mathematicians - especially by the leading ones, because it is too important. But additionally also some kind of public relations should be done - again by mathematicians. Who else can do this? Scientific journalism which concerns mathematics hardly exists, because people just don't know the subject and how to reflect upon it. And also the popular literature - does it exist at all?... In my opinion, since "What is Mathematics" by Courant and Robbins, not really much has been written by good people (with very few exceptions like "The Mathematical Experience" and few other books by Davis and Hersh, by Martin Gardner and Ian Stewart. But how many people do they reach?) It is not only the students in high school that should have some possibility to approach mathematics (to develop mathematical abilities) but also people in general, because it is a part of the culture and a part of the society. Nowadays mathematics appears to be something like an unknown culture. It should be spread out somehow. And I think the main means for this is to be able somehow to acquire credits for doing this. If it is done in a good way there should be some kind of real credits - both financial and moral or institutional, which means that such work should receive a high acceptance from the community..

A. JAFFE:

I really don't want to talk but maybe I'm forced to. I have a very simple thing to say. I've listened to this discussion and found it very interesting but for me the most important thing about mathematics is - it goes back to what Sergei Novikov said before - that there are so many wonderful, innovative, interesting and bright people who do mathematics. And I think mathematics will survive and flourish in the future if these bright mathematically inclined people would encourage doing mathematics. I don't think we can predict exactly what they will do, but I think if the bright people go off into other areas then mathematics will decay and if bright people continue to work in mathematics we'll have a good future.

V. ZAKHAROV:

I'd like to express some crucial optimism about the future of mathematics just appealing to my personal experience. When I started my scientific work as a physicist in the Budker Institute of Nuclear Physics, it was at the beginning of the 60s. I remember very well the typical background of even a bright physicist in mathematics. This background was relatively poor and people knew the mathematics of the last century, the top of this being special functions, theory of Bessel functions or hypergeometric functions, this was the top. And during this time the background of physicists in mathematics increased enormously and first we studied group representation because it was necessary for developing the quantum field theory, then much of topology, then such things like inverse scattering and lot of dynamical systems. I do not speak about differential geometry. So if we look at thc problem from the point of view of how to persuade the authorities to give money we can use this argument. The amount of mathematics which is involved in applied problems increased enormously in the last decades and it will increase for sure because of a lot of very important problems. I spoke about this, about such simple problems like how to calculate the gradient of pressure in a big tube when the fluid flows through this tube (and is unresolved so far), and many other simple questions which arise in nature. It's a challenge for mathematicians and this is the point. And so I think that mathematics has a big future in prospective application to the real world. But I see two possible obstacles. One is a traditional obstacle - it's a narrowness of people, their very narrow education. And I completely agree with Sergei Novikov that it's very important for people who are bright and have a good mathematical talent, to study more natural sciences to start building up a bridge from their knowledge to applied knowledge, because sometimes the ignorance of mathematicians is just unbelievable, in some simple physics. And I faced this situation and I was really astonished. The first is a traditional danger, it's not new, it existed through the whole this century. Another danger is new. It's connected with the development of computers because of the capacity of computers, the progress in enormous speed. It produced the temptation that any problem can be solved just by using a powerful enough computer. That's not true, of course, but the people who do this have a good chance of persuading the authorities that they are the right people to solve the problem because they have the strongest computers, and the authorities accept such arguments. This is again an experimental fact. So we should probably be more critical. And there are a lot of scientific journals now which are just filled with garbage, with a lot of made using computers. And the main thing, a lot of very nice pictures done on the color printer. But it's really very difficult to understand what it's about and what's the final conclusion of an article. It's the most typical thing. And real grants are going to these people and much more than to pure mathematics. And we should have this in mind. So mostly I agree that the main conclusion is that mathematicians should be optimistic about the possible applications. I could show immediately several very nice examples of how it could be, but it's a different story. So, first, we should be optimists and, second, people should try to be as broad as possible in their education, in the sense of including [in their] education, physics and other natural sciences. So people should be critical of the people who just imitate science. This is a pity but is more or less common place. I agree that the education of young people should be oriented in alignment with these principles.

T. GOWERS:

All this talk about practical applications of mathematics makes me feel just a little bit uncomfortable because nothing that I ever prove has any practical application whatsoever or I can't even imagine that it ever will have any practical application. So one of the topics that was supposed to be in the discussion was not just applications but image of mathematics. And I think it's important, not just that we should admit to ourselves that 90% of papers or possibly more in pure mathematics are completely useless to, say, engineers or biologists or the most theoretical physicists. But also this is not necessarily too bad a thing and I think we should think a little bit about just the sheer intellectual excitement of mathematics independently of its practical application. I mean this is not always an easy thing to justify to people who are giving us money, but actually I think it's very important because it really does cover a large... maybe it's not 90%, but it's a very, very significant proportion of mathematics that we all love, not doing practical applications. And we really must convey that this is worth doing as well. And if we concentrate too much on saying... on a practical application, then it might be some danger that people would say: "Oh, well, could you please justify mathematics from a practical point of view" and then maybe try to cut out some part of mathematics that can be justified on other grounds. So that's just the point I mean.

R. COIFMAN:

I think there is a miscommunication here. Practical applications are just one aspect of a pluralistic society. We are talking about mathematical content.

T. GOWERS:

Yes, and what I'm saying, I agree with that, so I'm just saying this is another aspect that we just haven't discussed at all, that it is an important part of the image of mathematics in the presentation of our image to the real outside world, that we should be confident that we have some intellectual justification.

S. KLAINERMAN:

I think it is possible find some unity between many of the points of view which were expressed here. First of all, it's clear that there are real distinctions between a physicist, an applied mathematician and a mathematician. As mathematicians we are conditioned to look at problems from the point of view of the deeper mathematical structure involved behind them. That doesn't mean, of course, that we should not look at problems, which come up from the real world. On the contrary. Though I agree with Misha [Gromov] that not all such problems can be solved there are nevertheless a few crucial ones which can. Certainly we have to come to terms with real world problems on our terms as mathematicians. I agree with Tim [Gowers] when he says that there is a certain, sort of, intellectual excitement which arises from looking at the mathematical structures behind problems. But this is not necessarily in contradiction with the fact that there are problems, natural problems in the physical world, which must have a deep mathematical structure, even though we may be ignorant of it at the present time. Take turbulence. It looks like an impossible problem right now but it will be solved. There are clear indications that the problem of turbulence is of such a type. I think it's a very good mathematical challenge for mathematicians, I mean for people like Tim [Gowers] who really perceive themselves as mathematicians and not as applied mathematicians or physicists. I think there are many examples like that. Here is another example which has to do with the protein-folding problem, mentioned before. It's clearly a fundamental problem in chemistry and biology as well. Here is an aspect of the problem, I have learned a few years ago from a chemist, which may catch the interest of mathematicians. The question is simply the following - you have a very complicated function, representing the energy, depending on a large number of dimensions which correspond to the many parameters involved in the physical structure of a protein. Chemists and biologists are interested to find the absolute minimum of that particular function. This particular chemist was looking for help from mathematicians and as he couldn't find it, came along with a solution, which he claims is very efficient. His idea is to consider to run the heat equation, in the configuration space of the large number of parameters involved in the problem, with initial data given by the function whose absolute minimum we want to calculate. According to him all the relative peaks of the function will dissipate very fast; only the absolute one will survive for a sufficiently long time. This is of course non-rigorous. I am not sure I quite believe it, but it's interesting from a sociological point of view. And then you go back and you find out exactly.

SOMEBODY:

The question definitely has been considered quite extensively.

S. KLAINERMAN:

Yes, I know. I'm just giving it as an example of a problem which could stimulate our mathematical imagination even though Misha [Gromov] does not agree. And I think there are many more problems like this. Finally I want to stress the fact that there are many unsolved questions of mathematical physics which have to do not with finding new equations, but deriving the consequences of well-known ones. There is no doubt that these problems have a deep mathematical structure even if at present time we may ignore how to use it. We should consider these problems from our point of view, as mathematicians. We don't have to ask the same questions physicists ask; we have to ask our own. And I firmly believe that if the problems are sufficiently interesting, they will generate good mathematics which will later have consequences in the real world. That's all.

V. MILMAN:

Because we should finish very quickly we will give one minute to a few people who wanted to say something.

A. GAMBURD:

I wanted to say a few words about pure mathematicians dealing with the real world, in particular about von Neumann. I think a unique and appealing feature of mathematics is that its relation with the real world has a double nature, or as Gromov's fish might put it, an amphibian character. (Apropos of funding: apparently fish became amphibian and entered the land in response to the uncertain patterns of rain and drought.) On the one hand, mathematics is a very successful escape from reality and has the aesthetic component that Professor Gowers talked about. On the other hand, mathematics works back to impact this very same reality and derives some of the best inspirations, even in the purest branches, from the natural sciences. Newton, mentioned repeatedly today, was an amphibian par excellence, and so was von Neumann. Until about 1940, the half-way point in his scientific life, von Neumann did first-rate pure mathematics; afterwards he turned to non-classical, new aspects of the applicability of mathematics in the real world, such as computers and game theory. (Incidentally, amphibians, when young have gills and are aquatic, but as adults are air-breathing and mainly terrestrial.) In the introduction to the "Theory of Games and Economic Behavior" von Neumann points out that the decisive phase of the application of mathematics to physics - Newton's creation of analytical mechanics- brought about, and can hardly be separated from, the discovery of calculus. Von Neumann then goes on to say that the complexity of problems in economics (and we might add many other fields mentioned today) is at least equal to that of problems in analytical mechanics. In his view, it is therefore to be expected - or feared - that to produce decisive success in these fields, mathematical discoveries of a stature comparable to that of calculus will be needed. Such discoveries would require pure mathematicians of Newton's or von Neumann's stature blazing new trails. But as Professor Fröhlich said earlier, the pressure to produce twenty papers a year (which is also felt by young people) tends to favor research in well-established fields.

N. ALON:

I think, incidentally, that the only person I know that produced twenty papers a year, every year, is Paul Erdös when he was eighty, so there is no point in talking about these young people who have to produce twenty papers a year. The way I understood Raphy [Coifman] (I think this was really his intention but you [Coifman] can tell me if you agree), is not that all of us should switch now and start to work only on applied problems, right? So, definitely most of us would keep doing what we have been doing all the time, and of course there is a lot of place for pure mathematics, but there are also a lot of very nice problems that are connected to the real world. That's the way I understand things and I think that to say outright that we don't even want to consider these problems, is good probably in order to provoke discussion but I don't think anybody can really mean it very seriously. But you can tell me if I'm wrong here; that's the way I interpret it.

So certainly we should keep doing what we are doing. We should also probably think about a way of explaining to the public, to non-mathematicians as well, why it is important. And you see that sometimes we have... even in this discussion, you see that we often have these problems of understanding what we are saying, right? Denis for example was asking to summarize the remark of somebody because it's too complicated to comprehend, and imagine what happens when we actually try to explain what we are doing to non-mathematicians. But probably this is also indeed an important ingredient, not only for funding but just for keeping the subject alive, to attract young people and to somehow keep it interesting.

A. WIGDERSON:

Time is up... Just maybe two points. I think that one thing that came to my understanding of what Raphy [Coifman] was saying at the beginning and I think what lots of people mean when they talk about doing mathematics in the real world is that we are converging to the definition of understanding a real world problem, when we have an efficient algorithm to get answers to the questions we are asking about, whatever complex system it is, in biology, in physics and so on. An efficient algorithm becomes the definition of understanding the structure of a complex problem. And I think that's a very important notion, and I think that's very important in mathematics. In fact when you start asking yourself: do you understand very basic questions in mathematics that have nothing to do with the real world, we know that (I don't know), that rational polynomials are a unique factorization domain. Did we understand this? I don't think we understood very much but now we know that in fact due to the work of Lovász and others, there is an efficient algorithm to factorize polynomials over the integers, and we now understand this problem much more and in fact it's much deeper, this understanding, and it goes beyond trivial theorems that we can prove, you know, to second year students. So I think this notion of understanding as providing an efficient algorithm is somehow a middle level between having just one or two or three unifying principles and between, say, having lots and lots of theories that explain different things.

D. KAZHDAN:

I cannot avoid just making a comparison that the definition which was given now, is very much in accordance with the definition at the end of the 19th century, before the works of Hilbert. If you look at the work of the English school on invariants, they looked for an explicit algorithm and they refused to accept Hilbert's proof of finiteness for rings of invariants or finiteness for basis of ideals exactly because proofs were not effective.

A. WIGDERSON:

I don't think you understood my point.

D. KAZHDAN:

It's quite possible, maybe you clarify it, but it sounds as a statement that we try to avoid arguments which we cannot make explicit. And I think the advances in mathematics starting from the beginning of the 20th century were exactly the realization of the possibility to go a long way using arguments which... in principle... Now some of these Hilbert's results can be made explicit, it's a very interesting possibility that we can go without making... calculate... explicit algorithms was interesting discovery. Maybe you [Wigderson] will clarify it.

R. COIFMAN:

This is a land of many religions, you know.

[HUBBUB]

Just a minute... It's OK, let's talk about it in the afternoon, but let me just try to summarize. If anything here came about, it's clear that this is a pluralistic society with different religions and we don't want to get into religious wars. The main issue really is how do we live together and continue surviving. And the danger at this point is that there is a transitional period in mathematics that is occurring. We really have to come to grips with it. I think this conference can have a major impact provided we do our homework and we each actually give some idea or some input into a definition of ourselves for the future. If we don't do that it'll be viewed as a failure. It'll make it even worse in terms of whatever, funding or support or understanding of society.

V. MILMAN:

If nobody wants to add anything to this discussion I think in two hours time we'll start with it's continuation. I have just one small last point to defend, in a sense, the importance of basic mathematics. I would like to say that one of the goals is to discover a new structure, to develop new ideas, a new form of thinking, and I believe Tim Gowers is still young enough to see how, surprisingly, the kind of new infinite-dimensional geometry he discovered will be applied. So this is my last point. And now you should take care of your luggage.

Transcribed by A. Scherbak

DISCUSSION on COMPUTER SCIENCE and DISCRETE MATHEMATICS

with introduction by M. Rabin

M. RABIN:

This is the afternoon for the discussion on theoretical computer science, computer science in general and discrete mathematics. I'm going to devote some opening remarks to computer science, then in the course of the discussion we will also ask people to make some comments about discrete mathematics and all of that is going to be intertwined with, I hope, a lively and productive discussion.

Computability 1935
Mathematical Machines 1935-58
Difficulty of Computations - Complexity 1958 - open
Efficient Algorithms 200 BC -
Use of Randomization 1975 -

Picture 1

I would like to start by giving you some kind of morphology and a very brief history of theoretical computer science, which can serve as a background to what we will say on the topic and also illustrate various points. So computer science is really an example of the process that people looking towards the future hoped to have in mathematics. Namely, in the 50's the whole field of computer technology arose and at the same time and in fact even before, the theory of computing was developed. This is a kind of applied mathematics and developed in the same way that a certain kind of mathematics, mainly continuous mathematics, was developed in order to deal with physical phenomena, mainly, let us say in the beginning, in classical physics, phenomena of mechanics and electricity and magnetism. Then the process of computing itself became the subject of mathematical study. We have some very useful and very influential results. So we really see here a new branch of applied mathematics. Speaking about some of the history and some of the achievements, the notion of computability itself was already introduced in 1935 with the work of Turing, Goedel, Church and was one of the foundations later of computer technology. And then also concurrently and going into the late 1950's and even beyond, various kinds of mathematical machines were defined. The best known is "Ardy" , a Turing machine that had, and this is an historical fact, a profound influence on the architecture of modern computers, actually there is a direct link between them. But also other machines, maybe less well-known, for example finite automata, so-called push-down store automata, which have a great significance both in linguistics, in the Chomsky theory of linguistics, and also in programming languages and even play a big role in the technology of compilation of programming languages. So looking at these abstract machines, there is again a direct link here between the theory and some very important applications.

Now, around 1958, computability and its mirror image undecidability were well understood not just in logic, but also in algebra and other fields; there were even undecidability results in topology, for example Markov. And now a final analysis of the realm of computable functions, i.e. in principle computable functions, was introduced; namely the computable functions starting with 1958 were classified according to the inherent difficulty of their computation. So computational tasks with a spectrum of solvable problems were introduced. Now I purposely mention the name first given to the field - difficulty of computation, rather than complexity because in a sense it is more appropriate, it talks about what you are really concerned with, namely the difficulty. You see complexity (and I'll come back to that topic later) deals with systems, which are very very complex in terms of having many components and especially interacting components and feedback mechanisms. So for example global climate is a complex system. Now, the difficulty of computational problems actually relays to very simple objects such as a number which is the product of two primes, and how difficult it is to factorize. The question is of cardinal importance. But the name "complexity" is a sort of come on, it's more sexy and so on, that's what we have now. And this, starting in 1958, is open. Now, we in essence arrive at the practice, that deals with the study of the computational process as an object, as material. And this is the waste field of efficient algorithms starting, and here I'm a little bit ignorant of exact dates, with Euclid's algorithm, which I think randomly assigned the 200 BC date (who knows when Euclid lived?) and going into the present. And the salient point of efficient algorithms and Euclid's algorithm is an outstanding example of that. You would assume from the definition of the greatest common divisor that you would need to know how to factorize in order to compute if you just look at the definition. And we know that even today, in 1999, factorization is a hard problem. And cryptographers will say: "Thank Heavens". But Euclid had a way, as we know and we could call it trivial, of very efficiently calculating greatest common divisors. And these efficient algorithms, which touch every aspect of computing whether they are in number theory, algebraic questions, combinatorics, graph theory, or what have you, languages, everywhere... data structures, which I'll mention further on, fast Fourier transform, signal processing, etc. Many of them are intellectual games in their own right, but many, many of those are also of the utmost practical significance. Within efficient algorithms there was the big surprise starting in 1975 that randomness, tossing coins can be used to solve problems, which before that and even now are inaccessible. This was really an enormous surprise and in many areas, for example, in protocols for security and other applications, randomness in fact is essential. There is a sharp distinction between randomization as used within algorithms, even though people use the name Carlo and Monte-Carlo methods because in the Monte-Carlo method, for example, in order to solve a potential equation or a diffusion equation you create a stochastic process that imitates, that behaves according to the equation that you are trying to solve, and then you perform very large scale experimentation, characterized by the use of many-many random bits. The randomized algorithms are very sparing in the number of bits that they use, and produce outstanding results.

Cryptography 1975–
Data Structures
Networks
Parallel/Distributed Systems

Picture 2

In 1975 modern cryptography started with the proposal of Diffie and Helman to use a public key, which I've explained in my lecture on cryptography and also had the bad fortune of missing it, can I ask others what that is? They had another very interesting point, which is sort of not explicitly stated in what they say. They proposed using well known mathematical functions and problems as the basis for cryptography. The cryptographic devices prior to that, such as the enigma machine and the data encryption standard, were a hotchpotch of boxes and processes, which did not provide you with any clear picture of what was going on. They first made the proposal to use a discrete lock and then came the proposal to use the presumed difficulty of factorization as the basis for cryptography. And then there are now proposals for using various latest shortest vector problems, etc. But the proposal to relate cryptography to well known mathematical problems is actually implicit perhaps even explicit in their paper. Now, cryptography, as I was explaining is not only the creation of secret codes, but is really the whole issue of having protocols, which is at least as important as the secret codes. And all of electronic commerce, all of Internet work and so on is actually dependent on and will be dependent on that technology, both on the secret codes and even more so on protocols. And all of that field has a very very nice mathematical structure.

Now, one should mention even though this is in some sense assumed under algorithms, that one should highlight data structures and especially very large data structures. So we know, and it is obvious, that modern life is dependent on extremely large so-called databases, that you have in banks, in, God forbid, internal revenue service, medical records, web-pages, all that data, which is floating around and is organized. And to access that data, search that data, determine it and so on, all of that depends on an understanding and on efficient algorithms for data structures. That is, a whole field, which is again mathematical combinatorial in nature and at the same time and even before is of essential practical importance and significance.

Then let me talk about the other component of modern communications plus computation complex, namely - networks. This is called the Internet, but I am talking about the underlying technology. So there are these enormous heterogeneous networks. Here there is a need for real-real understanding and again these are one of these very complex systems in the sense of actual complexity, not complexity theory, but systems, which have millions of components and where real theory and deeper understanding is required in order to bring feasibility and practicality. Let me mention to you one of the issues that is involved. We are all familiar with the frustration of trying to get something over the Internet and the delays that result. Now, the delays are not mainly a consequence of the communication lines. The lines have very high so-called bandwidths and these bandwidths will greatly increase with sophisticated fiber optics. The problem with the Internet is that this data (think about it as pages) is hopping from one note to another note. There isn't a direct connection created between you awhile that directly plugs you into the data base from which you want to draw the information. But it is hops from note to note, to note, to note. And in those notes you have so-called buffers (I hope I am not becoming too down to earth for people here, especially after lunch). Obviously you need temporary storage where those pages sit while they are being moved and while decisions are made on how to route them further. And the failures are buffer overflows. In telephony we used to consider questions of loads in terms of Markov processes. And that's how the telephone operators were able to provide sufficient flexibility with whatever, and also switching in order to route telephone conversations. This situation is much more complex among other things because of feedback phenomena. Namely if you fail to get the page then you immediately ask for the same page again or maybe the system itself generates the same page sending it in the same or may be in a different route. Usually this disaster is self-compounding and we don't have an understanding of these processes. Here we have to talk about two facets of this. First of all, a better understanding of the stochastic processes involved there. And then also mathematical, not technical, inventiveness on how to do that routing and again randomization and ideas of randomization obviously coming, how to do that routing in a way that smoothers out those loads. And that is a very active and by no means completely understood area. So that's another example.

Now, my list [Picture 2] is by no means comprehensive but I want to mention parallel and distributed systems and again the mathematical problems associated with them. So of course people build parallel computers. But the idea of parallelism is clear; that even though we have very powerful processors, very large memories and so on, the needs of very large scale computations, the needs of the consumption of instructions is so prodigious (for example, if you want to do weather prediction) that a single processor is not going to do it. The idea is to harness together hundreds or maybe thousands of processors in order to in unison solve this big problem. Now, this cannot be done pragmatically. In fact theory has had a profound influence on the technology of parallelism in several ways. First of all there were pure mathematical theorems, and I'll mention only one of those just for your enjoyment, a theorem, which essentially has no practical application but is just intellectually attractive. Namely consider the task of calculating, say, the determinant of a very very large system of equations. Say, ten thousand equations, a hundred thousand equations in a hundred thousand variables. Now, there are things like that, that arise. Now, if you think about the way we know how to do it then, say, Gaussian elimination (there are other methods but say Gaussian elimination) in essence is fairly eminently sequential. You can somewhat parallelize it (it's an ugly name but it's a used term)... While you are subtracting multiples of the first row from the second and the third and so on, you can have several processors working on it together. A surprising theorem due to Csansky says that you can do it in parallel; if the matrix is $n \times n$, using many processors, you can do it in time $\log^2 n$. So that is not obvious at all, how you cut it down from n, or my proposal that the total number of operations is n^3. So if you do n routes in parallel you cut it down to parallel time and square. Unfortunately despite the great attractiveness of Csansky you would may have to use n^6 processors. So we see that this is not so. However, the theoretical study has had a profound influence on the technology of parallelism and let me mention here one aspect of it, which also has to do a little bit with networks, namely in defining the connectivity, the various schemes for connectivity within these parallel computers. So if you have n processors, such as ten thousand processors, you want all of them to somehow share. So you would say: "Yes, I will connect every pair of those", well, ten thousand squared and so on, we are somehow always getting stuck on the same number - 50 million lines. It's similar to the private versus public key. So 50 million lines of direct connection are impossible. First of all computer scientists invented networks that did it in stages and are about $n \log n$ in size... And there are many proposals and in the deepest aspects of that very advanced work by Noga Alon and others; Sarnak, Lubotsky and Philips entered into it constructing very very efficient networks for tasks such as that. Also, not just Noga, Milman as well of course in the first paper. And Margulis, yes, thank you. You have to put that down to the fact that, you know, I'm sort of immersed in... I think by now we've mentioned everybody and really - thank you. I mean this is unbelievably beautiful, unbelievably beautiful work.

So just a final word about distributed systems. Distributed systems are something likely the Internet. The parallel computers are so-called strongly coupled. The whole parallel computer is in the same room, I mean with the advance of technology it could be in one shoebox, with a network, which connects all components and so on. The distributed systems are distributed world-wide and there are various things you want to do, especially in the area of protocols where they have to cooperate in various ways. And there are sophisticated algorithms that permit that in protocols, randomization is again very heavily used. That is a whole field in itself. Sometimes large distributed systems also act as almost a global parallel computer. So for example the factorization I spoke of, of a 140 digit number by using the general algebraic number field, I think, several thousand computers working cooperatively as some kind of a distributed system. That is but one example but let me come back to the question of the Internet and the large databases. Those very large data bases are already distributed because each university has home pages and data about the people at that university etc., etc. Sometimes you get the page from some central server, sometimes you go directly or you are being sent there, and there are issues of how to define and manage and what algorithms govern and make these distributed, say, database systems possible.

Now, that is going to be more or less the end of my introduction. There are things, which I've left out obviously, for example computational geometry, computer graphics related to computational geometry, and of course many other fields - computational algebra, number theory related to cryptography. The field is extremely wide. But as a bottom line I want to say that this is a real success story that young people at the time, saw the challenge, rose to the challenge, had wonderful opportunities. It's always to be, when you are one of the early people in the field, that there are gold mines to be discovered whereas later generations have to work much harder to get things, which are in many instances much less dramatic and fundamental, so that there are also drawbacks of being an early person. One humorist said: "The early Christians got the fattest lions." So the people who chose those careers are very often certainly in the 60's met with something between indifference and hostility within they are academics, say, math departments. But that is really... I mean these are small details, which don't matter at all in the large picture.

This is it thus far. Now, what would you like? Would you like us to start with some preliminary remarks on discrete mathematics and then open it for general discussion or should we have some general discussion right away and then move for a change of pace to discrete mathematics? Second choice? OK.

M. GROMOV:

Some talks were not conclusive... Sasha [Razborov] did not conclude his talk... Maybe he can say a couple of words and then we can discuss it.

M. RABIN:

OK, so we can do that. I would call upon everybody to be brief, may be I have already transgressed but... Where is Razborov? By popular demand (of one person)... Please... Do you [Razborov] need some transparencies?

A. RAZBOROV:

It is not so easy to talk about these things for obvious reasons: we still have to get to Tel-Aviv tonight. I have to be brief but what I wanted to say is mostly on my slides. Actually why don't we postpone this short lecture. I will go and bring my slides from the suitcase and will be back. And meanwhile you'll do something else.

M. RABIN:

All right. The shortest and best understood talk... words in this conference. Anybody to speak may be?

V. MILMAN:

I just would like to have a brief idea of what you said in your talk about this possibility to transfer information from A to B, say, that A is able to prove to B that he sent and to prove to C that he did not send. You said this but just the idea of both such possibilities in the same place.

M. RABIN:

OK. That's a wonderful question obviously planted by me. If we have time I'll tell you a joke about planted questions, but we don't. You want me to start with the joke? So I'll start with the joke, also of course into the microphone. This I've heard as a true story from my brother who used to teach at Oxford University before he transferred to Jerusalem. So they have debating clubs where there are also lectures and then there are discussions. Somebody was invited to give a talk at one of those clubs and he arranged with a friend two questions or number of questions. He gave his lecture and then asked: "Who has any question?" This guy raises his hand as a question. The speaker gives him a demonstrating answer. Everybody laughs. "Another question?" "Yes, I have another question." He didn't see, you know, which way it was going. The same thing is being absolutely decimated. Everybody laughs. So he became a little bit sorry and he said: "George, I don't quite remember what the third question we agreed upon was."

Now let me describe how you do deniable authentication. And these are those mind games. As I said also about the Internet and routing and so on, it's not just a question of analyzing a mathematical situation, which like in nature, is forced upon you. You can exercise your inventiveness and hopefully make it easy.

So here (Picture 3) we have the sender and the receiver. And the sender sends a message. It could be encrypted but this is not essential. This is

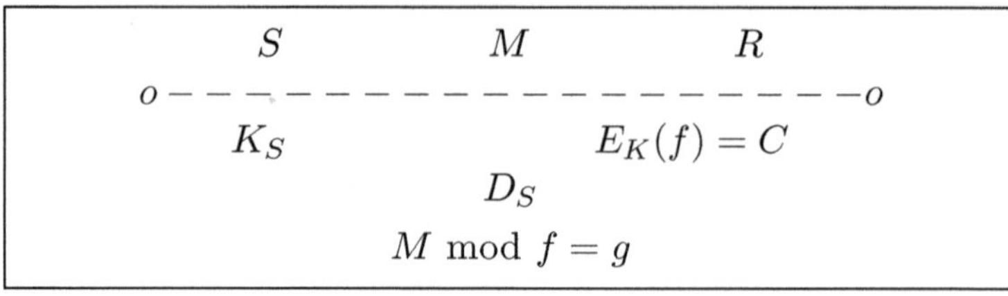

Picture 3

not essential to what we are saying. Now, the sender has a public key because the way he identifies himself or herself is with a public key K_S. And of course the sender also has a private key. And the public keys are in the directory. Now, when the message is received the receiver sends a question about this message encrypted by this public key. Now, what is the question? The question is the following: he sends an irreducible polynomial that he chose randomly, say, an irreducible polynomial of a field F_2 and what he wants to have in return... the message is of course a sequence of bits, say, a hundred thousand bits. It can be viewed as a polynomial of degree one hundred thousand. So the return is going to be M *mod* f. The question is how does S know f? Because it was able to decrypt it by use of the private key. So this is some polynomial f, say, of degree two hundred. Now, without knowing what the polynomial is, the probability of somebody giving the right residue is about 1/200. The sender who is really doing this knows that the only person who could send the correct answer with a very high probability, an overwhelming high probability ... must be the sender who has the private key. That's for him proof that M was sent by the sender. Now, how does deniability come in? Deniability... Yes, please.

V. MILMAN:

Is M the original message S to R? M was the original message and then it was transformed to g, wasn't it?

M. RABIN:

Yes. And then you perform the computation on it, the computation M mod f. In M mod f, f is a random irreducible polynomial chosen by R on that occasion. The fact that whoever is out there - there is no face, there is no handshake - whoever is out there was able to send this. That is proof that the original sender is actually S because he agreed to do it. How does deniability come into it? How does deniability come? Now you are getting to the highest order of mind games. The inquisitor... of course, if he completely places himself instead of R, is sitting there, then he is R and he also gets confirmation like R. But the inquisitor can do the following. He sits on the sidelines but tells R to send C. So he chooses f, encrypts it and says to the receiver: "When you get the message send this cybertext". Then in that case S will not be able to deny because the inquisitor will know that somebody decoded this C. So what S does is when he gets the question C he asks for a proof of knowledge of the plain text. I mentioned that as a new algorithm. So when he receives something encrypted by his public key he can ask the sender to prove that he is the sender (in this case R becomes the sender) that R knows what is inside there. What is the significance of the fact that he has proof that he knows. First of all there is a question: What does it mean to prove that you know? This sounds, you know, like knowledge in the sense of Descartes or Socrates. No, no, no, it has a very precise meaning. Namely, that actually R could reconstruct the contents, the hidden contents on his own very quickly. Why does this now make the authentication deniable? For the following reason: the inquisitor then comes and asks everybody to reveal their secret keys, so even D_S has to, it doesn't matter. What S is going to say is: "R, this exchange and the proof of knowledge did not happen between me, S, and R, the receiver, the voting booth". It rather happened between R and an accomplice, and what R really did was to decide about the M, decide about the f, and his accomplice by another line - this is what the inquisitor listens to, but they have another line or pre-arrangement - tell him what f is, and then the correct thing is returned. So the behavior in the situation where everything that the inquisitor saw could be simulated by R and then the accomplice and therefore S can say: "I didn't do that". It's in a sense like if you can prove that there is somebody who can perfectly, perfectly, but perfectly forge your signature then you can disown your own checks because you say: "There are maybe ten people who can completely forge it, so I didn't do it." So you see that we are talking here about fairly complicated mind games, but if you make the full analysis you see that it is compelling. So on the one hand you had end by a very simple protocol; the only part, which is not simple is that when R is sending C, which is the encryption of the question to S, he (R) can prove knowledge of the cybertext, of the plain text, of the secret inside, he can prove it and then you have the question how this is miracle achieved. In fact the proof is very simple in essence, I mean, at least it's very efficient and in itself is also done completely in the clear. So now R is the sender - any sender for any encrypted message using our methods can prove the knowledge of the plain text, he does it clearly and yet absolutely nothing is revealed, and these are components of this solution. But all together practical. So that's one of those situations where in the long run the wonders of cryptography, so to speak, are going to help in real privacy protection in certain situations.

OK, now, I think we have... Sasha [Razborov].

A. RAZBOROV:

I will just show you my last slides because the real reason why I couldn't show them during my official talk was that they were designed for a discussion rather than for a talk. So as you can see they are even arranged in the form of several theses and what I wanted and what I can say on the matter is already written there so I will simply give a few comments.

feasible proof of (a variant of) $P \neq NP$? A personal motivation
Serious attack on the $P? = NP$
question has never ceased since its formulation 30 years ago
it's a general feeling that $P \neq NP$ will be one of the central
open problems determining the development of mathematics
at least in the next century (cf. [Smale 98])
whenever an important open problem is hard it is always
helpful to accompany the attacks on the problem itself
by a meta-mathematical analysis

Picture 4

Well, the "SP and NP problem", which is probably at the heart of the modern theoretical computer science was discovered around 30 years ago and of course when mathematicians discover some problem, which is obviously very important, they try to solve it and if the problem is really important then, at least in somebody's opinion, it must resist any attempts to solve it because if you solve an important problem in 5 or 10 minutes then, no matter how important it was, it immediately ceases to be important. And there were actually several lines of attack on this problem. I have to skip some early attack on it via diagonalization techniques. Roughly, the point is that a very similar in spirit problem existed in the pure computability theory and it has what you would call now a trivial solution. It's the Post problem and it may be not trivial but it has an easy solution which is just a reformulation of this problem at a different level. So people were very enthusiastic about P vs. NP and they tried to solve it using some old-fashioned techniques, which, as we would say now, relativize and it turned out that they don't work. Some other techniques were discovered instead. And probably what people were trying to do during several last years was to solve this problem using lower bounds for Boolean circuits, which is well known to be almost equivalent reformulation of P vs. NP. Boolean circuits are very simple gadgets. They really were used at least at the earliest stages of technology for designing computers and it seemed easy to prove that some trivial device cannot compute some easy functions; it comes to very easy things and it can be explained to any schoolboy in half an hour. Nevertheless the progress was very limited, right? Well, this thesis can be discussed and...

M. RABIN:

Some people are curious about the first line - feasible proofs. You have some idea...

M. GROMOV:

But he is going to give the idea of definition of feasible proofs...

A. RAZBOROV:

Right. During my talk on feasible proofs I meant proofs in one or another logical system, which are feasible in the sense mathematical logicians would use this word. Today I would like to say that feasible proofs are just proofs which use some limited means. I will explain later what I mean.

Returning to P vs. NP, enough was said about this in my talk. The only thing I have to add is that also this approach (to prove lower bounds for Boolean circuits, which is an easy looking and combinatorial task) has not worked so far. Still there is a number of strong partial results along these lines, which somehow helped us to understand things better and I would like to propose for our discussion the following point. When we meet such a situation, when we have sufficiently many partial results proved with this or that machinery, but the "big" problem we are trying to solve resists any attacks, it is always helpful to look at its independence at least from this limited set of methods. In my personal opinion it is not very important whether we will be able to prove independence or we will not be able to do it. What is important is that no matter what will be the outcome, it really will advance our understanding of this problem.

M. RABIN:

What you propose is to define a certain limited logic... To define a limited logic and to show that within this limited logical system the question whether P is equal to NP is independent? That's that you mean? So what is that limited logic? Is it stronger or weaker?

A. RAZBOROV:

So far this meant two things. The first thing was already discussed during my talk, which is simply first order theory or propositional proof systems, and we are interested in short propositional proofs. So these are things feasible in a logical sense... Let me recall that "short" means polynomial not in n but in the number of input strings (2^n). And why is it natural and important? It's exactly explained by the chain of partial results about Boolean circuits we already have because all these results are feasible in this sense.

[Answering somebody's question:] Yes, n is the number of variables, and indeed in many respects it is more logical to look at the things exactly in the parameters you are talking about. But the trouble is that propositional proofs of size polynomial in n are too weak. You will have problems even with formalizing the statements in this system, and you will have even more serious problems with formalizing known proof techniques. Actually, most likely they do not formalize there at all (although we cannot prove it rigorously yet).

The second thing is natural proofs. We (with Steven Rudich) developed them and tried to approach our problem from a slightly different side. Namely, we forgot about logics, I mean for a moment. We simply looked at the normal proofs and tried to see what they have in common. And it turned out that it is possible to distill several (actually, three) simple properties. The first property is... Well, it may be a little bit technical for our discussion, let me just say that one property is combinatorial, and one computational (the third property simply states the soundness of the method). They are very natural and we thoroughly checked that all known results really fall into this framework. And it turned out that even in this framework we already can prove non-trivial things about independence. What is most interesting about all this activity, so far, is that here quite surprisingly (at least for myself), we entered quite a different world, namely, the world Mike Rabin was talking about.

Currently, the most promising approach to the $N? = NP$ question is to show lower bounds on the size of general Boolean circuits computing explicit Boolean functions. Many strong partial results were obtained along these lines. All these lower bound proofs are: (1) NATURAL (2) FEASIBLE the mathematical concept of poly-time computability is also underlying the meta-mathematical properties of these proof systems.

Picture 5

Namely, when we prove independence from natural proofs, we cannot do it unconditionally. We have to use some cryptographic assumptions like Alain [Connes] was talking about and Mike [Rabin] also. I should admit that the most difficult task in all this business is to calculate the parity of negations in front of the word "prove" you are using. You try to construct an algorithm to solve NP, then you try to prove that it does not exist (lower bounds), then you try to prove that some system cannot prove it does not exist (independence), etc. But after it is done it results in something absolutely amazing. Namely, in order to show independence (two negations) we have to use the cryptographic assumption, which is of lower-bounds type (one negation). So, it goes exactly the opposite way around, right? It was really strange and...

M. GROMOV:

Can you explain...

A. RAZBOROV:

With pleasure, and I think this will be the last thing I want to say because it is really a great illustration to the point I already made. So no matter where this will eventually lead us, it already has led us to some understanding of the nature of our difficulties, and here it is. There is a strong pseudo-random generator, the Goldreich, Goldwasser, Micali generator. And we can think of it as a generator which constructs a pseudo-random Boolean function. This is a Boolean function which on the one hand looks like a random one, and on the other hand, when viewed as an algorithmical task, can be easily computed with probability one. And the informal message which we already got doing our stuff (and I do expect and hope that there will be more messages of this kind) is this: look, you are going to show that there are no polynomial time algorithms for NP-problems, right? No polynomial time circuits, let me put it like this (I said above that this is more or less the same). But here is some collection, some bunch of easy Boolean functions provided by the GGM generator, which by definition look random. In fact they look very random. So if you can think of some argument proving that SATISFIABILITY is not doable by poly-size circuits then just think of how your argument is going to refute the easier fact that SATISFIABILITY is in the image of this generator. When you just try some obvious combinatorial ideas of proving lower bounds on this target, you immediately see how naive and meager they look in this context... I think that's what I was going to say. I'm sorry if it was not very intelligent, I was not prepared.

M. RABIN:

Thank you very much. So we may well have some reactions in general to computer science, theoretical computer science. I must say that I forgot one very important slide here. It's just going to take a second but it is really very important. So we had that sequence of developments. First of all theoretical computer science as the mathematical study of the process of computation. Then we had theoretical and also applied theoretical computer science constructing efficient algorithms, dealing with various problems of great practical importance. Now, the thing I forgot is the outreach and that is again in the spirit of what Raphy [Coifman] was saying. So now it turns out that the methodologies... not now, in the past decade... that the methodologies have a bearing on brain research. Not the actual physiology but trying to understand how the brain thinks. Outreach to computational biology in various forms and outreach to quantum computations. And for example Peter Shor couldn't have done what he did in constructing or proposing the quantum computation or factorization without a real understanding of the algorithms like Fourier transform and other things, that is, implementing and index calculations and so on, that is, implementing or hoping to implement by using of the quantum computer. And these are just a few examples.

So within the university (and that's the case which we also made at Harvard for the development of computer science), within the university, not just university, in intellectual life, theoretical computer science, computer science has an outreach to many other topics. Methodologies from computer science can be used in literature or in historical research and so on. It's becoming ever wider but maybe also more interesting. So I forgot

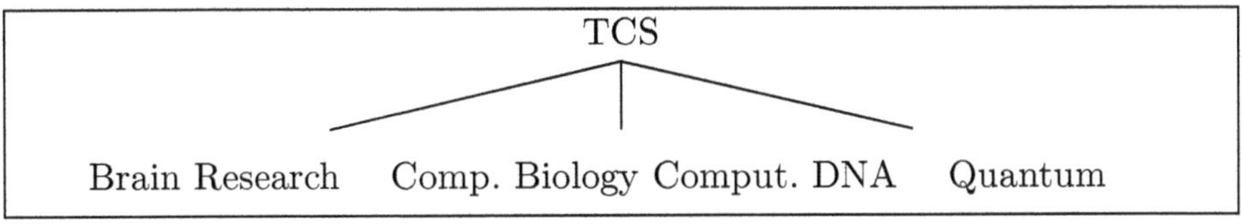

Picture 6

that and here it is. I invite comments of any sort.

S. KLAINERMAN:

I just wonder if it won't be useful to first hear some short special statements or otherwise we won't have time.

L. LOVÁSZ:

What I thought of commenting on is that discrete math here is sort of bunched together with theoretical computer science, and my first thought was that this is really doing some injustice to the field, because discrete math is a branch of mathematics and there are outstanding people in discrete mathematics who never think about computing. My second thought was, on the other hand, that this is actually a very good thing for the subject: we are sitting here with computer scientists (I don't know where I count myself), and it is one culture in some sense. We understand each other better than (unfortunately) some of the pure mathematicians. I think it is a little bit similar to the situation that people in analysis don't mind being put together with theoretical physicists. It's really a very closely tied subject and this brings me back to the question that Avi [Wigderson] raised at the end of the previous discussion session. I started out as a discrete mathematician doing graph theory and extreme graphs and this sort of things, and then when I first went to the United States in 72-73 I learned about computational complexity theory. It was immediately clear that this theory gives a lot of information about the kind of questions I was thinking about.

Let me add here one idea to Avi's question. So Avi suggested that if we look at some notion or concept we should ask: can this be efficiently computed or how efficiently can it be computed; and that question will lead us in many cases to deep structural understanding of the subject. Now, there is another thing here, which is called under the (probably not very fortunate) name of "non-determinism" in computing. What it means is, I would say, provability and refutability. If we study some property, how easy is it to give a certificate that the property is there or a certificate that it's not there? If you pick a number how easy is it to certify that it is a composite number? This is obvious, you just give a factorization. Is it easy to certify that it is a prime? As it turns out you can do it, but it's much trickier, you have to basically exhibit a primitive root. It turned out that in discrete math there were some basic questions, which people were trying to solve, like, given a graph, does it have perfect matching (a set of edges which matches up all the nodes)? There was a beautiful answer to this in the theorem of Tutte. But there is another question: does it have a Hamilton cycle (a cycle which goes through all the nodes). This is a very similar question. Its pictures look almost almost the same, but that turned out to be much more difficult. And computational complexity theory all of a sudden gave an explanation of this: one of these was in P and the other one was NP-complete. So if we assume the hypothesis that P is not equal to NP, then there is a clear explanation of why we had so much more success in answering one of the questions and why the other question was so difficult. This is not an isolated example. It turned out that for the next decade or so there was a big movement of trying to understand all sorts of questions in discrete math just asking these questions: is it easily computable, or is it easily refutable, or provable (certifiable). And in all cases it led to results, which go way beyond just graph theory and discrete math. As an example you can look at the question of linear programming, which is a very basic question in mathematics. Given a system of linear inequalities, is it solvable? It turned out that this is in both NP and co-NP, in other words, it's certifiable and refutable. This goes back to classical theorems from the last century by Fourier and Farkas. But it was not known at that time that it is polynomial time solvable. Then in the late seventies Khachiyan's algorithm was found (of course, as everything, it had various predecessors). A little bit later on it was observed that from the practical point of view, there were all sorts of troubles with the Khachiyan algorithm; it was very inefficient. So one asked, can you do it practically efficiently, and then Karmarkar's algorithm was discovered. Karmarkar's algorithm and other linear programming algorithms like the simplex method, are the core of many fantastic programs that solve systems of inequalities with millions of variables and millions of constraints, huge things. But also it became the core of a very interesting development of the theory of convex bodies by Nesterov and Nemirovsky, which is algorithmically motivated but is a structural study of convexity and as far as I can see brings a new aspect, a new way of looking at geometric properties of convex bodies.

So I think that to be packaged together with theoretical computer science means that discrete mathematics is in a happy state, that it, has an application area, which feeds exciting questions, and not only questions, but deep insights into the field. Now, of course to the third thought...

M. RABIN:

The third side of the coin.

L. LOVÁSZ:

The third side of the coin is that there are a lot of questions in the real world, which are discrete, or at least have a very heavy discrete element. Protein folding was mentioned here. Clearly the basic question is: you have an entirely discrete object which is just a sequence of amino acids. Now, how do you predict what it will do? It will curl up and it will act in very strange ways. It can be the venom of a snake or it can be an enzyme that helps your digestion and lots of other things. So how does it happen and what makes it happen? Of course, there are similar questions about the the DNA and so on.

Again, we were talking here about theoretical physics. Probably if you go to some other group of people, then they will be talking about an entirely different type of theoretical physics, which is much more discrete in nature, like statistical mechanics or other areas, Feynman graphs and so on. So to conclude, the third aspect of discrete mathematics is that there is a lot of discreteness out there. I think it's a little bit of a cultural thing I will try to talk more about this in my talk, so I don't want to say anything that I'll repeat. But it's cultural and I think it will be useful if people will be able to integrate both these sides.

M. RABIN:

Thank you very much. Before I call on Noga [Alon] I want to take issue with the question whether non-determinism is an appropriate or not appropriate name. I think it's a very appropriate name, and I introduced it into computation.

[LAUGHTER]

And now, you know, Shakespeare said: "What is in a name?" and so on, I think that part of its success was because the name was catchy... But that's just a historical comment. Noga [Alon], another discrete person.

N. ALON:

I'll try to talk discretely here. I guess Laszlo [Lovász] said almost everything I wanted to say about the connection between discrete mathematics and computer science. It's kind of obvious that theoretical computer science is one of the main reasons, maybe, for a lot of the development in discrete mathematics in, let's say, the last 20 years and some of the reasons we heard now. I wanted to mention or to add two more points. One is the connection between discrete mathematics and probability. Probability is clearly one of the big things that happened in discrete mathematics in the last 20 years. Maybe it started before that but it's mostly evident in recent years. And this is a connection that goes, like the connection with theoretical computer science, both ways. There are combinatorial methods that actually give new results in probability, for example, if you look at proofs of correlation inequalities, FKG-type inequalities used in percolation, these are definitely combinatorial in nature. And yet they give results in probability. And of course in the other direction there is what is called the probabilistic method, which started with Erdös and probably also with Shannon, where the most striking examples are when we use probabilistic reasoning to prove totally deterministic statements. This exists in other fields as well, but in discrete mathematics it's probably more striking than in other areas. And then of course there are also some subjects that are in-between, like the study of random graphs or random objects, or percolation. Sometimes it is difficult even to say if this is part of combinatorics or part of probability or something in-between. So definitely this connection is a very important connection and I think it will keep being so. Now, one more unrelated topic that I wanted to mention (and we touched upon this on Friday also), is again not something that exists only in discrete mathematics, but it exists in pure mathematics as well and it is connected to computer science, and that's the issue of computer-assisted proofs. Again, I think that they appear more in discrete mathematics, but they appear in other parts of mathematics as well, and what I mean here is not proofs where you gain some intuition by doing computer experiments, which is also very important, but I mean really proofs that even after you found the proof we at the moment do not know to check that it's valid without using computers. Still we cannot write the proof in a way that can be checked without using a computer. There are more and more examples like this and I guess it's more or less a safe prediction (although of course different people may think differently), that we would have more and more proofs of this type just because the tool is there. Is this good? Sometimes indeed when we prove, say, a result in pure mathematics we want to see what the reason is, and a reason usually means a short proof or a proof that we can understand and we can read. But inevitably it seems that there are short statements that have only long proofs; in fact there are also some formal ways to prove statements of this type. And it is very natural ("natural" is a bad word, we say, but it is reasonable) to believe that, indeed, since the tool is there, we are likely to use it. So there would be these proofs where we first do all the clever things and then we still have to do some computation that we have to do in ten hours of computer time. I think it would be interesting to also hear what people think about this. That is: Are these proofs desirable in pure mathematics? Should we avoid them? Would they exist in the future or would they vanish?

M. RABIN:

Fine. Maybe we will come back to the last extremely interesting point of Noga's. For example the work of Zeilberger proving various identities, say, about binomial coefficients and other mathematical functions. But I would like now to open... we've heard about computer science, theoretical computer science, discrete mathematics. Any comments?

S. KLAINERMAN:

Just to make a short comment about computer-assisted proofs. I am just trying to say that also in mathematics there are certain results, which are not very conceptual, which require a lot of steps, ten steps that have to be verified and are they satisfactory? I think the answer is... I mean the problem which is not very interesting and the result has kind of disappeared, but it is very important for the development of a certain field. There would be other proofs later on, which would be more conceptual. So I think that the same probably applies to this issue of computer-assisted proofs.

R. COIFMAN:

The question is what you think are the challenges that actually face you? I mean you mentioned biological brain computation. There are possibly completely different ways of processing information or computing if you wish, which don't follow the standard model that you have. They will promote challenges. How many explorations are happening in that direction? Or you really don't know much about the field? That's what I'm basically asking.

M. RABIN:

Well, it shouldn't really be a dialogue, say, between us because we want to open it, so just one word. So quantum computing for example is another way of doing computation. And it's wide open - there could be others. And with respect to brain computation - that of course comes to artificial intelligence, which on purpose, out of politeness, I didn't mention. But there is really a question we can do, may be eventually we will do artificial intelligence in one of two ways, maybe we'll really understand how the brain solves problems and how this piece of meat does remember and knows that we don't know Clinton's telephone number. So we may find other ways of doing it or we may understand how the brain does it or both. And nobody can say this is too much in the future but I'm sure it will be dealt with. Yes?

R. HADAMI:

My feeling is that there is a tendency to separate computer science from pure mathematics and to call it a separate arena of activity. And maybe the reason for not being able to properly attack the questions such as NP not being equal to P is because of this separation. Maybe you ask the question not in a very correct stage of mathematical technology. For instance, I don't know anything about it either, but the Weil Conjecture in mathematics needed, you know, a great framework of all the Grothendieck workings and constructions before anyone could even ask the question properly. Maybe you haven't asked the question properly and you have to develop a correct framework to ask the question in and then answer it. And this is pure mathematics and you need to be a great mathematician in order to do it. My feeling is that computer science is very down to earth, has stayed very down to earth, has stayed with all these basic tools of nice great mathematics - very bright minds but very low-level tools. And maybe this is one of the objects of this area, I think.

V. MILMAN:

The future will prove this very soon.

A. WIGDERSON:

Maybe the answer to Raphy [Coifman] was a little bit short... You may be totally right and there is sufficiently high mathematics, I mean the previous NP problem. In fact if you really followed what Sasha [Razborov] was saying in his proposal he just mentioned exactly this. I mean these natural proofs, what he calls natural proofs or the fragments of Peano arithmetic that he defined, just capture the methods that we have used so far in order to prove law bounds and Boolean circuits. They got us to the place where we stand. And if indeed the question is independent of the theories it means we need tools that are... you know, use heavier mathematics at least in the logical sense, OK? So that's why we invite all of you to work on this problem that uses [laughter] topology and dynamics and geometry and, you know whatever. That's on this point. The same statement may be said about the Riemannian hypothesis, Poincaré conjecture. You know maybe it's not solved yet because we didn't use the right tools. It's sort of an obvious statement about any problem that is still unsolved. I think it's really interesting - the path that was taken by Ajtai and Razborov and others to actually, you know, interrogate ourselves and not just say: "OK, let's invent more and morc tools and try to attack it from this". Not just trying to have more ideas but to examine the methods we've used so far to encapsulate them into a formal system and argue about this - whether it's sufficient or not. I don't know if this was done about the other major open questions.

R. HADAMI:

Why do you always say that this is a faculty of mathematics and computer science, not a faculty of mathematics and number theory?

A. WIGDERSON:

I don't care about names, I think this is beside the point. If you equate computer science with great mathematics I think you are a little short in numbers but anyway, let me say just a couple of words. I think this topic probably deserves a discussion of its own but the question was that there are other modes and models of computing and we have our Turing machine. Do we go somewhere else? And the answer is yes. I think that we now understand that we have tools to argue about models of computation, to define them, to study their strength and their connections to other models. In fact Mike [Rabin] mentioned some of them when he talked about distributed computing, parallel computing, and random computing.

And two words about some of the people who are trying to understand the brain from a computational point of view. I think it's extremely interesting when people have theories about memory, about association, visual systems, about things we do really try to understand, whether we can build a model of the brain, computational, not physiological or psychological, but computational, and say: "Oh, here is a computational model and look - it can do these things." And I think there are these really interesting works in this direction... Some of us are getting to this research, sure.

[HUBBUB]

A. RAZBOROV:

I just wanted to add a couple of short sentences. The situation in computer science is not something unique in the sense that people there have different opinions. I respond to the last points [concerning the tendency to separate computer science from pure mathematics]. One can get an experience like you are describing but believe me there are many people - and you can count on myself for example - who really share these two points with mathematicians that computer science should not be separated from mathematics and that whenever we feel that new mathematical, deep mathematical ideas are needed for our problems we immediately bring them in and we learn new things and there are many examples of this sort. So that's the only thing I wanted to add.

P. JONES:

I'd just like to say a little bit from the perspective of an analyst here, which is that I come here and I see some sort of tower of Babel, that you have various communities that should be talking very directly to each other and should be able to understand each other and there is the basic ability. And I think we shouldn't have too high an opinion of ourselves, that this confusion arrives not at some high level but at some extremely basic level. And just as an example of the failure of the analysts to communicate this clearly. Many people believe that what the Fourier analysts do is continuous arguments. But the whole of modern Fourier analysis is based on the fact that continuous arguments fail. So what we do is discreet arguments and these discreet arguments of course have deep analogies in theoretical computer science. This is a whole area that deserves to be studied.

V. MILMAN:

Perhaps only one word about something which was said here about computer supporting, I mean, solving problems with computer assisted proofs. I just think it depends on the question. For some question, when really the most important thing is to know the answer, Riemannian hypothesis is an example - it absolutely no matter how you come to it, needs a hundred or more hours of computing work to do this. But for some other questions the main point is in a different place. And very often exactly the problem where the point is in a different place, is solved with computer assistance. And this is where this contradiction lies. So for some problems you should know the answer, for some you want to understand what structure and ideas lie behind the simply formulated problem not to be able to have a simple answer. And here a computer assisted proof is just missing the goal, you know, someone wants to have a record instead of to understand what is going on.

N. ALON:

Just a very short "but". Suppose that these questions do not have proofs that are not computer assisted. Suppose that the only possible proof must use a computer.

V. MILMAN:

But it's always in the system of a given structure you already have in mathematics. It is always. Your answer is always in the special existing structure of all these axioms of structure you discovered. So you need to discover the next structure. It is all in the given framework; whatever you said is correct.

A. CONNES:

Just a question to what you [Wigderson] were saying. In your talk you mentioned permanents and determinants and so on. But I didn't get the details or the reformulation of the question. So could you give some details? I mean just formulated in such a way that... Just to make it clear. What the problem means? Otherwise I don't understand it.

A. WIGDERSON:

One of these mathematical definitions that everybody is familiar with. So I don't have to say anything, I will just put it again and you can... Give me one minute to find it.

L. LOVÁSZ:

Just a word regarding this question... I mean in connection with good questions and bad questions. No, I'm pointing to the chair in front of you. [Laughter.] I think that there is some virtue in discovering questions when you all of a sudden stand up and say: "Oh, how can we not know the answer to..." "P is not equal to NP" is one of those questions. Of course you couldn't have asked this question a hundred years ago. Or whether the determinant can be specialized to get the permanent.

A. WIGDERSON:

Basically the fact that the permanent is an interesting function is a non-trivial discovery of Valiant and in some sense it's a complete problem like the NP complete problem, only harder. But the nice thing about it is that it's not something like satisfiability where you are talking about Boolean formulas, which are ugly objects and they are discreet, and you understand that you don't understand what they are. Here you are talking about a polynomial that's so nice and looks so much like the determinant.

A. CONNES:

You did not say exactly what the operations are.

A. WIGDERSON:

In general what you are interested in is to take a projection, take one polynomial in some, let's say, real space or over any field that you like. OK, we take a polynomial with many variables over this field and a projection just means that we look at how this polynomial looks in an affine subspace of smaller dimension. That's it. And of course there is another polynomial of fewer variables, which has the dimension that you were down to. Right? So the two points I made were: is the determinant (or in fact every efficiently computable polynomial, not just the determinant), a polynomial projection of the permanent, of maybe, the square? That's Valiant's theory. In this sense the permanent is complete. The permanent is complete in this very simple sense of a projection to an affine subspace that is smaller than what you have started with. OK? And in the other direction: the permanent is also a restriction of the determinant. But the only proof we have so far is that it's exponential; the degree is exponentially higher. Whether you can do it in a polynomial, just with a polynomial increasing, is essentially equivalent to the P versus NP question... Well, let's say, if the field is a finite field and basically it's equivalent, up to notions whether Turing machines are different than circuits... It is equivalent and the reason it's equivalent and not just implying P different from NP, is that not only the permanence is complete in the sense that I explained but the determinant is also complete in another sense. Everything you have to do is to show a formula for it, you can encode as a small determinant. It's equivalent. So you can forget about computation and really work on this.

N. ALON:

It's similar to work on any other NP complete problem, only this looks well-defined...

M. RABIN:

Maybe I'll make a further comment on that and on your remark. There is no doubt that P versus NP, whether they are equal or not, is a centrally important problem with very far reaching consequences either way. And in relation to... or when it comes to our failure to settle it, I'm a little bit in sympathy with you [Wigderson], not, you know, about the elementary nature of theoretical computer science but that one of two things could happen. Either somebody will find an efficient algorithm for NP complete problems, so this is probably going to be very, very clever and involve new ideas in the same way that, say, Karmarkar was a departure and Khachian before, but maybe Karmarkar was a departure for what was done before. But we are probably not ready for a negative result. And part of the difficulty is that the problem is sort of too particular. So if you'll try to take any of these particular questions such as satisfiability or three-color ability or Hamiltonian circuit, they are sort of grainy. And what you would require would be some higher level of abstraction. So for example the problems of topology, of what was called combinatorial topology, were fairly inaccessible until there was a systematic introduction of various associated invariants, groups of various kinds, homomorphisms and isomorphisms between groups and so on. And we are lacking that kind of structure in the problem at hand and in computation in general. Computation seems to be sort of particularly dependent in some sense on models, on machines and so on. So maybe the projections (I doubt it), maybe these algebraic projections will be the sufficiently high-level abstraction, which is going to lead us, say, to the P not equal to NP proof, but we are probably far away, and because we are far away it's also senseless to try to predict. I would say that if the problem is solved in the direction of P=NP, it could happen in the coming decade. I would be much less willing to make a prediction if P is not equal to NP when a proof will actually arise.

M. GROMOV:

Is there what is called independence?

M. RABIN:

I don't believe in that.

L. LOVÁSZ:

It cannot be proved. I mean if we have a proof that it's independent, then it proves the statement that there is no polynomial time algorithm for a Hamiltonian cycle. The point is that that's a recognizable object at least if you define polynomial time properly.

M. RABIN:

Not quite, because there may be a polynomial time algorithm and you won't be able to prove that it does it.

L. LOVÁSZ:

Yes, but I assume that an algorithm comes with a proof of its running time bound. You can formalize polynomial time, for example, in terms of a program with a bounded depth of cycles and then there is a formal definition and if you use this formal definition, then it's recognizable.

M. RABIN:

But I would say that we shouldn't go far afield, you know, in the same talk and if you consider the difficulty of the question and the time it took to settle it you might claim that Fermat's last theorem is independent and similarly for the Riemannian hypothesis. I don't believe that to be the case.

N. ALON:

Well, Fermat's last theorem really cannot be independent because if it is false all you can write down a counterexample.

M. RABIN:

Yes, but it could be true without being provable. And we know it's true. [Laughter.] I want to get some other reactions. And by the way our time is running very short. But I invite, before Lovász, any other comments on this. Or does anybody want to relate what we are discussing now to Raphy's [Coifman] discussion in the morning? So you'll have the last word. A heavy responsibility.

L. LOVÁSZ:

I just wanted to mention an analogy to $P = NP$. There is the question, say, whether you can trisect an angle, which was raised in ancient times and was an entirely well defined question. The answer was suspected of being "no" but, of course, there was just no means of proving this. And then mathematics developed and developed and real numbers were invented and algebra was invented. And eventually now you can actually prove it quite easily. In a special class in high school, you can essentially give the proof that an angle cannot be trisected. So I believe that the situation is similar here.

M. RABIN:

I think that's a wonderful example. So I'm having the last word. It's a wonderful example in the following sense: that concepts, which seem to have nothing to do with the question at hand, namely the degree of an algebraic extension over Q in this case, and the fact that the algebraic extensions by ruler and compass constructions have degree 2^n, and 3 happens not to divide 2^n. So the Greeks couldn't do it, Newton couldn't do it and so on. But Gauss could do it. Yes. OK, thank you very much.

Transcribed by A. Scherbak

GAFA, Geom. funct. anal.
Special Volume – GAFA2000, 978 – 983
1016-443X/00/S20978-6 $ 1.50+0.20/0

GAFA Geometric And Functional Analysis

REFLECTIONS ON THE DEVELOPMENT OF MATHEMATICS IN THE 20TH CENTURY

D. Kazhdan

Let us compare mathematics as it was at the beginning of the century with contemporary mathematics. What has happened in the past hundred years?

To be sure, we have numerous new results and any attempt to list the most important results is bound to be incomplete. So let us ask our question differently: What are the basic changes in mathematical intuition or what questions are natural for us but could not be imagined at the beginning of the century?

There are some areas which have seen immense progress but where we do not find much change in mathematical intuition. For example I think that the sentence of Poincaré written at the turn of the century, "Analysis profits by geometric considerations, as it profits by the problems it is obliged to solve in order to satisfy the requirements of physics", adequately describes our contemporary understanding of analysis. Therefore I will not talk about the development of Analysis in this century. Rather, I will choose topics which in my opinion represent the basic shifts in mathematical perspective. Of course I can only present my personal views and different mathematicians will see the mathematical landscape in a completely different light.

Also, I prefer not to start from a discussion of particular mathematical achievements. Instead let us begin by considering the old question: "How is mathematics *possible*?" One of the possible interpretations of this question is, "How are we mathematicians able to perform our work?"

One of the main themes of 19th century mathematics was to "make mathematics rigorous." At the beginning of this century, therefore, the question, "How is mathematics *possible*?", might have been interpreted as the twofold directive:

1. set up a formalism adequate for mathematical reasoning and prove that such a formalism does not lead to contradictions, and
2. show that any question can be resolved.

The development of mathematics in the 20th century banished any hope for such a "naive" understanding. Godel has shown that we can never be sure that our framework, our chosen system of axioms does not lead to contradictions. Moreover we now know that any [sufficiently rich] system of axioms is incomplete. In other words if we are working in a framework of a sufficiently rich system of axioms then either our system leads to a contradiction or we meet statements about which we will never have anything to say. That is, we will neither be able to prove these statements, nor to disprove them, nor to show that we cannot either prove or disprove them. At first glance Godel seems to have signed the death sentence for mathematics. One would expect the unsolvable questions to jump at us in large numbers. If that were really the case, we would never be sure whether it makes sense to try seriously to solve difficult problems; surely, then, mathematics would come to a halt.

Fortunately, reality is very different. Aside from some very specific areas, we seem rarely to run into questions which we cannot settle and even in these areas we are sometimes able to prove that the questions we can't answer are "independent", that is, we know that we can neither prove nor disprove these statements. In view of this, we now give the old question, "How is mathematics possible?", a new interpretation: What is the mechanism which so often leads us to ask "meaningful" questions, i.e., questions which can be resolved?

I do not think that anyone has even an inkling of where to look for an answer to this. But I think that our ability to avoid the prognostication that might be suggested by Godel's theorem is related to the well-known but surprising observation that it is easier to solve a more general problem than a specific one. You see, there is a big difference between generalizations in mathematics and generalizations in social studies. In the case of social studies we pay for any generalization by being forced to accept an increasing number of counterexamples. In contrast, in mathematics where exceptions are not allowed, the existence of a sufficiently general statement to which we cannot find counterexamples is a strong indication that the statement is provable. [For example many people thought that the Fermat conjecture could neither be proved nor disproved nor shown to be undecidable. But immediately after Frey realized that the Fermat conjecture follows from the much more general Taniyama–Weil conjecture it became "clear" that Fermat's conjecture would be solved.]

We can also ask: "How is *mathematics* possible?" or, "Why doesn't mathematics split into a number of unrelated disciplines?" When one reads writings from the turn of the century one sees that the explosion of mathematics was seen to be the main problem which could destroy the unity of mathematics. Even then there was no mathematician who could follow all the developments; mathematics threatened to become a bunch of unrelated disciplines. Poincaré writes: "An attempt is made to cut it in pieces – to specialize. Too great a movement in this direction constitutes a serious obstacle to the progress of science." How could unity be preserved?

A choice of an answer to this question depends greatly on the answer to the first question: "How is mathematics *possible*?"

A "formal" interpretation of the first question" represents a very specific understanding of the structure of mathematics whereby logical structure takes on primary importance. This interpretation suggests Hilbert's one explanation for the unity – the main uniting force comes from the common structure: the logic of proofs.

On the other hand, Poincaré, for whom mathematics is characterized by the "economy of thought", writes that the unity of mathematics will be preserved by unexpected concurrencies as mathematics progresses.

We see now that both Hilbert and Poincaré are right – mathematics was able to preserve the unity during the multifaceted development of the 20th century and this unity is due both to the structural clarity and the immense number of unexpected connections between different areas of mathematics.

Actually the question, "How is mathematics *possible*?", was already asked by Kant who understood it as the question, "*How* is mathematics possible?" Kant saw the existence of mathematics as a proof for the existence of pure intuition. Mathematics for Kant was Euclidean. Such an understanding of mathematics does not correspond to everyday experience which teaches that some statements which are "intuitively clear" to one mathematician could be "counter-intuitive" to another. As Poincaré described beautifully in his article "Mathematical Discovery", an unexpected immediate illumination sometimes comes after a long and often seemingly unproductive period.

In other words mathematical intuition is not a natural phenomenon, is not given at birth, but develops throughout one's lifetime. I prefer to discuss the change in intuition of the mathematical community rather than follow the development of intuition of a particular mathematician [the topic of the article "Mathematical Discovery].

I think that the most drastic change in mathematical intuition came from the development of algebra. At the end of the previous century it was possible to subdivide mathematics into Algebra and Analysis, which contained Geometry, and these two areas were almost independent. At the end of this century we find ourselves in the position where the majority of achievements in Analysis and Geometry are, at least partially, based on the development of algebraic intuition. It is very characteristic that such a brilliant mathematician as Pontriagin dropped mathematics after the appearance of the post-war French school which was based on new algebraic intuition. This new understanding that the analysis of different algebraic structures is central for the development of mathematics found the most striking expression in the development of the category theory. I do not think that it would be possible to explain the basics of the category theory to any mathematician of the last century. The reason is that the theory of categories is "too simple". This theory, which originated in the forties, is based on a drastic shift of perspective: instead of studying the logic of the properties of mathematical objects the category theory studies the logic of relations. The category theory is perhaps the first serious extension of Aristotelian logic. In Aristotelian logic all the statements are "absolutely trivial" but in spite of this triviality Aristotelian logic is the backbone of all sciences. Analogously all the basic statements of the category theory are absolutely trivial but this logic of relations is the basis for a big chunk of modern mathematics. It is very significant that the first paper on the category theory was rejected by a first-rate mathematical journal for lack of content.

How does this new way of thinking change mathematical reality? It is impossible to describe the full picture while standing on one foot but I can give two applications of this new way of thinking. The first application is the possibility of constructing "ideal" objects which are completely defined in terms of their relations with the previously known objects. The second advantage coming from the category theory is the possibility of seeing familiar mathematical objects as "materializations" or, if you wish, shades of the more elaborate and structured objects. For example, much of the recent progress in representation theory is based on the understanding that, in a number of cases, functions are "materializations" of more elaborate algebraic-geometric objects.

The third topic I want to discuss is the change in structure of the interrelation between mathematics and physics. There were two different stages to this change.

In the first stage which started already at the beginning of this century, physicists realized that they needed mathematics not only as a tool to solve their problems, but also as a language to formulate laws of physics. Both the relativity theory and quantum mechanics rely on "modern" mathematics for the formulation of "physical" reality. There is no way to explain some of the most basic problems of contemporary physics to people who do not have an extensive mathematical background. But in this first stage, we still find a familiar structure to the relation between mathematics and physics when mathematics is used by physicists as a tool for the formulation and solution of their problems. The second stage, which started 20 years ago, brought a reversal of roles. In the last two decades of this century, we have had an increasing number of examples of applications of physics to mathematics. These applications are primarily in the form of conjectures which relate mathematical problems that were viewed by mathematicians as having nothing in common. How is this possible? In many physical theories there is a way to express physical quantities in terms of a functional integral. Since the functional integral does not have a rigorous definition such expressions do not have any meaning for mathematicians. Imagine now that the problem we consider depends on a parameter [say energy] and in the case when the energy is either very high or very low, there is a way to approximate the corresponding functional integrals by conventional mathematical expressions. These conventional expressions for the case of low and high energy are very different and we obtain two different rigorous expressions for the physical quantities – one from the analysis of the case when the energy is high and the other when it is low. From the point of view of physics both expressions are specializations of the original functional integral. Therefore "physical intuition" implies that these two different expressions coincide. On the other hand there is no obvious mathematical explanation for such a coincidence.

The existence of mathematical consequences of physical theories leads to the situation where mathematics plays a role of experimental physics for some branches of theoretical physics. It has become either impossible or too expensive to check the validity of some physical theories by experiment. Instead the validity of a physical theory is "confirmed" by the correctness of the mathematical predictions which can be deduced from this theory.

The last topic I want to discuss is the appearance of computer science. As a result of this development, mathematicians realized that it is not sufficient to ask whether a particular problem is solvable, but one should also inquire whether it can be solved in a reasonable amount of time. Computer scientists defined "reasonable" questions as such questions where you can check the correctness of an answer in a short [=polynomial] time. On the other hand one can consider a more restricted group of questions which can be solved in a short time. The basic problem of computer science is whether these two groups are really different, whether P≠NP. At first glance it is "clear" that P≠NP, that there are many ways to ask "reasonable" questions which are difficult to solve. But as we have already discussed, mathematical problems have a strong tendency to be solved in a relatively short time. Really, if a solution to a particular mathematical problem would take an exponentially long time we would never be able to solve such a problem. So either P=NP or we, mathematicians, are somehow able to choose very special "solvable" questions. Therefore we can restate the question, "Why are we mathematicians able to perform our work?", in a stronger form. We can ask: "What is the mechanism which leads us to ask questions which can be solved and can be solved in real time?"

David Kazhdan, Department of Mathematics, Harvard University, Cambridge, MA 02138, USA kazhdan@math.harvard.edu

MIX
Papier aus verantwortungsvollen Quellen
Paper from responsible sources
FSC® C105338

If you have any concerns about our products,
you can contact us on
ProductSafety@springernature.com

In case Publisher is established outside the EU,
the EU authorized representative is:
Springer Nature Customer Service Center GmbH
Europaplatz 3, 69115 Heidelberg, Germany

Printed by Libri Plureos GmbH
in Hamburg, Germany